Frontiers in Medicinal Chemistry
(Volume 7)

Edited by

Atta-ur-Rahman, *FRS*
Kings College
University of Cambridge
Cambridge
UK

Allen B. Reitz
Fox Chase Chemical Diversity Center, Inc.
Doylestown, PA 18902
USA

Iqbal Choudhary
H.E.J. Research Institute of Chemistry
University of Karachi, Karachi 75270
Pakistan

&

Jizhou Wang
U.S. Food and Drug Adminstration
Silver Spring, MD
USA

CONTENTS

CHAPTERS

PREFACE

This current Volume 7 of *Frontiers in Medicinal Chemistry* is comprised of 11 chapters written by the appropriate experts in their fields of study, focusing on topics of high important in the fields of medicinal chemistry, early-stage drug discovery, target validation, and chemical biology.

The first six manuscripts are focused on topics of interest to diseases of the central nervous system. The first by Kaur *et al.* is on antioxidants and neuroprotection, the second by Munoz-Torrero is on acetylcholinesterase inhibitors for Alzheimer's disease, the third by Ferrari-Toninelli *et al.* covers dopamine agonists for Parkinson's disease, the fourth by Marassiti *et al.* highlights the role of oxytocin in psychiatry, the fifth by Fernandez-Busquets *et al.* pertains to the many proteins for which inappropriate conformation contributes to pathology such as Alzheimer's, and the sixth by Loizzo *et al.* covers natural-product based cholinesterase inhibitors, also for Alzheimer's disease. The next three reviews are in the area of infectious disease research, the first by Abbat *et al.* on enzyme inhibitors for the treatment of malaria, the second by Fabbri *et al.* on bacterial protein toxins, and the third by Sciutto *et al.* on a vaccine approach to cysticercosis. The next review is on liver fibrosis by Zardi, and the final chapter is by Jiang *et al.* on regenerative engineering.

Thanks are given to the staff of Bentham Science Publishers for their support and guidance, particularly Mr. Mahmood Alam and Dr. Matthew Honan.

Atta-ur-Rahman, FRS
Kings College
University of Cambridge
Cambridge
UK

Allen B. Reitz
Fox Chase Chemical Diversity Center, Inc.
Doylestown, PA 18902
USA

Iqbal Choudhary
H.E.J. Research Institute of Chemistry
University of Karachi, Karachi 75270
Pakistan

_&_

Jizhou Wang
U.S. Food and Drug Adminstration
Silver Spring, MD
USA

List of Contributors

Alessia Fabbri	Department of Therapeutic Research and Medicines Evaluation, Istituto Superiore di Sanità, Viale Regina Elena, 299.00161 Roma Italy
Antonella Del Brocco	Department of Therapeutic Research and Medicines Evaluation, Istituto Superiore di Sanità, Viale Regina Elena, 299.00161 Roma Italy
Carla Fiorentini	Department of Therapeutic Research and Medicines Evaluation, Istituto Superiore di Sanità, Viale Regina Elena, 299.00161 Roma Italy
Carlos Larralde	Instituto de Investigaciones Biomédicas, UNAM, Universidad National Autonoma de Marico, Mexico
Cato T. Laurencin	The Raymond and Beverly Sackler Center for Biomedical, Biological, Physical and Engineering Sciences, University of Connecticut Health Center, Farmington, CT 06030, USA
Charanjit Kaur	Department of Anatomy, Yong Loo Lin School of Medicine, Blk MD10, 4 Medical Drive, National University of Singapore, Singapore 117597
Daniela Uberti	Department of Pharmaceutical Sciences, University of Piemonte Orientale "A Avogadro", Novara, Italy
Diego Muñoz-Torrero	Laboratori de Química Farmacèutica (Unitat Associada al CSIC), Facultat de Farmàcia, and Institut de Biomedicina (IBUB), Universitat de Barcelona, Av. Joan XXIII 27-31, E-08028-Barcelona, Spain
Donatella Marazziti	Dipartimento di Medicina Clinica e Sperimentale, Section of Psychiatry, University of Pisa, Pisa, Italy
Edda Sciutto	Instituto de Investigaciones Biomédicas, UNAM, Universidad National Autonoma de Marico, Mexico
Eng-Ang Ling	Department of Anatomy, Yong Loo Lin School of Medicine, Blk MD10, 4 Medical Drive, National University of Singapore, Singapore 117597
Enrico Maria Zardi	Department of Clinical Medicine, University "Campus Bio-Medico", Rome, Italy
Federica Menichini	AIFA Agenzia Italiana Del Farmaco, *via* del Tritone 181, 00187 Roma, Italy
Francesca Rosadi	Department of Therapeutic Research and Medicines Evaluation, Istituto Superiore di Sanità, Viale Regina Elena, 299.00161 Roma Italy
Francesco Menichini	Department of Pharmacy Health and Nutritional Sciences, University of Calabria, I-87036 Rende (CS), Italy
Gabriela Rosas	Facultad de Medicina, Universidad Autónoma del Estado de Morelos, Cuernavaca, Morelos, México

Giovanna Cenini	Department of Pharmaceutical Sciences, University of Piemonte Orientale "A Avogadro", Novara, Italy
Giulia Ballan	Department of Therapeutic Research and Medicines Evaluation, Istituto Superiore di Sanità, Viale Regina Elena, 299.00161 Roma Italy
Giulia Ferrari-Toninelli	Department of Pharmaceutical Sciences, University of Piemonte Orientale "A Avogadro", Novara, Italy
Giuseppina Maccarinelli	Department of Pharmaceutical Sciences, University of Piemonte Orientale "A Avogadro", Novara, Italy
Gladis Fragoso	Instituto de Investigaciones Biomédicas, UNAM, Universidad National Autonoma de Marico, Mexico
Gurugirijha Rathnasamy	Department of Anatomy, Yong Loo Lin School of Medicine, Blk MD10, 4 Medical Drive, National University of Singapore, Singapore 117597
Lakshmi S. Nair	The Raymond and Beverly Sackler Center for Biomedical, Biological, Physical and Engineering Sciences, University of Connecticut Health Center, Farmington, CT 06030, USA
Liliana Dell'Osso	Dipartimento di Medicina Clinica e Sperimentale, Section of Psychiatry, University of Pisa, Pisa, Italy
Mariagrazia Grilli	Department of Molecular and Translational Medicine, University of Brescia, Brescia, Italy
Mario Catena Dell'Osso	Dipartimento di Medicina Clinica e Sperimentale, Section of Psychiatry, University of Pisa, Pisa, Italy
Marisela Hernández	Instituto de Investigaciones Biomédicas, UNAM, Universidad National Autonoma de Marico, Mexico
Maurizio Memo	Department of Molecular and Translational Medicine, University of Brescia, Brescia, Italy
Monica R. Loizzo	Department of Pharmacy Health and Nutritional Sciences, University of Calabria, I-87036 Rende (CS), Italy
Natalia S. de Groot	Medical Research Council Laboratory of Molecular Biology, Francis Crick Avenue, Cambridge, CB2 0QH United Kingdom
Paola Bettinsoli	Department of Pharmaceutical Sciences, University of Piemonte Orientale "A Avogadro", Novara, Italy
Prasad V. Bharatam	Department of Medicinal Chemistry, National Institute of Pharmaceutical Education and Research (NIPER)
Rosa Tundis	Department of Pharmacy Health and Nutritional Sciences, University of Calabria, I-87036 Rende (CS), Italy
Salvador Ventura	Institut de Biotecnologia i Biomedicina, Universitat Autònoma de Departament de Bioquímica i Biologia Molecular, Facultat de

Biociències and Barcelona, E-08193 Bellaterra, Spain

Sangamesh G. Kumbar The Raymond and Beverly Sackler Center for Biomedical, Biological, Physical and Engineering Sciences, University of Connecticut Health Center, Farmington, CT 06030, USA

Sara A. Bonini Department of Pharmaceutical Sciences, University of Piemonte Orientale "A Avogadro", Novara, Italy

Sara Travaglione Department of Therapeutic Research and Medicines Evaluation, Istituto Superiore di Sanità, Viale Regina Elena, 299.00161 Roma Italy

Sergio Rosales-Mendoza Laboratorio de Biofarmacéuticos Recombinantes, Facultad de Ciencias Químicas, Universidad Autónoma de San Luis Potosí, México

Sheenu Abbat Department of Pharmacoinformatics, National Institute of Pharmaceutical Education and Research (NIPER), S.A.S Nagar-160 062, Punjab, India

Shweta Bhagat Department of Medicinal Chemistry, National Institute of Pharmaceutical Education and Research (NIPER)

Stefano Baroni Dipartimento di Medicina Clinica e Sperimentale, Section of Psychiatry, University of Pisa, Pisa, Italy

Stefano Loizzo Department of Therapeutic Research and Medicines Evaluation, Istituto Superiore di Sanità, Viale Regina Elena, 299.00161 Roma Italy

Tao Jiang The Raymond and Beverly Sackler Center for Biomedical, Biological, Physical and Engineering Sciences, University of Connecticut Health Center, Farmington, CT 06030, USA

Valentina Falaschi Dipartimento di Medicina Clinica e Sperimentale, Section of Psychiatry, University of Pisa, Pisa, Italy

Xavier Fernàndez-Busquets Nanomalaria Group, Institute for Bioengineering of Catalonia (IBEC), Baldiri Reixac 10-12, ES-08028 Barcelona, Spain; Barcelona Centre for International Health Research (CRESIB, Hospital Clínic-Universitat de Barcelona), Rosselló 149-153, ES-08036 Barcelona, Spain

CHAPTER 1

Antioxidants and Neuroprotection in the Brain and Retina

Charanjit Kaur, Gurugirijha Rathnasamy and Eng-Ang Ling*

Department of Anatomy, Yong Loo Lin School of Medicine, Blk MD10, 4 Medical Drive, National University of Singapore, Singapore 117597

Abstract: Oxidative stress is implicated in the pathogenesis of a number of neurological disorders such as Alzheimer's disease (AD), Parkinson's disease (PD), multiple sclerosis and stroke in the adult as well as in conditions such as periventricular white matter damage in the neonatal brain. It has also been linked to the disruption of blood brain barrier (BBB) in hypoxic-ischemic injury. Both experimental and clinical results have shown that antioxidants such as melatonin, a neurohormone synthesized and secreted by the pineal gland and edaravone (3-methyl-1-phenyl-2-pyrazolin-5-one), a newly developed drug, are effective in reducing oxidative stress and are promising neuroprotectants in reducing brain damage. Indeed, the neuroprotective effects of melatonin in many central nervous system (CNS) disease conditions such as amyotrophic lateral sclerosis, PD, AD, ischemic injury, neuropsychiatric disorders and head injury are well documented. Melatonin affords protection to the BBB in hypoxic conditions by suppressing the production of vascular endothelial growth factor and nitric oxide which are known to increase vascular permeability. The protective effects of melatonin against hypoxic damage have also been demonstrated in newborn animals whereby it attenuated damage in different areas of the brain. Furthermore, exogenous administration of melatonin in newborn animals effectively enhanced the surface receptors and antigens on the macrophages/microglia in the CNS indicating its immunoregulatory actions. Not only did melatonin prevent neuronal loss in the brain, but also protected the retinal neurons in ocular diseases such as glaucoma, diabetic retinopathy and age related macular degeneration. Another anti-oxidant, edaravone has been shown to reduce oxidative stress, edema, infarct volume, inflammation and apoptosis following ischemic injury of the brain in the adult as well as decrease free radical production in the neonatal brain following hypoxic-ischemic insult. It can counteract toxicity from activated microglia. This review summarizes the clinical and experimental data highlighting the therapeutic potential of melatonin and edaravone in neuroprotection in various disorders of the CNS.

Keywords: Activated microglia, antioxidants, blood brain barrier disruption, edaravone, excitotoxicity, glutathione, hypoxia, inflammation, lipid peroxidation,

***Corresponding author Eng-Ang Ling:** Department of Anatomy, Yong Loo Lin School of Medicine, Blk MD10, 4 Medical Drive, National University of Singapore, Singapore 117597; Tel: 65-65163203; E-mail: antlea@nus.edu.sg

Atta-ur-Rahman, Allen B. Reitz, M. Iqbal Choudhary and Jizhou Wang (Eds.)

malondialdehyde, melatonin, mitochondrial dysfunction, neurodegeneration, neuronal death, nitric oxide, oligodendrocytes death, oxidative stress, reactive oxygen species, retinal ganglion cell, vascular endothelial growth factor.

INTRODUCTION

Free radicals are chemical compounds containing one or more unpaired electrons in their outer orbits. Free radicals such as reactive oxygen species (ROS) are generated predominantly during cellular respiration and normal metabolism. The formation of ROS is balanced under normal conditions as they are neutralized or eliminated by antioxidants. However, under pathological conditions such as brain injuries and stroke, there is increased formation of free radicals resulting in the disruption of balance between their formation and the ability of cells to defend against them. This is known as oxidative stress. Free radicals can cause damage to lipids, proteins and DNA, resulting in their oxidation and causing cell death by necrosis or apoptosis. Certain organs such as the brain are predisposed to higher levels of oxidative stress as there is an imbalance between the amount of free radicals generated due to its high metabolic activity and low levels of endogenous antioxidants. Examples of ROS include hydroxyl ($OH^{\bullet}$), superoxide ($O_2^{\bullet-}$), hydrogen peroxide (H_2O_2), singlet oxygen ($^{-1}O_2$) and hypochloric acid (HOCl). Besides the ROS, reactive nitrogen species (RNS) such as nitrous oxide ($NO^{\bullet}$), peroxynitrite ($OONO^{-}$) and peroxynitrous acid (ONOOH) can facilitate nitrosylation reactions and cause damage to cells. A range of defense mechanisms exist in the tissues by which the cells are protected from the harmful effects of free radicals *e.g.* the presence of antioxidant enzymes such as superoxide dismutase (SOD), glutathione peroxidase (GPx), alpha lipoic acid (LA), catalase (CAT) and coenzyme Q10. SOD transforms harmful free radicals into hydrogen peroxide whereas CAT, an antioxidant enzyme, along with GPx converts the hydrogen peroxide into water and oxygen. LA, a cofactor for mitochondrial alpha-ketoacid dehydrogenase complexes, is a natural endogenous antioxidant.

Several endogenous antioxidant defenses are also present in the brain which prevents oxidative damage by ROS [1-5]. Neurons have lower levels of antioxidants such as glutathione (GSH) and GSH-related enzymes than glia [6-13]. *In vitro* studies have provided evidence that due to their higher antioxidant defences, glial cells play a critical role in protecting the neurons from oxidative damage [9, 14-17].

OXIDATIVE STRESS AND CELLULAR FUNCTIONS

ROS and RNS play a dual role as they have both deleterious and beneficial actions; moderate levels acting as signals to promote cell proliferation and survival, whereas increased production induces cell death [18, 19].

NO and ROS act as signaling molecules for many physiological functions including regulation of vascular tone, monitoring of oxygen tension in the control of ventilation and erythropoietin production, and signal transduction from membrane receptors in various physiological processes [20]. ROS are involved in modulation of many cellular processes from short-term ion-channel activation [21-24] to gene transcription and expression [25-27]. ROS are also involved in synaptic transmission [24, 28-34] and are known to mediate nonsynaptic communication between neurons and glia [35]. High ROS and NO concentration, generated in response to decreased activity of mitochondrial electron transport chain, could induce apoptotic cell death in various cell types [36-41] by mediating the release of mitochondrial cytochrome c into the cytosol [42].

MECHANISMS OF NEURONAL DEATH IN NEURODEGENERATIVE DISORDERS

Neuronal loss in neurodegenerative diseases is caused by a number of mechanisms such as excitotoxicity, disruption of blood brain barrier (BBB) and inflammation. Many studies have reported that these mechanisms are related to oxidative stress. Since the human brain represents only 2% of the body weight but accounts for 20% of the total adult body oxygen utilization [43], it generates more ROS than any other tissue [44]. The high content of polyunsaturated fatty acids, non-heme iron and ascorbic acid in the brain coupled with a relatively inefficient enzymatic antioxidative defense system make it particularly susceptible to damage by free radicals.

Loss of neurons and decline in cognitive functions occurs in neurodegenerative diseases including Alzheimer's disease (AD), Parkinson's disease (PD), Huntington's chorea and amyotrophic lateral sclerosis among many others [45]. Neuronal loss has also been reported in stroke, multiple sclerosis as well as in aging. Besides neuronal death, oligodendrocyte death might contribute to the pathology of neurodegenerative conditions such as multiple sclerosis [46-48], ischemic injury [49, 50], AD [51-54] and aging [55]. The involvement of oxidative stress in neuronal death [56] in the pathogenesis of these neurodegenerative disorders is well documented. Neurons are particularly vulnerable to oxidative stress as they have low antioxidant defences. Oligodendrocytes are also highly susceptible to damage induced by free radicals as they are iron containing cells [57] and have reduced level of glutathione (GSH), glutathione peroxidase and mitochondrial manganese superoxide dismutase [58].

AD, one of the major causes of dementia, is characterized by the deposition of the amyloid β (Aβ) peptide and microtubule associated protein tau. Abnormal tau phosphorylation has been reported to play a critical role in the pathogenesis of AD [59]. A central role of ROS in the pathogenesis of AD has been supported [60, 61]. Aβ has the capacity to interact with transition metals generating redox-active ions, which precipitate lipid peroxidation and cellular oxidative stress [62]. Mediation of oxidant toxicity by Aβ has also been implicated in tau phosphorylation [63-65]. Besides oxidative stress, mitochondrial dysfunction [66] and disruption of BBB have been described to occur in AD.

Aβ homeostasis and regulation of Aβ transport is maintained by the BBB *via* two main receptors, the low density lipoprotein receptor related protein 1 expressed on the brain capillary endothelium [67] and the receptor for advanced glycation end products [68] expressed on neurons, glial cells and endothelial cells [69]. These receptors have the ability to transport circulating free Aβ into the CNS [70] and brain-derived Aβ into the blood [71]. Following transport at the BBB, the Aβ is rapidly taken up by neurons inducing cellular stress [70].

PD is characterized by loss of dopaminergic neurons in the substantia nigra and the presence of the fibrillar cytoplasmic aggregates of α-synuclein, the Lewy bodies, in surviving neurons [72]. Free radicals and mitochondrial dysfunction have been implicated in the pathogenesis of PD [73]. In addition to the neuronal loss, pre- and postsynaptic neurochemical changes also occur [74]. The α-synuclein protein participates in the modulation of dopamine amounts at nerve terminals as it participates in synthesis, storage, release, and uptake of dopamine [75, 76] and also regulates the activity of the dopamine transporter (DAT) [77]. Oxidative stress, demonstrated in PD patients [78, 79], has been consistently associated with the loss of dopaminergic neurons in PD which are known to produce abundant quantities of ROS [80] involved in caspase activation and consequent apoptosis [81]. Increased levels of ROS are also associated with dopamine metabolism [82] and it has been reported that dopamine-associated oxidative stress may contribute to an inflammatory reaction seen in PD [83]. Impaired function of the BBB has been suggested to cause or accelerate progress of the disease [84] as absorption or metabolism of putative PD toxins, and their elimination across the BBB may be faulty [85].

Multiple sclerosis is an autoimmune disease characterized by demyelinating lesions. Inflammation, infiltration of inflammatory cells and activation of microglia have been reported as factors responsible for neural tissue damage [86]. Relatively high concentrations of NO have been reported to be produced by microglia, macrophages and other cells at sites of inflammation in multiple sclerosis [87-89]. This high level of NO reacts with superoxide to generate

$OONO^-$ [88, 90, 91] which is implicated in damaging myelin and axons [92]. Injection of a $OONO^-$ donor, 3-morpholinosydnonimine, into the corpus callosum of rats produced axonal damage and demyelination similar to that seen in multiple sclerosis. These lesions were suppressed by administration of the $OONO^-$ scavenger, uric acid [92]. Disruption of BBB has been reported to contribute to disease progression in multiple sclerosis [93, 94].

In hypoxic-ischemic insults, one of the earliest mechanisms postulated to explain CNS tissue damage was the generation of free radicals [95]. It is known that production of $O_2^{\bullet-}$ and NO by excitotoxic mechanisms results in formation of $OONO^-$ and $OH^{\bullet}$ [96]. $O_2^{\bullet-}$ participates in $OONO^-$ formation [97] and is also involved in the iron-catalyzed Haber-Weiss reaction [98] which causes H_2O_2 to be converted to OH. $OONO^-$ and OH have the potential to react with and damage lipids, proteins and DNA. Suppression of antioxidant enyzmes also adds to oxidative stress in conditions such as stroke. Mean plasma levels of nonenzymatic antioxidants and antioxidant enzyme activities have been reported to be lower in stroke patients compared to healthy controls [99]. Hypoxia also causes increased permeability of the vascular endothelium at the BBB resulting in leakage of plasma proteins and subsequent edema of the brain [100, 101]. The breakdown of the BBB following the hypoxic exposure may be attributed to a remarkable increase in vascular endothelial growth factor (VEGF) production [102].

OXIDATIVE STRESS AND THE BLOOD BRAIN BARRIER

Integrity of the BBB, formed by the capillary endothelial cells with tight junctions (TJ) between them, is important for the homeostatic regulation of the brain microenvironment [103] and maintenance of the immune-privileged status of the brain by restricting the entry of T lymphocytes. BBB disruption is an early phenomenon in many conditions such as multiple sclerosis [104] and AD [105]. Leakage of plasma proteins [104] due to endothelial TJ abnormality is widespread in the white matter in multiple sclerosis [93, 94, 106].

In vivo and *in vitro* studies have reported that free radicals cause disruption of BBB [107, 108]. ROS contribute to dysfunction of the BBB by altering the distribution or causing the disappearance of tight junction proteins claudin-5 and occludin [109]. Protein tyrosine kinase (PTK) signaling pathway may link oxidative stress, matrix metalloproteinases (MMPs), and loss of BBB function [110]. It has been reported that ROS-activated PTK pathway phosphorylates the TJ proteins (occludin, claudin-5, and ZO-1) at tyrosine residues [111]. Loss of brain endothelium stability and increase in BBB permeability *via* degradation of basement membrane proteins by activated MMPs has also been reported to be mediated by ROS [111-114]. It has been suggested that $OONO^-$ may be a direct trigger for increasing BBB permeability in CNS inflammation [115]. The involvement of $OONO^-$ in inducing BBB damage is evidenced by many studies which report that uric acid, a product of purine metabolism and a known scavenger of $OONO^-$, inactivates $OONO^-$ and reduces BBB permeability and immune cell invasion into the CNS [116, 117]. Lipid peroxidation has been reported to increase capillary permeability and cause formation of oedema [118].

OXIDATIVE STRESS AND INFLAMMATION

Inflammation is associated with multiple sclerosis, brain injuries, infections, AD, stroke, hypoxia-ischemia as well as with brain aging. Both microglia and astrocytes are activated in response to injury or infection and produce pro-inflammatory mediators such as tumor necrosis factor (TNF-α) and interleukins-1 and 6 (IL-1 and IL-6). Experimental studies on animal models producing inflammation in the brain with lipopolysaccaride (LPS) have shown the expression of TNF-α and interleukins [119-121]. Microglia have been reported to be the main cell type expressing IL-6 [122], IL-1α, IL-1β [123] and TNF-α. Melatonin has been shown to reverse the inflammatory response by suppressing the production of inflammatory cytokines. Activated glia is reported to be associated with Aβ peptide deposits in the brain of AD patients as a result of chronic inflammation [124]. The involvement of inflammation and activated microglia in PD was initially suggested by McGeer *et al.* [125] and later reported in many other studies [126-128]. Elevated levels TNF-α and IL1β in the brain and cerebrospinal fluid (CSF) of PD patients have been demonstrated by many studies [129-133]. Microglia and astrocytes are stimulated by oxidative stress to secrete proinflammatory cytokines such as TNF-α [134, 135] which causes neuronal caspase-8 activation [136].

OXIDATIVE STRESS AND EXCITOTOXICITY

Excitotoxic cell death is a common pathologic event occurring in the CNS in a number of neurologic diseases and insults including multiple sclerosis, stroke, and traumatic injury [137-140]. Glutamate induces and exacerbates cell death by activating ionotropic and group I mGluRs [141, 142]. Accumulation of excess glutamate in the extracellular spaces leads to persistent activation of glutamate receptors resulting in excitotoxic degeneration of the neurons and glial cells [143-146]. Higher concentrations of extracellular glutamate have been reported in the serum or CSF of patients suffering from multiple sclerosis [147-149] and this has been hypothesized to be caused by enhanced release from activated macrophages/microglia or alterations in the expression of glutamate transporters [150]. Glutamate concentration in the CSF and brain tissue of adult patients and children suffering from traumatic brain injury [151-156] was also found to be higher than normal subjects. Massive increase in levels of glutamate (> 300 times the normal levels) was detected in the extracellular fluid in the brain tissue in stroke [157].

Glutamate toxicity leads to increased production of free radicals such as superoxide or hydroxyl radicals [158-163] which impair plasma membrane ionic transport systems [164] and results in glutathione depletion [165]. Glutathione is a major cellular antioxidant which protects the cells against oxidative stress [166-168] and its depletion results in an increase in ROS generation [169, 170]. Glutathione depletion in oligodendrocytes leading to their death by free radical mediated mechanisms has been described to be caused by glutamate [171]. Developing oligodendrocytes were reported to be vulnerable to intracellular glutathione depletion [172, 173] and subsequent accumulation of ROS [171, 172, 174].

EXOGENOUS ANTIOXIDANTS AND THE ADULT BRAIN

Melatonin

Melatonin (5-methoxy-N-acetyltryptamine), a derivative of the amino acid tryptophan (Fig. **1**), is primarily secreted by the pineal gland and acts as an antioxidant [56, 175, 176], a biologic modulator of mood, sleep, sexual behaviour, reproductive alterations and circadian rhythms [177, 178] and has immunoregulatory actions [179, 180]. Its anticonvulsant [181-183], anti-aging [184] and oncostatic effects against a variety of tumours [185-191] have been widely reported. It is also known to alleviate a variety of circadian rhythm-related disorders, including jet lag, and shift work [192-196]. Many studies have reported the neuroprotective ability of melatonin in PD [175, 197], AD [175, 197, 198], amyotrophic lateral sclerosis [199], neural trauma [200, 201] and ischemia/reperfusion injury [202-205]. It has also been shown to reduce vascular leakage in the brain when the BBB is disrupted [102].

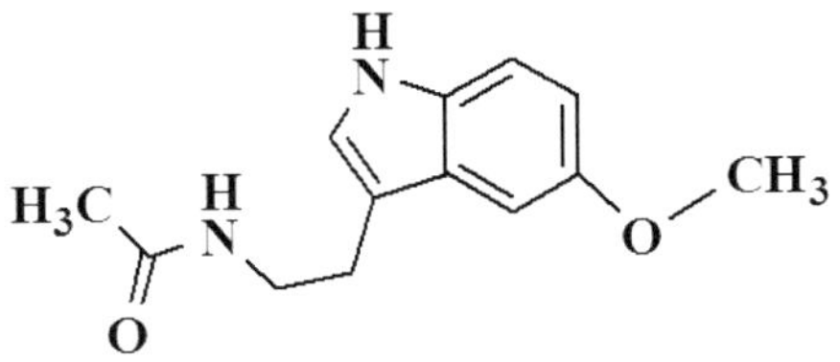

Figure 1: Structure of Melatonin.

Melatonin acts as an antioxidant by its ability to scavenge both ROS and RNS and also its ability to stimulate the activity of antioxidative enzymes [206]. It can remove the OH, H_2O_2, $^1O_2$, HOCl, $OONO^-$, ONOOH [176, 207] and reduce lipid peroxidation [208]. Aldehydes such as malondialdehyde (MDA) and 4-hydroxy-2-nonenal [4- HNE) formed during lipid peroxidation cause tissue damage by augmenting free radical events [209]. Melatonin interacts with MDA to protect against cellular damage induced by oxidative stress [210]. The activity and expression of antioxidant enzymes, such as SOD, CAT, and GPx and glutathione reductase has been shown to be increased by melatonin [211-215]. It has been reported to downregulate the prooxidant enzymes such as inducible nitric oxide synthases and, hence, production of NO [102, 216]. It also reduces excitotoxicity by reducing the release of glutamate (Kaur *et al.*, unpublished observations) and inflammation by decreasing production of inflammatory cytokines [217].

Melatonin has been reported to inhibit apoptosis of neurons induced by A*β* and suppress the the formation of A*β* itself from amyloid *β* precursor protein (APP) [65]. Aβ induced activation of microglial nicotinamide adenine dinucleotide phosphate (NADPH) oxidase has been reported as a source of ROS in AD [218]. It was recently reported that melatonin was able to decrease the cellular ROS level by inhibiting the activity of NADPH oxidase [219]. It also reduced tau hyperphosphorylation and MDA levels [220]. In experimentally produced AD, melatonin was able to decrease expression of proinflammatory IL-1 α [124].

Free radicals generated from both mitochondrial impairment and dopamine metabolism contribute to oxidative stress and are considered to play a critical role in pathogenesis of PD [221]. Inflammation, excitotoxicity and excessive NO generation are other contributing factors in the progression of disease [222, 223]. Melatonin has been shown to reduce neuronal loss in experimental models of PD [74, 224] by preventing α-synuclein aggregation, downregulating DAT, inhibiting dopamine autoxidation as well as free radical-mediated dopamine oxidation [74, 225]. It has been reported to reduce formation of OH and increase SOD activity in mice treated with parkinsonian neurotoxin, 1-methyl-4-phenyl-1, 2, 3, 6-tetrahydropyridine [226]. However, some studies have reported that melatonin administration exacerbates motor impairment in PD patients [227, 228].

Oxidative stress has been considered to play a pivotal role in hypoxic-ischemic brain damage. Melatonin is known to reduce oxidative damage during ischemia and reperfusion [229], abolish the increase in MDA levels [230] and preserves mitochondrial function during ischemia and reperfusion. Disturbance of mitochondrial respiration during ischemia has been reported to result in excessive generation of ROS [231]. Preservation of mitochondrial function by melatonin suppresses mitochondrial oxidative stress [232]. Melatonin treatment is also known to prevent the production of iNOS involved in the inflammatory process [233, 234]. Increased iNOS expression occurs in many tissues in response to hypoxia [235] resulting in increased production of NO. NO can function as a neurotoxin by being a free radical as well as a precursor of $OONO^-$, a potent oxidant, a nitrating agent for protein modification and by increasing intranuclear Ca^{2+} influx [236]. Recent studies have reported significant reduction in NO production following melatonin administration in hypoxic rats [102, 216].

Melatonin's effectiveness in reducing edema formation in ischemic animals was evident [237] with its ability to decrease the permeability of the disrupted BBB. Reduction in increased vascular permeability in hypoxic/ischemic conditions [238, 239] occurs through modulation of endothelial–neutrophil interactions [240], and antioxidant actions [241-244]. This could also be due to its suppressive effect on the tissue concentration of VEGF [102, 245] resulting in reduced vascular permeability and edema in many parts of the CNS [102]. Suppression of NO and aquaporin 4 (AQ4) production by melatonin has also been reported to occur in hypoxic brains [102, 246]. AQ4 is the astroglial water channel which facilitates water movement into and out of brain parenchyma [247] and has been implicated in the production of edema in pathological conditions.

Edaravone

Edaravone (3-methyl-1-1phenyl-2-pyrazolin-5-one, MCI-186, Fig. **2**) is a newly developed free radical scavenger, with neuroprotective effects in cerebral ischemia [248, 249]. It has been approved in Japan for treating acute brain infarction [250]. Edaravone has been reported to exert antioxidant effects as it can quench ·OH and inhibit ·OH-dependent and ·OH-independent lipid peroxidation [248, 251]. It has been shown to reduce the increase of hydroxyl radical and superoxide anion level in several models of cerebral ischemia [252, 253].

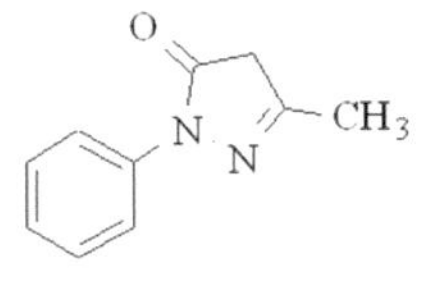

Figure 2: Structure of Edaravone (3-methyl-1-1phynyl-2-pyrazolin-5-one).

ROS and Ca^{2+} overload during ischemia/reperfusion also result in opening of the mitochondrial permeability transition pore (PTP), a non-specific pore in the inner membrane of the mitochondria, leading to cell damage [254]. Edaravone has been reported to attenuate significantly Ca^{2+} induced swelling in rat brain mitochondria [254]. The neuroprotective effects of edaravone were reported to be mediated *via* its antioxidant effect through suppression of lipid peroxidation and oxidative DNA damage in a mouse stroke model [255]. It is also known to exert anti-inflammatory effect by inhibition of microglial activity and inhibition of peroxynitrite production through reduction of iNOS activity [255]. Edaravone has also been shown to protect ischemic neurons from apoptosis *via* suppressing the gene expression of the Fas/FasL signaling pathway [256, 257] and through a Bax/Bcl-2 dependent anti-apoptotic mechanism [258]. The protective effect of edaravone in hypoxic/ischemic injury has also been attributed to inhibition of endoplasmic stress and subsequent apoptotic signaling pathway [259].

Edaravone has been also reported to suppress brain edema in hypoxic/ischemic conditions in animal models [260, 261] and patients with acute cerebral ischemia [262, 263]. This effect was attributed to edaravone's ability to inhibit VEGF expression in astrocytes [261].

Besides inhibiting ROS generation, edaravone has also been shown to reduce ROS-induced inflammatory reaction in stroke [264]. Oxidative stress activates nuclear-factor kappa B (NF-κB) and several mediators of inflammation such as iNOS, cytokines, cyclooxygenase-2 which result in delayed damage to the ischemic area in stroke [265-272]. Edaravone has been reported to reduce iNOS expression and suppress the increase in oxidative markers, such as 4-hydroxy-nonenal (4-HNE)-modified protein as well as neutrophil activation in ischemic injury [25, 156, 273].

Exogenous Antioxidants and the Developing Brain

The developing brain has been described to be more susceptible than the adult brain to oxidative stress occurring in many conditions such as hydrocephalus [274] and hypoxic ischemic injury [275]. The antioxidant defenses in the developing brain are also low as it possesses lower levels of the GPx and CAT [276]. Overproduction of NO, lipid peroxidation [245, 277, 278] and DNA damage in response to asphyxia or hypoxia/ischemia has been reported. Activation of microglial cells, inflammation and cell death occurs in the brain white matter in the asphyxiated fetal brain [279]. Increased expression of pro-inflammatory cytokines such as TNF-α, IL-1β and IL-18 by activated microglia in hypoxic/ischemic injury in the neonatal rat [280-282] was recently reported to be one of the mechanisms responsible underlying deaths of oligodendrocytes [283].

In addition to the above, increased vascular permeability has also been reported in the developing brain in many regions in hypoxic injury [102, 245, 284] and this was attributed to increased production of VEGF and NO. Cerebral edema was evidenced by marked swelling of astrocytes associated with the blood vessels in the hypoxic brains [245]. Melatonin was effective in reducing MDA and nitrite/nitrate levels in the blood of asphyxiated newborns [285] as well as VEGF and NO production in the hypoxic brain [245, 284]. Melatonin has also been reported to decrease BBB permeability, increase glutathione content and reduce NO levels in the choroid plexus of the infantile rats with kaolin-induced hydrocephalus [286]. Maternal administration of melatonin has been reported to abrogate oxidative damage to the fetal brain in asphyxia [278] by minimizing inflammation and cell death [279].

Trauma is known to trigger apoptotic neuronal death in the developing brain [287]. Oxidative stress, derangement in cerebral blood flow, excitotoxicity and inflammation described as a result of traumatic damage in infants and children [288] may be some of the factors underlying cell death. Melatonin has proven to be useful in counteracting oxidative damage and reducing cell death in traumatic brain injury in the immature brain [289, 290]. Maternally-administered lipopolysaccharide has been reported to increase the level of TNF-α protein in fetal brain [291-293] and maternal administration of melatonin significantly decreased LPS induced enhanced TNF-α production in fetal brain [293].

Edaravone has also been shown to have neuroprotective effects in the developing brain. It suppressed the production of nitric oxide, lipid peroxidation and formation of free radicals during hypoxic-ischemic insult and was able to attenuate neuronal damage [294-296]. Maternal administration of edaravone has been shown to reduce hydroxyl radical formation and inhibit fetal brain injury resulting from umbilical cord occlusion [297, 298].

Exogenous Antioxidants and the Adult Retina

During embryonic development the retina develops as an outgrowth of the brain. Similar to the brain, the retina and retinal neurons, especially the retinal ganglion cells (RGCs), are highly vulnerable to damage due to hypoxia-ischemia, oxidative stress, excitotoxicity and inflammation. In sight threatening ocular diseases such as glaucoma, diabetic retinopathy, age-related macular degeneration and retinitis pigmentosa; several factors including excitotoxicity [299, 300], NO [300], inflammation [301] and increased vascular leakage [302] have been reported to mediate death of RGCs. Nevertheless, all these factors have been shown to cause apoptosis of RGCs by promoting free radical generation. This suggests that oxidative stress plays a key role in the death of the RGCs in various ocular pathologies [303-305]. Consistent with this there were elevated levels of free radicals in the retina, in conditions such as diabetic retinopathy [304], retinal ischemia [305] and glaucoma [306]. Just like the brain, the retina is also sensitive to high levels of glutamate [307] which is known to exert its action specifically through the N-methyl-d-aspartate (NMDA) receptors [307]. In the above mentioned ocular disorders, activation of NMDA receptors has been demonstrated to mediate death of RGCs. However, this was demonstrated to occur *via* generation of free radicals such as NO and superoxide [308]. Furthermore, the increased level of free radical NO was associated with blood retinal barrier (BRB) disruption [302]. Similarly, pro-inflammatory cytokine TNF-α was reported to induce free radical generation in cultured retinal pigment epithelial cells [309]. As mentioned above the elevated levels of free radicals result in peroxidation of the lipid rich retinal membranes and DNA damage, leading to increased levels of MDA and 8-hydroxy-2′-deoxyguanosine (8-OHdG) respectively [304]. Several anti-oxidants have been suggested to have therapeutic potential in ameliorating death of RGCs in various ocular diseases. However their clinical efficacy warrants further investigation.

Melatonin and its derivatives have been shown to have a therapeutic potential in treating ocular diseases such as glaucoma [310, 311], retinal ischemia [312], age-related macular degeneration [313] and diabetic retinopathy [314]. Melatonin has been shown to exert its antioxidant effect on the retina by scavenging free radicals and NO or by inhibiting their formation [315, 316]. Furthermore, in glaucomatous retina, melatonin has been shown to reduce the death of RGCs by decreasing the intraocular pressure [317]. Daily intake of melatonin, approximately 3mg for three months before bed time, has been shown to delay the onset of age-related macular degeneration [313]. In ischemic retina, hypoxia inducible factor (HIF)-1α induced NO production was suggested to cause detrimental effects on RGCs. However, melatonin was shown to increase the survival of RGCs in ischemic retina by inhibiting the increased expression of HIF-1α [312]. In rats with type 2 diabetes mellitus, apart from preventing oxidative stress and reduction in CAT activity in the retina, melatonin was able to decrease the NOS activity and TNF-α level [318]. Moreover, melatonin was shown to protect the integrity of BRB by decreasing VEGF and NO levels in hypoxic rat retina [302].

Edaravone has been reported to lower MDA levels and raise the activity of superoxide dismutase in ischemic retinal tissues [319]. In addition, it was reported to attenuate apoptosis of retinal neurons induced by either ischemia or NMDA [319, 320]. In oxygen-glucose deprived RGC cultures edaravone reduced the levels of toxic free radicals such as hydrogen peroxide, superoxide and hydroxyl radicals [320]. In light induced photoreceptor degeneration models, edaravone protected the photoreceptor cells from free radicals and this was evident with the reduced expression of 8-OHdG [321].

Exogenous Antioxidants and the Developing Retina

The developing neural retina is also affected by various insults such as hypoxia-ischemia, bright light exposure, hyperoxia and asphyxia [322-324] leading to death of RGCs. Our studies on developing hypoxic retina showed a widespread presence of apoptotic and necrotic cells, swollen Müller cell processes, and increased vascular leakage. In hypoxic immature retina there was increased expression of glutamate receptors and glutamate levels [325, 326]. The activation of glutamate receptors following a hypoxic insult resulted in increased intracellular ROS production within the RGCs [326]. Along with this, activated retinal microglial cells were actively involved in inflammatory response in the hypoxic immature retina [327]. Moreover, oxidative stress was evident with increased lipid peroxidation and reduced glutathione content [328]. The apoptosis of RGCs was evident with increased expression of cytosolic cytochrome-c and caspase-3 in them following hypoxic insult [325, 328].

However, these changes were attenuated in retinas of hypoxic neonatal rats treated with melatonin [324]. In cultures of RGCs exposed to hypoxia and treated with melatonin, there was reduced expression of cytosolic cytochrome-c and caspase-3 when compared with hypoxic RGCs not treated with melatonin [328]. Moreover, melatonin effectively reduced the production of cytokines in hypoxic retinal microglial cultures. Melatonin when administered to hypoxic neonatal rats resulted in reduced number of apoptotic and necrotic cells and reduced the swelling of Müller cell processes [324]. Concomitantly there was increased glutathione content and reduced lipid peroxidation in retinas of hypoxic rats treated with melatonin [328]. In hypoxic neonatal retina, melatonin was shown to protect the BRB integrity by down-regulating the expression of VEGF and its receptors Flt-1 and FlK-1, which are implicated in causing BRB disruption [324, 328]. Associated with this there was reduced leakage of intraperitoneally or intravenously administered tracers such as rhodamine isothiocyanate and horseradish peroxidase from the retinal and hyaloid vessels in the retina of hypoxic rats administered with melatonin [324].

Although there is no investigation available on the beneficial effects of edaravone in neonatal retinal injury, edaravone could be of therapeutic interest owing to its anti-oxidant nature. Similar to its beneficial effect in the developing brain, it is possible that edaravone could abrogate free radical mediated reactions in the immature retina too.

CONCLUSION

Several experimental and clinical studies have reported the therapeutic potential of melatonin in alleviation of oxidative stress and inflammation in many CNS conditions such as AD, PD and stroke in the adult brain. Melatonin reduces damage to the developing brain in hypoxic/ischemic or traumatic brain injury by similar mechanisms as in the adult brain. This neuroprotective property of melatonin was also observed in both the adult and developing retina. Another antioxidant drug, edaravone, has shown promising results due to its effectiveness in reducing lipid peroxidation as well as inflammation in ischemic injury of the brain and has proved to be beneficial in reducing the infarct volume in stroke patients in clinical trials in Japan. Keeping the beneficial effects of melatonin in view, it merits consideration as a potential therapeutic agent for treatment of CNS disorders of different etiologies in adults, infants and children. Future research on other beneficial properties of edaravone besides amelioration of oxidative stress and inflammation and its involvement in various cellular signaling events is warranted to explore its full neuroprotective potential.

ACKNOWLEDGEMENTS

This study was supported by a research grant (R181-000-0098-112) from the National University of Singapore.

CONFLICT OF INTEREST

The authors confirm that this chapter contents have no conflict of interest.

DISCLOSURE

This is an updated report of our earlier review article "Antioxidants and neuroprotection in the adult and developing central nervous system (review)" by Kaur C and Ling EA. Curr Med Chem. 2008;15(29):3068-80.

ABBREVIATIONS

AD	=	Alzheimer's disease
AQ4	=	Aquaporin 4
BBB	=	Blood-brain barrier
BRB	=	Blood-retinal barrier
CAT	=	Catalase
CNS	=	Central nervous system
CSF	=	Cerebrospinal fluid
DAT	=	Dopamine receptor
GPx	=	Glutathione peroxidase
GSH	=	Glutathione
HIF-1α	=	Hypoxia inducible factor-1α
H_2O_2	=	Hydrogen peroxide
OH°	=	Hydroxyl radical
HOCl	=	Hypochloric acid
8-OHdG	=	8-hydroxy-2′-deoxyguanosine
LA	=	Lipoicacid
LPS	=	Lipopolysaccharides
MDA	=	Malondialdehyde
MMPs	=	Matrix metalloproteinases
NADPH	=	Nicotinamide adenine dinucleotide phosphate oxidase

NMDA = N-methyl-d-aspartate

NO = Nitric oxide

NO° = Nitrous oxide

$OONO^{-}$ = Peroxynitrite

ONOOH = Peroxynitrous acid

PD = Parkinson's disease

PTK = Protein tyrosine kinase

RGCs = Retinal ganglion cells

ROS = Reactive oxygen species

RNS = Reactive nitrogen species

$^{-}1O_2$ = Singlet oxygen

SOD = Superoxide dismutase

TJ = Tight junction

VEGF = Vascular endothelial growth factor

REFERENCES

[1] Yu BP. Cellular defenses against damage from reactive oxygen species. Physiol rev. 1994;74(1):139-62.

[2] Cohen G. Enzymatic/nonenzymatic sources of oxyradicals and regulation of antioxidant defenses. Ann N Y Acad Sci. 1994;738:8-14.

[3] Meister A. Glutathione-ascorbic acid antioxidant system in animals. J Biol Chem. 1994;269(13):9397-400.

[4] Rice ME. Ascorbate regulation and its neuroprotective role in the brain. Trends Neurosci. 2000;23(5):209-16.

[5] Dringen R, Pawlowski PG, Hirrlinger J. Peroxide detoxification by brain cells. J Neurosci Res. 2005;79(1-2):157-65.

[6] Slivka A, Mytilineou C, Cohen G. Histochemical evaluation of glutathione in brain. Brain Res. 1987;409(2):275-84.

[7] Raps SP, Lai JC, Hertz L, Cooper AJ. Glutathione is present in high concentrations in cultured astrocytes but not in cultured neurons. Brain Res. 1989;493(2):398-401.

[8] Makar TK, Nedergaard M, Preuss A, Gelbard AS, Perumal AS, Cooper AJ. Vitamin E, ascorbate, glutathione, glutathione disulfide, and enzymes of glutathione metabolism in cultures of chick

astrocytes and neurons: evidence that astrocytes play an important role in antioxidative processes in the brain. J Neurochem. 1994;62(1):45-53.
[9] Desagher S, Glowinski J, Premont J. Astrocytes protect neurons from hydrogen peroxide toxicity. J Neurosci. 1996;16(8):2553-62.
[10] Trepanier G, Furling D, Puymirat J, Mirault ME. Immunocytochemical localization of seleno-glutathione peroxidase in the adult mouse brain. Neuroscience. 1996;75(1):231-43.
[11] Peuchen S, Bolanos JP, Heales SJ, Almeida A, Duchen MR, Clark JB. Interrelationships between astrocyte function, oxidative stress and antioxidant status within the central nervous system. Prog Neurobiol. 1997;52(4):261-81.
[12] Rice ME, Russo-Menna I. Differential compartmentalization of brain ascorbate and glutathione between neurons and glia. Neuroscience. 1998;82(4):1213-23.
[13] Dringen R, Hirrlinger J. Glutathione pathways in the brain. Biol Chem. 2003;384(4):505-16.
[14] Drukarch B, Schepens E, Jongenelen CA, Stoof JC, Langeveld CH. Astrocyte-mediated enhancement of neuronal survival is abolished by glutathione deficiency. Brain Res. 1997;770(1-2):123-30.
[15] Drukarch B, Schepens E, Stoof JC, Langeveld CH, Van Muiswinkel FL. Astrocyte-enhanced neuronal survival is mediated by scavenging of extracellular reactive oxygen species. Free Radic Biol Med. 1998;25(2):217-20.
[16] Tanaka J, Toku K, Zhang B, Ishihara K, Sakanaka M, Maeda N. Astrocytes prevent neuronal death induced by reactive oxygen and nitrogen species. Glia. 1999;28(2):85-96.
[17] Dringen R, Kussmaul L, Gutterer JM, Hirrlinger J, Hamprecht B. The glutathione system of peroxide detoxification is less efficient in neurons than in astroglial cells. J Neurochem. 1999;72(6):2523-30.
[18] Valko M, Leibfritz D, Moncol J, Cronin MT, Mazur M, Telser J. Free radicals and antioxidants in normal physiological functions and human disease. Int J Biochem Cell Biol. 2007;39(1):44-84.
[19] Trachootham D, Lu W, Ogasawara MA, Nilsa RD, Huang P. Redox regulation of cell survival. Antioxid Redox Signal. 2008;10(8):1343-74.
[20] Droge W. Free radicals in the physiological control of cell function. Physiol rev. 2002;82(1):47-95.
[21] Seutin V, Scuvee-Moreau J, Massotte L, Dresse A. Hydrogen peroxide hyperpolarizes rat CA1 pyramidal neurons by inducing an increase in potassium conductance. Brain Res. 1995;683(2):275-8.
[22] Krippeit-Drews P, Kramer C, Welker S, Lang F, Ammon HP, Drews G. Interference of H2O2 with stimulus-secretion coupling in mouse pancreatic beta-cells. The Journal of physiology. 1999;514 (Pt 2):471-81.
[23] Avshalumov MV, Chen BT, Koos T, Tepper JM, Rice ME. Endogenous hydrogen peroxide regulates the excitability of midbrain dopamine neurons *via* ATP-sensitive potassium channels. J Neurosci. 2005;25(17):4222-31.
[24] Avshalumov MV, Rice ME. Activation of ATP-sensitive K+ (K(ATP)) channels by H2O2 underlies glutamate-dependent inhibition of striatal dopamine release. Proc Natl Acad Sci U S A. 2003;100(20):11729-34.
[25] Suzuki YJ, Forman HJ, Sevanian A. Oxidants as stimulators of signal transduction. Free Radic Biol Med. 1997;22(1-2):269-85.
[26] Haddad JJ. Antioxidant and prooxidant mechanisms in the regulation of redox(y)-sensitive transcription factors. Cell Signal. 2002;14(11):879-97.
[27] Esposito F, Ammendola R, Faraonio R, Russo T, Cimino F. Redox control of signal transduction, gene expression and cellular senescence. Neurochem Res. 2004;29(3):617-28.
[28] Pellmar TC. Peroxide alters neuronal excitability in the CA1 region of guinea-pig hippocampus *in vitro*. Neuroscience. 1987;23(2):447-56.
[29] Pellmar TC. Use of brain slices in the study of free-radical actions. Journal of neuroscience methods. 1995;59(1):93-8.
[30] Avshalumov MV, Rice ME. NMDA receptor activation mediates hydrogen peroxide-induced pathophysiology in rat hippocampal slices. Journal of neurophysiology. 2002;87(6):2896-903.
[31] Chen BT, Avshalumov MV, Rice ME. H2O2 is a novel, endogenous modulator of synaptic dopamine release. J. Neurophysiol. 2001;85(6):2468-76.
[32] Chen BT, Avshalumov MV, Rice ME. Modulation of somatodendritic dopamine release by endogenous H(2)O(2): susceptibility in substantia nigra but resistance in VTA. J. Neurophysiol. 2002;87(2):1155-8.

[33] Avshalumov MV, Chen BT, Rice ME. Mechanisms underlying H(2)O(2)-mediated inhibition of synaptic transmission in rat hippocampal slices. Brain Res. 2000;882(1-2):86-94.
[34] Avshalumov MV, Chen BT, Marshall SP, Pena DM, Rice ME. Glutamate-dependent inhibition of dopamine release in striatum is mediated by a new diffusible messenger, H_2O_2. J Neurosci. 2003;23(7):2744-50.
[35] Atkins CM, Sweatt JD. Reactive oxygen species mediate activity-dependent neuron-glia signaling in output fibers of the hippocampus. J Neurosci. 1999;19(17):7241-8.
[36] Slater AF, Stefan C, Nobel I, van den Dobbelsteen DJ, Orrenius S. Signalling mechanisms and oxidative stress in apoptosis. Toxicology letters. 1995;82-83:149-53.
[37] Adamson DC, Wildemann B, Sasaki M, Glass JD, McArthur JC, Christov VI, *et al.* Immunologic NO synthase: elevation in severe AIDS dementia and induction by HIV-1 gp41. Science. 1996;274:1917-21.
[38] De Groot CJ, Ruuls SR, Theeuwes JW, Dijkstra CD, Van der Valk P. Immunocytochemical characterization of the expression of inducible and constitutive isoforms of nitric oxide synthase in demyelinating multiple sclerosis lesions. J Neuropathol Exp Neurol. 1997;56(1):10-20.
[39] Thippeswamy T, Morris R. Cyclic guanosine 3',5'-monophosphate-mediated neuroprotection by nitric oxide in dissociated cultures of rat dorsal root ganglion neurones. Brain Res. 1997;774(1-2):116-22.
[40] Albina JE, Reichner JS. Role of nitric oxide in mediation of macrophage cytotoxicity and apoptosis. Cancer metastasis reviews. 1998;17(1):39-53.
[41] Dumont A, Hehner SP, Hofmann TG, Ueffing M, Droge W, Schmitz ML. Hydrogen peroxide-induced apoptosis is CD95-independent, requires the release of mitochondria-derived reactive oxygen species and the activation of NF-kappaB. Oncogene. 1999;18(3):747-57.
[42] Ushmorov A, Ratter F, Lehmann V, Droge W, Schirrmacher V, Umansky V. Nitric-oxide-induced apoptosis in human leukemic lines requires mitochondrial lipid degradation and cytochrome C release. Blood. 1999;93(7):2342-52.
[43] Clarke DD, Sokoloff L. Circulation and energy metabolism of the brain. In: Siegel G, Agranoff B, Albers RW, Molinoff P, Eds. Basic Neurochemistry: Molecular, Cellular, and Medical Aspects. 5th ed. New York: Raven Press 1994; pp. 645-680.
[44] Reiter RJ, Tan DX, Leon J, Kilic U, Kilic E. When melatonin gets on your nerves: its beneficial actions in experimental models of stroke. Exp. Biol. Med.. 2005;230:104-17.
[45] Martin JB. Molecular basis of the neurodegenerative disorders. N Engl J Med. 1999;340(25):1970-80.
[46] Boccaccio GL, Steinman L. Multiple sclerosis: from a myelin point of view. J Neurosci Res. 1996;45(6):647-54.
[47] Lucchinetti CF, Bruck W, Rodriguez M, Lassmann H. Distinct patterns of multiple sclerosis pathology indicates heterogeneity on pathogenesis. Brain Pathol. 1996;6(3):259-74.
[48] Benjamins JA, Nedelkoska L, George EB. Protection of mature oligodendrocytes by inhibitors of caspases and calpains. Neurochem Res. 2003;28(1):143-52.
[49] Mandai K, Matsumoto M, Kitagawa K, Matsushita K, Ohtsuki T, Mabuchi T, *et al.* Ischemic damage and subsequent proliferation of oligodendrocytes in focal cerebral ischemia. Neuroscience. 1997;77(3):849-61.
[50] Mabuchi T, Kitagawa K, Ohtsuki T, Kuwabara K, Yagita Y, Yanagihara T, *et al.* Contribution of microglia/macrophages to expansion of infarction and response of oligodendrocytes after focal cerebral ischemia in rats. Stroke. 2000;31(7):1735-43.
[51] Lee JT, Xu J, Lee JM, Ku G, Han X, Yang DI, *et al.* Amyloid-beta peptide induces oligodendrocyte death by activating the neutral sphingomyelinase-ceramide pathway. J Cell Biol. 2004;164(1):123-31.
[52] Whitman GT, Cotman CW. Oligodendrocyte degeneration in AD. Neurobiol Aging. 2004;25(1):33-6.
[53] Xu WL, Qiu CX, Wahlin A, Winblad B, Fratiglioni L. Diabetes mellitus and risk of dementia in the Kungsholmen project: a 6-year follow-up study. Neurology. 2004;63(7):1181-6.
[54] Roth AD, Ramirez G, Alarcon R, Von Bernhardi R. Oligodendrocytes damage in Alzheimer's disease: beta amyloid toxicity and inflammation. Biol Res. 2005;38(4):381-7.
[55] Bartzokis G. Age-related myelin breakdown: a developmental model of cognitive decline and Alzheimer's disease. Neurobiol Aging. 2004;25(1):5-18; author reply 49-62.
[56] Reiter RJ. Oxidative damage in the central nervous system: protection by melatonin. Prog Neurobiol. 1998;56(3):359-84.

[57] Connor JR, Pavlick G, Karli D, Menzies SL, Palmer C. A histochemical study of iron-positive cells in the developing rat brain. J Comp Neurol. 1995;355:111-23.

[58] Bernhard MC, Junker E, Hettinger A, Lauterburg BH. Time course of total cysteine, glutathione and homocysteine in plasma of patients with chronic hepatitis C treated with interferon-alpha with and without supplementation with N-acetylcysteine. Journal of hepatology. 1998;28(5):751-5.

[59] Vickers JC, Dickson TC, Adlard PA, Saunders HL, King CE, McCormack G. The cause of neuronal degeneration in Alzheimer's disease. Prog Neurobiol. 2000;60(2):139-65.

[60] Smith MA, Sayre LM, Monnier VM, Perry G. Radical AGEing in Alzheimer's disease. Trends Neurosci. 1995;18(4):172-6.

[61] Markesbery WR. Oxidative stress hypothesis in Alzheimer's disease. Free Radic Biol Med. 1997;23(1):134-47.

[62] Rottkamp CA, Raina AK, Zhu X, Gaier E, Bush AI, Atwood CS, *et al.* Redox-active iron mediates amyloid-beta toxicity. Free Radic Biol Med. 2001;30(4):447-50.

[63] Busciglio J, Lorenzo A, Yeh J, Yankner BA. beta-amyloid fibrils induce tau phosphorylation and loss of microtubule binding. Neuron. 1995;14(4):879-88.

[64] Pratico D, Uryu K, Leight S, Trojanoswki JQ, Lee VM. Increased lipid peroxidation precedes amyloid plaque formation in an animal model of Alzheimer amyloidosis. J Neurosci. 2001;21(12):4183-7.

[65] Srinivasan V. Melatonin oxidative stress and neurodegenerative diseases. Indian journal of experimental biology. 2002;40(6):668-79.

[66] Takuma K, Yan SS, Stern DM, Yamada K. Mitochondrial dysfunction, endoplasmic reticulum stress, and apoptosis in Alzheimer's disease. J Pharmacol Sci. 2005;97(3):312-6.

[67] Wang X, Lee SR, Arai K, Lee SR, Tsuji K, Rebeck GW, *et al.* Lipoprotein receptor-mediated induction of matrix metalloproteinase by tissue plasminogen activator. Nat Med. 2003;9(10):1313-7.

[68] Deane R, Zlokovic BV. Role of the blood-brain barrier in the pathogenesis of Alzheimer's disease. Current Alzheimer research. 2007;4(2):191-7.

[69] Lue LF, Yan SD, Stern DM, Walker DG. Preventing activation of receptor for advanced glycation endproducts in Alzheimer's disease. Current drug targets CNS and neurological disorders. 2005;4(3):249-66.

[70] Deane R, Du Yan S, Submamaryan RK, LaRue B, Jovanovic S, Hogg E, *et al.* RAGE mediates amyloid-beta peptide transport across the blood-brain barrier and accumulation in brain. Nat Med. 2003;9(7):907-13.

[71] Shibata M, Yamada S, Kumar SR, Calero M, Bading J, Frangione B, *et al.* Clearance of Alzheimer's amyloid-ss(1-40) peptide from brain by LDL receptor-related protein-1 at the blood-brain barrier. J Clin Invest. 2000;106(12):1489-99.

[72] Spillantini MG, Crowther RA, Jakes R, Hasegawa M, Goedert M. alpha-Synuclein in filamentous inclusions of Lewy bodies from Parkinson's disease and dementia with lewy bodies. Proc Natl Acad Sci U S A. 1998;95(11):6469-73.

[73] Jenner P, Olanow CW. Understanding cell death in Parkinson's disease. Ann Neurol. 1998;44(3 Suppl 1):S72-84.

[74] Lin CH, Huang JY, Ching CH, Chuang JI. Melatonin reduces the neuronal loss, downregulation of dopamine transporter, and upregulation of D2 receptor in rotenone-induced parkinsonian rats. J Pineal Res. 2008;44(2):205-13.

[75] Sidhu A, Wersinger C, Vernier P. alpha-Synuclein regulation of the dopaminergic transporter: a possible role in the pathogenesis of Parkinson's disease. FEBS Lett. 2004;565(1-3):1-5.

[76] Yu S, Ueda K, Chan P. Alpha-synuclein and dopamine metabolism. Mol Neurobiol. 2005;31(1-3):243-54.

[77] Lee FJ, Liu F, Pristupa ZB, Niznik HB. Direct binding and functional coupling of alpha-synuclein to the dopamine transporters accelerate dopamine-induced apoptosis. FASEB J. 2001;15(6):916-26.

[78] Halliwell B, Gutteridge JMC. Oxidative stress in PD. In: Halliwell B, Gutteridge JMC, Eds. Free Radicals in Biology and Medicine. New York: Oxford University Press 1999; pp.744-758

[79] Waragai M, Wei J, Fujita M, Nakai M, Ho GJ, Masliah E, *et al.* Increased level of DJ-1 in the cerebrospinal fluids of sporadic Parkinson's disease. Biochem Biophys Res Commun. 2006;345(3):967-72.

[80] Zhang J, Perry G, Smith MA, Robertson D, Olson SJ, Graham DG. Parkinson's disease is associated with oxidative damage to cytoplasmic DNA and RNA in substantia nigra neurons. Am J Pathol. 1999;154:1423-29.

[81] Friedlander RM. Apoptosis and caspases in neurodegenerative diseases. N Engl J Med. 2003;348(14):1365-75.

[82] Gotz ME, Kunig G, Riederer P, Youdim MB. Oxidative stress: free radical production in neural degeneration. Pharmacology & therapeutics. 1994;63(1):37-122.

[83] Rowe DB, Le W, Smith RG, Appel SH. Antibodies from patients with Parkinson's disease react with protein modified by dopamine oxidation. J Neurosci Res. 1998;53(5):551-8.

[84] Kortekaas R, Leenders KL, van Oostrom JC, Vaalburg W, Bart J, Willemsen AT, *et al.* Blood-brain barrier dysfunction in parkinsonian midbrain *in vivo*. Ann Neurol. 2005;57(2):176-9.

[85] Zlokovic BV. The blood-brain barrier in health and chronic neurodegenerative disorders. Neuron. 2008;57(2):178-201.

[86] Gonsette RE. Oxidative stress and excitotoxicity: a therapeutic issue in multiple sclerosis? Mult Scler. 2008;14(1):22-34.

[87] Bo L, Mork S, Kong PA, Nyland H, Pardo CA, Trapp BD. Detection of MHC class II-antigens on macrophages and microglia, but not on astrocytes and endothelia in active multiple sclerosis lesions. J Neuroimmunol. 1994;51(2):135-46.

[88] Cross AH, Manning PT, Keeling RM, Schmidt RE, Misko TP. Peroxynitrite formation within the central nervous system in active multiple sclerosis. J Neuroimmunol. 1998;88(1-2):45-56.

[89] Hill KE, Zollinger LV, Watt HE, Carlson NG, Rose JW. Inducible nitric oxide synthase in chronic active multiple sclerosis plaques: distribution, cellular expression and association with myelin damage. J Neuroimmunol. 2004;151(1-2):171-9.

[90] Oleszak EL, Zaczynska E, Bhattacharjee M, Butunoi C, Legido A, Katsetos CD. Inducible nitric oxide synthase and nitrotyrosine are found in monocytes/macrophages and/or astrocytes in acute, but not in chronic, multiple sclerosis. Clin Diagn Lab Immunol. 1998;5(4):438-45.

[91] Liu JS, Zhao ML, Brosnan CF, Lee SC. Expression ofinducible nitric oxide synthase and nitrotyrosine in multiple sclerosis lesions. Am J Pathol. 2001;158:2057-66.

[92] Touil T, Deloire-Grassin MS, Vital C, Petry KG, Brochet B. *In vivo* damage of CNS myelin and axons induced by peroxynitrite. Neuroreport. 2001;12(16):3637-44.

[93] Kirk J, Plumb J, Mirakhur M, McQuaid S. Tight junctional abnormality in multiple sclerosis white matter affects all calibres of vessel and is associated with blood-brain barrier leakage and active demyelination. The Journal of pathology. 2003;201(2):319-27.

[94] Leech S, Kirk J, Plumb J, McQuaid S. Persistent endothelial abnormalities and blood-brain barrier leak in primary and secondary progressive multiple sclerosis. Neuropathol Appl Neurobio. 2007;33(1):86-98.

[95] Flamm ES, Demopoulos HB, Seligman ML, Poser RG, Ransohoff J. Free radicals in cerebral ischemia. Stroke. 1978;9(5):445-7.

[96] Warner DS, Sheng H, Batinic-Haberle I. Oxidants, antioxidants and the ischemic brain. The Journal of experimental biology. 2004;207(Pt 18):3221-31.

[97] Beckman JS, Beckman TW, Chen J, Marshall PA, Freeman BA. Apparent hydroxyl radical production by peroxynitrite: implications for endothelial injury from nitric oxide and superoxide. Proc Natl Acad Sci U S A. 1990;87(4):1620-4.

[98] Liochev SI, Fridovich I. The Haber-Weiss cycle -- 70 years later: an alternative view. Redox report : communications in free radical research. 2002;7(1):55-7; author reply 9-60.

[99] Cherubini A, Ruggiero C, Polidori MC, Mecocci P. Potential markers of oxidative stress in stroke. Free Radic Biol Med. 2005;39(7):841-52.

[100] Tanno H, Nockels RP, Pitts LH, Noble LJ. Breakdown of the blood-brain barrier after fluid percussive brain injury in the rat. Part 1: Distribution and time course of protein extravasation. J Neurotrauma. 1992;9(1):21-32.

[101] Plateel M, Dehouck MP, Torpier G, Cecchelli R, Teissier E. Hypoxia increases the susceptibility to oxidant stress and the permeability of the blood-brain barrier endothelial cell monolayer. J Neurochem. 1995;65(5):2138-45.

[102] Kaur C, Sivakumar V, Zhang Y, Ling EA. Hypoxia-induced astrocytic reaction and increased vascular permeability in the rat cerebellum. Glia. 2006;54:826-39.

[103] Abbott NJ. Astrocyte-endothelial interactions and blood-brain barrier permeability. J Anat. 2002;200(6):629-38.

[104] van Horssen J, Brink BP, de Vries HE, van der Valk P, Bo L. The blood-brain barrier in cortical multiple sclerosis lesions. J Neuropathol Exp Neurol. 2007;66(4):321-8.

[105] Bowman GL, Kaye JA, Moore M, Waichunas D, Carlson NE, Quinn JF. Blood-brain barrier impairment in Alzheimer disease: stability and functional significance. Neurology. 2007;68(21):1809-14.

[106] Plumb J, McQuaid S, Mirakhur M, Kirk J. Abnormal endothelial tight junctions in active lesions and normal-appearing white matter in multiple sclerosis. Brain Pathol. 2002;12(2):154-69.

[107] Blasig IE, Mertsch K, Haseloff RF. Nitronyl nitroxides, a novel group of protective agents against oxidative stress in endothelial cells forming the blood-brain barrier. Neuropharmacology. 2002;43(6):1006-14.

[108] Parathath SR, Gravanis I, Tsirka SE. Nitric oxide synthase isoforms undertake unique roles during excitotoxicity. Stroke. 2007;38(6):1938-45.

[109] Schreibelt G, Kooij G, Reijerkerk A, van Doorn R, Gringhuis SI, van der Pol S, *et al.* Reactive oxygen species alter brain endothelial tight junction dynamics *via* RhoA, PI3 kinase, and PKB signaling. FASEB J. 2007;21(13):3666-76.

[110] Nakashima I, Kato M, Akhand AA, Suzuki H, Takeda K, Hossain K, *et al.* Redox-linked signal transduction pathways for protein tyrosine kinase activation. Antioxid Redox Signal. 2002;4(3):517-31.

[111] Haorah J, Ramirez SH, Schall K, Smith D, Pandya R, Persidsky Y. Oxidative stress activates protein tyrosine kinase and matrix metalloproteinases leading to blood-brain barrier dysfunction. J Neurochem. 2007;101(2):566-76.

[112] Gasche Y, Fujimura M, Morita-Fujimura Y, Copin JC, Kawase M, Massengale J, *et al.* Early appearance of activated matrix metalloproteinase-9 after focal cerebral ischemia in mice: a possible role in blood-brain barrier dysfunction. J Cereb Blood Flow Metab. 1999;19(9):1020-8.

[113] Kim GW, Gasche Y, Grzeschik S, Copin JC, Maier CM, Chan PH. Neurodegeneration in striatum induced by the mitochondrial toxin 3-nitropropionic acid: role of matrix metalloproteinase-9 in early blood-brain barrier disruption? J Neurosci. 2003;23(25):8733-42.

[114] Machado LS, Kozak A, Ergul A, Hess DC, Borlongan CV, Fagan SC. Delayed minocycline inhibits ischemia-activated matrix metalloproteinases 2 and 9 after experimental stroke. BMC Neurosci. 2006;7:56.

[115] Phares TW, Fabis MJ, Brimer CM, Kean RB, Hooper DC. A peroxynitrite-dependent pathway is responsible for blood-brain barrier permeability changes during a central nervous system inflammatory response: TNF-alpha is neither necessary nor sufficient. J Immunol. 2007;178(11):7334-43.

[116] Hooper DC, Scott GS, Zborek A, Mikheeva T, Kean RB, Koprowski H, *et al.* Uric acid, a peroxynitrite scavenger, inhibits CNS inflammation, blood-CNS barrier permeability changes, and tissue damage in a mouse model of multiple sclerosis. FASEB J. 2000;14(5):691-8.

[117] Kean RB, Spitsin SV, Mikheeva T, Scott GS, Hooper DC. The peroxynitrite scavenger uric acid prevents inflammatory cell invasion into the central nervous system in experimental allergic encephalomyelitis through maintenance of blood-central nervous system barrier integrity. J Immunol. 2000;165(11):6511-8.

[118] Das BS, Mohanty S, Mishra SK, Patnaik JK, Satpathy SK, Mohanty D, *et al.* Increased cerebrospinal fluid protein and lipid peroxidation products in patients with cerebral malaria. Trans R Soc Trop Med Hyg. 1991;85(6):733-4.

[119] Ban E, Haour F, Lenstra R. Brain interleukin 1 gene expression induced by peripheral lipopolysaccharide administration. Cytokine. 1992;4(1):48-54.

[120] Gatti S, Bartfai T. Induction of tumor necrosis factor-alpha mRNA in the brain after peripheral endotoxin treatment: comparison with interleukin-1 family and interleukin-6. Brain Res. 1993;624(1-2):291-4.

[121] Sharman K, Sharman E, Campbell A, Bondy SC. Reduced basal levels and enhanced LPS response of IL-6 mRNA in aged mice. The journals of gerontology Series A, Biological sciences and medical sciences. 2001;56(12):B520-3.

[122] Ye SM, Johnson RW. Increased interleukin-6 expression by microglia from brain of aged mice. J Neuroimmunol. 1999;93(1-2):139-48.
[123] Yu WH, Go L, Guinn BA, Fraser PE, Westaway D, McLaurin J. Phenotypic and functional changes in glial cells as a function of age. Neurobiol Aging. 2002;23(1):105-15.
[124] Shen Y, Zhang G, Liu L, Xu S. Suppressive effects of melatonin on amyloid-beta-induced glial activation in rat hippocampus. Archives of medical research. 2007;38(3):284-90.
[125] McGeer PL, Itagaki S, Boyes BE, McGeer EG. Reactive microglia are positive for HLA-DR in the substantia nigra of Parkinson's and Alzheimer's disease brains. Neurology. 1988;38(8):1285-91.
[126] Sawada M, Imamura K, Nagatsu T. Role of cytokines in inflammatory process in Parkinson's disease. J Neural Transm Suppl. 2006(70):373-81.
[127] Whitton PS. Inflammation as a causative factor in the aetiology of Parkinson's disease. Br J Pharmacol. 2007;150(8):963-76.
[128] Wilms H, Zecca L, Rosenstiel P, Sievers J, Deuschl G, Lucius R. Inflammation in Parkinson's diseases and other neurodegenerative diseases: cause and therapeutic implications. Curr Pharm Des. 2007;13(18):1925-8.
[129] Mogi M, Harada M, Kondo T, Riederer P, Inagaki H, Minami M, *et al.* Interleukin-1 beta, interleukin-6, epidermal growth factor and transforming growth factor-alpha are elevated in the brain from parkinsonian patients. Neurosci Lett. 1994;180(2):147-50.
[130] Blum-Degen D, Muller T, Kuhn W, Gerlach M, Przuntek H, Riederer P. Interleukin-1 beta and interleukin-6 are elevated in the cerebrospinal fluid of Alzheimer's and *de novo* Parkinson's disease patients. Neurosci Lett. 1995;202(1-2):17-20.
[131] Muller T, Blum-Degen D, Przuntek H, Kuhn W. Interleukin-6 levels in cerebrospinal fluid inversely correlate to severity of Parkinson's disease. Acta Neurol Scand. 1998;98(2):142-4.
[132] Nagatsu T. Parkinson's disease: changes in apoptosis-related factors suggesting possible gene therapy. J Neural Transm. 2002;109(5-6):731-45.
[133] Nagatsu T, Sawada M. Cellular and molecular mechanisms of Parkinson's disease: neurotoxins, causative genes, and inflammatory cytokines. Cell Mol Neurobiol. 2006;26(4-6):781-802.
[134] Pawate S, Shen Q, Fan F, Bhat NR. Redox regulation of glial inflammatory response to lipopolysaccharide and interferongamma. J Neurosci Res. 2004;77(4):540-51.
[135] Yang CS, Lee HM, Lee JY, Kim JA, Lee SJ, Shin DM, *et al.* Reactive oxygen species and p47phox activation are essential for the Mycobacterium tuberculosis-induced pro-inflammatory response in murine microglia. J Neuroinflammation. 2007;4:27.
[136] Shibata N, Kobayashi M. [The role for oxidative stress in neurodegenerative diseases]. Brain and nerve = Shinkei kenkyu no shinpo. 2008;60(2):157-70.
[137] Li XJ, Gu J, Pan BS, Sun FY. Effect of melatonin on production of hydroxyl radical and lactate dehydrogenase during hypoxia in rat cortical slices. Zhongguo yao li xue bao = Acta pharmacologica Sinica. 1999;20(3):201-5.
[138] Fern R, Moller T. Rapid ischemic cell death in immature oligodendrocytes: a fatal glutamate release feedback loop. J Neurosci. 2000;20(1):34-42.
[139] Tekkok SB, Goldberg MP. Ampa/kainate receptor activation mediates hypoxic oligodendrocyte death and axonal injury in cerebral white matter. J Neurosci. 2001;21(12):4237-48.
[140] Goldberg MP, Ransom BR. New light on white matter. Stroke. 2003;34(2):330-2.
[141] Allen JW, Vicini S, Faden AI. Exacerbation of neuronal cell death by activation of group I metabotropic glutamate receptors: role of NMDA receptors and arachidonic acid release. Exp Neurol. 2001;169(2):449-60.
[142] Hilton GD, Nunez JL, Bambrick L, Thompson SM, McCarthy MM. Glutamate-mediated excitotoxicity in neonatal hippocampal neurons is mediated by mGluR-induced release of Ca++ from intracellular stores and is prevented by estradiol. Eur J Neurosci. 2006;24(11):3008-16.
[143] Haas J, Erdo SL. Quisqualate-induced excitotoxic death of glial cells: transient vulnerability of cultured astrocytes. Glia. 1991;4(1):111-4.
[144] Yoshioka A, Hardy M, Younkin DP, Grinspan JB, Stern JL, Pleasure D. Alpha-amino-3-hydroxy-5-methyl-4-isoxazolepropionate (AMPA) receptors mediate excitotoxicity in the oligodendroglial lineage. J Neurochem. 1995;64(6):2442-8.
[145] Matute C, Sanchez-Gomez MV, Martinez-Millan L, Miledi R. Glutamate receptor-mediated toxicity in optic nerve oligodendrocytes. Proc Natl Acad Sci U S A. 1997;94(16):8830-5.

[146] McDonald JW, Althomsons SP, Hyrc KL, Choi DW, Goldberg MP. Oligodendrocytes from forebrain are highly vulnerable to AMPA/kainate receptor-mediated excitotoxicity. Nat Med. 1998;4(3):291-7.

[147] Westall FC, Hawkins A, Ellison GW, Myers LW. Abnormal glutamic acid metabolism in multiple sclerosis. J Neurol Sci. 1980;47(3):353-64.

[148] Stover JF, Pleines UE, Morganti-Kossmann MC, Kossmann T, Lowitzsch K, Kempski OS. Neurotransmitters in cerebrospinal fluid reflect pathological activity. Eur J Clin Invest. 1997;27(12):1038-43.

[149] Sarchielli P, Greco L, Floridi A, Gallai V. Excitatory amino acids and multiple sclerosis: evidence from cerebrospinal fluid. Arch Neurol. 2003;60(8):1082-8.

[150] Werner P, Pitt D, Raine CS. Multiple sclerosis: altered glutamate homeostasis in lesions correlates with oligodendrocyte and axonal damage. Ann Neurol. 2001;50(2):169-80.

[151] Baker AJ, Moulton RJ, MacMillan VH, Shedden PM. Excitatory amino acids in cerebrospinal fluid following traumatic brain injury in humans. J Neurosurg. 1993;79(3):369-72.

[152] Palmer AM, Marion DW, Botscheller ML, Bowen DM, DeKosky ST. Increased transmitter amino acid concentration in human ventricular CSF after brain trauma. Neuroreport. 1994;6(1):153-6.

[153] Zauner A, Bullock R, Kuta AJ, Woodward J, Young HF. Glutamate release and cerebral blood flow after severe human head injury. Acta neurochirurgica Supplement. 1996;67:40-4.

[154] Bullock R, Zauner A, Woodward JJ, Myseros J, Choi SC, Ward JD, *et al.* Factors affecting excitatory amino acid release following severe human head injury. J Neurosurg. 1998;89(4):507-18.

[155] Ruppel RA, Kochanek PM, Adelson PD, Rose ME, Wisniewski SR, Bell MJ, *et al.* Excitatory amino acid concentrations in ventricular cerebrospinal fluid after severe traumatic brain injury in infants and children: the role of child abuse. J Pediatr. 2001;138(1):18-25.

[156] Zhang J, Johnston G, Stebler B, Keller ET. Hydrogen peroxide activates NFkappaB and the interleukin-6 promoter through NFkappaB-inducing kinase. Antioxid Redox Signal. 2001;3(3):493-504.

[157] Bullock R, Zauner A, Woodward J, Young HF. Massive persistent release of excitatory amino acids following human occlusive stroke. Stroke. 1995;26(11):2187-9.

[158] Lafon-Cazal M, Pietri S, Culcasi M, Bockaert J. NMDA-dependent superoxide production and neurotoxicity. Nature. 1993;364(6437):535-7.

[159] Dugan LL, Sensi SL, Canzoniero LM, Handran SD, Rothman SM, Lin TS, *et al.* Mitochondrial production of reactive oxygen species in cortical neurons following exposure to N-methyl-D-aspartate. J Neurosci. 1995;15(10):6377-88.

[160] Reynolds IJ, Hastings TG. Glutamate induces the production of reactive oxygen species in cultured forebrain neurons following NMDA receptor activation. J Neurosci. 1995;15(5 Pt 1):3318-27.

[161] Bindokas VP, Jordan J, Lee CC, Miller RJ. Superoxide production in rat hippocampal neurons: selective imaging with hydroethidine. J Neurosci. 1996;16(4):1324-36.

[162] Perez Velazquez JL, Frantseva MV, Carlen PL. *In vitro* ischemia promotes glutamate-mediated free radical generation and intracellular calcium accumulation in hippocampal pyramidal neurons. J Neurosci. 1997;17(23):9085-94.

[163] Sengpiel B, Preis E, Krieglstein J, Prehn JH. NMDA-induced superoxide production and neurotoxicity in cultured rat hippocampal neurons: role of mitochondria. Eur J Neurosci. 1998;10(5):1903-10.

[164] Kourie JI. Interaction of reactive oxygen species with ion transport mechanisms. Am J Physiol. 1998;275(1 Pt 1):C1-24.

[165] Ratan RR, Murphy TH, Baraban JM. Oxidative stress induces apoptosis in embryonic cortical neurons. J Neurochem. 1994;62(1):376-9.

[166] Meister A, Anderson ME. Glutathione. Annu Rev Biochem. 1983;52:711-60.

[167] Mizui T, Kinouchi H, Chan PH. Depletion of brain glutathione by buthionine sulfoximine enhances cerebral ischemic injury in rats. Am J Physiol. 1992;262(2 Pt 2):H313-7.

[168] Bobyn PJ, Franklin JL, Wall CM, Thornhill JA, Juurlink BH, Paterson PG. The effects of dietary sulfur amino acid deficiency on rat brain glutathione concentration and neural damage in global hemispheric hypoxia-ischemia. Nutr Neurosci. 2002;5(6):407-16.

[169] Coyle JT, Puttfarcken P. Oxidative stress, glutamate, and neurodegenerative disorders. Science. 1993;262(5134):689-95.

[170] Tan S, Sagara Y, Liu Y, Maher P, Schubert D. The regulation of reactive oxygen species production during programmed cell death. J Cell Biol. 1998;141(6):1423-32.
[171] Oka A, Belliveau MJ, Rosenberg PA, Volpe JJ. Vulnerability of oligodendroglia to glutamate: pharmacology, mechanisms, and prevention. J Neurosci. 1993;13(4):1441-53.
[172] Back SA, Gan X, Li Y, Rosenberg PA, Volpe JJ. Maturation-dependent vulnerability of oligodendrocytes to oxidative stress-induced death caused by glutathione depletion. J Neurosci. 1998;18(16):6241-53.
[173] Back SA, Han BH, Luo NL, Chricton CA, Xanthoudakis S, Tam J, *et al.* Selective vulnerability of late oligodendrocyte progenitors to hypoxia-ischemia. J Neurosci. 2002;22(2):455-63.
[174] Yonezawa M, Back SA, Gan X, Rosenberg PA, Volpe JJ. Cystine deprivation induces oligodendroglial death: rescue by free radical scavengers and by a diffusible glial factor. J Neurochem. 1996;67(2):566-73.
[175] Reiter RJ, Carneiro RC, Oh CS. Melatonin in relation to cellular antioxidative defense mechanisms. Hormone and metabolic research = Hormon- und Stoffwechselforschung = Hormones et metabolisme. 1997;29(8):363-72.
[176] Reiter RJ, Tan DX, Mayo JC, Sainz RM, Leon J, Czarnocki Z. Melatonin as an antioxidant: biochemical mechanisms and pathophysiological implications in humans. Acta Biochim Pol. 2003;50(4):1129-46.
[177] Beyer CE, Steketee JD, Saphier D. Antioxidant properties of melatonin--an emerging mystery. Biochem Pharmacol. 1998;56(10):1265-72.
[178] Nowak JZ, Zawilska JB. Melatonin and its physiological and therapeutic properties. Pharm World Sci. 1998;20(1):18-27.
[179] Haldar C, Singh R, Guchhait P. Relationship between the annual rhythms in melatonin and immune system status in the tropical palm squirrel, Funambulus pennanti. Chronobiol Int. 2001;18(1):61-9.
[180] Maestroni GJ. The immunotherapeutic potential of melatonin. Expert opinion on investigational drugs. 2001;10(3):467-76.
[181] Turgut M, Uyanikgil Y, Ates U, Baka M, Yurtseven ME. Pinealectomy stimulates and exogenous melatonin inhibits harmful effects of epileptiform activity during pregnancy in the hippocampus of newborn rats: an immunohistochemical study. Child's nervous system : ChNS : official journal of the International Society for Pediatric Neurosurgery. 2006;22(5):481-8.
[182] Yahyavi-Firouz-Abadi N, Tahsili-Fahadan P, Riazi K, Ghahremani MH, Dehpour AR. Involvement of nitric oxide pathway in the acute anticonvulsant effect of melatonin in mice. Epilepsy Res. 2006;68(2):103-13.
[183] Yildirim M, Marangoz C. Anticonvulsant effects of melatonin on penicillin-induced epileptiform activity in rats. Brain Res. 2006;1099(1):183-8.
[184] Reiter RJ. Pineal function during aging: attenuation of the melatonin rhythm and its neurobiological consequences. Acta Neurobiol Exp (Wars). 1994;54 Suppl:31-9.
[185] Anisimov VN, Popovich IG, Zabezhinski MA. Melatonin and colon carcinogenesis: I. Inhibitory effect of melatonin on development of intestinal tumors induced by 1,2-dimethylhydrazine in rats. Carcinogenesis. 1997;18(8):1549-53.
[186] Gilad E, Laufer M, Matzkin H, Zisapel N. Melatonin receptors in PC3 human prostate tumor cells. J Pineal Res. 1999;26(4):211-20.
[187] Petranka J, Baldwin W, Biermann J, Jayadev S, Barrett JC, Murphy E. The oncostatic action of melatonin in an ovarian carcinoma cell line. J Pineal Res. 1999;26(3):129-36.
[188] Kanishi Y, Kobayashi Y, Noda S, Ishizuka B, Saito K. Differential growth inhibitory effect of melatonin on two endometrial cancer cell lines. J Pineal Res. 2000;28(4):227-33.
[189] Siu SW, Lau KW, Tam PC, Shiu SY. Melatonin and prostate cancer cell proliferation: interplay with castration, epidermal growth factor, and androgen sensitivity. Prostate. 2002;52(2):106-22.
[190] Kiefer T, Ram PT, Yuan L, Hill SM. Melatonin inhibits estrogen receptor transactivation and cAMP levels in breast cancer cells. Breast Cancer Res Treat. 2002;71(1):37-45.
[191] Lissoni P, Chilelli M, Villa S, Cerizza L, Tancini G. Five years survival in metastatic non-small cell lung cancer patients treated with chemotherapy alone or chemotherapy and melatonin: a randomized trial. J Pineal Res. 2003;35(1):12-5.
[192] Croughs RJ, de Bruin TW. Melatonin and jet lag. Neth J Med. 1996;49(4):164-6.

[193] Arendt J, Skene DJ, Middleton B, Lockley SW, Deacon S. Efficacy of melatonin treatment in jet lag, shift work, and blindness. J Biol Rhythms. 1997;12(6):604-17.

[194] Arendt J. Jet-lag. Lancet. 1998;351(9098):293-4.

[195] Manfredini R, Manfredini F, Conconi F. Standard melatonin intake and circadian rhythms of elite athletes after a transmeridian flight. J Int Med Res. 2000;28(4):182-6.

[196] Parry BL. Jet lag: minimizing it's effects with critically timed bright light and melatonin administration. J Mol Microbiol Biotechnol. 2002;4(5):463-6.

[197] Reiter RJ. Oxidative processes and antioxidative defense mechanisms in the aging brain. FASEB J. 1995;9(7):526-33.

[198] Pappolla MA, Chyan YJ, Poeggeler B, Frangione B, Wilson G, Ghiso J, *et al.* An assessment of the antioxidant and the antiamyloidogenic properties of melatonin: implications for Alzheimer's disease. J Neural Transm. 2000;107(2):203-31.

[199] Jacob S, Poeggeler B, Weishaupt JH, Siren AL, Hardeland R, Bahr M, *et al.* Melatonin as a candidate compound for neuroprotection in amyotrophic lateral sclerosis (ALS): high tolerability of daily oral melatonin administration in ALS patients. J Pineal Res. 2002;33(3):186-7.

[200] Beni SM, Kohen R, Reiter RJ, Tan DX, Shohami E. Melatonin-induced neuroprotection after closed head injury is associated with increased brain antioxidants and attenuated late-phase activation of NF-kappaB and AP-1. FASEB J. 2004;18(1):149-51.

[201] Maldonado MD, Murillo-Cabezas F, Terron MP, Flores LJ, Tan DX, Manchester LC, *et al.* The potential of melatonin in reducing morbidity-mortality after craniocerebral trauma. J Pineal Res. 2007;42(1):1-11.

[202] Carrillo-Vico A, Calvo JR, Abreu P, Lardone PJ, Garcia-Maurino S, Reiter RJ, *et al.* Evidence of melatonin synthesis by human lymphocytes and its physiological significance: possible role as intracrine, autocrine, and/or paracrine substance. FASEB J. 2004;18(3):537-9.

[203] Cuzzocrea S, Reiter RJ. Pharmacological action of melatonin in shock, inflammation and ischemia/reperfusion injury. Eur J Pharmacol. 2001;426(1-2):1-10.

[204] Cheung RT. The utility of melatonin in reducing cerebral damage resulting from ischemia and reperfusion. J Pineal Res. 2003;34(3):153-60.

[205] Chen TY, Lee MY, Chen HY, Kuo YL, Lin SC, Wu TS, *et al.* Melatonin attenuates the postischemic increase in blood-brain barrier permeability and decreases hemorrhagic transformation of tissue-plasminogen activator therapy following ischemic stroke in mice. J Pineal Res. 2006;40(3):242-50.

[206] Reiter RJ, Tan DX, Manchester LC, Tamura H. Melatonin defeats neurally-derived free radicals and reduces the associated neuromorphological and neurobehavioral damage. J Physiol Pharmacol. 2007;58 Suppl 6:5-22.

[207] Reiter RJ, Tan DX, Manchester LC, Qi W. Biochemical reactivity of melatonin with reactive oxygen and nitrogen species: a review of the evidence. Cell Biochem Biophys. 2001;34(2):237-56.

[208] Kacmaz A, User EY, Sehirli AO, Tilki M, Ozkan S, Sener G. Protective effect of melatonin against ischemia/reperfusion-induced oxidative remote organ injury in the rat. Surgery today. 2005;35(9):744-50.

[209] Esterbauer H, Schaur RJ, Zollner H. Chemistry and biochemistry of 4-hydroxynonenal, malonaldehyde and related aldehydes. Free Radic Biol Med. 1991;11(1):81-128.

[210] Li G, Li L, Yin D. A novel observation: melatonin's interaction with malondiadehyde. Neuro endocrinol letters. 2005;26(1):61-6.

[211] Pablos MI, Agapito MT, Gutierrez R, Recio JM, Reiter RJ, Barlow-Walden L, *et al.* Melatonin stimulates the activity of the detoxifying enzyme glutathione peroxidase in several tissues of chicks. J Pineal Res. 1995;19(3):111-5.

[212] Ozturk G, Coskun S, Erbas D, Hasanoglu E. The effect of melatonin on liver superoxide dismutase activity, serum nitrate and thyroid hormone levels. Jpn J Physiol. 2000;50(1):149-53.

[213] Meki AR, Hussein AA. Melatonin reduces oxidative stress induced by ochratoxin A in rat liver and kidney. Comp Biochem Physiol C Toxicol Pharmacol. 2001;130(3):305-13.

[214] Reiter RJ, Tan DX, Gitto E, Sainz RM, Mayo JC, Leon J, *et al.* Pharmacological utility of melatonin in reducing oxidative cellular and molecular damage. Polish journal of pharmacology. 2004;56(2):159-70.

[215] Subramanian P, Mirunalini S, Pandi-Perumal SR, Trakht I, Cardinali DP. Melatonin treatment improves the antioxidant status and decreases lipid content in brain and liver of rats. Eur J Pharmacol. 2007;571(2-3):116-9.
[216] Kaur C, Sivakumar V, Lu J, Ling EA. Increased vascular permeability and nitric oxide production in response to hypoxia in the pineal gland. J Pineal Res. 2007;42(4):338-49.
[217] Omurtag GZ, Tozan A, Sehirli AO, Sener G. Melatonin protects against endosulfan-induced oxidative tissue damage in rats. J Pineal Res. 2008;44(4):432-8.
[218] Wilkinson BL, Landreth GE. The microglial NADPH oxidase complex as a source of oxidative stress in Alzheimer's disease. J Neuroinflammation. 2006;3:30.
[219] Zhou J, Zhang S, Zhao X, Wei T. Melatonin impairs NADPH oxidase assembly and decreases superoxide anion production in microglia exposed to amyloid-beta1-42. J Pineal Res. 2008;45(2):157-65.
[220] Wang DL, Ling ZQ, Cao FY, Zhu LQ, Wang JZ. Melatonin attenuates isoproterenol-induced protein kinase A overactivation and tau hyperphosphorylation in rat brain. J Pineal Res. 2004;37(1):11-6.
[221] Mayo JC, Sainz RM, Tan DX, Antolin I, Rodriguez C, Reiter RJ. Melatonin and Parkinson's disease. Endocrine. 2005;27(2):169-78.
[222] Beal MF. Mitochondria, oxidative damage, and inflammation in Parkinson's disease. Ann N Y Acad Sci. 2003;991:120-31.
[223] Hald A, Lotharius J. Oxidative stress and inflammation in Parkinson's disease: is there a causal link? Exp Neurol. 2005;193(2):279-90.
[224] Sharma R, McMillan CR, Tenn CC, Niles LP. Physiological neuroprotection by melatonin in a 6-hydroxydopamine model of Parkinson's disease. Brain Res. 2006;1068(1):230-6.
[225] Rocchitta G, Migheli R, Mura MP, Esposito G, Marchetti B, Desole MS, *et al.* Role of endogenous melatonin in the oxidative homeostasis of the extracellular striatal compartment: a microdialysis study in PC12 cells *in vitro* and in the striatum of freely moving rats. J Pineal Res. 2005;39(4):409-18.
[226] Thomas B, Mohanakumar KP. Melatonin protects against oxidative stress caused by 1-methyl-4-phenyl-1,2,3,6-tetrahydropyridine in the mouse nigrostriatum. J Pineal Res. 2004;36(1):25-32.
[227] Willis GL. The role of ML-23 and other melatonin analogues in the treatment and management of Parkinson's disease. Drug News Perspect. 2005;18(7):437-44.
[228] Willis GL, Robertson AD. Recovery of experimental Parkinson's disease with the melatonin analogues ML-23 and S-20928 in a chronic, bilateral 6-OHDA model: a new mechanism involving antagonism of the melatonin receptor. Pharmacol Biochem Behav. 2004;79(3):413-29.
[229] Tan DX, Manchester LC, Reiter RJ, Qi WB, Karbownik M, Calvo JR. Significance of melatonin in antioxidative defense system: reactions and products. Biol Signals Recept. 2000;9(3-4):137-59.
[230] Tutunculer F, Eskiocak S, Basaran UN, Ekuklu G, Ayvaz S, Vatansever U. The protective role of melatonin in experimental hypoxic brain damage. Pediatrics international : official journal of the Japan Pediatric Society. 2005;47(4):434-9.
[231] Bai J, Cederbaum AI. Mitochondrial catalase and oxidative injury. Biol Signals Recept. 2001;10(3-4):189-99.
[232] Acuna-Castroviejo D, Escames G, Rodriguez MI, Lopez LC. Melatonin role in the mitochondrial function. Frontiers in bioscience : a journal and virtual library. 2007;12:947-63.
[233] Alonso M, Collado PS, Gonzalez-Gallego J. Melatonin inhibits the expression of the inducible isoform of nitric oxide synthase and nuclear factor kappa B activation in rat skeletal muscle. J Pineal Res. 2006;41(1):8-14.
[234] Escames G, Lopez LC, Ortiz F, Ros E, Acuna-Castroviejo D. Age-dependent lipopolysaccharide-induced iNOS expression and multiorgan failure in rats: effects of melatonin treatment. Exp Gerontol. 2006;41(11):1165-73.
[235] Jung M, Drapier JC, Weidenbach H, Renia L, Oliveira L, Wang A, *et al.* Effects of hepatocellular iron imbalance on nitric oxide and reactive oxygen intermediates production in a model of sepsis. J hepatol. 2000;33(3):387-94.
[236] Mishra OP, Zubrow AB, Ashraf QM, Delivoria-Papadopoulos M. Nuclear Ca(++)-influx, Ca (++)/calmodulin-dependent protein kinase IV activity and CREB protein phosphorylation during post-hypoxic reoxygenation in neuronal nuclei of newborn piglets: the role of nitric oxide. Neurochem Res. 2006;31(12):1463-71.

[237] Kondoh T, Uneyama H, Nishino H, Torii K. Melatonin reduces cerebral edema formation caused by transient forebrain ischemia in rats. Life Sci. 2002;72(4-5):583-90.
[238] Zhang ZG, Zhang L, Jiang Q, Zhang R, Davies K, Powers C, *et al.* VEGF enhances angiogenesis and promotes blood-brain barrier leakage in the ischemic brain. J Clin Invest. 2000;106(7):829-38.
[239] Zhang Z, Chopp M. Vascular endothelial growth factor and angiopoietins in focal cerebral ischemia. Trends Cardiovasc Med. 2002;12(2):62-6.
[240] Lotufo CM, Lopes C, Dubocovich ML, Farsky SH, Markus RP. Melatonin and N-acetylserotonin inhibit leukocyte rolling and adhesion to rat microcirculation. Eur J Pharmacol. 2001;430(2-3):351-7.
[241] Cuzzocrea S, Zingarelli B, Gilad E, Hake P, Salzman AL, Szabo C. Protective effect of melatonin in carrageenan-induced models of local inflammation: relationship to its inhibitory effect on nitric oxide production and its peroxynitrite scavenging activity. J Pineal Res. 1997;23(2):106-16.
[242] Cuzzocrea S, Costantino G, Mazzon E, Caputi AP. Regulation of prostaglandin production in carrageenan-induced pleurisy by melatonin. J Pineal Res. 1999;27(1):9-14.
[243] Bilici D, Akpinar E, Kiziltunc A. Protective effect of melatonin in carrageenan-induced acute local inflammation. Pharmacol Res. 2002;46(2):133-9.
[244] El-Shenawy SM, Abdel-Salam OM, Baiuomy AR, El-Batran S, Arbid MS. Studies on the anti-inflammatory and anti-nociceptive effects of melatonin in the rat. Pharmacol Res. 2002;46(3):235-43.
[245] Kaur C, Sivakumar V, Lu J, Tang FR, Ling EA. Melatonin attenuates hypoxia-induced ultrastructural changes and increased vascular permeability in the developing hippocampus. Brain Pathol. 2008;18(4):533-47.
[246] Cuzzocrea S, Costantino G, Gitto E, Mazzon E, Fulia F, Serraino I, *et al.* Protective effects of melatonin in ischemic brain injury. J Pineal Res. 2000;29(4):217-27.
[247] Papadopoulos MC, Verkman AS. Aquaporin-4 gene disruption in mice reduces brain swelling and mortality in pneumococcal meningitis. J Biol Chem. 2005;280(14):13906-12.
[248] Watanabe T, Yuki S, Egawa M, Nishi H. Protective effects of MCI-186 on cerebral ischemia: possible involvement of free radical scavenging and antioxidant actions. J Pharmacol Exp Ther. 1994;268(3):1597-604.
[249] Yoneda Y, Uehara T, Yamasaki H, Kita Y, Tabuchi M, Mori E. Hospital-based study of the care and cost of acute ischemic stroke in Japan. Stroke. 2003;34(3):718-24.
[250] Watanabe T, Tahara M, Todo S. The novel antioxidant edaravone: from bench to bedside. Cardiovasc Ther. 2008;26(2):101-14.
[251] Yamamoto T, Yuki S, Watanabe T, Mitsuka M, Saito KI, Kogure K. Delayed neuronal death prevented by inhibition of increased hydroxyl radical formation in a transient cerebral ischemia. Brain Res. 1997;762(1-2):240-2.
[252] Hoehn B, Yenari MA, Sapolsky RM, Steinberg GK. Glutathione peroxidase overexpression inhibits cytochrome C release and proapoptotic mediators to protect neurons from experimental stroke. Stroke. 2003;34(10):2489-94.
[253] Okatani Y, Wakatsuki A, Kaneda C. Melatonin increases activities of glutathione peroxidase and superoxide dismutase in fetal rat brain. J Pineal Res. 2000;28(2):89-96.
[254] Takayasu Y, Nakaki J, Kawasaki T, Koda K, Ago Y, Baba A, *et al.* Edaravone, a radical scavenger, inhibits mitochondrial permeability transition pore in rat brain. J Pharmacol Sci. 2007;103(4):434-7.
[255] Zhang N, Komine-Kobayashi M, Tanaka R, Liu M, Mizuno Y, Urabe T. Edaravone reduces early accumulation of oxidative products and sequential inflammatory responses after transient focal ischemia in mice brain. Stroke. 2005;36(10):2220-5.
[256] Suzuki K, Kazui T, Terada H, Umemura K, Ikeda Y, Bashar AH, *et al.* Experimental study on the protective effects of edaravone against ischemic spinal cord injury. J Thorac Cardiovasc Surg. 2005;130(6):1586-92.
[257] Xiao B, Bi FF, Hu YQ, Tian FF, Wu ZG, Mujlli HM, *et al.* Edaravone neuroprotection effected by suppressing the gene expression of the Fas signal pathway following transient focal ischemia in rats. Neurotox Res. 2007;12(3):155-62.
[258] Amemiya S, Kamiya T, Nito C, Inaba T, Kato K, Ueda M, *et al.* Anti-apoptotic and neuroprotective effects of edaravone following transient focal ischemia in rats. Eur J Pharmacol. 2005;516(2):125-30.
[259] Qi X, Okuma Y, Hosoi T, Nomura Y. Edaravone protects against hypoxia/ischemia-induced endoplasmic reticulum dysfunction. J Pharmacol Exp Ther. 2004;311(1):388-93.

[260] Abe M, Reiter RJ, Orhii PB, Hara M, Poeggeler B. Inhibitory effect of melatonin on cataract formation in newborn rats: evidence for an antioxidative role for melatonin. J Pineal Res. 1994;17(2):94-100.

[261] Ishikawa A, Yoshida H, Metoki N, Toki T, Imaizumi T, Matsumiya T, *et al.* Edaravone inhibits the expression of vascular endothelial growth factor in human astrocytes exposed to hypoxia. Neurosci Res. 2007;59(4):406-12.

[262] Houkin K, Nakayama N, Kamada K, Noujou T, Abe H, Kashiwaba T. Neuroprotective effect of the free radical scavenger MCI-186 in patients with cerebral infarction: clinical evaluation using magnetic resonance imaging and spectroscopy. Journal of stroke and cerebrovascular diseases : the official journal of National Stroke Association. 1998;7(5):315-22.

[263] Mitsumori K, Sakuragi M, Kitami K, Koyanagi I, Murata J. The effect of MCI-186 on regional cerebral blood flow in patients with acute cerebral infarction. Ther Res. 1998;19:553-65.

[264] Barone FC, Feuerstein GZ. Inflammatory mediators and stroke: new opportunities for novel therapeutics. J Cereb Blood Flow Metab. 1999;19(8):819-34.

[265] Baird AE, Benfield A, Schlaug G, Siewert B, Lovblad KO, Edelman RR, *et al.* Enlargement of human cerebral ischemic lesion volumes measured by diffusion-weighted magnetic resonance imaging. Ann Neurol. 1997;41(5):581-9.

[266] Matsui T, Mori T, Tateishi N, Kagamiishi Y, Satoh S, Katsube N, *et al.* Astrocytic activation and delayed infarct expansion after permanent focal ischemia in rats. Part I: enhanced astrocytic synthesis of s-100beta in the periinfarct area precedes delayed infarct expansion. J Cereb Blood Flow Metab. 2002;22(6):711-22.

[267] Bemeur C, Ste-Marie L, Desjardins P, Vachon L, Butterworth RF, Hazell AS, *et al.* Dehydroascorbic acid normalizes several markers of oxidative stress and inflammation in acute hyperglycemic focal cerebral ischemia in the rat. Neurochem Int. 2005;46(5):399-407.

[268] Block F, Dihne M, Loos M. Inflammation in areas of remote changes following focal brain lesion. Prog Neurobiol. 2005;75(5):342-65.

[269] Bramlett HM, Dietrich WD. Pathophysiology of cerebral ischemia and brain trauma: similarities and differences. J Cereb Blood Flow Metab. 2004;24(2):133-50.

[270] Khan M, Sekhon B, Jatana M, Giri S, Gilg AG, Sekhon C, *et al.* Administration of N-acetylcysteine after focal cerebral ischemia protects brain and reduces inflammation in a rat model of experimental stroke. J Neurosci Res. 2004;76(4):519-27.

[271] Kim JB, Yu YM, Kim SW, Lee JK. Anti-inflammatory mechanism is involved in ethyl pyruvate-mediated efficacious neuroprotection in the postischemic brain. Brain Res. 2005;1060(1-2):188-92.

[272] Wen Y, Yang S, Liu R, Perez E, Yi KD, Koulen P, *et al.* Estrogen attenuates nuclear factor-kappa B activation induced by transient cerebral ischemia. Brain Res. 2004;1008(2):147-54.

[273] Takahashi G, Sakurai M, Abe K, Itoyama Y, Tabayashi K. MCI-186 prevents spinal cord damage and affects enzyme levels of nitric oxide synthase and Cu/Zn superoxide dismutase after transient ischemia in rabbits. J Thorac Cardiovasc Surg. 2003;126(5):1461-6.

[274] Socci DJ, Bjugstad KB, Jones HC, Pattisapu JV, Arendash GW. Evidence that oxidative stress is associated with the pathophysiology of inherited hydrocephalus in the H-Tx rat model. Exp Neurol. 1999;155(1):109-17.

[275] Lee EJ, Lee MY, Chen HY, Hsu YS, Wu TS, Chen ST, *et al.* Melatonin attenuates gray and white matter damage in a mouse model of transient focal cerebral ischemia. J Pineal Res. 2005;38(1):42-52.

[276] Pan Z, Perez-Polo R. Role of nerve growth factor in oxidant homeostasis: glutathione metabolism. J Neurochem. 1993;61(5):1713-21.

[277] Kinsman S, Eslin D, Jones H, Lange H, McAllister JP. Induction of neuronal nitric oxide synthase during the late stage of congenital hydrocephalus in the HTX rat. Society for Neuroscience Abstracts 1997; 23: 1874.

[278] Miller SL, Yan EB, Castillo-Melendez M, Jenkin G, Walker DW. Melatonin provides neuroprotection in the late-gestation fetal sheep brain in response to umbilical cord occlusion. Dev Neurosci. 2005;27(2-4):200-10.

[279] Welin AK, Svedin P, Lapatto R, Sultan B, Hagberg H, Gressens P, *et al.* Melatonin reduces inflammation and cell death in white matter in the mid-gestation fetal sheep following umbilical cord occlusion. Pediatr Res. 2007;61(2):153-8.

[280] Hedtjarn M, Leverin AL, Eriksson K, Blomgren K, Mallard C, Hagberg H. Interleukin-18 involvement in hypoxic-ischemic brain injury. J Neurosci. 2002;22(14):5910-9.
[281] Hedtjarn M, Mallard C, Hagberg H. Inflammatory gene profiling in the developing mouse brain after hypoxia-ischemia. J Cereb Blood Flow Metab. 2004;24(12):1333-51.
[282] Hedtjarn M, Mallard C, Arvidsson P, Hagberg H. White matter injury in the immature brain: role of interleukin-18. Neurosci Lett. 2005;373(1):16-20.
[283] Deng Y, Lu J, Sivakumar V, Ling EA, Kaur C. Amoeboid microglia in the periventricular white matter induce oligodendrocyte damage through expression of proinflammatory cytokines *via* MAP kinase signaling pathway in hypoxic neonatal rats. Brain Pathol. 2008;18(3):387-400.
[284] Sivakumar V, Lu J, Ling EA, Kaur C. Vascular endothelial growth factor and nitric oxide production in response to hypoxia in the choroid plexus in neonatal brain. Brain Pathol. 2008;18(1):71-85.
[285] Fulia F, Gitto E, Cuzzocrea S, Reiter RJ, Dugo L, Gitto P, *et al.* Increased levels of malondialdehyde and nitrite/nitrate in the blood of asphyxiated newborns: reduction by melatonin. J Pineal Res. 2001;31(4):343-9.
[286] Turgut M, Erdogan S, Ergin K, Serter M. Melatonin ameliorates blood-brain barrier permeability, glutathione, and nitric oxide levels in the choroid plexus of the infantile rats with kaolin-induced hydrocephalus. Brain Res. 2007;1175:117-25.
[287] Felderhoff-Mueser U, Sifringer M, Pesditschek S, Kuckuck H, Moysich A, Bittigau P, *et al.* Pathways leading to apoptotic neurodegeneration following trauma to the developing rat brain. Neurobiol Dis. 2002;11(2):231-45.
[288] Bayir H, Kochanek PM, Clark RS. Traumatic brain injury in infants and children: mechanisms of secondary damage and treatment in the intensive care unit. Crit Care Clin. 2003;19(3):529-49.
[289] Ozdemir D, Tugyan K, Uysal N, Sonmez U, Sonmez A, Acikgoz O, *et al.* Protective effect of melatonin against head trauma-induced hippocampal damage and spatial memory deficits in immature rats. Neurosci Lett. 2005;385(3):234-9.
[290] Ozdemir D, Uysal N, Gonenc S, Acikgoz O, Sonmez A, Topcu A, *et al.* Effect of melatonin on brain oxidative damage induced by traumatic brain injury in immature rats. Physiol Res. 2005;54(6):631-7.
[291] Urakubo A, Jarskog LF, Lieberman JA, Gilmore JH. Prenatal exposure to maternal infection alters cytokine expression in the placenta, amniotic fluid, and fetal brain. Schizophr Res. 2001;47(1):27-36.
[292] Ning H, Wang H, Zhao L, Zhang C, Li XY, Chen YH, *et al.* Maternally-administered lipopolysaccharide (LPS) increases tumor necrosis factor alpha in fetal liver and fetal brain: its suppression by low-dose LPS pretreatment. Toxicology letters. 2008;176(1):13-9.
[293] Xu DX, Wang H, Ning H, Zhao L, Chen YH. Maternally administered melatonin differentially regulates lipopolysaccharide-induced proinflammatory and anti-inflammatory cytokines in maternal serum, amniotic fluid, fetal liver, and fetal brain. J Pineal Res. 2007;43(1):74-9.
[294] Ikeda T, Xia YX, Kaneko M, Sameshima H, Ikenoue T. Effect of the free radical scavenger, 3-methyl-1-phenyl-2-pyrazolin-5-one (MCI-186), on hypoxia-ischemia-induced brain injury in neonatal rats. Neurosci Lett. 2002;329(1):33-6.
[295] Noor JI, Ikeda T, Ueda Y, Ikenoue T. A free radical scavenger, edaravone, inhibits lipid peroxidation and the production of nitric oxide in hypoxic-ischemic brain damage of neonatal rats. Am J Obstet Gynecol. 2005;193(5):1703-8.
[296] Noor JI, Ueda Y, Ikeda T, Ikenoue T. Edaravone inhibits lipid peroxidation in neonatal hypoxic-ischemic rats: an *in vivo* microdialysis study. Neurosci Lett. 2007;414(1):5-9.
[297] Masaoka N, Nakajima Y, Watanabe M, Hayakawa Y, Nagaishi M, Yamamoto T, *et al.* MCI-186 administered to the maternal circulation inhibits fetal brain injury resulting from total umbilical cord occlusion in the chronically instrumented fetal lamb. J Matern Fetal Neonatal Med. 2006;19(10):625-31.
[298] Nakajima Y, Masaoka N, Hayakawa Y, Watanabe M, Nagaishi M, Sakata H, *et al.* The production of hydroxyl radicals in the fetal lamb brain resulting from occlusion of the umbilical circulation and the transplacental effect of MCI-186 to inhibit hydroxyl radical production. Pediatr Res. 2006;59(2):216-20.
[299] Salt TE, Cordeiro MF. Glutamate excitotoxicity in glaucoma: throwing the baby out with the bathwater? Eye (Lond). 2006;20(6):730-1; author reply 1-2.
[300] Nucci C, Tartaglione R, Rombola L, Morrone LA, Fazzi E, Bagetta G. Neurochemical evidence to implicate elevated glutamate in the mechanisms of high intraocular pressure (IOP)-induced retinal ganglion cell death in rat. Neurotoxicology. 2005;26(5):935-41.

[301] Whitcup SM, Nussenblatt RB, Lightman SL, Hollander DA. Inflammation in retinal disease. International journal of inflammation. 2013;2013:724648.
[302] Kaur C, Sivakumar V, Yong Z, Lu J, Foulds WS, Ling EA. Blood-retinal barrier disruption and ultrastructural changes in the hypoxic retina in adult rats: the beneficial effect of melatonin administration. The Journal of pathology. 2007;212(4):429-39.
[303] Izzotti A, Bagnis A, Sacca SC. The role of oxidative stress in glaucoma. Mutation research. 2006;612(2):105-14.
[304] Kowluru RA, Chan PS. Oxidative stress and diabetic retinopathy. Experimental diabetes research. 2007;2007:43603.
[305] Kuriyama H, Waki M, Nakagawa M, Tsuda M. Involvement of oxygen free radicals in experimental retinal ischemia and the selective vulnerability of retinal damage. Ophthalmic research. 2001;33(4):196-202.
[306] Oshida E, Matsumoto Y, Arai K. Free radicals in the aqueous humor of patients with glaucoma. Clin Ophthalmol. 2010;4:653-60.
[307] Sucher NJ, Lipton SA, Dreyer EB. Molecular basis of glutamate toxicity in retinal ganglion cells. Vision research. 1997;37(24):3483-93.
[308] El-Remessy AB, Khalil IE, Matragoon S, Abou-Mohamed G, Tsai NJ, Roon P, *et al.* Neuroprotective effect of (-)Delta9-tetrahydrocannabinol and cannabidiol in N-methyl-D-aspartate-induced retinal neurotoxicity: involvement of peroxynitrite. Am J Pathol. 2003;163(5):1997-2008.
[309] Yang D, Elner SG, Bian ZM, Till GO, Petty HR, Elner VM. Pro-inflammatory cytokines increase reactive oxygen species through mitochondria and NADPH oxidase in cultured RPE cells. Exp Eye Res. 2007;85(4):462-72.
[310] Lundmark PO, Pandi-Perumal SR, Srinivasan V, Cardinali DP, Rosenstein RE. Melatonin in the eye: implications for glaucoma. Exp Eye Res. 2007;84(6):1021-30.
[311] Belforte NA, Moreno MC, de Zavalia N, Sande PH, Chianelli MS, Keller Sarmiento MI, *et al.* Melatonin: a novel neuroprotectant for the treatment of glaucoma. J Pineal Res. 2010;48(4):353-64.
[312] Park SW, Lee HS, Sung MS, Kim SJ. The effect of melatonin on retinal ganglion cell survival in ischemic retina. Chonnam medical journal. 2012;48(2):116-22.
[313] Yi C, Pan X, Yan H, Guo M, Pierpaoli W. Effects of melatonin in age-related macular degeneration. Ann N Y Acad Sci. 2005;1057:384-92.
[314] Jiang T, Chang Q, Zhao Z, Yan S, Wang L, Cai J, *et al.* Melatonin-mediated cytoprotection against hyperglycemic injury in Muller cells. PloS one. 2012;7(12):e50661.
[315] Siu AW, Maldonado M, Sanchez-Hidalgo M, Tan DX, Reiter RJ. Protective effects of melatonin in experimental free radical-related ocular diseases. J Pineal Res. 2006;40(2):101-9.
[316] Saenz DA, Turjanski AG, Sacca GB, Marti M, Doctorovich F, Sarmiento MI, *et al.* Physiological concentrations of melatonin inhibit the nitridergic pathway in the Syrian hamster retina. J Pineal Res. 2002;33(1):31-6.
[317] Crooke A, Colligris B, Pintor J. Update in glaucoma medicinal chemistry: emerging evidence for the importance of melatonin analogues. Curr Med Chem. 2012;19(21):3508-22.
[318] Salido EM, Bordone M, De Laurentiis A, Chianelli M, Keller Sarmiento MI, Dorfman D, *et al.* Therapeutic efficacy of melatonin in reducing retinal damage in an experimental model of early type 2 diabetes in rats. J Pineal Res. 2013;54(2):179-89.
[319] Song Y, Gong YY, Xie ZG, Li CH, Gu Q, Wu XW. Edaravone (MCI-186), a free radical scavenger, attenuates retinal ischemia/reperfusion injury in rats. Acta Pharmacol Sin. 2008;29(7):823-8.
[320] Inokuchi Y, Imai S, Nakajima Y, Shimazawa M, Aihara M, Araie M, *et al.* Edaravone, a free radical scavenger, protects against retinal damage *in vitro* and *in vivo*. J Pharmacol Exp Ther. 2009;329(2):687-98.
[321] Imai S, Inokuchi Y, Nakamura S, Tsuruma K, Shimazawa M, Hara H. Systemic administration of a free radical scavenger, edaravone, protects against light-induced photoreceptor degeneration in the mouse retina. Eur J Pharmacol. 2010;642(1-3):77-85.
[322] Lawwill T, Crockett S, Currier G. Retinal damage secondary to chronic light exposure, thresholds and mechanisms. Documenta ophthalmologica Advances in ophthalmology. 1977;44(2):379-402.
[323] Chemtob S, Roy MS, Abran D, Fernandez H, Varma DR. Prevention of postasphyxial increase in lipid peroxides and retinal function deterioration in the newborn pig by inhibition of cyclooxygenase activity and free radical generation. Pediatr Res. 1993;33(4 Pt 1):336-40.

[324] Kaur C, Sivakumar V, Foulds WS, Luu CD, Ling EA. Cellular and vascular changes in the retina of neonatal rats after an acute exposure to hypoxia. Invest Ophthalmol Vis Sci. 2009;50(11):5364-74.

[325] Kaur C, Sivakumar V, Foulds WS, Luu CD, Ling EA. Hypoxia-induced activation of N-methyl-D-aspartate receptors causes retinal ganglion cell death in the neonatal retina. J Neuropathol Exp Neurol. 2012;71(4):330-47.

[326] Sivakumar V, Foulds WS, Luu CD, Ling EA, Kaur C. Hypoxia-induced retinal ganglion cell damage through activation of AMPA receptors and the neuroprotective effects of DNQX. Exp Eye Res. 2013;109:83-97.

[327] Sivakumar V, Foulds WS, Luu CD, Ling EA, Kaur C. Retinal ganglion cell death is induced by microglia derived pro-inflammatory cytokines in the hypoxic neonatal retina. The Journal of pathology. 2011;224(2):245-60.

[328] Kaur C, Sivakumar V, Robinson R, Foulds WS, Luu CD, Ling EA. Neuroprotective effect of melatonin against hypoxia-induced retinal ganglion cell death in neonatal rats. J Pineal Res. 2013;54(2):190-206.

CHAPTER 2

Acetylcholinesterase Inhibitors as Disease-Modifying Therapies for Alzheimer's Disease

Diego Muñoz-Torrero*

Laboratori de Química Farmacèutica (Unitat Associada al CSIC), Facultat de Farmàcia, and Institut de Biomedicina (IBUB), Universitat de Barcelona, Av. Joan XXIII 27-31, E-08028-Barcelona, Spain

Abstract: The therapeutic arsenal for the treatment of Alzheimer's disease (AD) remains confined to a group of four inhibitors of AChE and one NMDA receptor antagonist, which are used to provide a relief of the very late symptoms of the dementia, *i.e.* the cognitive and functional decline. In line with the growing body of evidence of the pivotal role of the β-amyloid peptide (Aβ) in the pathogenesis of AD, alternative classes of drugs targeting mainly the formation or the aggregation of Aβ are actively pursued by the pharmaceutical industry, as they could positively modify the course of AD, stopping or slowing down disease progression. While mostly amyloid-directed drug candidates are being scrutinized in the past decades as disease-modifying drugs, mounting preclinical and clinical evidence is pointing towards a disease-modifying role also for currently marketed anti-Alzheimer AChE inhibitors (AChEIs), particularly for donepezil. In this review, the neuroprotective effects exhibited by currently commercialized AChEIs will be briefly discussed, together with the secondary mechanisms through which they could exert such effects. This review will focus also on particular classes of AChEIs, namely dual binding site AChEIs, which are being purposely designed to target Aβ aggregation and/or other biological targets that contribute to AD pathogenesis, thus constituting very promising disease-modifying anti-Alzheimer drug candidates.

Keywords: Acetylcholinesterase inhibitors, AChE periperal site, AChE-Aβ interaction, AP2238 derivatives, Aβ aggregation, bis-tacrines, cannabinoid-based Aβ aggregation inhibitors, carbamate-based AChEIs, disease-modifying anti-Alzheimer drugs, dual site binding, huperzine B dimers, lipocrine, marketed AChEIs, memoquin, multifunctional drugs, neuroprotection, piperidine-based AChEIs, polyamine-based AChEIs, pyridinium-based AChEIs, tacrine-based hybrids.

***Corresponding author Diego Muñoz-Torrero:** Laboratori de Química Farmacèutica, Facultat de Farmàcia, Universitat de Barcelona, Av. Joan XXIII 27-31, E-08028-Barcelona, Spain; Tel: +34-93-4024533; Fax: +34-93-4035941; E-mail: dmunoztorrero@ub.edu
Dedicated to Prof. Pelayo Camps.

Atta-ur-Rahman, Allen B. Reitz, M. Iqbal Choudhary and Jizhou Wang (Eds.)

1. INTRODUCTION

Alzheimer's disease (AD), the leading cause of dementia in elderly people, is already approaching epidemic proportions, affecting millions of people worldwide [1]. Its prevalence is expected to increase dramatically during the next decades, along with the increase of average life expectancy and the aging of the population [1]. Unfortunately, to date an effective curative or preventive therapy still remains elusive. Thus, this devastating disorder constitutes one the most important unmet needs in modern medicine.

The two most characteristic neuropathological hallmarks of AD are the senile plaques and the neurofibrillary tangles, which are mainly composed of β-amyloid peptide (Aβ) and hyperphosphorylated tau protein, respectively. These two proteins seem to be at the root of the pathogenesis of the disease [2]. Some debate exists about which of them is the trigger and which one is a consequence of the neurodegenerative process of AD [3] or even about if both of them are lesions representing the effect rather than the cause of the disease [4]. However, so far the most prevailing hypothesis on the cause of AD is the so-called amyloid cascade hypothesis, which points to Aβ as the initiating molecule in AD [5-9]. This hypothesis establishes that increased production, aggregation and accumulation of Aβ in the brain triggers a cascade of neurotoxic events which eventually leads to a widespread neuronal degeneration and to the clinical symptomatology of the dementia. In view of this scenario, it is obvious that Aβ constitutes the main biological target to be addressed in order to prevent, delay or stop the neurodegenerative process of AD, *i.e.* in order to positively modify AD progression. As a consequence, in the last years most of the academic and industry research efforts to develop disease-modifying anti-Alzheimer drugs are focusing on the targets involved upstream in the neurotoxic cascade of the disease, namely Aβ formation, aggregation, clearance or metabolism [10-14]. Indeed, more than 40% of the anti-Alzheimer drug candidates currently undergoing clinical trials target Aβ aggregation/amyloid plaque formation and Aβ formation [12]. The clinical development of some promising *disease-modifying* labelled anti-Alzheimer drug candidates such as tramiprosate (**1**, Fig. (**1**)), a glycosaminoglycan mimetic that binds to soluble Aβ, thereby inhibiting its aggregation and amyloid plaque formation [15], and (*R*)-flurbiprofen (**2**, Fig. (**1**)), a modulator of the activity of γ-secretase [16], which, together with β-secretase, is involved in Aβ formation from the amyloid precursor protein (APP), was recently discontinued by efficacy issues [17]. Other Aβ-directed drug candidates, including some active or passive immunization protocols involving either second generation Aβ vaccines with improved safety profile relative to the first generation vaccine AN-1792 or anti-Aβ monoclonal antibodies are still under clinical development [10, 18].

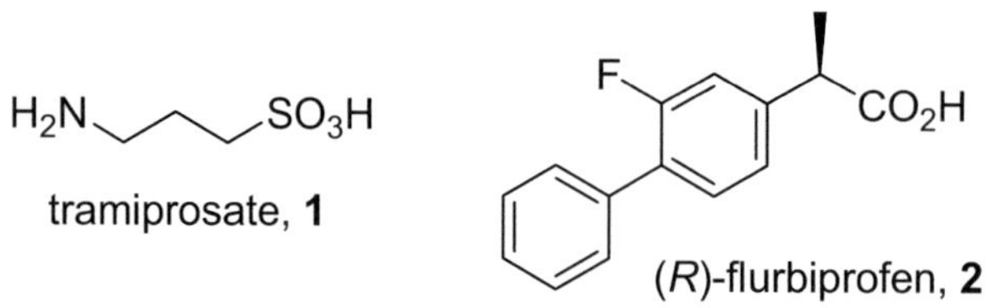

Figure 1: Disease-modifying anti-Alzheimer drug candidates recently withdrawn from clinical trials.

While these amyloid-directed disease-modifying anti-Alzheimer drug candidates are under development, the current therapeutic arsenal for AD still remains dominated by a group of drugs which were conceptually developed to treat the very late symptoms of dementia, which, in turn, arise from the deficit in central nervous system (CNS) of different neurotransmitters, particularly acetylcholine (ACh), which are being biosynthesized and released from the neurons which are dying due to the neurodegeneration. According to the so-called cholinergic hypothesis of AD, most of the cognitive and functional decline in AD patients is a direct consequence of a marked cholinergic deficit in CNS, and, consequently, the use of cholinomimetic agents, able to compensate for this deficit should allow alleviation of the symptomatology of the disease. Indeed, with the sole exception of the recently approved glutamate NMDA receptor antagonist memantine (**7**, Fig. (**2**)), the other four commercialized anti-Alzheimer drugs, namely tacrine (**3**, Fig. (**2**)), donepezil (**4**), rivastigmine (**5**), and galantamine (**6**) are a group of cholinomimetic agents with the pharmacological profile of inhibitors of the enzyme acetylcholinesterase (AChE). The enzyme AChE is responsible for the hydrolysis of the neurotransmitter ACh at the synaptic cleft. The four commercialized AChE inhibitors (AChEIs) inhibit this hydrolytic process, thereby leading to increased levels of ACh at the synapse, compensating for the central cholinergic deficit characteristic of AD, and alleviating its symptomatology.

Apart from the inhibition of the hydrolysis of ACh, these compounds have been shown to exhibit a number of biological effects which are clearly beyond the symptomatic improvement of cognition and function and are suggestive of the fact that currently commercialized AChEIs could positively modify AD progression, thus seriously challenging the prevailing view that these drugs are merely a symptomatic treatment for AD. Herein, the evidences of neuroprotective effects in the currently commercialized anti-Alzheimer AChEIs are briefly presented, together with the potential mechanisms through which these effects could be exerted and the rationalization from a structural point of view of the most important of these mechanisms, *i.e.* those addressing Aβ formation and particularly Aβ aggregation as the earliest events of the neurotoxic cascade of AD.

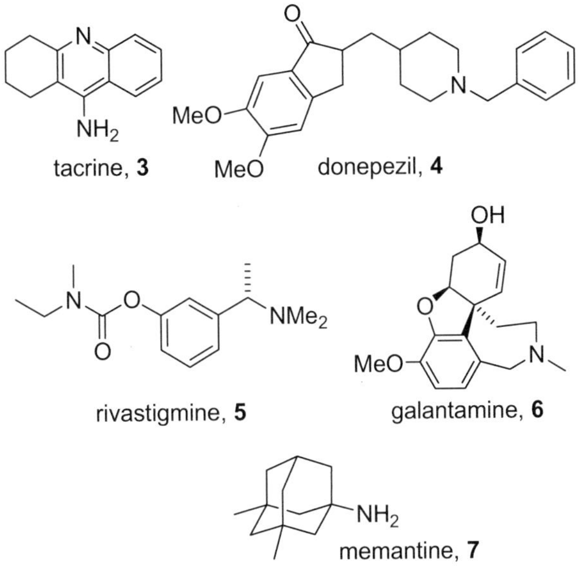

Figure 2: Currently commercialized anti-Alzheimer drugs.

In the last decade, the consideration of the structural determinants allowing interference on Aβ aggregation by AChEIs has driven in great part the development of a novel class of AChEI, the so-called dual binding site inhibitors. The firstly developed structural families of dual binding site AChEIs have been the subject of several reviews in the last years [19-28]. However, until recently, the determination of the Aβ antiaggregating effect of this kind of compounds was rather rare. In this review, it is presented the rational design, structure-activity relationship studies, as well as the pharmacological profile of those structural families of dual binding site AChEIs for which the effect on Aβ aggregation and in some cases on other targets involved in the neurotoxic cascade have been proved, thus constituting very promising amyloid- and cholinergic-directed disease-modifying anti-Alzheimer drug candidates.

2. CURRENTLY COMMERCIALIZED ANTI-ALZHEIMER AChEIs AS DISEASE-MODIFYING DRUGS

Over the past few years numerous preclinical, radiologic and clinical studies have evidenced a number of neuroprotective effects associated to the use of AChEIs that challenge the prevailing view of these compounds just as palliative drugs, and suggest that these drugs could have an impact on the natural history of AD [reviewed in 29-35].

A large body of evidence coming from preclinical studies has shown that the four marketed anti-Alzheimer AChEIs, and particularly donepezil, with which most studies have been carried out, exhibit different neuroprotective effects both *in vitro* and *in vivo* against a variety of neurotoxic insults. These compounds display neuroprotection against glutamate- [30, 31, 36-40] or NMDA- [41] induced toxicity normally in primary cultures of rat cortical neurons, against Aβ-induced toxicity in primary cultures of rat septal cholinergic neurons [42, 43], SH-SY5Y cells [44] or mice [45], and against other neurotoxic agents such as hydrogen peroxide [46] or carbon monoxide [47]. Some neuroprotective effects of these drugs in oxygen-glucose deprivation models of neuronal injury have been also reported [30, 36, 48, 49].

Analogously, there is also growing evidence from clinical trials suggestive of a positive disease-modifying role for AChEIs. Usually, the longer the patient preserves cognition and function, the longer the delay to nursing home placement, the most clinically relevant end point associated with AD progression. Indeed, AChEIs may afford sustained benefits over time, a considerable number of AD patients showing long-lasting cognitive stabilization after AChEI treatment, which can not be solely explained on the basis of a merely symptomatic effect [34]. A recent study on available long-term clinical data from patients remaining on treatment with rivastigmine or donepezil for up to 5 years or with galantamine for up to 4 years has suggested that AChEIs may delay, although not stop, cognitive and functional decline [50]. Also, specifically designed clinical studies with AChEIs have shown that these drugs can positively alter the course of AD, as indicated by the delay in admission to a nursing home (delaying end point design), by the preservation of benefits after withdrawal of the AChEI, resulting in less deterioration in cognitive function relative to placebo-treated patients (withdrawal design), and by the preserved advantage in AChEI-treated AD patients relative to a group of patients initially randomized to placebo and taking the AChEI in an extension phase of the study (randomized start design) [33].

The neuroprotective effect of AChEIs during clinical trials has also been demonstrated by means of neuroradiological techniques such as magnetic resonance imaging (MRI), proton magnetic resonance spectroscopy (MRS) and positron emission tomography (PET) [51]. MRI-detectable volumetric changes in the medial temporal lobe or in the whole brain due to the progressive brain atrophy in AD patients constitute valid biomarkers of pathologic progression of AD [52]. Thus, quantification of the rate of hippocampal atrophy over time has been used to test the disease-modifying effects of AChEIs [53]. Indeed, the most rigorous evidence for the neuroprotective role of AChEIs comes from MRI studies which showed that two different groups of AD patients receiving donepezil underwent a significantly lower degree of hippocampal atrophy than the untreated groups [54, 55].

Although all the currently commercialized AChEIs have been shown to exert to a greater or lesser extent neuroprotective effects, neither are all of them active in the same assays nor their potencies as neuroprotectants correlate with their potencies as AChEIs. Thus, it has become apparent that AChEIs mediate their neuroprotective effect through mechanisms other than inhibition of the catalytic activity of AChE. The knowledge of the mechanism of action responsible for the neuroprotective effects of AChEIs is of great interest, especially for the design and development of new drugs with the same mode of action and improved potency.

Different mechanisms have been proposed through which AChEIs could exert their neuroprotective effects, including blockade of the glutamate-induced excitotoxicity through antagonism of NMDA receptors [56], overexpression of the antiapoptotic protein bcl-2 following stimulation of $\alpha7$ and $\alpha4\beta2$ nicotinic receptors [31, 32, 39, 40, 44], mobilization of calcium pools from the intracellular endoplasmic reticulum following activation of σ_1 receptors [45, 47, 57], or upregulation of the neuroprotective splice variants of AChE (readthrough form, AChE-R) [29].

However, the most interesting way to modify AD progression would involve an interference in the earliest events of the neurotoxic cascade, namely in Aβ formation and aggregation. Indeed, this kind of mechanisms has been demonstrated for AChEIs [reviewed in 32, 58-62].

Some interconnections between the cholinergic system and APP metabolism have been established. It is known that carbachol, a non-selective muscarinic agonist, increases the secretion of the non-amyloidogenic fragment sAPPα in cells expressing M_1 and M_3 muscarinic receptors, but not in cells expressing M_2 and M_4 receptors [63]. This effect is particularly interesting taking into account that formation of sAPPα involves the proteolytic cleavage of APP by α-secretase. Because the APP processing *via* α- or β-secretase are mutually exclusive, the activation of the α-secretase pathway involved in this effect precludes formation of Aβ. Direct activation of muscarinic M_1 and M_3 receptors by selective muscarinic agonists or indirect activation thereof by AChEIs seems to trigger a downstream signal transduction cascade involving protein kinase C (PKC), which in turn can stimulate the non-amyloidogenic processing of APP. The effects of AChEIs on the levels of sAPPα differ among cell types and depend on the specific drug, the duration of treatment and tested dose [58]. The increase in the levels of sAPPα induced by tacrine and donepezil has been demonstrated in several studies *in vitro* [64, 65]. Moreover, some changes in APP processing have been observed in AD patients short-term treated with donepezil, which seems to be able to restore the balance between α- and β-secretase activities, altered in AD pathology [66, 67].

Regarding the effect of AChEIs on APP processing, these drugs may not operate through a single mechanism. As above mentioned, AChEIs could act through a cholinergic pathway by increasing the levels of the neurotransmitter ACh, which in turn would stimulate muscarinic M_1 receptors and the PKC pathway. In addition to PKC, other pathways can be activated downstream of the cholinergic receptor that can modulate APP processing, such as tyrosine kinase, and MAP kinase [60]. Moreover, AChEIs could interfer APP processing through other mechanisms such as regulation of its synthesis or turnover acting at transcriptional or post-transcriptional stages, maturation and sorting acting at post-translational stages, or even inhibition of APP processing enzymes such as BACE-1 (β-secretase) and γ-secretase [58], *i.e.* the enzymes involved in the formation of Aβ. Indeed, donepezil has been recently reported to exhibit a potent inhibitory activity of BACE-1, with an IC_{50} of 170 nM, while tacrine, rivastigmine and galantamine were inactive at 3-5 μM concentrations [68, 69].

Overall, most of the above-mentioned mechanisms of neuroprotection by AChEIs depend on their action on biological targets other than the enzyme AChE, and therefore are very difficult to rationalize from a structural point of view, which would be necessary for the design of novel drugs hitting these specific targets.

The situation is completely different when an additional mechanism of neuroprotection by AChEIs is considered, namely their interference with Aβ aggregation. Thus, consideration of this mechanism, which derives from some findings of Inestrosa one decade ago, has allowed the rationalization in structural terms of the apparent superiority of donepezil relative to the other commercialized anti-Alzheimer AChEIs in most of the reported studies of neuroprotection, and, more interestingly, the rational design of novel classes of specific AChEIs structurally armed to target Aβ aggregation.

Inestrosa found that the enzyme AChE, which colocalizes with Aβ in the senile plaques, can bind to Aβ, accelerating its aggregation into amyloid fibrils as en early event during the aggregation process of Aβ [70, 71]. The Aβ proaggregating effect of AChE was reported to be independent of the source and structural polymorphism of the enzyme. Conversely, other proteins which also colocalize with Aβ in the senile plaques, such as butyrylcholinesterase (BChE) were found not to be able to significantly affect Aβ aggregation [71]. Not only accelerates AChE the aggregation of Aβ, but it also increases Aβ neurotoxicity. The stable macromolecular complex formed upon binding of AChE to Aβ is more neurotoxic than Aβ fibrils alone both *in vitro* (PC12 cells and primary retina cells) [72, 73] and *in vivo* (rat hippocampus) [74]. In the latter studies, the hippocampal injection of AChE-Aβ complexes resulted in the appearance of some features reminiscent of AD-like lesions in rat brain. The animals injected with the AChE-Aβ complexes formed bigger amyloid deposits than those formed in animals injected with Aβ alone, which led to a stronger neuronal cell loss [74].

Very interestingly, the results obtained by Inestrosa were independently confirmed by Brimijoin in *in vivo* studies with doubly transgenic mice expressing human APP (hAPP) and human AChE (hAChE) in brain. Simultaneous expression of hAPP (and therefore Aβ) and hAChE resulted in a clear earlier onset of amyloid plaque formation in cerebral cortex relative to singly transgenic animals expressing only hAPP (30-50% sooner) [75]. These plaques were associated with a variety of markers indicative of classic mature AD lesions, including macrophage and astrocyte antigens and AChE itself. Moreover, amyloid burden rose with age, remaining always higher in the doubly transgenic animals than in the singly transgenic littermates. Even more interestingly, Brimijoin and Soreq found that the doubly transgenic mice exhibited increased AD-like pathology, the increase of brain amyloid burden closely correlating with a decline of working memory [76, 77].

The results obtained by Inestrosa and Brimijoin clearly raised the notion that those particular classes of AChEIs able to disrupt the AChE-Aβ interaction could affect the basic disease process in AD [77]. Inestrosa found that AChE binds to Aβ through a hydrophobic environment close to the so-called peripheral site of the enzyme [78]. Indeed, propidium, a specific inhibitor known to block the peripheral site of AChE, at 50 μM inhibited by 75% the AChE-induced Aβ aggregation, while the active site inhibitor edrophonium at 100 μM had no effect on Aβ aggregation [71]. In an independent work, Andrisano obtained similar results: at 100 μM, propidium was found to inhibit by 82% the AChE-induced Aβ aggregation, while edrophonium was inactive (unless otherwise stated, the antiaggregating effects reported throughout the text refer to the $A\beta_{1-40}$ peptide) [79].

The peripheral site of AChE is placed at the mouth of a narrow gorge, about 20 Å deep, at the bottom of which is buried the active site where the hydrolysis of ACh takes place [80]. In principle, the four commercialized AChEIs were developed to inhibit the hydrolysis of the neurotransmitter, and, therefore they were expected to interact at the active site of the enzyme. In the event, tacrine, rivastigmine and galantamine do clearly interact within the active site of AChE as shown in the X-ray tridimensional structures of their complexes with *Torpedo californica* AChE (*Tc*AChE) solved by Sussman and Silman [81-83], while a completely different mode of interaction is displayed by donepezil, whose molecule spans the whole length of the active site gorge, simultaneously interacting with the tryptophan residues characteristic of the active and peripheral sites of AChE (Trp84 and Trp279, respectively) [84] (Fig. (**3**)). Thus, donepezil is the sole commercialized AChEI exhibiting this dual site binding.

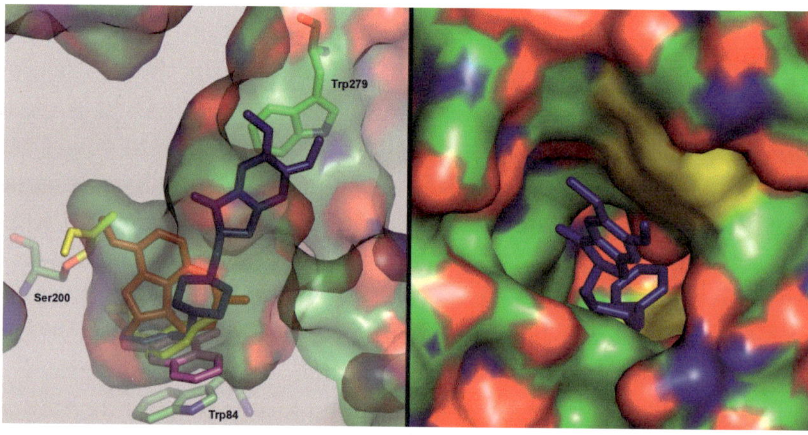

Figure 3: Binding mode of the currently commercialized anti-Alzheimer AChEIs with *Tc*AChE. **a)** Comparison of the binding modes of tacrine (PDB code 1ACJ, colored in magenta), rivastigmine (PDB code 1GQR, colored in yellow), galantamine (PDB code 1DX6, colored in orange), and donepezil (PDB code 1EVE, colored in blue). Trp84/Ser200 and Trp279 have been included as reference residues of the active site and peripheral site, respectively. **b)** Another view of the binding mode of donepezil (colored in blue) along the AChE gorge. The surfaces of Trp84 (at the bottom) and Trp279 (at the entrance) have been colored in yellow.

According to the results obtained by Inestrosa, the enzyme AChE would occupy a place upstream in the neurotoxic cascade of AD. Consequently, those inhibitors able to block the recognition zone of Aβ within AChE, *i.e.* the peripheral site, would be expected to block the AChE-induced Aβ aggregation, and therefore the cascade of neurodegeneration from the beginning. Indeed, as previously mentioned, the peripheral site inhibitor propidium exhibits an outstanding Aβ antiaggregating effect *in vitro*. However, its cationic character precludes its entrance in CNS, and, therefore, its use as an anti-Alzheimer agent. Of course, this is not an issue in the established anti-Alzheimer drug donepezil, which is also able to interact with the peripheral site of AChE, and to inhibit the AChE-induced aggregation of Aβ, though to a lesser extent than propidium (22% at 100 μM) [79], but to a greater extent than marketed active site inhibitors (Table **1**). Worthy of note, the bisquaternary ligand decamethonium, which, as donepezil, is able to simultaneously interact with both the active and peripheral sites of AChE, exhibits a similar Aβ antiaggregating effect (25% at 100 μM) under the same assay conditions [79].

Likely, some of the neuroprotective effects found for donepezil must be related to its moderate ability to block Aβ aggregation due to its dual binding site character. Indeed, some studies have directly linked the neuroprotective effects exerted by donepezil against Aβ in primary cultures of rat septal neurons with its ability to inhibit Aβ aggregation [42, 43]. Moreover, the clearly higher inhibitory potency of donepezil relative to the rest of commercialized AChEIs (Table **1**) could be ascribed to the increased number of contact points with the biological target due to its dual binding site character. As previously mentioned, the outstanding pharmacological profile of donepezil includes a surprisingly potent inhibitory activity toward BACE-1, a kind of effect shared by other dual binding site AChEIs (see below) but not by the other marketed anti-Alzheimer AChEIs (Table **1**).

Table 1: Inhibitory activities of currently commercialized anti-Alzheimer AChEIs toward hAChE, hAChE-induced $A\beta_{1-40}$ aggregation and BACE-1

Drug	hAChE (IC_{50} nM)	AChE-Induced Aβ aggregation (% inhibition)	BACE-1 (IC_{50} nM)
Donepezil	23[a]	22[b]	170[c]
Tacrine	424[a]	7[b]	not active[c]
Rivastigmine	3030[a]	< 5[a]	not active[c]
Galantamine	2010[a]	18[a]	not active[c]

[a] Data from ref. [85]. [b] Data from ref. [79]. [c] Data from ref. [68, 69], tacrine, rivastigmine and galantamine were not active at 4, 3 and 5 μM, respectively.

Thus, donepezil exhibits a unique pharmacological profile encompassing potent inhibition of hAChE, and significant ability to block both formation and aggregation of Aβ. The latter effects endow donepezil with potential to interfere upstream in the neurotoxic cascade of AD, and thus to positively modify AD progression.

3. DUAL BINDING SITE AChEIs WITH Aβ ANTIAGGREGATING EFFECTS

As previously mentioned donepezil is the sole commercialized dual binding site AChEI, although it was not designed as such. Indeed, donepezil was approved by FDA in 1996 for its ability to block the catalytic activity of AChE, but it was not until 1999 when the X-ray tridimensional structure of its complex with *Tc*AChE was solved [84], thereby establishing its dual site binding mode with the enzyme. Previously, bisquaternary AChEIs such as decamethonium were believed to derive their enhanced potency relative to similar monoquaternary ligands, from their ability to span the active site and a second "anionic" site, the peripheral site, about 14 Å distant from the active site [86]. In 1993, Sussman identified the residue Trp279, near the top of the *Tc*AChE active site gorge, as a part of the peripheral "anionic" site, and established a dual site binding mode for the bisquaternary ligand decamethonium [81].

The findings of Inestrosa and Brimijoin and the particular architecture of the enzyme AChE, in which the active and peripheral sites were close enough as to be simultaneously spanned by a single not too large molecule, have driven the rational design of novel classes of dual binding site AChEIs. In 1996, the first family of purposely designed dual binding site AChEIs was developed by Pang [87], and extended in a collaborative work with Carlier and Han [reviewed in 20, 21] and independently by Campiani [88-90] and Hu [91]. Soon after, other groups developed new families of dual binding site AChEIs which have been the subject of several reviews in the last years [19-28]. In this section, only those families of dual binding site AChEIs which have been shown to exhibit Aβ antiaggregating effects and/or other neuroprotective effects will be considered.

The pioneering work of Pang, which was contemporaneous to the findings of Inestrosa, focused on dimers of tacrine which were designed with the main aim of increasing AChE inhibitory potency and selectivity [87]. Indeed, the most potent of Pang's compounds, the so-called *bis*(7)-tacrine (**8**, Fig. (**4**)), turned out to be 150-fold more potent than the parent compound tacrine as inhibitor of rat brain AChE and 250-fold more selective toward AChE than BChE [92]. The increase in inhibitory potency due to the dual site binding is even more dramatic when the source of the enzyme is hAChE, *bis*(7)-tacrine being more than 500-fold more potent than tacrine, although only 7-fold more selective toward hAChE than toward hBChE [93].

The structure-activity relationships (SAR) data regarding AChE inhibitory activity within the family of bis-tacrines can be briefly summarized as follows (Fig. (**4**)):

i) The optimal tether length is that equivalent to 7 methylene groups, although penta-, hexa-, and octa- to deca-methylene linkers also allow some dual site binding [92]. ii) Substitution of one of the tacrine units by basic amino groups of different hydrophobicity such as amino, dimethylamino, 4-pyridylamino, and 4-quinolinylamino leads to a 7-200-fold decrease in AChE inhibitory activity relative to *bis*(7)-tacrine, **8** [94]. iii) Introduction of a chlorine atom at position 6 of the two tacrine units increases 3-fold the AChE inhibitory activity relative to **8** [91], while introduction of two chlorine atoms at positions 6 and 8 of one or both units of tacrine was detrimental for the activity (5-115-fold decrease relative to **8**) [88, 89]. iv) Replacement of the central methylene group of the linker of **8** by a protonatable methylamino group leads to a 20-fold increase in AChE inhibitory potency [89], but substitution of the methyl group attached to the amino function by larger alkyl groups such as ethyl or 2-hydroxyethyl or by an allyl substituent leads to a slightly increased or even to a decreased activity relative to **8** [90]. Interestingly, the substitution of the methyl group attached to this nitrogen atom of the tether by an acetyl group, which renders neutral the initially basic nitrogen atom, leads to a dramatic decrease of AChE inhibitory activity (1150-fold relative to **8**, and 25000-fold relative to the methylamino-substituted analog), while keeping a potent low nanomolar BChE inhibitory activity [89, 90]. v) Substitution of the exocyclic amino group of one unit of tacrine by a sulfur atom or a methylene group leads to a sharp decrease of AChE inhibitory activity (260- and 25-fold, respectively), reversing the selectivity toward BChE in the first case [90].

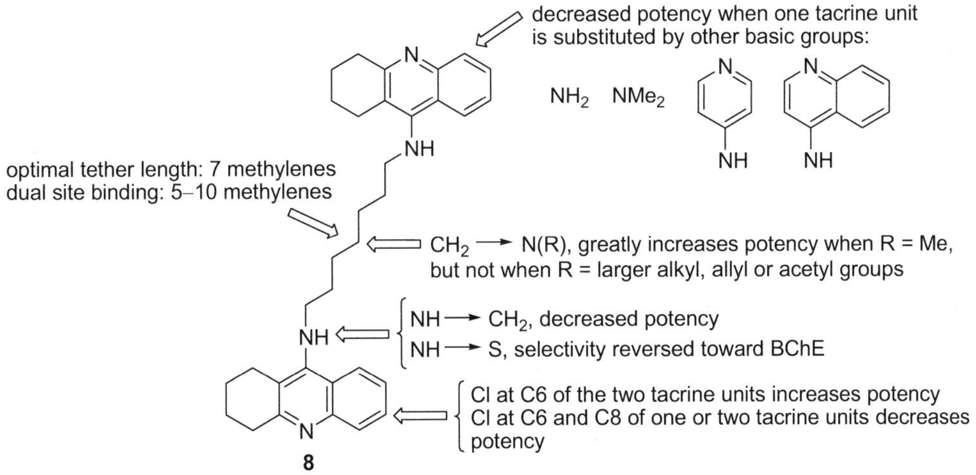

8

Figure 4: Schematic structure-activity relationships regarding AChE inhibitory activity within the family of bis-tacrines.

Bis(7)-tacrine has emerged as a very promising anti-Alzheimer drug candidate, not only for its high AChE inhibitory activity, but also for its wide range of neuroprotective effects against a variety of neuronal injury models. *Bis*(7)-tacrine protects against apoptosis induced by ischemia in primary cultured cortical mouse astrocytes [95], by hydrogen peroxide in rat pheochromocytoma PC12 cells [96], by glutamate in primary cultured cerebellar granule neurons [97], and by Aβ in primary cultured rat cortical neurons [98]. The mechanisms involved in the neuroprotective effects of *bis*(7)-tacrine against glutamate-induced excitotoxicity have been thoroughly studied by Han. *Bis*(7)-tacrine exhibits an antagonistic action on the glutamate NMDA receptor in rat cortical membranes [97, 99-101], acting at the MK-801-binding site with an affinity similar to that of the anti-Alzheimer drug memantine [97, 99]. However, *bis*(7)-tacrine is much more potent than memantine in reducing glutamate-induced excitotoxicity in cerebellar granule neurons (about 80-fold more potent) [99], thus suggesting that this compound could act on some intracellular biological targets which mediate the downstream signal transduction of glutamate. Indeed, it has been shown that the neuroprotective effects of *bis*(7)-tacrine against the glutamate-induced excitotoxicity involve a multi-target synergistic mechanism. Apart from a blockade of NMDA receptors, *bis*(7)-tacrine can selectively inhibit neuronal nitric oxide synthase, an enzyme which is linked to the NMDA receptor and activated by glutamate-induced calcium overinflux, thus leading to an excessive generation of nitric oxide, a mediator of glutamate- and also Aβ-mediated neuronal cell death [99, 102]. Another mechanism that contributes to the neuroprotective effects of *bis*(7)-tacrine against the glutamate-induced excitotoxicity seems to be the prevention or reversal of the mitochondrial dysfunction, which constitutes a primary event in glutamate excitotoxicity. Thus, it has been found that post-treatment with *bis*(7)-tacrine prevented the neuronal cell death and also the decline of mitochondrial membrane potential and the decrease of ATP production induced by glutamate, and more importantly, it rescued the mitochondria from the changes in the expression pattern of a number of proteins involved in energy metabolism, oxidative stress and apoptosis [103].

The above mentioned mechanisms of neuroprotection by *bis*(7)-tacrine have not been related to its dual site binding character, that otherwise has been clearly established by resolution of the X-ray crystal structure of its *Tc*AChE complex [104]. Indeed, the determination of the Aβ antiaggregating effect of the firstly developed dual binding site AChEI has remained a gap until recently, that Bolognesi reported a 68% inhibition of the AChE-induced Aβ aggregation by *bis*(7)-tacrine at 100 μM [93].

Not only can *bis*(7)-tacrine interfere upstream in the neurotoxic cascade of AD by inhibiting Aβ aggregation, but, as previously reported for donepezil [68, 69], this compound is also able to reduce Aβ formation *in vitro* by directly inhibiting BACE-1 in a concentration-dependent manner, exhibiting an IC_{50} value of 7.5 μM [105]. Interestingly, in Neuro2a APPswe neuroblastoma cells the BACE-1 inhibitory activity of *bis*(7)-tacrine was accompanied by a slight activation of α-secretase [105].

In summary, *bis*(7)-tacrine exhibits a unique pharmacological profile encompassing a number of different disease-modifying effects upstream and downstream the neurotoxic cascade of AD.

Tacrine is the active site AChEI which most extensively has been incorporated in the structure of dual binding site AChEIs, likely as a consequence of the ease of synthesis of tacrine moeties and its ability to interact with both the active and peripheral sites of AChE. Indeed, apart from bis-tacrines a large number of families of dual binding site AChEIs are based on tacrine [93, 106-116], though just a few of them have been tested for their ability to block the AChE-induced Aβ aggregation.

Melchiorre and Bolognesi developed the new dual binding site AChEI **9** (Fig. (**5**)), composed of a unit of tacrine as the active site interacting moiety and the phenantridinium motif of propidium as the peripheral site interacting moiety, connecting both units through a triamine linker [109]. This hybridization resulted in a 20000- and 300-fold increase in hAChE inhibitory activity relative to the parent compounds propidium and tacrine, respectively. Also, hybrid **9** turned out to be a potent inhibitor of the AChE-induced Aβ aggregation, exhibiting a 95% inhibition at 100 μM, with an IC_{50} value of 14 μM, values that are similar to those of propidium (82% inhibition and IC_{50} of 13 μM) (Table **2**) [109]. The outstanding activities of **9** led to consider that the length of the linker was close to optimal, so no further extension of this particular family has been reported.

Martínez developed a new family of tacrine-based dual binding site AChEIs which are the most potent inhibitors of AChE-induced Aβ aggregation and among the most potent inhibitors of hAChE so far reported [112]. On the basis of its pharmacological, structural, and electronic properties, an indol system was chosen as the peripheral site interacting unit of this novel structural class, which was linked to the active-site interacting tacrine unit through an amide-containing oligomethylene linker. By structural modification of the linker (length of the linker, position of the amide functionality, incorporation of a protonatable methylamino group within the oligomethylene chain) and substitution at the indol and tetrahydroacridine systems, a large structural family was developed (compounds **10-14** are shown in Fig. (**5**) as selected examples), from which the following SAR data, about bovine AChE inhibitory activity, were derived: i) The position of the amide group within the linker is of utmost importance. The optimal location of the amide group implies a separation from the indole system of 2 methylene groups, as in hybrids **10** and **11** (Fig. (**5**)). Shortening of this distance to one methylene group, lengthening to 3 methylene groups or to an oxyethylene group, or direct attachment of the amide function to the indol system result in a clear drop of AChE inhibitory activity (in general around 10-fold decrease, but it can be up to 2300-fold). The additional interactions gained by the amide functionality at its optimal location, namely hydrogen bonds between the amide NH group and the hydroxy group of Tyr330 and between the amide carbonyl oxygen atom and the hydroxy group of Tyr121, account for the higher inhibitory potency of those hybrids with properly located amide group. ii) Conformational restriction of the linker by introduction of a C-C double bond between the amide functionality and the indol system leads to a 900-fold decreased activity. iii) The optimal separation of 2 methylene groups between the amide group and the indol system, must be accompanied by an additional separation of 6 or 7 methylene groups between the amide group and the tacrine moiety, as in hybrids **10** and **11**. Shortening of this distance to 5 methyelene groups or lengthening to 8-10 methylene groups results in lower AChE inhibitory potency (8-1000-fold). iv) Substitution of a methylene group by a basic methylamino group within the oligomethylene chain which connects the amide functionality and the tacrine moiety results in a 50-fold drop in activity. v) Introduction of a chlorine atom at position 6 of the tacrine unit leads to the optimal activity (20-50-fold increased activity relative to the counterparts unsubstituted at the tacrine moiety). vi) Introduction of substituents (cyano, bromo) at the indol system leads to a decreased activity (3-35-fold).

The most potent compouds of this novel family, compounds **10-14** (Fig. (**5**)), exhibited a highly potent inhibitory activity toward both bovine and human AChE, with IC_{50} values in the picomolar range, and also toward human BChE (hBChE), with IC_{50} values in the low nanomolar range (Table **2**). Very interestingly, most of these hybrids turned out to be more potent than propidium as inhibitors of the AChE-induced Aβ aggregation, exhibiting percentages of inhibition of 82-99% at 100 μM, with IC_{50} values in the low micromolar range (Table **2**). Besides inhibition of the AChE-induced Aβ aggregation, these compounds inhibited the spontaneous Aβ aggregation (without AChE) by 49-65%

at 100 μM, an effect also shared by propidium (46% inhibition) but not by donepezil, decamethonium or tacrine, which were found inactive in this assay. Moreover, these compounds take on added value that they possess a wide therapeutic safety range, as evidenced by their lack of cytotoxicity in human neuroblastoma SH-SY5Y cells, thus constituting very promising disease-modifying anti-Alzheimer drug candidates [112, 117].

Table 2: Inhibitory activities of tacrine-based dual binding site AChEIs toward hAChE, hBChE, and hAChE-induced $A\beta_{1-40}$ aggregation

Compound	hAChE (IC_{50} nM)	hBChE (IC_{50} nM)	AChE-Induced Aβ aggregation (% inhibition)
8[a]	0.81	5.66	68
9[b]	1.55	64	95[c]
10[d]	0.07	2.9	98[c]
11[d]	0.02	0.1	96[c]
12[d]	0.12	12	84[c]
13[d]	0.23	1.7	82[c]
14[d]	0.13	3.2	99[c]
Tacrine[a]	424	46	7
Propidium[b]	32300	13200	82[c]

[a] Data from ref. [93]. [b] Data from ref. [109]. [c] IC_{50} values for hybrids **9-14** and propidium 14, 2, 6, 7, 15, 2, and 13 μM, respectively. [d] Data from ref. [112].

Several questions around BChE have been a matter of controversy in the last decade, particularly the role of this enzyme in the treatment of AD and the existence or not of a peripheral site also in BChE. Initially, inhibition of BChE was regarded as an undesirable effect since it was thought to be related to the occurrence of peripheral cholinergic side-effects. Moreover, the higher AChE *versus* BChE inhibitory activity of the first tacrine-based homo- and heterodimers, which was then considered an added value, was initially ascribed to the lack of a peripheral binding site in BChE [87, 94]. However, both selective and non-selective AChEIs seem to produce qualitatively similar adverse effects, that, consequently, do not appear to result from selective inhibition of BChE [118, 119]. Moreover, recent evidences have shown that in advanced AD patients, AChE activity is greatly reduced in specific brain regions, while BChE activity increases, perhaps partly compensating for AChE decrease [120]. The increasing role of BChE in the hydrolysis of ACh as the ratio AChE/BChE gradually decreases in these patients suggests that inhibition of BChE might be valuable in the search for anti-Alzheimer agents [120, 121]. Indeed, it has been suggested that dual inhibition of AChE and BChE could increase the efficacy of treatment and broaden the indications [121] or even to confer additional neuroprotective benefits [122]. Thus, it has been reported that dual cholinesterase inhibition over a period of 20 weeks in AD patients may decrease the rate of cortical atrophy, as evidenced by MRI studies [122]. Recently, several X-ray crystal structures of truncated or full-length BChE have been reported [123-126]. As in AChE, the active site of BChE is buried at the bottom of a deep and narrow gorge. The main differences between both enzymes seem to reside in the absence of several aromatic residues at the peripheral site, at the gorge entrance and in the acyl pocket of BChE relative to AChE [127]. However, several aromatic residues such as Tyr332 or Phe278, located at the entrance of the gorge, have been related to the peripheral site of BChE [89, 90, 128, 129]. These aromatic residues could be responsible for π-π interactions with aromatic moieties of the above mentioned tacrine-based dual binding site AChEIs [89, 90]. Indeed, although these compounds are rather selective for AChE inhibition, the clearly higher BChE inhibitory activity of most of them relative to tacrine seems to agree with the existence of the peripheral site in this enzyme.

In a program directed to the synthesis of novel dual binding site AChEIs based on a polyamine backbone, additionally endowed with M_2 muscarinic antagonistic activity, Melchiorre developed one decade ago caproctamine (**15**, Fig. (**6**)), as an interesting lead compound for AD treatment [130]. A large family of polyamine-based AChEIs has been developed by structural modification of this lead [130-132], from which the following SAR data regarding hAChE inhibitory activity resulted: i) The *o*-methoxybenzyl groups are essential for a good AChE inhibitory activity. Their removal results in a dramatic drop of activity, what is easily understandable taking into account that these moieties are responsible for the interactions with the characteristic tryptophan residues at the active and peripheral sites of AChE (see below) [130]. Moreover, removal or replacement of the methoxy substituents by other groups of different lipophilicity and electronic character or isomerization to the *meta* or *para* positions leads to a decreased activity. It was found that lipophilicity of the substituents at these benzene rings had no effect on activity, while their electronic properties played a pivotal role. Indeed, the AChE inhibitory activity of these compounds was dependent on the extent of protonation of the benzylamino nitrogen atoms, whose basicity is increased by electron-donating groups at the benzene ring [131]. ii) Demethylation of the benzylamino nitrogen atoms results in a 2-fold decrease of AChE inhibitory activity [130], while replacement of the methyl group by larger alkyl groups leads to a higher potency, which was found optimal for the ethyl-substituted derivative of caproctamine, **16** (10-fold more potent) [131]. iii) The *N*-methyl-substituted amido groups of the lead are essential for AChE inhibitory activity. Deoxygenation or replacement by methylenoxy groups results in 3-180 and 10-fold decreased activity, respectively [130-132]. Also, *N*-demethylation of the amido groups leads to a 6-fold drop of activity [130]. iv) Variation of the length of the inner spacer between 5 and 8 methylene groups has no significant effect on AChE inhibitory activity, as it is the case for the substitution of two central methylenes by a disulfide bridge [130]. Conversely, substitution of the flexible octamethylene chain of **15** and **16** by a conformationally constrained dipiperidine moiety, as in **17** and **18**, leads to a 2-6-fold increased activity relative to **16** [132], while substitution by a dianiline moiety, as in **19**, results in a 2-fold decreased activity [132].

Molecular dynamics simulation studies confirmed the dual site binding of the lead **15** to AChE. Despite the noticeable length of the caproctamine molecule, it seems that it can fold in such a way as to place the active and peripheral site interacting units, *i.e.* the *o*-methoxybenzylamino groups, at an appropriate distance to simultaneously reach Trp84 and Trp279 residues [130].

However, contrary to the expectations for a dual binding site AChEI, the inhibitory effect of both caproctamine and its ethyl-analog **16** on the AChE-induced Aβ aggregation has been recently found not to be significant (Table **3**) [132]. Conversely, the conformationally constrained analogs, particularly **17** and **19**, exhibit a moderate Aβ antiaggregating activity (Table **3**) [132].

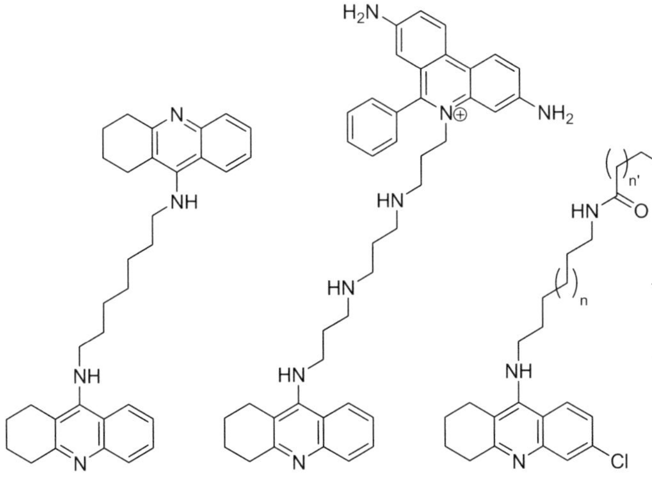

	n	n'	R
10	1	1	H
11	2	1	H
12	2	0	H
13	2	0	Br
14	0	2	H

8 **9**

Figure 5: Tacrine-based dual binding site AChEIs with Aβ antiaggregating effects.

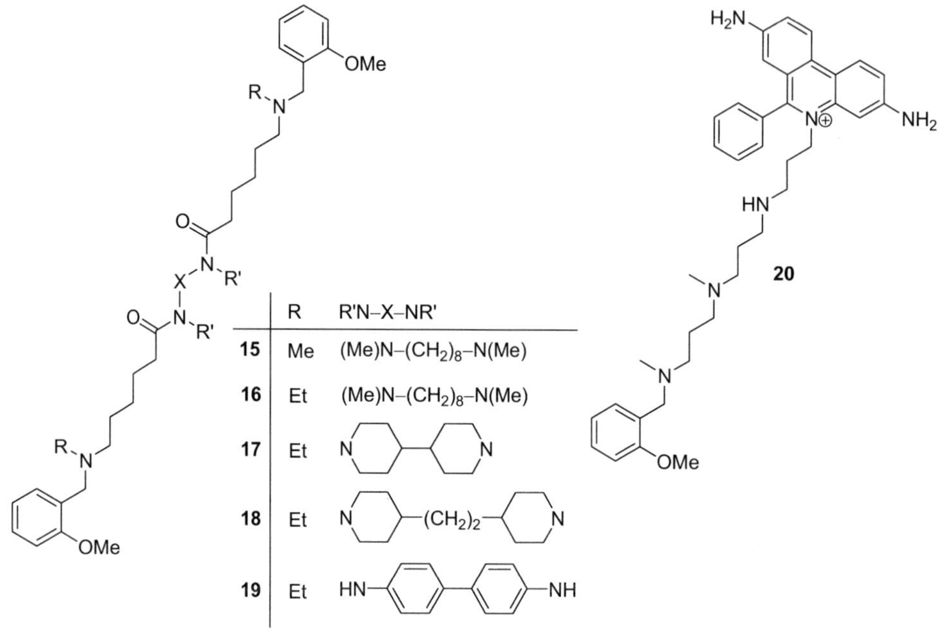

Figure 6: Polyamine-based dual binding site AChEIs with Aβ antiaggregating effects.

Table 3: Inhibitory activities of polyamine-based dual binding site AChEIs toward hAChE and hAChE-induced $A\beta_{1-40}$ aggregation

Compound	hAChE (IC_{50} nM)	AChE-Induced Aβ aggregation (% inhibition)
15[a]	170	< 5
16[a]	19	< 5
17[a]	3	41
18[a]	7	15
19[a]	36	35
20[b]	952	79[c]

[a] Data from ref. [132]. [b] Data from ref. [109]. [c] IC_{50} value 18 μM.

Another polyamine-based dual binding site AChEI has been developed by Bolognesi and Melchiorre, by combination of a propidium moiety, as in compound **9** (Fig. (**5**)), with the *N*-methyl-*o*-methoxybenzylamino moiety of caproctamine [109]. The resulting hybrid compound, **20** (Fig. (**6**)), although more potent hAChE inhibitor than propidium, is significantly less potent than both caproctamine and especially the propidium-tacrine hybrid **9**. Likely, the best suitability of the tacrine moiety of **9** relative to the caproctamine-derived moiety of **20** to interact with the enzyme active site could account for the higher AChE inhibitory activity of **9** relative to **20** [109]. Although an insufficient length of the linker, which would preclude a correct dual site binding, could be also responsible for the low AChE inhibitory activity of hybrid **20**, its dual binding site character was confirmed by determination of its Aβ antiaggregating activity. Thus, hybrid **20** was found to inhibit by 79% the AChE-induced Aβ aggregation at 100 μM, being clearly more potent in this assay than the other polyamine-based inhibitors, although slighly less potent than **9** [109].

In sharp contrast to tacrine-based dual binding site AChEIs, these polyamine-based compounds are weak inhibitors of hBChE, exhibiting IC_{50} values in the micromolar range. However, some of them **(15-17)** take on added value a significant muscarinic M_2 antagonistic activity, with K_b values ranging from 400 to 2000 nM [130-132].

As previously mentioned, the firstly developed classes of dual binding site AChEIs either were designed just to improve AChE inhibitory activity as in the case of *bis*(7)-tacrine or in spite of the fact that their potential to block the AChE-induced Aβ aggregation was recognized, this effect was not assessed. It has not been until relatively recently that this kind of effect has been measured for some dual binding site AChEIs. Indeed, in 2003 Valenti reported the first direct evidence of the Aβ antiaggregating effect of a purposely designed dual binding site AChEI [133]. Thus, on the basis of docking simulations AP2238 (**21**, Fig. (**7**)) was rationally designed by combination of three structural motifs suited for interaction with the active and peripheral sites and also with the aromatic residues which are lining the wall of the active-site gorge as a third recognition zone within the enzyme. For the interaction with the AChE active and peripheral sites, the *N*-benzylamino moiety present in many known AChEIs such as donepezil [23, 134], and a synthetically accessible coumarin system were chosen, respectively, while a *p*-phenylene moiety was selected for the interaction with mid-gorge aromatic residues [133]. Indeed, π-π stacking interactions of the aromatic rings of the active and peripheral site-interacting units of AP2238 with Trp86 and Trp286 (numbering of residues corresponding to hAChE, equivalent to Trp84 and Trp279 in *Tc*AChE) were predicted through the docking simulations, while π-π and OH-π interactions were predicted between the *p*-phenylene linker and the aromatic residues Tyr341 and Tyr124 [133]. Moreover, the docking model suggested that AP2238 contacted the peripheral site of AChE better than donepezil, a fact, which was confirmed by its greater inhibitory effect on the AChE-induced Aβ aggregation relative to donepezil (35% inhibition at 100 μM, Table **4**). The moderately high hAChE inhibitory activity of this compound was also indicative of its dual site binding character.

In an attempt to improve the moderate potency of AP2238 toward both hAChE and hAChE-induced Aβ aggregation, Piazzi has recently reported the synthesis of a large series of AP2238 derivatives [135], from which the following SAR data regarding hAChE inhibitory activity were derived: i) Introduction of a substituent (nitro, methyl or methoxy) in the benzene ring of the benzylamino moiety causes a 3-500-fold drop of activity, with the sole exception of the introduction of an *o*-methoxy substituent, the resulting compound **22** (Fig. (**7**)) being nearly equipotent to AP2238 (Table **4**). The *N*-methyl-substituted *o*-methoxybenzylamino moiety of **22** is identical to that of caproctamine (**15**). Therefore, the suitability of the *o*-methoxy-substitution in AP2238 derivatives could be analogously related to its positive effect on the basicity of the nitrogen atom of this group, and, consequently, on its extent of protonation. ii) Also, as reported for caproctamine analogs, replacement of the methyl group on the nitrogen atom of the benzylamino moiety by an ethyl group (compound **23**, AP2243) leads to a higher potency (2.5-fold), while replacement by a 2-hydroxyethyl group (compound **24**) has no effect on activity (Table **4**). iii) Isomerization of the linker from *p*-phenylene to *m*-phenylene, as in compound **25**, leads to a 2-5-fold decrease in AChE inhibitory activity. iv) The best substitution pattern for the peripheral site interacting unit, involves the presence of an intact 6,7-dimethoxy-substituted coumarin moiety. Replacement of this heterocyclic unit by other methoxy-substituted oxygenated heteroaromatic moieties such as xanthen-9-one, flavone or isoflavone leads to a dramatic drop of activity (30-1800-fold). Moreover, monomethoxy-substitution at position 6 or 7 causes a 1.5-4- and 10-20-fold decreased AChE inhibitory activity. A similar trend was found for monosubstitution at position 6 with nitro and amino groups.

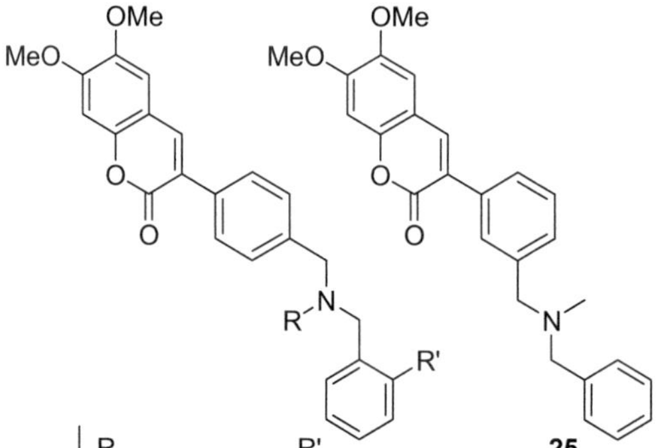

25

	R	R'
21	Me	H
22	Me	MeO
23	Et	H
24	2-hydroxyethyl	H

Figure 7: AP2238 and related dual binding site AChEIs with Aβ antiaggregating effects.

All AP2238 derivatives turned out to be weak inhibitors of hBChE, with IC_{50} values in the micromolar range. As found for the AChE inhibitory activity, the modest Aβ antiaggregating effect of the lead AP2238 was not generally improved in its analogs, which exhibited lower percentages of inhibition (Table **4**), including the monosubstituted 6-methoxy-, 7-methoxy, 6-nitro- and 6-amino-derivatives of AP2238 (< 5% inhibition, not shown in Table **4**). Only AP2243 (**23**) exhibited a nearly identical Aβ antiaggregating effect than AP2238 [135].

Table 4: Inhibitory activities of AP2238 and derivatives toward hAChE and hAChE-induced $A\beta_{1-40}$ aggregation[a]

Compound	hAChE (IC_{50} nM)	AChE-Induced Aβ aggregation (% inhibition)
21	45	35
22	62	15
23	18	38
24	49	6
25	202	25

[a] Data from ref. [135].

The vast majority of the dual binding site AChEIs so far developed are reversible, and, consequently, do not modify covalently the catalytic serine residue. A similar situation is observed with the currently commercialized AChEIs, among which only the carbamate rivastigmine covalently modifies by carbamoylation this serine residue. Recently, Recanatini has developed a novel family of dual binding site AChEIs which bear a carbamate functionality in their active site interacting unit [136], using as the lead compounds the subnanomolar carbamate inhibitors xanthostigmine (**26**, Fig. (**8**)) and the seven-methylene homolog thereof, **27** [137, 138]. In spite of the significant length of the xanthostigmine molecule, docking simulations showed that its tricyclic ring was not able to reach Trp279 at the peripheral site, interacting instead with several aromatic residues of the active-site gorge [137]. In a further attempt to contact the peripheral site, the alkoxy chain between the heteroaryl moiety and the tertiary amine of xanthostigmine was lengthened from 3 up to 10 methylene groups, obtaining the best AChE inhibitory activity for the seven-methylene linked derivative, **27** [138]. However, no evidence was given about the dual site binding character of **27**, which, in the event, was equipotent to **26** as hAChE inhibitor (Table **5**). Indeed, neither **26** nor **27** were able to significantly inhibit the AChE-induced aggregation of Aβ, as it could be expected for a dual binding site AChEI, exhibiting percentages of inhibition lower than 10% at 100 μM (Table **5**) [136]. These results are similar to those found for the carbamate active site inhibitor rivastigmine (Table **1**), but quite different from those obtained with physostigmine (30% inhibition) [79, 85]. In the latter case, formation of a transient AChE-physostigmine intermediate involving the peripheral site has been proposed to explain its relatively high Aβ antiaggregating effect [136].

In order to ensure the reaching of Trp279 at the peripheral site, and thus, the dual site binding, Recanatini designed a novel family of carbamate-based compounds, keeping the *N*-methyl-*N*-[(3-methylcarbamoyloxyphenyl)methyl]amino moiety and the tri- or heptamethylene alkoxy spacer chain of **26** or **27**, but replacing the tricyclic xanthenone system of these leads by different bicyclic benzocycloalkanone ring systems, to which different sterically encumbering π-electron-rich arylidene groups were attached [136]. The SAR data related to the hAChE inhibitory activity of this family can be briefly summarized as follows: i) The length of the alkoxy linker which confers an optimal activity is that equivalent to 7 methylene groups, the heptamethylene-linked derivatives being 2-6-fold more potent than their trimethylene-linked counterparts. ii) The bulky arylidene groups attached at position 2 of the benzocycloalkanone ring system turned out to be of pivotal importance. Indeed, removal of this arylidene group leads to a 20-3000-fold drop of activity. Regarding the nature of this arylidene moiety, the best activity was observed for those compounds bearing a 3,4,5-trimethoxyphenyl as the aryl group of the arylidene moiety, these compounds being nearly 5-fold more potent than those containing 1-naphthyl, 2-naphthyl or phenyl, 15-fold more potent than those containing 3,4-dichlorophenyl, 85-fold more potent than those containing 3,5-dichlorophenyl, and 150-fold more potent than those containing 9-anthryl. iii) The benzocycloalkanone ring system conferring an optimal AChE inhibitory activity turned out to be the 3-benzofuranone system present in compounds **28-34** (Fig. (**8**)). These compounds are 4-, 10- and 20-fold more potent than their counterparts bearing 1-indanone, 3,4-dihydro-2*H*-naphthalen-1-one and 4-chromenone systems, respectively.

Overall, the best compound of the series, **28**, is still slighly less potent hAChE inhibitor than the leads **26** and **27** (Table **5**). Although these results could lead to question the dual site binding character of the novel derivatives, this aspect was confirmed by docking studies, and, more importantly through determination of their inhibitory activity on the AChE-induced Aβ aggregation. All the compounds of the series, including all the substitution patterns above mentioned, exhibited a remarkable percentage of inhibition of Aβ aggregation at 100 μM, which ranged from 29% to 67% [136]. Even a synthetic intermediate used in the preparation of

29 lacking the carbamate functionality, which of course was essentially inactive toward AChE, exhibited an outstanding 50% inhibition of the AChE-induced Aβ aggregation. Since all the compounds of the overall series were tested for their Aβ antiaggregating effect, some SAR data could also be derived for this activity in this structural class. Even though no great differences of Aβ antiaggregating activity were observed among the different compounds of the series, it could be concluded that the best substitution pattern conferring an optimal Aβ antiaggregating effect involves the presence of a heptamethylene spacer chain, a 1-naphthyl or 9-anthryl group in the arylidene moiety, and a 3-benzofuranone system [136]. The extended planar conjugated structure resulting from the increased extension of the conjugated π-system by attachment of the arylidene moiety to the benzocycloalkanone system in these compouds relative to the leads **26** and **27** seems to have a role in their interference with the AChE-Aβ interaction at the peripheral site. Indeed, a benzofuranone derivative lacking the arylidene moety of compounds **28-34** was found to exhibit a significantly lower Aβ antiaggregating effect (18% inhibition).

Moreover, in spite of being clearly selective for hAChE inhibition, all the compounds of the series turned out to be rather potent inhibitors of hBChE, with IC_{50} values in the nanomolar range (Table **5**).

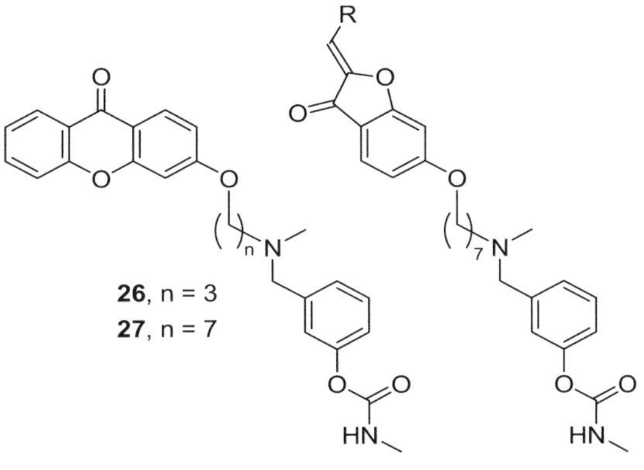

	R
28	3,4,5-trimethoxyphenyl
29	1-naphthyl
30	2-naphthyl
31	9-anthryl
32	phenyl
33	3,4-dichlorophenyl
34	3,5-dichlorophenyl

Figure 8: Carbamate-based dual binding site AChEIs with Aβ antiaggregating effects.

Table 5: Inhibitory activities of carbamate-based dual binding site AChEIs toward hAChE, hBChE, and hAChE-induced Aβ aggregation[a]

Compound	hAChE (IC_{50} nM)	hBChE (IC_{50} nM)	AChE-Induced Aβ aggregation (% inhibition)
26	0.3	48	< 10
27	0.3	17	< 10
28	0.5	136	41
29	2.8	40	67
30	2.1	75	53
31	79	761	65
32	2	58	43
33	8	95	49
34	45	130	48

[a] Data corresponding to selected compounds from ref. [136].

Kwon has recently developed a novel class of piperidine-based dual binding site AChEIs bearing aromatic groups, namely benzoate and benzhydryl, at both extremes of the molecule to allow the interaction with the aromatic rings of the tryptophan residues of both binding sites of AChE [139]. Selected compounds of this family are shown in Fig. (**9**). The SAR data related to their *Electrophorus electricus* AChE (*Ee*AChE) inhibitory activity can be summarized as follows: i) The benzhydryl group of these compounds is very important for the AChE inhibitory activity. Its replacement by a benzyl group causes, in general, the loss of the activity. ii) The presence of a chlorine atom at position 4 of the benzene ring of the benzoate moiety leads to the optimal activity. Its removal causes a 4-fold drop of activity. Its replacement by a fluorine atom, branched alkyl substituents or a nitro group leads to a 3-20-fold decreased activity. Also, its isomerization to position 3 leads to a 10-fold lower activity. iii) Insertion of a NH group between the ester carbonyl group and the benzene ring, thus providing a carbamate functionality, leads to a 8-fold diminished activity.

Even though none of these compounds exhibited a high inhibitory potency toward *Ee*AChE, with IC_{50} values in the micromolar range in most cases (Table **6**), their dual site binding character was established on the basis of docking simulations and determination of their inhibitory activity on the AChE-induced Aβ aggregation. Thus, molecular modeling studies suggested that the benzoate moiety would be placed at the AChE active site for interaction with Trp84, while one of the benzene rings of the benzhydryl moiety would establish π-stacking interactions with Trp279 at the peripheral site. The increased flexibility of a benzyl group relative to the benzhydryl group could account for the weaker interaction with Trp279 of benzyl-substituted analogs, and, therefore, for their lower AChE inhibitory and Aβ antiaggregating effects [139]. The benzhydryl-substituted compounds of this series, as compounds **35-38** exhibited a remarkable inhibitory activity on the AChE-induced $A\beta_{1-42}$ aggregation both at 100 μM and even at 1 μM [139]. Interestingly, as previously found in compounds **10-14** (Fig. (**5**)), these piperidine-based dual binding site AChEIs are able to inhibit the spontaneous Aβ aggregation (in the absence of AChE), exhibiting IC_{50} values up to 20 μM and being more potent than tacrine and donepezil.

All of these compounds turned out to be weak inhibitors of BChE too, with IC_{50} values in the micromolar range, the benzyhydryl- and benzyl-substituted compounds being selective for AChE inhibition and for BChE inhibition, respectively.

Interestingly enough, some of these compounds exhibited a neuroprotective effect against the toxicity induced by $A\beta_{1-42}$ in human neuroblastoma IMR-32 cells [139]. Compounds **35** and **37** were the most potent, affording 91% and 85% of cell viability, respectively. In this assay, donepezil showed a similar degree of neuroprotection (83% survival), while tacrine not only did not afford any protection, but it caused some cytotoxicity instead.

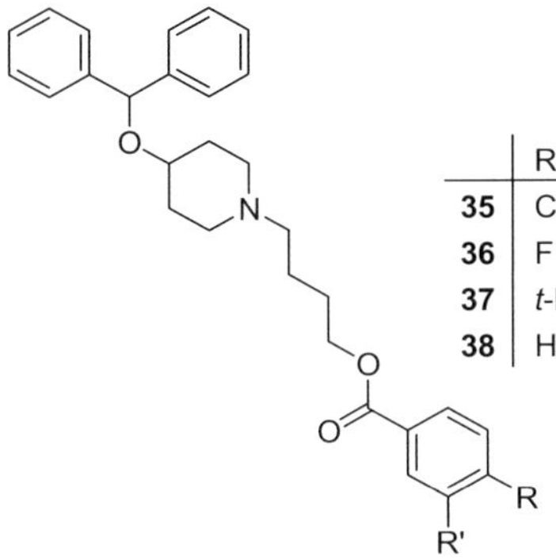

	R	R'
35	Cl	H
36	F	H
37	*t*-Bu	H
38	H	F

Figure 9: Piperidine-based dual binding site AChEIs with Aβ antiaggregating effects.

Table 6: Inhibitory activities of piperidine-based dual binding site AChEIs toward *Ee*AChE, and *Ee*AChE-induced $A\beta_{1-42}$ aggregation[a]

Compound	*Ee*AChE (IC_{50} nM)	AChE-Induced Aβ aggregation (% inhibition)[b]
35	320	56/42
36	840	42/28
37	2410	72/61
38	7720	52/41

[a] Data corresponding to selected compounds from ref. [139].[b] Inhibition of the AChE-induced $A\beta_{1-42}$ aggregation at 100 μM and 1 μM.

The potential to block the spontaneous aggregation of Aβ has also been reported in a short family of pyridinium- and bispyridinium-based dual binding site AChEIs developed by Holzgrabe [140-142]. Even though these compounds turned out to be rather weak inhibitors of *Ee*AChE, with IC_{50} values in the micromolar range, some of them exhibited a significant inhibitory activity on the spontaneous aggregation of both $A\beta_{1\text{-}40}$ and $A\beta_{1\text{-}42}$ at 100 μM. Thus, pyridinium-based compounds **39** and **40** and bispyridinium-based compound **41** (Fig. (**10**)) inhibited $A\beta_{1\text{-}40}$ aggregation by 60%, 32%, and 55%, respectively, and $A\beta_{1\text{-}42}$ aggregation by 14%, 21%, and 61%, respectively [142].

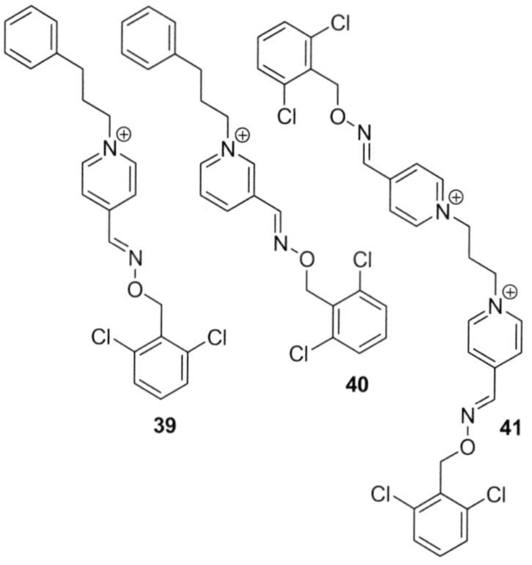

Figure 10: Pyridinium- and bispyridinium-based dual binding site AChEIs with Aβ antiaggregating effects.

As previously mentioned, the main pharmacological effect of dual binding site AChEIs, *i.e.* the inhibitory activity on the AChE-induced Aβ aggregation, arises from the blockade of the peripheral site of the enzyme.

Using a combination of a computational docking model and experimental screening Janda identified five peripheral site AChEIs among a library of compounds designed to disrupt protein-protein interactions, which turned out to be potent inhibitors of AChE-induced Aβ aggregation (compounds **42-46**, Fig. (**11**)) [143]. Docking simulations and competition assays with propidium indicated that these compounds were able to interact with the peripheral site of AChE, even with higher docking energy than propidium. In the event, all of these compounds exhibited essentially the same Aβ antiaggregating effect than propidium, with percentages of inhibition of the AChE-induced Aβ aggregation averaging 78% at 100 μM. As propidium, these compounds showed a weak *Ee*AChE inhibitory activity, with K_i values in the micromolar range [143].

Some cannabinoids have been found to exhibit neuroprotective effects against Aβ-induced toxicity, likely by regulating downstream events in the neurotoxic cascade of AD [144, 145]. Recently, Janda reported that cannabinoids have also potential to act upstream in the neurotoxic cascade of AD, namely on AChE-induced Aβ aggregation [146]. Thus, the widely explored cannabinoid Δ^9-tetrahydrocannabinol (THC), *i.e.* the active component of marijuana (*Cannabis sativa*) (**47**, Fig. (**11**)), was found to be able to inhibit hAChE-induced Aβ aggregation and *Ee*AChE, as well. Both effects seem to be mediated by binding of THC to the peripheral site of AChE, as inferred from docking simulations and competition studies with propidium.

The AChE inhibitory activity of THC was in the micromolar range, but its Aβ antiaggregating effect was superior to that of propidium. Indeed, at 50 μM THC completely blocked the AChE-induced Aβ aggregation, while with propidium the effect was not complete [146]. Moreover, the highly lipophilic nature of THC is also a clear advantage over propidium, whose cationic nature precludes its potential use for AD treatment.

The outstanding Aβ antiaggregating effect of compounds **42-47** should prompt the rational design of novel drug candidates targeting the peripheral site of AChE [147].

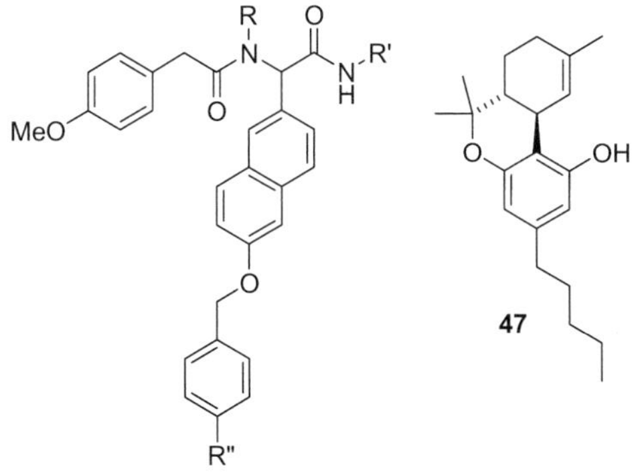

47

	R	R'	R"
42	*p*-methoxybenzyl	cyclohexyl	MeO
43	*p*-methoxybenzyl	benzyl	MeO
44	*n*-Bu	cyclohexyl	MeO
45	*n*-Bu	benzyl	MeO
46	2-methoxyethyl	cyclohexyl	H

Figure 11: Peripheral site AChEIs with Aβ antiaggregating effects.

Apart from compounds herein described, other families of dual binding site AChEIs have been developed, whose members display potent AChE inhibitory activity, and, in most cases, also potent BChE inhibitory activity [114-116, 148-150]. Even some of them, namely a dimer of the natural product huperzine B developed by Bai, exhibited significant neuroprotective effects against Aβ-induced cytotoxicity and against oxygen-glucose deprivation in PC12 cells, and ameliorated Aβ-induced redox disequilibrium and the staurosporine-induced apoptosis in PC12 cells [150]. Unfortunately, in these studies the potential Aβ antiaggregating effects of the novel dual binding site AChEIs were not assessed. As pointed out elsewhere [85, 143], the evaluation of this effect for a series of compounds can be certainly expensive, the overall cost being inaccessible for most of the academic laboratories, but it refers to the probably most important biological effect attainable by these compounds. It has been reported that dual site binding may not be a sufficient condition to ensure a high Aβ antiaggregating action [85, 132]. Indeed, great differences can be observed among the Aβ-antiaggregating effects of different classes of dual binding site AChEIs. Therefore, a systematic evaluation of this effect in the novel families of dual binding site AChEIs would be highly valuable in order to shed light on the structural determinants for a high Aβ antiaggregating action, thus driving the rational design of improved compounds [151].

Of note, very recently, several novel structural classes of dual binding site AChEIs with Aβ anti-aggregating properties have been reviewed [152].

4. MULTIFUNCTIONAL DUAL BINDING SITE AChEIs

It is becoming increasingly clearer that the use of a single drug hitting a single biological target may not be sufficient to efficiently tackle complex diseases which are polyetiological in origin, as it is the case for neurodegenerative diseases in general and for AD in particular [153]. The consideration of these diseases as multifactorial has had immediate implications for therapeutic strategies. Thus, the use of multiple drugs directed toward multiple targets has been suggested as the most efficient way to treat these diseases [154]. Indeed, combination therapies are being included in clinical trials.

An emerging alternative approach is based on the use of a single drug able to modulate multiple targets identified for a particular disease. Among the potential multifunctional agents, some naturally occurring compounds have been claimed to be very promising for AD treatment [155-159]. However, this approach normally involves the use of synthetic hybrid compounds, rationally designed to incorporate into a single framework at least two different pharmacophoric moieties [160-164]. The use of multifunctional drugs should be advantageous over the classical polypharmacy, since it reduces the risk of drug-drug interactions and simplifies pharmacokinetic and pharmacodynamic studies, manufacture, formulation, clinical trial designs, and also the therapeutic regimens, thus resulting in a greater patient compliance. However, one of the main issues associated with the use of such multifunctional drugs resides in the difficult optimization of potencies at two separate targets with a single drug [162].

The design of multifunctional drugs addressing neurodegenerative diseases has been recently reviewed by Cavalli, Bolognesi and Melchiorre, and Youdim and Van der Schyf [28, 165-167].

In most cases, at least one of the pharmacophoric moieties of multifunctional drugs is directly related to an AChEI [28]. Particularly, as it has been already suggested [28], dual binding site AChEIs themselves could be considered multifunctional anti-Alzheimer drug candidates, in virtue of their ability to modulate at least two events of the neurotoxic cascade, namely Aβ aggregation and the central cholinergic deficit. However, the search for multipotency in anti-Alzheimer drug candidates has gone beyond the effects afforded by the dual site binding. Thus, novel multifunctional anti-Alzheimer agents have been purposely designed by combination of a moiety derived from a known active site AChEI with a different pharmacophoric moiety, structurally suited to simultaneously interact with the AChE peripheral site thus affording dual site binding, and separately with a second biological target involved in the neurotoxic cascade of AD. Another approach, very recently exploited, consists of the incorporation of additional functionalities into the structure of a classical dual binding site AChEI, as to allow it to address an additional biological target other than AChE.

The first approach was efficiently used by Melchiorre in the development of lipocrine (**48**, Fig. (**12**)), a molecule wich combines a unit of 6-chlorotacrine with a unit of the universal antioxidant lipoic acid, and therefore conceived to address AChE-mediated events and oxidative stress, one of the downstream processes of the neurotoxic cascade of AD [69, 110, 168]. A short family of hybrids bearing these or related structural motifs was developed, from which the following SAR data regarding hAChE inhibitory activity derived: i) Lipocrine was the most potent compound of the series. Shortening of the oligomethylene chain between the two nitrogen atoms from 3 to 2 methylenes or lengthening to 4-7 methylenes results in nearly 15- and 5-fold drop of activity, respectively, while attachment of the lateral chain at position 3 of the tacrine unit instead of at the exocyclic amino group leads to a dramatic decrease of potency. ii) Removal of the chlorine atom of **48** at position 6 of the tacrine unit leads to a 30-fold decreased activity, while substitution of the tacrine unit by a 2-chloro-6,7-dimethoxyquinazoline system (compound **50**, Fig. (**12**)) or by a caproctamine-like *N*-ethyl-*o*-methoxybenzylamino moiety or other alkoxybenzylamino groups results in a dramatic loss of activity. iii) Substitution of the dithiolane ring of **48** by a 2-thienyl ring (compound **49**) leads to a 10-fold decrease of activity. iv) Racemic lipocrine and its (*R*)-enantiomer were equipotent.

Lipocrine and some analogs turned out to be potent inhibitors of hBChE, with IC_{50} values in the low nanomolar range. Lipocrine was selective for AChE inhibition while in the analogs lacking the chlorine atom in the tacrine unit the selectivity was reversed.

Most dual binding site AChEIs bear an aromatic system and/or a quaternary or protonatable nitrogen atom in the peripheral site interacting unit. Even though this is not the case for lipocrine, it has been argued that the dithiolane ring of lipocrine could interact with the AChE peripheral site [28, 168]. Indeed, kinetic experiments evidenced the dual site binding character of lipocrine, which consequently was found to be able to inhibit the AChE-induced Aβ aggregation. Both, lipocrine and its (*R*)-enantiomer turned out to be the most potent Aβ antiaggregating compounds of the series, exhibiting percentages of inhibition around 65% at 100 μM and IC_{50} values around 45 μM (Table 7). Compounds **49** and **50** exhibited lower, but still significant, Aβ antiaggregating effects, while some derivatives bearing methoxybenzylamino groups instead of the tacrine unit were weak inhibitors of the AChE-induced Aβ aggregation (9-17% inhibition) [69].

The effects of lipocrine on its second biological target, namely, the oxidative stress, were also very interesting. Thus, lipocrine produced a strong inhibitory effect on the *tert*-butyl hydroperoxide-induced formation of reactive oxygen species (ROS) in SH-SY5Y cells (64% inhibition at 50 μM), being in this effect more potent than the parent lipoic acid.

Moreover, as found in other dual binding site AChEIs, lipocrine turned out to be a potent inhibitor of BACE-1 (IC_{50} 70 nM) [69]. Thus, lipocrine exhibits a unique pharmacological profile encompassing ability to inhibit both Aβ formation and aggregation, hAChE activity and also the oxidative damage.

Another example of incorporation of an antioxidant moiety into the framework of a dual binding site AChEI was recently reported by Rodríguez-Franco, who combined a tacrine related moiety with a unit related to the antioxidant melatonin [169]. This molecular hybridization resulted in a series of subnanomolar inhibitors of hAChE, the most potent of which, compound **51** (Fig. (**12**)), exhibiting an IC_{50}

value of 8 pM. The SAR data regarding AChE inhibitory activity within this family could be summarized as follows: i) The introduction of chlorine atoms at the tacrine unit of these compounds produced surprising results. Thus, the presence of a chlorine atom at position 6 of the tacrine unit leads to a 5-fold improvement of both bovine and human AChE inhibitory activity, but introduction of the chlorine atom at position 8 decreases potency by 15-fold, and 6,8-dichlorosubstitution leads to a 2-10-fold drop of bovine AChE inhibitory activity but to a 10-fold increased hAChE inhibitory activity relative to the 6-chlorosubstituted counterparts. ii) Shortening of the oligomethylene chain between the tacrine amino and the carbonyl groups of **51** from 6 to 5 methylenes leads to a 2-10-fold decrease of bovine AChE inhibitory activity. iii) Introduction of methoxy or hydroxy substituents at position 5 of the indol moiety leads to 2-6- and 10-fold decreased bovine AChE inhibitory activity, respectively.

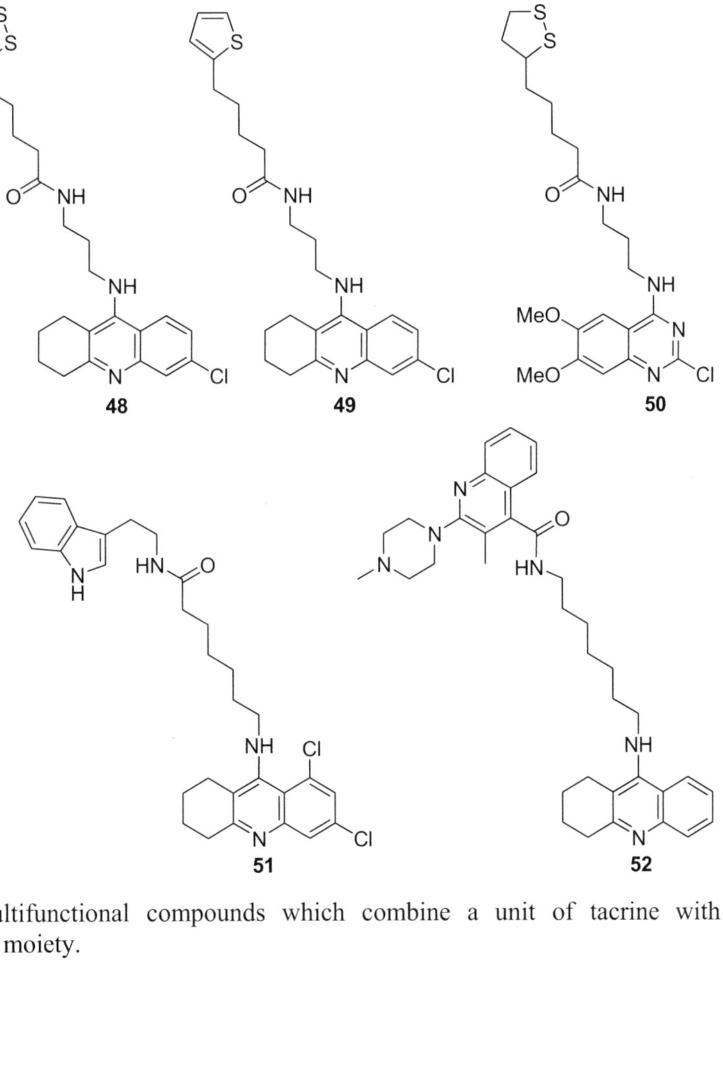

Figure 12: Multifunctional compounds which combine a unit of tacrine with a different pharmacophoric moiety.

Table 7: Inhibitory activities of lipocrine and related compounds toward hAChE, hBChE and hAChE-induced $A\beta_{1-40}$ aggregation[a]

Compound	hAChE (IC_{50} nM)	hBChE (IC_{50} nM)	AChE-Induced Aβ aggregation (% inhibition)
rac-**48**	0.25	11	62
(*R*)-**48**	0.23	not det.	68
49	2.7	31	24
50	inactive	not det.	32

[a] Data corresponding to selected compounds from ref. [69].

Worthy of note, the indol-unsubstituted compounds of this series are isomers of the indol-tacrine hybrids **10-14** (Fig. (**5**)), bearing an inverse amide group. The dual site binding of the tacrine-melatonin hybrids to AChE has been recently suggested on the basis of molecular modeling studies [170].

Although selective for hAChE inhibition, these compounds exhibited potent hBChE inhibitory activities, with IC_{50} in the low nanomolar range. Moreover, due to the presence of the melatonin moiety, these compounds exhibit an interesting oxygen radical absorbance capacity (2-4 trolox equivalents), being thus even more potent than the parent melatonin [169, 170].

In a study directed to delineate the geometry of the serotonin 5-HT_3 receptor binding pockets, Cappelli designed hybrid **52** (Fig. (**12**)), which combines a tacrine unit and a quipazine-related quinoline moiety with affinity for 5-HT_3 receptors [171]. Indeed, a computational analysis of **52** suggested that its tacrine unit interacts with a subsite located at the boundary of the 5-HT_3 receptor extracellular domain, which could be comparable to the peripheral site of AChE. Regarding its interaction with AChE, docking simulations suggested its dual site binding, by placing the tacrine unit at the active site of the enzyme and the quinoline moiety at the peripheral site. As a result, compound **52** was found to be a potent inhibitor of hAChE (IC_{50} = 4 nM), and also hBChE (IC_{50} = 40 nM), and a high affinity ligand of the rat 5-HT_3 receptor (K_i = 6 nM), a profile which seems suitable for restoring the normal cholinergic tone in AD patients, due to the modulation of the tonic inhibitory control on cortical ACh release by 5-HT_3 receptors [171]. Unfortunately, its potential Aβ antiaggregating effect has not been assessed.

Some classical dual binding site AChEIs, namely *bis*(7)-tacrine (**8**, Figs. (**4**) and (**5**)), AP2238 (**21**, Fig. (**7**)) and caproctamine (**15**, Fig. (**6**)), have recently been used as templates to build up novel multifunctional drugs, through incorporation of additional functionalization to endow the resulting molecules with the ability to hit biological targets other than AChE.

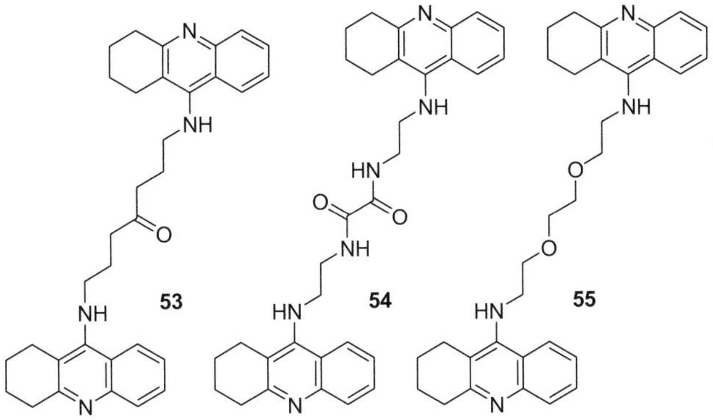

Figure 13: Multifunctional anti-Alzheimer compounds designed from *bis*(7)-tacrine.

Table 8: Inhibitory activities of multifunctional anti-Alzheimer compounds designed from *bis*(7)-tacrine toward hAChE, hBChE, and hAChE-induced Aβ_{1-40} aggregation[a]

Compound	hAChE (IC_{50} nM)	hBChE (IC_{50} nM)	AChE-Induced Aβ aggregation (% inhibition)
53	1.83	11	76
54	6.65	5.44	54
55	20	35	62
8	0.81	5.66	68

[a] Data from ref. [93].

Bolognesi has recently developed some derivatives of *bis*(7)-tacrine by introduction in its heptylene linker of suitable groups as to allow chelation of metal ions such as copper or iron [93], which seem to play a crucial role both in the aggregation of Aβ and in mediating the oxidative stress mechanism of Aβ neurotoxicity by induction of ROS generation [172, 173]. Indeed, the resulting bis-tacrines **53-55** (Fig. (**13**)) bearing ketone carbonyl, oxalamide or ethylenedioxy groups within the linker exhibited the desired additional property of acting as metal chelators, as inferred from changes in their electronic spectra in the presence of Fe^{2+} and Cu^{2+}, although at the expense of a decrease in their hAChE inhibitory activities relative to the parent **8** (2-25-fold) (Table **8**) [93]. Docking simulations confirmed the dual site binding of hybrids **53-55** to AChE, what accounts for their high inhibitory potency toward both hAChE and hAChE-induced Aβ aggregation, the latter effect being in the range of that of the parent **8**. Their lower hAChE inhibitory activity relative to **8** was then suggested to be due to their higher desolvation penalty as a consequence of their increased polarity [93]. As also found in other dual binding site AChEIs, these compounds turned out to be potent inhibitors of hBChE, with IC_{50} values in the low nanomolar range.

In a program directed to develop multifunctional anti-Alzheimer compounds targeting AChE and BACE-1, Piazzi modified the structure of AP2243 (**23**, Fig. (**7**)), the *N*-ethyl analog of the dual binding site AChEI AP2238 (**21**), by substitution of the coumarin methoxy substituents by other groups, namely dihalophenylalkylamidic functions [174], present in known BACE-1 inhibitors [175]. As found in *bis*(7)-tacrine-based multifunctional compounds, introduction of the alternative substituents endowed the resulting hybrids (selected examples shown in Fig. (**14**)) with the additional activity (BACE-1 inhibition), although at the expense of a lower hAChE inhibitory activity relative to the parent AP2243 (**23**), much more dramatic in this case (10-400-fold lower) (Table **9**) [174]. Even though it has been reported that these compounds exhibit a binding mode similar to that of the parent compound, their clearly lower AChE inhibitory activity seem to indicate that the bulky substituents on the coumaring ring preclude a proper penetration into the enzyme gorge [174]. Unfortunately, the effect of the alternative substitution on the moderate Aβ antiaggregating activity of the parent compound was not assessed.

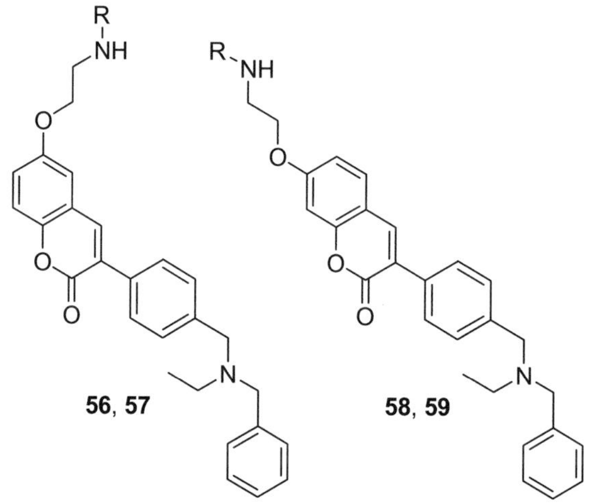

56, R = 3,5-difluorophenylacetyl
57, R = 3,5-difluorophenylacryloyl
58, R = 3,5-difluorophenylacetyl
59, R = 3-(3,4-difluorophenyl)propanoyl

Figure 14: Multifunctional anti-Alzheimer compounds designed from AP2243.

Table 9: Inhibitory activities of multifunctional anti-Alzheimer compounds designed from AP2243 toward hAChE, and BACE-1[a]

Compound	hAChE (IC_{50} nM)	BACE-1 (IC_{50} nM)
56	551	149
57	7160	99
58	181	150
59	327	114
24	18	≈ 238[b]

[a] Data corresponding to selected compounds from ref. [174]. [b] 42% Inhibition at 202 nM.

Melchiorre and Bolognesi have recently developed one of the most promising multifunctional anti-Alzheimer drug candidates by structural modification of the dual binding site AChEI caproctamine (**15**, Fig. (**6**)) by incorporation in the linker of a functionality suited to confer additional radical scavenger activity, namely the 1,4-benzoquinone moiety of the natural antioxidant coenzyme Q10 [85, 176-178]. Not only exhibited the resulting compound, memoquin (**60**, Fig. (**15**)), the additional antioxidant activity, but it turned out to be more potent than caproctamine and its *N*-ethyl analog **16** (Fig. (**6**)) toward hAChE inhibition (100- and 10-fold, respectively), toward hBChE inhitibion (8- and 1.5-fold, respectively), and toward hAChE-induced Aβ aggregation (> 17-fold higher percentages of inhibition at 100 μM) (Table **10**) [85, 178]. Worthy of note, molecular dynamics simulations confirmed the dual site binding of memoquin. Moreover, memoquin also inhibited *in vitro* by 96% the spontaneous aggregation of $A\beta_{1-42}$ when tested at an equimolar concentration of 50 μM, and it was found to potently inhibit BACE-1, with an IC_{50} of 108 nM [178]. Regarding its antioxidant effect, it was found that memoquin was able *in vitro* to decrease the formation of free radicals, and to inhibit *tert*-butyl hydroperoxide-induced ROS generation in SH-SY5Y cells. The latter effect was remarkably increased after chemical induction of the enzyme NAD(P)H:quinone oxidoreductase, an enzyme which catalyzes the reduction of quinones to their antioxidant active hydroquinone form, and which is upregulated in AD patients in response to oxidative stress. With the purpose of tuning the effect of memoquin on its multiple targets, a large series of differently substituted derivatives was prepared [85]. Some selected examples are shown in Fig. (**15**). The *o*-methoxybenzyl groups and the ethyl-substituted benzylamino nitrogen atoms are essential for an optimal AChE inhibitory activity as reported for caproctamine analogs (see above). Indeed, the SAR affecting these parts of the molecule of memoquin are very similar to those reported for caproctamine analogs. Moreover, replacement of the methoxyphenyl moiety by different heteroaromatic rings (furan, thiophene, pyridine) leads to a clear drop of potency (40-125-fold). The SAR for hAChE inhibition affecting other parts of the lead molecule can be briefly summarized as follows: i) Introduction of substituents at positions 2 and 5 of the quinone ring leads to a decreased hAChE inhibitory activity (2-80-fold), especially for bulky substituents such as phenyl groups, while replacement of the quinone ring by a benzene ring causes a 30-fold decrease of activity. ii) Shortening of the hexamethylene spacer chain between nitrogen atoms of **60** to 2-5 methylenes or lengthening to 7 methylenes causes a 1.5-40- and 5-fold drop of activity, respectively. iii) Lengthening of the chain between methoxyphenyl rings and basic nitrogen atoms from 1 to 2 methylenes leads to a 10-fold decrease of activity. iv) Replacement of the methylene group of the spacer chain between nitrogen atoms closer to the basic nitrogen atoms by a carbonyl group led to a dramatic loss of potency (7500-fold).

All of the compounds of the series were rather weak inhibitors of hBChE, with micromolar IC_{50} values. The Aβ antiaggregating effect of these compounds was much less sensitive to the structural changes. Thus, the inhibitory activity on the AChE-induced $A\beta_{1-40}$ aggregation was in the range 70-90% for most of the tested compounds, irrespective of the substitution or replacement at the quinone ring, shortening of the spacer chain between nitrogen atoms or substitution on the basic nitrogen atoms. Only, introduction of bulky phenyl substituents at the quinone ring or replacement of the aromatic methoxy groups by nitro groups produced a more pronounced loss of activity (around 2-fold), while again, the replacement of the methylene group closer to the basic nitrogen atoms by a carbonyl group led to a nearly complete loss of activity. Moreover, all of the compounds exhibited a certain inhibitory effect on the spontaneous aggregation of $A\beta_{1-42}$, most of them exhibiting a percentage of inhibition in the range 50-70% at 10 μM. Again most of the structural changes had no or little effect on this activity. The greater changes were observed upon shortening of the spacer chain between nitrogen atoms to 2-4 methylenes, introduction of the carbonyl group close to the basic nitrogen atoms, substitution of the aromatic methoxy groups by methyl, chloro or nitro substituents, or replacement of the methoxyphenyl ring by heteroaromatic rings, which caused modest decreases in the percentages of inhibition (2-5-fold) [85].

As previously found for memoquin, its derivative **61** was able to protect SH-SY5Y cells expressing high levels of NAD(P)H:quinone oxidoreductase against oxidative stress-evoked ROS generation, although to a lesser extent than memoquin [85].

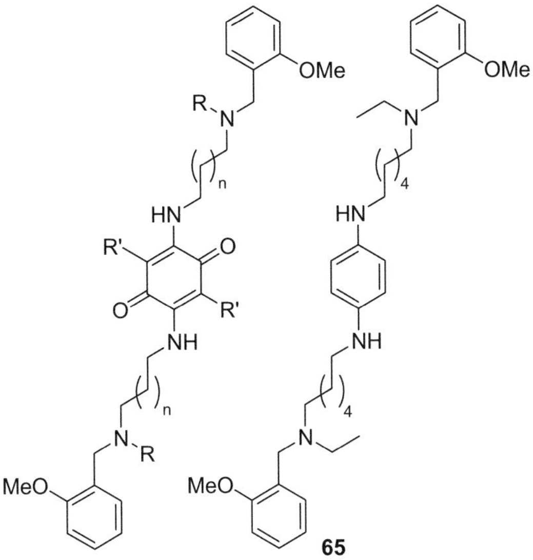

65

	R	R'	n
60	Et	H	4
61	Et	Br	4
62	Et	phenyl	4
63	Et	H	1
64	H	H	4

Figure 15: Multifunctional anti-Alzheimer compounds designed from caproctamine.

Table 10: Inhibitory activities of multifunctional anti-Alzheimer compounds designed from caproctamine toward hAChE, hAChE-induced $A\beta_{1-40}$ aggregation, and spontaneous $A\beta_{1-42}$ aggregation[a]

Compound	hAChE (IC_{50} nM)	AChE-Induced Aβ aggregation (% inhibition)	Spontaneous Aβ aggregation (% inhibition)
60	1.55	87	67
61	2.66	84	64
62	123	46	68
63	2.25	91	32
64	123	90	70
65	43	72	57
15	170	< 5	20

[a] Data corresponding to selected compounds from ref. [85].

In the past few years, the search for multifunctional anti-Alzheimer drug candidates has been an area of very active research (for recent reviews see [179-

191]). In most cases, a moiety with known AChE inhibitory activity has been linked to another pharmacophoric moiety, which was to provide interesting complementary activities such as NMDA receptor antagonist activity [192-195], metal-chelation properties [196-200], MAO inhibitory activity [196, 197, 201-203], or, prominently, antioxidant activity [205-216].

5. *IN VIVO* RELEVANCE OF THE Aβ ANTIAGGREGATING EFFECT OF DUAL BINDING SITE AChEIs

The most interesting property of dual binding site AChEIs is their *in vitro* proven ability to inhibit the AChE-induced aggregation of Aβ, because of the derived potential to interfere upstream in the neurotoxic cascade of AD, and, thus to positively modify AD progression. However, some concerns may exist about the extrapolation of the outstanding *in vitro* Aβ antiaggregating effects of these compounds to the physiological conditions.

Most of these assays are perfomed using high concentrations of AChE, Aβ and inhibitor. On the one hand, the concentrations of AChE and Aβ are much higher than those present in the brain, but they are necessary to accelerate the aggregation process up to a reasonable extent for analytical purposes. On the other hand, the concentration of inhibitor in these assays is usually 100 μM, and therefore, much higher than those required for AChE inhibition, which are in the nanomolar or even in the picomolar range. However, as pointed out elsewhere [79], these high concentration values should be normalized in order to compare the inhibition data. Thus, if one considers the inhibitor/AChE concentration ratio in both the Ellman's assay for determination of the AChE inhibitory activity and in the thioflavin T-based fluorometric assay for determination of the inhibitory activity on the AChE-induced Aβ aggregation, the resulting values are indeed of the same magnitude. Thus, it would seem reasonable that similar amounts of a given inhibitor might lead to both inhibitory activities.

More interestingly, the proof-of-concept of the therapeutic usefulness of the *in vitro* Aβ antiaggregating effect of dual binding site AChEIs has been recently obtained in *in vivo* studies.

Indeed, memoquin has exhibited very interesting neuroprotective effects in AD11 anti-NGF transgenic mice, a comprehensive animal model for AD which expresses recombinant antibodies against nerve growth factor [217, 218]. These animals develop a progressive neurodegenerative phenotype, displaying a full set of hallmarks of AD including widespread neuronal loss, formation of Aβ plaques, intracellular accumulation of hyperphosphorylated tau, severe cholinergic deficit, and behavioral impairment linked to the progression of neurodegeneration, among others. Regarding Aβ accumulation, this peptide appears in hippocampal dystrophic neurites at 6 months of age, and at the age of 15 months they show a clear accumulation of extracellular Aβ deposits [219, 220]. Memoquin treatment delayed the Aβ expression and significantly reduced Aβ deposits in 15-months-old AD11 mice [178]. Moreover, memoquin rescued the cholinergic deficit in basal forebrain and completely or partially prevented tau hyperphosphorylation at the age of 2-6- and 15-months, respectively. Moreover, these neuroprotective effects of memoquin were accompanied by its ability to rescue behavioral deficits linked to attention and memory. Recently, additional behavioral studies with memoquin have been carried out in two different mouse models, in which amnesia was induced either with scopolamine or with Aβ, administered intraperitoneally or intracerebroventricularly, respectively [221]. In these studies, memoquin, at a dose of 15 mg/kg, was able to rescue spatial, episodic, aversive, short and long-term memory in both mouse models. Taking into account the *in vivo* neuroprotective and cognitive-enhancing effects of memoquin, as well as its good oral bioavailabilitity, ability to cross the blood-brain barrier and a favorable safety profile [178, 221], it is clear that it represents a very promising disease-modifying anti-Alzheimer drug candidate, prone to be subjected to clinical testing.

NP-61 (formerly NP-0361) is a dual binding site AChEI developed by Martínez at the company Noscira (formerly Neuropharma, S.A.) [222], which has been shown to exhibit a picomolar inhibitory activity of AChE and to inhibit the AChE-induced Aβ aggregation *in vitro* with an IC_{50} value one order of magnitude lower than that of propidium. Regarding its *in vivo* effects, when orally administered for 3 months to hAPP transgenic mice (Swedish and London mutation), NP-61 decreased the levels of soluble and aggregated $A\beta_{1-40}$ and $A\beta_{1-42}$ and reduced the number, surface area and size of amyloid plaques in cortex and hippocampus [223, 224]. Moreover, it was determined that the reduced brain amyloid plaque load resulted in a significantly increased cognition (spatial learning and memory in the Morris water maze test). The ability of NP-61 to revert the AD-like neurodegenerative phenotype in transgenic mice as well as the good safety results demonstrated in regulatory toxicological studies were clearly suggestive of its promising therapeutic potential for clinical application. Indeed, the regulatory authorities of the United Kingdom and the corresponding Ethics Committee approved the first clinical trial with NP-61 in 2007.

Other multifunctional dual binding site or peripheral site AChEIs endowed with potent *in vitro* inhibitory activities against AChE and Aβ aggregation, as well as with other complementary effects such as antioxidant activity (*i.e.* a tacrine-melatonin hybrid [225]), metal chelating properties (*i.e.* a tacrine-8-hydroxyquinoline hybrid [226]), MAO-B inhibitory activity (*i.e.* a 9-anthraldehyde pyrazoline derivative [227]), or tau anti-aggregating properties (*i.e.* a tacrine-huprine hybrid [228]) have also shown beneficial effects in APP/PS1 mice or scopolamine- and Aβ-induced amnesia mouse models, in terms of cognition and/or amyloid pathology.

Thus, the proof-of-concept performed in animal models with these dual binding site or peripheral site AChEIs and the fact that the first example of this kind of compounds entered clinical trials are a clear evidence of the promising role of these compounds as disease-modifying agents for AD treatment.

CONCLUSION

The multiple and increasing evidence of different neuroprotective effects by the currently marketed AChEIs makes untenable the notion that these drugs only afford a symptomatic relief of the disease. As the evidences of neuroprotection are overwhelming the simplistic consideration of their symptomatic role, it is acquiring increasing importance the belief that these drugs should be used in the prodromal or very early stages of AD.

Neuroprotection mediated by AChEIs likely occur *via* multiple pathways, and different mechanisms might be in operation, depending on the nature of the neurotoxic insult. The higher these mechanisms can interfere in the neurotoxic cascade of AD, the stronger the potential of the drug to posively modify the progress of neurodegeneration. Unlike the rest of commercialized AChEIs, donepezil is able to inhibit *in vitro* the two earliest events of the neurotoxic cascade of AD, namely Aβ formation as well as Aβ aggregation, which likely accounts for the apparent superiority of donepezil in the neuroprotection studies with these drugs. The interference of donepezil and the other AChEIs in Aβ formation, either by stimulation of the α-secretase pathway or by inhibition of β-secretase, BACE-1 (in the sole case of donepezil), is related to their indirect activation of the cholinergic system and downstream signal transduction cascades or, in the case of BACE-1 inhibition is difficult to rationalize from a structural point of view. Conversely, the unique ability of donepezil to block Aβ aggregation can be ascribed to its chemical structure, and the resulting particular mode of interaction with AChE, by spanning the entire length of the active site gorge and simultaneously reaching both the active and peripheral binding sites of the enzyme. Particularly, according to Inestrosa's findings, the blockade of the peripheral site of AChE accounts for the Aβ antiaggregating effect of donepezil.

Novel families of AChEIs with the same binding mode of donepezil, *i.e.* dual binding site AChEIs, and to a lesser extent peripheral site AChEIs, have been purposely designed in the last decade with the main aim of achieving a potent Aβ antiaggregating effect, apart from a highly potent AChE inhibitory activity, and in some cases, also other complementary actions arising from hitting other biological targets involved in AD neurotoxic cascade.

Inestrosa and Brimijoin demonstrated that AChE, upon binding to Aβ through its peripheral site, accelerates Aβ aggregation *in vitro*, amyloid plaque formation *in vivo*, the latter effect resulting in a significant cognitive impairment in animal models. Now, dual binding site AChEIs have been shown to be able to inhibit *in vitro* the AChE-induced Aβ aggregation, to significantly reduce *in vivo* the amyloid burden, and to rescue behavioral and memory deficits due to the amyloid plaque load. Thus, dual binding site AChEIs constitute very promising anti-Alzheimer disease-modifying drug candidates, which should contribute, together with the amyloid-directed drug candidates currently undergoing clinical trials, to fulfil one of the most important unmet needs of modern medicine, namely the possibility of preventing, stopping or delaying the neurodegenerative process of AD.

ACKNOWLEDGEMENTS

The author gratefully acknowledge Dr. Axel Bidon-Chanal from the Departament de Fisicoquímica of the Universitat de Barcelona (Spain) for providing the figure of the superimposed 3D X-ray crystal structures of the complexes *Tc*AChE-marketed AChEIs and Dr. Manuela Bartolini from the Department of Pharmaceutical Sciences of the University of Bologna (Italy) for helpful scientific discussions.

CONFLICT OF INTEREST

The author confirms that this chapter contents have no conflict of interest.

DISCLOSURE

This is an updated report of the earlier review article by D. Muñoz-Torrero "Acetylcholinesterase inhibitors as disease-modifying therapies for Alzheimer's disease (review)". Curr Med Chem 2008; 15(24): 2433-55.

ABBREVIATIONS

Aβ	=	β-Amyloid peptide
ACh	=	Acetylcholine
AChE	=	Acetylcholinesterase
AChEI	=	Acetylcholinesterase inhibitor
AD	=	Alzheimer's disease
APP	=	Amyloid precursor protein
BChE	=	Butyrylcholinesterase
CNS	=	Central nervous system
*Ee*AChE	=	*Electrophorus electricus* AChE
hAChE	=	Human acetylcholinesterase
hAPP	=	Human amyloid precursor protein
hBChE	=	Human butyrylcholinesterase
PKC	=	Protein kinase C
ROS	=	Reactive oxygen species
SAR	=	Structure-activity relationships
*Tc*AChE	=	*Torpedo californica* acetylcholinesterase
THC	=	Δ^9-Tetrahydrocannabinol

REFERENCES

[1] Wimo A, Prince M. World Alzheimer Report 2010. The global economic impact of dementia; Alzheimer's Disease International. Available from: http://www.alz.co.uk.

[2] Klafki H-W, Staufenbiel M, Kornhuber J, Wiltfang J. Therapeutic approaches to Alzheimer's disease. Brain 2006; 129: 2840-55.

[3] Schönheit B, Zarski R, Ohm TG. Spatial and temporal relationships between plaques and tangles in Alzheimer-pathology. Neurobiol Aging 2004; 25: 697-711.

[4] Castellani RJ, Zhu X, Lee H-G, Moreira PI, Perry G, Smith MA. Neuropathology and treatment of Alzheimer disease: did we lose the forest for the trees? Expert Rev Neurother 2007; 7: 473-85.
[5] Hardy J, Allsop D. Amyloid deposition as the central event in the aetiology of Alzheimer's disease. Trends Pharm Sci 1991; 12: 383-8.
[6] Selkoe DJ. The molecular pathology of Alzheimer's disease. Neuron 1991; 6: 487-98.
[7] Hardy JA, Higgins GA. Alzheimer's disease: the amyloid cascade hypothesis. Science 1992; 256: 184-5.
[8] Hardy J, Selkoe DJ. The amyloid hypothesis of Alzheimer's disease: progress and problems on the road to therapeutics. Science 2002; 297: 353-6.
[9] Hardy J. Has the amyloid cascade hypothesis for Alzheimer's disease been proved? Curr Alzheimer Res 2006; 3: 71-3.
[10] Melnikova I. Therapies for Alzheimer's disease. Nat Rev Drug Discovery 2007; 6: 341-2.
[11] Siemers ER, Dean RA, Demattos R, May PC. New pathways in drug discovery for Alzheimer's disease. Curr Neurol Neurosci Rep 2006; 6: 372-8.
[12] Skovronsky DM, Lee VM-Y, Trojanowski JQ. Neurodegenerative diseases: new concepts of pathogenesis and their therapeutic implications. Annu Rev Pathol; 2006; 1: 151-70.
[13] Saido TC, Iwata N. Metabolism of amyloid β peptide and pathogenesis of Alzheimer's disease: towards presymptomatic diagnosis, prevention and therapy. Neurosci Res 2006; 54: 235-53.
[14] Tschäpe J-A, Hartmann T. Therapeutic perspectives in Alzheimer's disease. Recent Pat CNS Drug Discovery 2006; 1: 119-27.
[15] Aisen PS, Gauthier S, Vellas B, *et al.* Alzhemed: a potential treatment for Alzheimer's disease. Curr Alzheimer Res 2007; 4: 473-8.
[16] Geerts H. Drug evaluation: (R)-flurbiprofen – an enantiomer of flurbiprofen for the treatment of Alzheimer's disease. IDrugs 2007; 10: 121-33.
[17] Sullivan MG. Tramiprosate falls short in Phase III Alzheimer's trial: unusually large placebo effect could be a recurring problem in studies that allow concomitant medications. Clin Psychiatry News 2007; 35: 27.
[18] Hawkes CA, McLaurin J. Immunotherapy as treatment for Alzheimer's disease. Expert Rev Neurother 2007; 7: 1535-48.
[19] Castro A, Martinez A. Peripheral and dual binding site acetylcholinesterase inhibitors: implications in treatment of Alzheimer's disease. Mini-Rev Med Chem 2001; 1: 267-72.
[20] Du D-M, Carlier PR. Development of bivalent acetylcholinesterase inhibitors as potential therapeutic drugs for Alzheimer's disease. Curr Pharm Des 2004; 10: 3141-56.
[21] Li WM, Kan KKW, Carlier PR, Pang YP, Han YF. East meets West in the search for Alzheimer's therapeutics – novel dimeric inhibitors from tacrine and huperzine A. Curr Alzheimer Res 2007; 4: 386-96.
[22] Recanatini M, Valenti P. Acetylcholinesterase inhibitors as a starting point towards improved Alzheimer's disease therapeutics. Curr Pharm Des 2004; 10: 3157-66.
[23] Muñoz-Torrero D, Camps P. Dimeric and hybrid anti-Alzheimer drug candidates. Curr Med Chem 2006; 13: 399-422.
[24] Castro A, Martinez A. Targeting beta-amyloid pathogenesis through acetylcholinesterase inhibitors. Curr Pharm Des 2006; 12: 4377-87.
[25] Holzgrabe U, Kapková P, Alptüzün V, Scheiber J, Kugelmann E. Targeting acetylcholinesterase to treat neurodegeneration. Expert Opin Ther Targets 2007; 11: 161-79.
[26] Musial A, Bajda M, Malawska B. Recent developments in cholinesterases inhibitors for Alzheimer's disease treatment. Curr Med Chem 2007; 14: 2654-79.
[27] Haviv H, Wong DM, Silman I, Sussman JL. Bivalent ligands derived from Huperzine A as acetylcholinesterase inhibitors. Curr Top Med Chem 2007; 7: 375-87.
[28] Cavalli A, Bolognesi ML, Minarini A, *et al.* Multi-target-directed ligands to combat neurodegenerative diseases. J Med Chem 2008; 51: 347-72.
[29] Francis PT, Nordberg A, Arnold SE. A preclinical view of cholinesterase inhibitors in neuroprotection: do they provide more than symptomatic benefits in Alzheimer's disease? Trends Pharmacol Sci 2005; 26: 104-11.
[30] Riepe MW. Cholinergic treatment: what are the early neuropathological targets? Eur J Neurol 2005; 12(Suppl 3): 3-9.

[31] Geerts H. Indicators of neuroprotection with galantamine. Brain Res Bull 2005; 64: 519-24.
[32] Nordberg A. Mechanisms behind the neuroprotective actions of cholinesterase inhibitors in Alzheimer disease. Alzheimer Dis Assoc Disord 2006; 20(Suppl 1): S12-S18.
[33] Mori E, Hashimoto M, Krishnan KR, Doraiswamy PM. What constitutes clinical evidence for neuroprotection in Alzheimer disease. Support for the cholinesterase inhibitors? Alzheimer Dis Assoc Disord 2006; 20(Suppl 1): S19-S26.
[34] Sabbagh MN, Farlow MR, Relkin N, Beach TG. Do cholinergic therapies have disease-modifying effects in Alzheimer's disease? Alzheimer's Dementia 2006; 2: 118-25.
[35] Relkin NR. Beyond symptomatic therapy: a re-examination of acetylcholinesterase inhibitors in Alzheimer's disease. Expert Rev Neurother 2007; 7: 735-48.
[36] Akaike A. Preclinical evidence of neuroprotection by cholinesterase inhibitors. Alzheimer Dis Assoc Disord 2006; 20(Suppl 1): S8-S11.
[37] Takada Y, Yonezawa A, Kume T, *et al.* Nicotinic acetylcholine receptor-mediated neuroprotection by donepezil against glutamate neurotoxicity in rat cortical neurons. J Pharmacol Exp Ther 2003; 306: 772-7.
[38] Akaike A, Katsuki H, Kume T, Takada-Takatori Y. Neuroprotective action of donepezil mediated by neuronal nicotinic receptors. Psychogeriatrics 2006; 6(Suppl 2): S47-S56.
[39] Takada-Takatori Y, Kume T, Sugimoto M, Katsui H, Sugimoto H, Akaike A. Acetylcholinesterase inhibitors used in treatment of Alzheimer's disease prevent glutamate neurotoxicity *via* nicotinic acetylcholine receptors and phosphatidylinositol 3-kinase cascade. Neuropharmacology 2006; 51: 474-86.
[40] Kihara T, Sawada H, Nakamizo T, *et al.* Galantamine modulates nicotinic receptor and blocks Aβ-enhanced glutamate toxicity. Biochem Biophys Res Commun 2004; 325: 976-82.
[41] Akasofu S, Kimura M, Kosasa T, Ogura H, Sawada K. Protective effect of donepezil in primary-cultured rat cortical neurons exposed to *N*-methyl-D-aspartate (NMDA) toxicity. Eur J Pharmacol 2006; 530: 215-22.
[42] Kimura M, Akasofu S, Ogura H, Sawada K. Protective effect of donepezil against Aβ(1-40) neurotoxicity in rat septal neurons. Brain Res 2005; 1047: 72-84.
[43] Kimura M, Komatsu H, Ogura H, Sawada K. Comparison of donepezil and memantine for protective effect against amyloid-beta(1–42) toxicity in rat septal neurons. Neurosci Lett 2005; 391: 17-21.
[44] Arias E, Gallego-Sandín S, Villarroya M, García AG, López MG. Unequal neuroprotection afforded by the acetylcholinesterase inhibitors galantamine, donepezil, and rivastigmine in SH-SY5Y neuroblastoma cells: Role of nicotinic receptors. J Pharmacol Exp Ther 2005; 315: 1346-53.
[45] Meunier J, Ieni J, Maurice T. The anti-amnesic and neuroprotective effects of donepezil against amyloid $\beta_{25\text{-}35}$ peptide-induced toxicity in mice involve an interaction with the σ_1 receptor. Br J Pharmacol 2006; 149: 998-1012.
[46] Zhang HY, Tang XC. Huperzine B, a novel acetylcholinesterase inhibitor, attenuates hydrogen peroxide induced injury in PC12 cells. Neurosci Lett 2000; 292: 41-4.
[47] Meunier J, Ieni J, Maurice T. Antiamnesic and neuroprotective effects of donepezil against learning impairments induced in mice by exposure to carbon monoxide gas. J Pharmacol Exp Ther 2006; 317: 1307-19.
[48] Zhou J, Fu Y, Tang XC. Huperzine A and donepezil protect rat pheochromocytoma cells against oxygen–glucose deprivation. Neurosci Lett 2001; 306: 53-6.
[49] Sobrado M, Roda JM, López MG, Egea J, García AG. Galantamine and memantine produce different degrees of neuroprotection in rat hippocampal slices subjected to oxygen–glucose deprivation. Neurosci Lett 2004; 365: 132-6.
[50] Bullock R, Dengiz A. Cognitive performance in patients with Alzheimer's disease receiving cholinesterase inhibitors for up to 5 years. Int J Clin Pract 2005; 59: 817-22.
[51] Modrego PJ. The effect of drugs for Alzheimer disease assessed by means of neuroradiological techniques. Curr Med Chem 2006; 13: 3417-24.
[52] Silbert LC, Quinn JF, Moore MM, *et al.* Changes in premorbid brain volumen predict Alzheimer's disease pathology. Neurology 2003; 61: 487-92.
[53] Mori E, Lee K, Yasuda M, *et al.* Accelerated hippocampal atrophy in Alzheimer's disease with apolipoprotein E epsilon4 allele. Ann Neurol 2002; 51: 209-14.

[54] Hashimoto M, Kazui H, Matsumoto K, Nakano Y, Yasuda M, Mori E. Does donepezil treatment slow the progression of hippocampal atrophy in patients with Alzheimer's disease? Am J Psychiatry 2005; 162: 676-82.
[55] Krishnan KRR, Charles HC, Doraiswamy PM, *et al.* Randomized, placebo-controlled trial of the effects of donepezil on neuronal markers and hippocampal volumes in Alzheimer's disease. Am J Psychiatry 2003; 160: 2003-11.
[56] Wang X-D, Chen X-Q, Yang H-H, Hu G-Y. Comparison of the effects of cholinesterase inhibitors on [^{3}H]MK-801 binding in rat cerebral cortex. Neurosci Lett 1999; 272: 21-4.
[57] Maurice T, Meunier J, Feng B, Ieni J, Monaghan DT. Interaction with σ_1 protein, but not *N*-methyl-D-aspartate receptor, is involved in the pharmacological activity of donepezil. J Pharmacol Exp Ther 2006; 317: 606-14.
[58] Lahiri DK, Rogers JT, Greig NH, Sambamurti K. Rationale for the development of cholinesterase inhibitors as anti-Alzheimer agents. Curr Pharm Des 2004; 10: 3111-9.
[59] Colombres M, Sagal JP, Inestrosa NC. An overview of the current and novel drugs for Alzheimer's disease with particular reference to anti-cholinesterase compounds. Curr Pharm Des 2004; 10: 3121-30.
[60] Racchi M, Mazzucchelli M, Porrello E, Lanni C, Govoni S. Acetylcholinesterase inhibitors: novel activities of old molecules. Pharmacol Res 2004; 50: 441-51.
[61] Verhoeff NPLG. Acetylcholinergic neurotransmission and β-amyloid cascade: implications for Alzheimer's disease. Expert Rev Neurother 2005; 5: 277-84.
[62] Bandyopadhyay S, Goldstein LE, Lahiri DK, Rogers JT. Role of the APP non-amyloidogenic signaling pathway and targeting α-secretase as an alternative drug target for treatment of Alzheimer's disease. Curr Med Chem 2007; 14: 2848-64.
[63] Nitsch RM, Slack BE, Wurtman RJ, Growdon JH. Release of Alzheimer amyloid precursor derivatives stimulated by activation of muscarinic acetylcholine receptors. Science 1992; 258: 304-7.
[64] Chong YH, Suh YH. Amyloidogenic processing of Alzheimer's amyloid precursor protein *in vitro* and its modulation by metal ions and tacrine. Life Sci 1996; 59: 545-57.
[65] Zimmermann M, Gardoni F, Marcello E, *et al.* Acetylcholinesterase inhibitors increase ADAM10 activity by promoting protein its trafficking in neuroblastoma cell lines. J Neurochem 2004; 90: 1489-99.
[66] Borroni B, Colciaghi F, Pastorino L, *et al.* Amyloid precursor protein in platelets of patients with Alzheimer disease: effect of acetylcholinesterase inhibitor treatment. Arch Neurol 2001; 58: 442-6.
[67] Zimmermann M, Borroni B, Cattabeni F, Padovani A, Di Luca M. Cholinesterase inhibitors influnce APP metabolism in Alzheimer disease patients. Neurobiol Dis 2005; 19: 237-42.
[68] Mancini F, Naldi M, Cavrini V, Andrisano V. Multiwell fluorometric and colorimetric microassays for the evaluation of beta-secretase (BACE-1) inhibitors. Anal Bioanal Chem 2007; 388: 1175-83.
[69] Rosini M, Andrisano V, Bartolini M, Melchiorre C. Organic compounds useful for the treatment of Alzheimer's disease, their use and method of preparation. WO 2006/080043 A2.
[70] Alvarez A, Bronfman F, Pérez, CA, Vicente M, Garrido J, Inestrosa NC. Acetylcholinesterase, a senile plaque component, affects the fibrillogenesis of amyloid-β-peptides. Neurosci Lett 1995; 201: 49-52.
[71] Inestrosa NC, Alvarez A, Pérez CA, *et al.* Acetylcholinesterase accelerates assembly of amyloid-β-peptides into Alzheimer's fibrils: Possible role of the peripheral site of the enzyme. Neuron 1996; 16: 881-91.
[72] Alvarez A, Alarcón R, Opazo C, *et al.* Stable complexes involving acetylcholinesterase and amyloid-β peptide change the biochemical properties of the enzyme and increase the neurotoxicity of Alzheimer's fibrils. J Neurosci 1998; 18: 3213-23.
[73] Muñoz F, Inestrosa NC. Neurotoxicity of acetylcholinesterase amyloid-β peptide aggregates is dependent on the type of Aβ peptide and the AChE concentration present in the complexes. FEBS Lett 1999; 450: 205-9.
[74] Reyes AE, Chacón MA, Dinamarca MC, Cerpa W, Morgan C, Inestrosa NC. Acetylcholinesterase-Aβ complexes are more toxic than Aβ fibril in rat hippocampus. Effect on rat β-amyloid aggregation, laminin expression, reactive astrocytosis, and neuronal cell loss. Am J Pathol 2004; 164: 2163-74.
[75] Rees T, Hammond PI, Soreq H, Younkin S, Brimijoin S. Acetylcholinesterase promotes beta-amyloid plaques in cerebral cortex. Neurobiol Aging 2003; 24: 777-87.

[76] Rees TM, Berson A, Sklan EH, *et al.* Memory deficits correlating with acetylcholinesterase splice shift and amyloid burden in doubly transgenic mice. Curr Alzheimer Res 2005; 2: 291-300.
[77] Rees TM, Brimijoin S. The role of acetylcholinesterase in the pathogenesis of Alzheimer's disease. Drugs Today 2005; 39: 75-83.
[78] De Ferrari GV, Canales MA, Shin I, Weiner LM, Silman I, Inestrosa NC. A structural motif of acetylcholinesterase that promotes amyloid β-peptide fibril formation. Biochemistry 2001; 40: 10447-57.
[79] Bartolini M, Bertucci C, Cavrini V, Andrisano V. β-Amyloid aggregation induced by human acetylcholinesterase: inhibition studies. Biochem Pharmacol 2003; 65: 407-16.
[80] Sussman JL, Harel M, Frolow F, *et al.* Atomic structure of acetylcholinesterase from Torpedo californica: A prototypic acetylcholine-binding protein. Science 1991; 253: 872-9.
[81] Harel M, Schalk I, Ehret-Sabatier L, *et al.* Quaternary ligand binding to aromatic residues in the active-site gorge of acetylcholinesterase. Proc Natl Acad Sci USA 1993; 90: 9031-5.
[82] Bar-On P, Millard CB, Harel M, *et al.* Kinetic and structural studies on the interaction of cholinesterases with the anti-Alzheimer drug rivastigmine. Biochemistry 2002; 41: 3555-64.
[83] Greenblatt HM, Kryger G, Lewis T, Silman I, Sussman J. Structure of acetylcholinesterase complexed with (–)-galanthamine at 2.3 Å resolution. FEBS Lett 1999; 463: 321-6.
[84] Kryger G, Silman I, Sussman JL. Structure of acetylcholinesterase complexed with E2020 (Aricept®): Implications for the design of new anti-Alzheimer drugs. Structure 1999; 7: 297-307.
[85] Bolognesi ML, Banzi R, Bartolini M, *et al.* Novel class of quinone-bearing polyamines as multi-target-directed ligands to combat Alzheimer's disease. J Med Chem 2007; 50: 4882-97.
[86] Mooser G, Sigman DS. Ligand binding properties of acetylcholinesterase determined with fluorescent probes. Biochemistry 1974; 13: 2299-307.
[87] Pang Y-P, Quiram P, Jelacic T, Hong F, Brimijoin S. Highly potent, selective, and low cost bis-tetrahydroaminacrine inhibitors of acetylcholinesterase. Steps toward novel drugs for treating Alzheimer's disease. J Biol Chem 1996; 271: 23646-9.
[88] Savini L, Campiani G, Gaeta A, *et al.* Novel and potent tacrine-related hetero- and homobivalent ligands for acetylcholinesterase and butyrylcholinesterase. Bioorg Med Chem Lett 2001; 11: 1779-82.
[89] Savini L, Gaeta A, Fattorusso C, *et al.* Specific targeting of acetylcholinesterase and butyrylcholinesterase recognition sites. Rational design of novel, selective, and highly potent cholinesterase inhibitors. J Med Chem 2003; 46: 1-4.
[90] Campiani G, Fattorusso C, Butini S, *et al.* Development of molecular probes for the identification of extra interaction sites in the mid-gorge and peripheral sites of butyrylcholinesterase (BuChE). Rational design of novel, selective, and highly potent BuChE inhibitors. J Med Chem 2005; 48: 1919-29.
[91] Hu M-K, Wu L-J, Hsiao G, Yen M-H. Homodimeric tacrine congeners as acetylcholinesterase inhibitors. J Med Chem 2002; 45: 2277-82.
[92] Carlier PR, Han YF, Chow E-SH, *et al.* Evaluation of short-tether bis-THA AChE inhibitors. A further test of the dual binding site hypothesis. Bioorg Med Chem 1999; 7: 351-7.
[93] Bolognesi ML, Cavalli A, Valgimigli L, *et al.* Multi-target-directed drug design strategy: from a dual binding site acetylcholinesterase inhibitor to a trifunctional compound against Alzheimer's disease. J Med Chem 2007; 50: 6446-9.
[94] Carlier PR, Chow ES-H, Han Y, Liu J, El Yazal J, Pang Y-P. Heterodimeric tacrine-based acetylcholinesterase inhibitors: investigating ligand-peripheral site interactions. J Med Chem 1999; 42: 4225-31.
[95] Han Y-F, Wu D-C, Xiao X-Q, *et al.* Protection against ischemic injury in primary cultured astrocytes of mouse cerebral cortex by *bis*(7)-tacrine, a novel anti-Alzheimer's agent. Neurosci Lett 2000; 288: 95-8. Corrigendum in: Wu D-C, Xiao X-Q, Ng AKY, *et al.* Neurosci Lett 2000; 290: 84.
[96] Xiao X-Q, Lee NTK, Carlier PR, Pang Y-P, Han Y-F. Bis(7)-tacrine, a promising anti-Alzheimer's agent, reduces hydrogen peroxide-induced injury in rat pheochromocytoma cells: comparison with tacrine. Neurosci Lett 2000; 290: 197-200.
[97] Li W, Pi R, Chan HHN, *et al.* Novel dimeric acetylcholinesterase inhibitor bis7-tacrine, but not donepezil, prevents glutamate-induced neuronal apoptosis by blocking N-methyl-D-aspartate receptors. J Biol Chem 2005; 280: 18179-88.

[98] Fu H, Li W, Lao Y, *et al.* Bis(7)-tacrine attenuates β amyloid-induced neuronal apoptosis by regulating L-type calcium channels. J Neurochem 2006; 98: 1400-10.

[99] Li W, Xue J, Niu C, *et al.* Synergistic neuroprotection by bis(7)-tacrine *via* concurrent blockade of N-methyl-D-aspartate receptors and neuronal nitric-oxide synthase. Mol Pharmacol 2007; 71: 1258-67.

[100] Yu Z-L, Carlier PR, Han Y-F. *Bis* (7)-tacrine, a potential AChE inhibitor, displaces [H3] dizocilpine binding to N-methyl-D-aspartate receptors. *IXth International Meeting on Cholinesterases*, Suzhou, China, 2007.

[101] Luo J, Li W, Liu Y, *et al.* Novel dimeric bis(7)-tacrine proton-dependently inhibits NMDA-activated currents. Biochem Biophys Res Commun 2007; 361: 505-9.

[102] Li W, Lee NTK, Fu H, *et al.* Neuroprotection *via* inhibition of nitric oxide synthase by bis(7)-tacrine. NeuroReport 2006; 17: 471-4.

[103] Fu H, Li W, Liu Y, *et al.* Mitochondrial proteomic analysis and characterization of the intracellular mechanisms of bis(7)-tacrine in protecting against glutamate-induced excitotoxicity in primary cultured neurons. J Proteome Res 2007; 6: 2435-46.

[104] Rydberg EH, Brumshtein B, Greenblatt HM, *et al.* Complexes of alkylene-linked tacrine dimers with *Torpedo californica* acetylcholinesterase: Binding of Bis(5)-tacrine produces a dramatic rearrangement in the active-site gorge. J Med Chem 2006; 49: 5491-500.

[105] Fu H, Li W, Luo J, *et al.* Promising anti-Alzheimer's dimer bis(7)-tacrine reduces β-amyloid generation by directly inhibiting BACE-1 activity. Biochem Biophys Res Commun 2008; 366: 631-6.

[106] Carlier PR, Du D-M, Han Y, Liu J, Pang Y-P. Potent, easily synthesized huperzine A-tacrine hybrid acetylcholinesterase inhibitors. Bioorg Med Chem Lett 1999; 9: 2335-8.

[107] Lewis WG, Green LG, Grynszpan F, *et al.* Click chemistry *in situ*: acetylcholinesterase as a reaction vessel for the selective assembly of a femtomolar inhibitor from an array of building blocks. Angew Chem Int Ed 2002; 41: 1053-7.

[108] Shao D, Zou C, Luo C, Tang X, Li Y. Synthesis and evaluation of tacrine-E2020 hybrids as acetylcholinesterase inhibitors for the treatment of Alzheimer's disease. Bioorg Med Chem Lett 2004; 14: 4639-42.

[109] Bolognesi ML, Andrisano V, Bartolini M, Banzi R, Melchiorre C. Propidium-based polyamine ligands as potent inhibitors of acetylcholinesterase and acetylcholinesterase-induced amyloid-β aggregation. J Med Chem 2005; 48: 24-7.

[110] Rosini M, Andrisano V, Bartolini M, *et al.* Rational approach to discover multipotent anti-Alzheimer drugs. J Med Chem 2005; 48: 360-3.

[111] Camps P, Formosa X, Muñoz-Torrero D, Petrignet J, Badia A, Clos MV. Synthesis and pharmacological evaluation of huprine-tacrine heterodimers: subnanomolar dual binding site acetylcholinesterase inhibitors. J Med Chem 2005; 48: 1701-4.

[112] Muñoz-Ruiz P, Rubio L, García-Palomero E, *et al.* Design, synthesis, and biological evaluation of dual binding site acetylcholinesterase inhibitors: new disease-modifying agents for Alzheimer's disease. J Med Chem 2005; 48: 7223-33.

[113] Alonso D, Dorronsoro I, Rubio L, *et al.* Donepezil-tacrine hybrid related derivatives as new dual binding site inhibitors of AChE. Bioorg Med Chem 2005; 13: 6588-97.

[114] Gemma S, Gabellieri E, Huleatt P, *et al.* Discovery of huperzine A-tacrine hybrids as potent inhibitors of human cholinesterases targeting their midgorge recognition sites. J Med Chem 2006; 49: 3421-5.

[115] Elsinghorst PW, González Tanarro CM, Gütschow M. Novel heterobivalent tacrine derivatives as cholinesterase inhibitors with notable selectivity toward butyrylcholinesterase. J Med Chem 2006; 49: 7540-4.

[116] Elsinghorst PW, Cieslik JS, Mohr K, Tränkle C, Gütschow M. First gallamine-tacrine hybrid: design and characterization at cholinesterases and the M_2 muscarinic receptor. J Med Chem 2007; 50: 5685-95.

[117] del Monte-Millán M, García-Palomero E, Valenzuela R, *et al.* Dual binding site acetylcholinesterase inhibitors: potential new disease-modifying agents for AD. J Mol Neurosci 2006; 30: 85-8.

[118] Weinstock M. Selectivity of cholinesterase inhibition. CNS Drugs 1999; 12: 307-23.

[119] Camps P, Muñoz-Torrero D. Cholinergic drugs in pharmacotherapy of Alzheimer's disease. Mini-Rev Med Chem 2002; 2: 11-25.

[120] Giacobini E. Cholinesterase inhibitors: new roles and therapeutic alternatives. Pharmacol Res 2004; 50: 433-40.

[121] Lane RM, Potkin SG, Enz A. Targeting acetylcholinesterase and butyrylcholinesterase in dementia. Int J Neuropsychopharmacol 2006; 9: 101-24.
[122] Venneri A, McGeown WJ, Shanks MF. Empirical evidence of neuroprotection by dual cholinesterase inhibition in Alzheimer's disease. Neuroreport 2005; 16: 107-10.
[123] Nachon F, Nicolet Y, Viguié N, Masson P, Fontecilla-Camps JC, Lockridge O. Engineering of a monomeric and low-glycosylated form of human butyrylcholinesterase: expression, purification, characterization and crystallization. Eur J Biochem 2002; 269: 630-7.
[124] Nicolet Y, Lockridge O, Masson P, Fontecilla-Camps JC, Nachon F. Crystal structure of human butyrylcholinesterase and of its complexes with substrate and products. J Biol Chem 2003; 278: 41141-7.
[125] Nachon F, Asojo OA, Borgstahl GEO, Masson P, Lockridge O. Role of water in aging of human butyrylcholinesterase inhibited by echothiophate: the crystal structure suggests two alternative mechanisms of aging. Biochemistry 2005; 44: 1154-62.
[126] Ngamelue MN, Homma K, Lockridge O, Asojo OA. Crystallization and X-ray structure of full-length recombinant human butyrylcholinesterase. Acta Cryst 2007; F63: 723-7.
[127] Loudwig S, Nicolet Y, Masson P, *et al.* Photoreversible inhibition of cholinesterases: catalytic serine-labeled caged butyrylcholinesterase. ChemBioChem 2003; 4: 762-7.
[128] Masson P, Froment M-T, Fort S, *et al.* Butyrylcholinesterase-catalyzed hydrolysis of N-methylindoxyl acetate: analysis of volume changes upon reaction and hysteretic behavior. Biochim Biophys Acta 2002; 1597: 229-43.
[129] Lin G, Chen G-H, Lu C-P, Yeh S-C. QSARs for peripheral anionic site of butyrylcholinesterase with inhibitions by 4-acyloxy-biphenyl-4′-N-butylcarbamates. QSAR Comb Sci 2005; 24: 943-52.
[130] Melchiorre C, Andrisano V, Bolognesi ML, *et al.* Acetylcholinesterase noncovalent inhibitors based on a polyamine backbone for potential use against Alzheimer's disease. J Med Chem 1998; 41: 4186-9.
[131] Tumiatti V, Rosini M, Bartolini M, *et al.* Structure-activity relationships of acetylcholinesterase noncovalent inhibitors based on a polyamine backbone. 2. Role of the substituents on the phenyl ring and nitrogen atoms of caproctamine. J Med Chem 2003; 46: 954-66.
[132] Tumiatti V, Andrisano V, Banzi R, *et al.* Structure-activity relationships of acetylcholinesterase noncovalent inhibitors based on a polyamine backbone. 3. Effect of replacing the inner polymethylene chain with cyclic moieties. J Med Chem 2004; 47: 6490-8.
[133] Piazzi L, Rampa A, Bisi A, *et al.* 3-(4-[[Benzyl(methyl)amino]methyl]phenyl)-6,7-dimethoxy-2*H*-2-chromenone (AP2238) inhibits both acetylcholinesterase and acetylcholinesterase-induced β-amyloid aggregation: a dual function lead for Alzheimer's disease therapy. J Med Chem 2003; 46: 2279-82.
[134] Sugimoto H, Yamanishi Y, Iimura Y, Kawakami Y. Donepezil hydrochloride (E2020) and other acetylcholinesterase inhibitors. Curr Med Chem 2000; 7: 303-39.
[135] Piazzi L, Cavalli A, Belluti F, *et al.* Extensive SAR and computational studies of 3-{4-[(benzylmethylamino)methyl]phenyl}-6,7-dimethoxy-2*H*-2-chromenone (AP2238) derivatives. J Med Chem 2007; 50: 4250-4.
[136] Belluti F, Rampa A, Piazzi, L, *et al.* Cholinesterase inhibitors: xanthostigmine derivatives blocking the acetylcholinesterase-induced β-amyloid aggregation. J Med Chem 2005; 48: 4444-56.
[137] Rampa A, Bisi A, Valenti P, *et al.* Acetylcholinesterase inhibitors: synthesis and structure-activity relationships of ω-[*N*-methyl-*N*-(3-alkylcarbamoyloxyphenyl)-methyl]aminoalkoxyheteroaryl derivatives. J Med Chem 1998; 41: 3976-86.
[138] Rampa A, Piazzi L, Belluti F, *et al.* Acetylcholinesterase inhibitors: SAR and kinetic studies on ω-[*N*-methyl-*N*-(3-alkylcarbamoyloxyphenyl)methyl]aminoalkoxyaryl derivatives. J Med Chem 2001; 44: 3810-20.
[139] Kwon YE, Park JY, No KT, *et al.* Synthesis, *in vitro* assay, and molecular modeling of new piperidine derivatives having dual inhibitory potency against acetylcholinesterase and $A\beta_{1-42}$ aggregation for Alzheimer's disease therapeutics. Bioorg Med Chem 2007; 15: 6596-607.
[140] Kapková P, Stiefl N, Sürig U, Engels B, Baumann K, Holzgrabe U. Synthesis, biological activity and docking studies of new acetylcholinesterase inhibitors of the bispyridinium type. Arch Pharm 2003; 336: 523-40.
[141] Alptüzün V, Kapková P, Baumann K, Erciyas E, Holzgrabe U. Synthesis and biological activity of pyridinium-type acetylcholinesterase inhibitors. J Pharm Pharmacol 2003; 55: 1397-404.

[142] Kapková P, Alptüzün V, Frey P, Erciyas E, Holzgrabe U. Search for dual function inhibitors for Alzheimer´s disease: synthesis and biological activity of acetylcholinesterase inhibitors of pyridinium-type and their Aβ fibril formation inhibition capacity. Bioorg Med Chem 2006; 14: 472-8.

[143] Dickerson TJ, Beuscher IV AE, Rogers CJ, *et al.* Discovery of acetylcholinesterase peripheral anionic site ligands through computational refinement of a directed library. Biochemistry 2005; 44: 14845-53.

[144] Iuvone T, Esposito G, Esposito R, Santamaria R, Di Rosa M, Izzo AA. Neuroprotective effect of cannabidiol, a non-psychoactive component from *Cannabis sativa*, on β-amyloid-induced toxicity in PC12 cells. J Neurochem 2004; 89: 134-41.

[145] Ramírez BG, Blázquez C, Gómez del Pulgar T, Guzmán M, de Ceballos ML. Prevention of Alzheimer's disease pathology by cannabinoids: neuroprotection mediated by blockade of microglial activation. J Neurosci 2005; 25: 1904-13.

[146] Eubanks LM, Rogers CJ, Beuscher IV AE, *et al.* A molecular link between the active component of marijuana and Alzheimer's disease pathology. Mol Pharmaceutics 2006; 3: 773-7.

[147] Johnson G, Moore SW. The peripheral anionic site of acetylcholinesterase: structure, functions and potential role in rational drug design. Curr Pharm Des 2006; 12: 217-25.

[148] Decker M. Homobivalent quinazolinimines as novel nanomolar inhibitors of cholinesterases with dirigible selectivity toward butyrylcholinesterase. J Med Chem 2006; 49: 5411-3.

[149] Sauvaître T, Barlier M, Herlem D, *et al.* New potent acetylcholinesterase inhibitors in the tetracyclic triterpene series. J Med Chem 2007; 50: 5311-23.

[150] He X-C, Feng S, Wang Z-F, *et al.* Study on dual-site inhibitors of acetylcholinesterase: Highly potent derivatives of bis- and bifunctional huperzine B. Bioorg Med Chem 2007; 15: 1394-408.

[151] Galdeano C, Viayna E, Arroyo P, *et al.* Structural determinants of the multifunctional profile of dual binding site acetylcholinesterase inhibitors as anti-Alzheimer agents. Curr Pharm Des 2010; 16: 2818-36.

[152] Viayna E, Sabate R, Muñoz-Torrero D. Dual inhibitors of β-amyloid aggregation and acetylcholinesterase as multi-target anti-Alzheimer drug candidates. Curr Top Med Chem 2013; 13: 1820-42.

[153] Viayna E, Sola I, Di Pietro O, Muñoz-Torrero D. Human disease and drug pharmacology, complex as real life. Curr Med Chem 2013; 20: 1623-34.

[154] Marlatt MW, Webber KM, Moreira PI, *et al.* Therapeutic opportunities in Alzheimer disease: one for all or all for one? Curr Med Chem 2005; 12: 1137-47.

[155] Zhang H-Y. One-compound-multiple-targets strategy to combat Alzheimer's disease. FEBS Lett 2005; 579: 5260-4.

[156] Shen L, Zhang H-Y, Ji H-F. A theoretical study on Cu(II)-chelating properties of curcumin and its implications for curcumin as a multipotent agent to combat Alzheimer's disease. J Mol Struct 2005; 757: 199-202.

[157] Ji H-F, Zhang H-Y. Theoretical evaluation of flavonoids as multipotent agents to combat Alzheimer's disease. J Mol Struct 2006; 767: 3-9.

[158] Zhang H-Y, Yang D-P, Ji H-F. Naturally occurring multipotent anti-Alzheimer's agents. Expert Opin Drug Discovery 2006; 1: 269-77.

[159] Zhang H-Y. Can food-derived multipotent agents reduce the risk of Alzheimer's disease? Trends Food Sci Technol 2007; 18: 492-5.

[160] Morphy R, Rankovic Z. Designed multiple ligands. An emerging drug discovery paradigm. J Med Chem 2005; 48: 6523-43.

[161] Morphy R, Rankovic Z. The physicochemical challenges of designing multiple ligands. J Med Chem 2006; 49: 4961-70.

[162] Zimmermann GR, Lehár J, Keith CT. Multi-target therapeutics: when the whole is greater than the sum of the parts. Drug Discovery Today 2007; 12: 34-42.

[163] Viegas-Junior C, Danuello A, da Silva Bolzani V, Barreiro EJ, Manssour Fraga CA. Molecular hybridization: a useful tool in the design of new drug prototypes. Curr Med Chem 2007; 14: 1829-52.

[164] Meunier B. Hybrid molecules with a dual mode of action: dream or reality? Acc Chem Res 2008; 41: 69-77.

[165] Youdim MBH, Buccafusco JJ. Multi-functional drugs for various CNS targets in the treatment of neurodegenerative disorders. Trends Pharmacol Sci 2005; 26: 27-35.

[166] Van der Schyf CJ, Geldenhuys WJ, Youdim MBH. Multifunctional drugs with different CNS targets for neuropsychiatric disorders. J Neurochem 2006; 99: 1033-48.
[167] Van der Schyf CJ, Mandel S, Geldenhuys WJ, *et al.* Novel multifunctional anti-Alzheimer drugs with various CNS neurotransmitter targets and neuroprotective moieties. Curr Alzheimer Res 2007; 4: 522-36.
[168] Bolognesi ML, Minarini A, Tumiatti V, Melchiorre C. Lipoic acid, a lead structure for multi-target-directed drugs for neurodegeneration. Mini-Rev Med Chem 2006; 6: 1269-74.
[169] Rodríguez-Franco MI, Fernández-Bachiller MI, Pérez C, Hernández-Ledesma B, Bartolomé B. Novel tacrine-melatonin hybrids as dual-acting drugs for Alzheimer disease, with improved acetylcholinesterase inhibitory and antioxidant properties. J Med Chem 2006; 49: 459-62.
[170] Fernández-Bachiller MI, Pérez C, Campillo NE, *et al.* Tacrine-melatonin hybrids as multifunctional agents for Alzheimer's disease, with cholinergic, antioxidant, and neuroprotective properties. ChemMedChem 2009; 4: 828-41.
[171] Cappelli A, Gallelli A, Manini M, *et al.* Further studies on the interaction of the 5-hydroxytryptamine$_3$ (5-HT$_3$) receptor with arylpiperazine ligands. Development of a new 5-HT$_3$ receptor ligand showing potent acetylcholinesterase inhibitory properties. J Med Chem 2005; 48: 3564-75.
[172] Blat D, Weiner L, Youdim MBH, Fridkin M. A novel iron-chelating derivative of the neuroprotective peptide NAPVSIPQ shows superior antioxidant and antineurodegenerative capabilities. J Med Chem 2008; 51: 126-34.
[173] Deraeve C, Boldron C, Maraval A, *et al.* Preparation and study of new poly-8-hydroxyquinoline chelators for an anti-Alzheimer strategy. Chem Eur J 2008; 14: 682-96.
[174] Piazzi L, Cavalli A, Colizzi F, *et al.* Multi-target-directed coumarin derivatives: hAChE and BACE1 inhibitors as potential anti-Alzheimer compounds. Bioorg Med Chem Lett 2008; 18: 423-6.
[175] Guo T, Hobbs DW. Development of BACE1 inhibitors for Alzheimer's disease. Curr Med Chem 2006; 13: 1811-29.
[176] Cavalli A, Bolognesi ML, Bartolini M, *et al.* Memoquin: a new therapeutic poly-agent for Alzheimer's disease. Neurobiol Aging 2004; 25(Suppl 2): S194.
[177] Capsoni S, Andrisano V, Bartolini M, *et al.* Memoquin, a novel multifunctional compound for alzheimer's disease: an up-date on preclinical studies. Alzheimer's Dementia 2006; 2(Suppl 1): S73.
[178] Cavalli A, Bolognesi ML, Capsoni S, *et al.* A small molecule targeting the multifactorial nature of Alzheimer's disease. Angew Chem Int Ed 2007; 46: 3689-92.
[179] Tranches Dias KS, Viegas C. Multi-target directed drugs: A modern approach for design of new drugs for the treatment of Alzheimer's disease. Curr Neuropharmacol 2014; 12: 239-55.
[180] Rodriguez Simoes MC, Dias Viegas FP, Soares Moreira M, *et al.* Donepezil: An important prototype to the design of new drug candidates for Alzheimer's disease. Mini-Rev Med Chem 2014; 14: 2-19.
[181] Bansal Y, Silakari O. Multifunctional compounds: Smart molecules for multifactorial diseases. Eur J Med Chem 2014; 76: 31-42.
[182] Geldenhuys WJ, Van der Schyf CJ. Rationally designed multi-targeted agents against neurodegenerative diseases. Curr Med Chem 2013; 20: 1662-72.
[183] Chen X, Decker M. Multi-target compounds acting in the central nervous system designed from natural products. Curr Med Chem 2013; 20: 1673-85.
[184] Russo P, Frustaci A, Del Bufalo A, Fini M, Cesario A. Multitarget drugs of plants origin acting on Alzheimer's disease. Curr Med Chem 2013; 20: 1686-93.
[185] Carreiras MC, Mendes E, Perry MJ, Francisco AP, Marco-Contelles J. The multifactorial nature of Alzheimer's disease for developing potential therapeutics. Curr Top Med Chem 2013; 13: 1745-70.
[186] Minarini A, Milelli A, Simoni E, *et al.* Multifunctional tacrine derivatives in Alzheimer's disease. Curr Top Med Chem 2013; 13: 1771-86.
[187] Rampa A, Belluti F, Gobbi S, Bisi A. Hybrid-based multi-target ligands for the treatment of Alzheimer's disease. Curr Top Med Chem 2011; 11: 2716-30.
[188] Egea J, de los Rios C. 1,8-naphthyridine derivatives as cholinesterases inhibitors and cell Ca^{2+} regulators, a multitarget strategy for Alzheimer's disease. Curr Top Med Chem 2011; 11: 2807-2823.
[189] Bajda M, Guzior N, Ignasik M, Malawska B. Multi-target-directed ligands in Alzheimer's disease treatment. Curr Med Chem 2011; 18: 4949-75.

[190] Kozurkova M, Hamulakova S, Gazova Z, Paulikova H, Kristian P. Neuroactive multifunctional tacrine congeners with cholinesterase, anti-amyloid aggregation and neuroprotective properties. Pharmaceuticals 2011; 4: 382-418.
[191] Bolognesi ML, Minarini A, Rosini M, Tumiatti V, Melchiorre C. From dual binding site acetylcholinesterase inhibitors to multi-target-directed ligands (MTDLs): A step forward in the treatment of Alzheimer's disease. Mini-Rev Med Chem 2008; 8: 960-7.
[192] Rosini M, Simoni E, Minarini A, Melchiorre C. Multi-target design strategies in the context of Alzheimer's disease: Acetylcholinesterase inhibition and NMDA receptor antagonism as the driving forces. Neurochem Res 2014; 39: 1914-23.
[193] Rosini M, Simoni E, Bartolini M, *et al.* The bivalent ligand approach as a tool for improving the *in vitro* anti-Alzheimer multitarget profile of dimebon. ChemMedChem 2013; 8: 1276-81.
[194] Simoni E, Daniele S, Bottegoni G, *et al.* Combining galantamine and memantine in multitargeted, new chemical entities potentially useful in Alzheimer's disease. J Med Chem 2012; 55: 9708-21.
[195] Rosini M, Simoni E, Bartolini M, *et al.* Inhibition of acetylcholinesterase, β-amyloid aggregation and NMDA receptors in Alzheimer's disease: a promising direction for the multi-target-directed ligands gold rush. J Med Chem 2008; 51: 4381-4.
[196] Wang L, Esteban G, Ojima M, *et al.* Donepezil + propargylamine + 8-hydroxyquinoline hybrids as new multifunctional metal-chelators, ChE and MAO inhibitors for the potential treatment of Alzheimer's disease. Eur J Med Chem 2014; 80: 543-61.
[197] Zheng H, Fridkin M, Youdim MBH. Novel chelators targeting cell cycle arrest, acetylcholinesterase, and monoamine oxidase for Alzheimer's therapy. Curr Drug Targets 2012; 13: 1089-106.
[198] Zheng H, Youdim MBH, Fridkin M. Selective acetylcholinesterase inhibitor activated by acetylcholinesterase releases an active chelator with neurorescuing and anti-amyloid activities. ACS Chem Neurosci 2010; 1: 737-46.
[199] Fernández-Bachiller MI, Pérez C, González-Muñoz GC, *et al.* Novel tacrine-8-hydroxyquinoline hybrids as multifunctional agents for the treatment of Alzheimer's disease, with neuroprotective, cholinergic, antioxidant, and copper-complexing properties. J Med Chem 2010; 53: 4927-37.
[200] Zheng H, Youdim MBH, Fridkin M. Site-activated multifunctional chelator with acetylcholinesterase and neuroprotective-neurorestorative moieties for Alzheimer's therapy. J Med Chem 2009; 52: 4095-8.
[201] Esteban G, Allan J, Samadi A, *et al.* Kinetic and structural analysis of the irreversible inhibition of human monoamine oxidases by ASS234, a multi-target compound designed for use in Alzheimer's disease. Biochim Biophys Acta 2014; 1844: 1104-10.
[202] Bautista-Aguilera OM, Esteban G, Bolea I, *et al.* Design, synthesis, pharmacological evaluation, QSAR analysis, molecular modeling and ADMET of novel donepezileindolyl hybrids as multipotent cholinesterase/monoamine oxidase inhibitors for the potential treatment of Alzheimer's disease. Eur J Med Chem 2014; 75: 82-95.
[203] Bolea I, Gella A, Unzeta M. Propargylamine-derived multitarget-directed ligands: fighting Alzheimer's disease with monoamine oxidase inhibitors. J Neural Transm 2013; 120: 893-902.
[204] Weinreb O, Amit T, Bar-Am O, Youdim MBH. Ladostigil: a novel multimodal neuroprotective drug with cholinesterase and brain-selective monoamine oxidase inhibitory activities for Alzheimer's disease treatment. Curr Drug Targets 2012; 13; 483-94.
[205] Rosini M, Simoni E, Milelli A, Minarini A, Melchiorre C. Oxidative stress in Alzheimer's disease: Are we connecting the dots? J Med Chem 2014; 57: 2821-31.
[206] Teixeira J, Silva T, Andrade PB, Borges F. Alzheimer's disease and antioxidant therapy: How long how far? Curr Med Chem 2013; 20: 2939-52.
[207] Firuzi O, Miri R, Tavakkoli M, Saso L. Antioxidant therapy: Current status and future prospects. Curr Med Chem 2011; 18: 3871-88.
[208] López-Iglesias B, Pérez C, Morales-García JA, *et al.* New melatonin-*N*,*N*-dibenzyl(*N*-methyl)amine hybrids: Potent neurogenic agents with antioxidant, cholinergic, and neuroprotective properties as innovative drugs for Alzheimer's disease. J Med Chem 2014: 57: 3773-85.
[209] Thiratmatrakul S, Yenjai C, Waiwut P, *et al.* Synthesis, biological evaluation and molecular modeling study of novel tacrine-carbazole hybrids as potential multifunctional agents for the treatment of Alzheimer's disease. Eur J Med Chem 2014; 75: 21-30.

[210] Qiang X, Sang Z, Yuan W, *et al.* Design, synthesis and evaluation of genistein-*O*-alkylbenzylamines as potential multifunctional agents for the treatment of Alzheimer's disease. Eur J Med Chem 2014; 76: 314-31.

[211] Kálai T, Altman R, Maezawa I, *et al.* Synthesis and functional survey of new tacrine analogs modified with nitroxides or their precursors. Eur J Med Chem 2014; 77: 343-50.

[212] Belluti F, De Simone A, Tarozzi A, *et al.* Fluorinated benzophenone derivatives: Balanced multipotent agents for Alzheimer's disease. Eur J Med Chem 2014; 78: 157-66.

[213] Pudlo M, Luzet V, Ismaïli L, Tomassoli I, Iutzeler A, Refouvelet B. Quinolone-benzylpiperidine derivatives as novel acetylcholinesterase inhibitor and antioxidant hybrids for Alzheimer's disease. Bioorg Med Chem 2014; 22: 2496-507.

[214] Bolognesi ML, Bergamini C, Fato R, Oiry J, Vasseur J-J, Smietana M. Synthesis of new lipoic acid conjugates and evaluation of their free radical scavenging and neuroprotective activities. Chem Biol Drug Des 2014; 83: 688-96.

[215] Luo Z, Sheng J, Sun Y, *et al.* Synthesis and evaluation of multi-target-directed ligands against Alzheimer's disease based on the fusion of donepezil and ebselen. J Med Chem 2013; 56: 9089-99.

[216] Fernández-Bachiller MI, Pérez C, Monjas L, Rademann J, Rodríguez-Franco MI. New tacrine-4-oxo-4*H*-chromene hybrids as multifunctional agents for the treatment of Alzheimer's disease, with cholinergic, antioxidant, and *β*-amyloid-reducing properties. J Med Chem 2012; 55: 1303-17.

[217] Capsoni S, Ugolini G, Comparini A, Ruberti F, Berardi N, Cattaneo A. Alzheimer-like neurodegeneration in aged antinerve growth factor transgenic mice. Proc Natl Acad Sci USA 2000; 97: 6826-31.

[218] Capsoni S, Cattaneo A. On the molecular basis linking Nerve Growth Factor (NGF) to Alzheimer's disease. Cell Mol Neurobiol 2006; 26: 619-33.

[219] Capsoni S, Giannotta S, Cattaneo A. Early events of an Alzheimer-like neurodegeneration in anti-nerve growth factor trasngenic mice. Brain Aging 2002; 2: 24-43.

[220] Capsoni S, Giannotta S, Cattaneo A. β-Amyloid plaques in a model for sporadic Alzheimer's disease based on transgenic anti-nerve growth factor antibodies. Mol Cell Neurosci 2002; 21: 15-28.

[221] Capurro V, Busquet P, Lopes JP, *et al.* Pharmacological characterization of memoquin, a multi-target compound for the treatment of Alzheimer's disease. PLoS ONE 2013; 8: e56870.

[222] Martinez A, Castro A. Novel cholinesterase inhibitors as future effective drugs for the treatment of Alzheimer's disease. Expert Opin Investig Drugs 2006; 15: 1-12.

[223] Vericat JA, Muñoz P, Windisch M, Hutter-Paier B, Medina M, Martinez A. NP-0361: a dual AChE inhibitor that interferes with the β-amyloid plaque charge load. 5th Neurobiology of Aging Conference, San Diego, USA, 2004.

[224] García-Palomero E, Muñoz P, Usan P, *et al.* Potent β-amyloid modulators. Neurodegener Dis 2008; 5: 153-156.

[225] Spuch C, Antequera D, Fernández-Bachiller MI, Rodríguez-Franco MI, Carro E. A new tacrine-melatonin hybrid reduces amyloid burden and behavioral deficits in a mouse model of Alzheimer's disease. Neurotox Res 2010; 17: 421-31.

[226] Antequera D, Bolos M, Spuch C, *et al.* Effects of a tacrine-8-hydroxyquinoline hybrid (IQM-622) on Aβ accummulation and cell death: Involvement of hippocampal neuronal loss in Alzheimer's disease. Neurobiol Dis 2012; 46: 682-91.

[227] Mishra N, Sasmal D, Singh KK. Attenuating $A\beta_{1-42}$-induced toxicity by a novel acetylcholinesterase inhibitor. Neuroscience 2013; 250: 309-19.

[228] Viayna E, Sola I, Bartolini M, *et al.* Synthesis and multitarget biological profiling of a novel family of rhein derivatives as disease-modifying anti-Alzheimer agents. J Med Chem 2014; 57: 2549-67.

CHAPTER 3

Dopamine Receptor Agonists for Protection and Repair in Parkinson's Disease: An Update

Giulia Ferrari-Toninelli[2], Sara A. Bonini[2], Paola Bettinsoli[2], Giuseppina Maccarinelli[2], Giovanna Cenini[2], Mariagrazia Grilli[1], Daniela Uberti[2] and Maurizio Memo[1,*]

[1]*Department of Molecular and Translational Medicine, University of Brescia, Brescia, Italy and* [2]*Department of Pharmaceutical Sciences, University of Piemonte Orientale "A Avogadro", Novara, Italy*

Abstract: Dopamine agonists have been shown to possess neuroprotective properties in different *in vitro* and *in vivo* experimental Parkinson's disease models, because their capability to counteract- neuronal cell death. Here we update the molecular evidence underlying the wide pharmacological spectrum of dopamine agonists currently used for treating Parkinson's disease patients. In particular, the mechanism of action of different dopamine agonists does not always appear to be restricted to the stimulation of selective dopamine receptor subtypes since at least some of these drugs are endowed with antioxidant, antiapoptotic or neurotrophic properties. These activities are molecule-specific and may contribute to the clinical efficacy of these drugs for the treatment of chronic and progressive neurodegenerative diseases in which oxidative injury and/or protein misfolding and aggregation exert a primary role. However, despite increasing number of experimental results confirm their neuroprotective effects, further studies are needed to definitively confirm dopamine agonists as disease-modifying agents.

Keywords: Alzheimer's disease, amyolid fibril, amyotrophic lateral sclerosis, apomorphine, bromocryptine, disease-modifying therapy, dopamine, dopamine receptor agonists, Free radicals, neurodegeneration, neurogenesis, neuroimaginig, neuroprotection, oxidative stress, Parkinson's disease, pergolide, pramipexole, protein aggregation, ropinirole, rotigotine, α-synuclein.

INTRODUCTION

Parkinson's disease (PD) is the most common neurodegenerative movement disorder. Approximately 2% of the population older than 65 years suffer from this slowly progressive neurodegenerative disease. More than 90% of PD cases are

***Corresponding author Maurizio Memo:** Department of Molecular and Translational Medicine, Viale Europa 11, 25124 Brescia, Italy; Tel: (+39) – 030 – 3717516; Fax: (+39) – 030 – 3717529; E-mail: maurizio.memo@unibs.it

Atta-ur-Rahman, Allen B. Reitz, M. Iqbal Choudhary and Jizhou Wang (Eds.)

sporadic. The primary cause of the disorder is the progressive loss of the pigmentated DAergic neurons in the substantia nigra *pars compacta* (SNpc) accompanied by the appearance of intracytoplasmic inclusions known as Lewy bodies, containing aggregated alpha synuclein. Despite significant progress has been made in controlling the symptoms of the disease, a neuroprotective or disease-modifying therapy that can slow or stop disease progression has not yet been obtained. To date, the etiopathogenesis of nigral DAergic neuron loss in PD is unknown. However, the presence of ongoing oxidative stress as the result of inefficacious antioxidant defence mechanisms and generation of radical oxygen species (ROS) in the SNpc of the parkinsonian brain are considered to be important pathogenic mechanisms [1-3]. It should be noted that part of these free radicals are inevitably produced by DA metabolism in the brain either by enzymes through the action of monoamine oxidase-B or by auto-oxidation [4]. Other sources of increased radical production may be endogenous neurotoxins occurring in the brain like tetrahydroisoquinolines or exogenously administered neurotoxins like the widely used herbicide paraquat which have similar neurochemical properties like the well-known neurotoxin, 1-methyl-4-phenyl-1,2,3,6-tetrahydropyridine (MPTP) [5-7]. Moreover, it has been suggested that PD could be associated with excitotoxicity and apoptosis [8] and with a loss of neurotrophic factors [9, 10]. Finally, protein aggregation and misfolding have emerged as important mechanisms not only in PD, with α-synuclein aggregating Lewy bodies, but also in many other neurodegenerative disorders, including as Alzheimer's disease (AD). Therefore, an effective anti-parkinsonian therapy should interfere with the overall mechanisms that finally lead to the progressive DAergic death in the CSN.

Levodopa combined with a peripheral DOPA-decarboxylase inhibitor and a COMT inhibitor is considered the therapy of choice for PD. Levodopa is nearly always effective, but has a high incidence of adverse effects with long term use, including response fluctuations (on/off phenomena) and dyskinesias. More recently DAergic agonists, acting directly at the receptor level, would be able to decrease the incidence of these motor complications [2, 3]. However it seems that these molecules are also endowed with the ability to modify the disease progression protecting neuron from degeneration.

Here we update the knowledge about the potential effects of dopaminergic agonists in restoring the impaired DA function as well as in preventing neurodegenerative processes because of thier additional pharmacological properties.

NEUROPROTECTIVE AND NEURORESTORATIVE PROPERTIES OF DA AGONISTS

The potential of DA agonists in inhibit intracellular death pathways and/or stimulate neuron regeneration may be a consequence of specific DA receptor stimulation or be completely independent from them. These additional pharmacological effects may enlarge the pharmacological spectrum of different DA agonists, contribute to slow down the progression of the neurodegenerative process, also involving sites of action distal from the nigro-striatal areas.

Antioxidant Activity

Increased oxidative stress is thought to be involved in nigral cell death, that is characteristic of PD. This oxidative stress may be further exacerbated by levodopa therapy. These mechanisms have been proven *in vitro* and in animal models, but their relevance in humans remains speculative [2, 3].

Most DA agonists have demonstrated protective properties in cell culture against a range of toxins, including DA, 6-hydroxydopamine (6-OH-DA), 1-metyl-4-phenylpyridinium (MPP^+) and hydroxy peroxide [2, 3, 11]. DA agonists also protect against toxin action *in vivo*, as shown in rodents receiving intrastriatal injection of 6-OH-DA or MPP^+. As reported in details in the following chapters, at cellular level, independent groups have demonstrated decreased free radical production and an amelioration of DA neuronal loss following DA agonist treatment.

Stimulation of Neurogenesis

The neurotransmitter and its receptors appear early during ontogenesis and affect cell proliferation in the embryonic germinal zones [12, 13]. Interestingly, the regulation of neural stem cells by dopamine persists in the restricted brain areas where neurogenesis occurs in adulthood, particularly in the subventricular zone (SVZ) within the lateral wall of the lateral ventricles and in the subgranular zone of the hippocampus. Ongoing adult neurogenesis is currently believed to be an important form of neural plasticity, enabling organisms to adapt to environmental changes and possibly influencing learning and memory throughout life. On the other hand promotion of adult neurogenesis may offer a potential approach for replacing neurons or neuritic networks that degenerate or lose function during aging or in neuropathological settings, including Parkinson's disease. This therapeuthic strategy is particularly intriguing since in post-mortem brains of individuals with Parkinson's disease the numbers of neural precursor cells in neurogenic regions are dramatically reduced [14]. In a very elegant study Hoglinger and co-authors [14] provided experimental evidence that the highly proliferative precursors cells located in the adult murine subependymal zone lining the lateral ventricle receive dopaminergic afferents. This innervation appears functionally relevant. Transient and bilateral dopaminergic denervation by administration of MPTP resulted in transient and bilateral decrease in the subependymal zone proliferation. Additionally, ablation of mesencephalic dopaminergic neurons of adult rats by stereotaxic injection of 6-OHDA into the nigrostriatal pathway resulted in the unilateral dopaminergic denervation of the subependymal zone as well as in the marked reduction in the number of proliferating precursors. Conversely, cell proliferation was restored by L-DOPA chronic infusion. Similar results were obtained in the hippocampal neurogenic region, which has been shown to receive a dopaminergic from the ventral tegmental area. More recently, the existence of a topographically organized projection from the SNpc to the SVZ was also demonstrated in primates [15], with the anteromedial SNpc projecting to the anteroventral SVZ and the posterolateral SNpc to the posterodorsal SVZ.

The effects on precursor cell proliferation elicited by dopamine are likely to be mediated by D2-like receptors, since D2 and D3 receptors are expressed on neural stem/progenitors cells. As recently summarized by Joyce and Millan [16], different D3 receptor-preferring agonists augment mitogenesis in the SVZ. In particular, Van Kapen and associates demonstrated induction of neurogenesis leading to the regeneration of DAergic pathways, suggesting that this effect may participate to the restoration of DAergic nigrostriatal pathway and locomotor activity in rat model of PD [17, 18]. However, species-specific differences of D3 receptors in regulating neurogenesis have been reported [19, 20]. A two-fold induction of cell proliferation in the SVZ and rostral migratory stream of the adult rats was demonstrated following icv administration of the dopamine D(3) receptor agonist, 7-hydroxy-N,N-di-n-propyl-2-aminotetralin (7-OH-DPAT) for 2 weeks [17]. The same drug had no effect on mice [20, 21]. More robust are the data on the involvement of D2 receptors in regulating neurogenesis, based on both the use of D2-null mice and D2 agonists/antagonists [22].

Stimulation of Neurotrophic Factors

One of the most convincing evidence for the role of neutrophic factors in PD is the study of Elsworth *et al.* [23]. They found that implantation of adeno-associated virus type 2 encoding gial derived neurotrophic factor (AAV2-GDNF) in the normal monkey caudate nucleus induced overexpression of GDNF, that persisted for at least 6 months after injection. In a 6-month within-animal controlled study, AAV2-GDNF enhanced the survival of fetal dopamine neurons by 4-fold, and increased the outgrowth of grafted fetal dopamine neurons by almost 3-fold in the caudate nucleus of MPTP-treated monkeys, compared with control grafts in the other caudate nucleus. GDNF is a potent neurotrophic factor that is crucial to the development, survival, and outgrowth of DA neurons [24, 25]. GDNF is highly expressed in the developing rat striatum, yet its concentration is relatively low in the adult brain [26-28]. Several studies in the 6-OH-DA lesioned striatum of rats have demonstrated improvements in survival and outgrowth of grafted fetal DA neurons when central injections of GDNF have been administered, or when GDNF overexpressing cells have been co-grafted to the striatum [29-35]. Other than GDNF, brain-derived neurotrophic factor (BDNF) is one of the major trophic factors for DAergic neurons [36-38]. Transplantation of modified fibroblasts or astrocytes that express BDNF into either the striatum or the midbrain attenuates 6-OH-DA-induced losses of nigrostriatal neurons [39, 40]. Also, BDNF can modulate dopaminergic neurotransmission in nigrostriatal neurons, as shown by elevated rotational behavior and increased turnover of dopamine in the striatum [41]. BDNF can promote functional recovery from 6-OHDA lesions following expression in striatal cells from an AAV vector [41]. These studies support the relevance of neurotrophic factors for a neurorestorative effect expected in the treatment of PD. As reported in the following chapters, some DA agonists have been found to stimulate either BDNF or GDNF in experimental cell cultures. These pharmacological effects may be useful to improve treatment for PD.

Anti-Fibrillary Activity

Several data suggest anti-fibrillary effects of DA agonists. The aggregation of β-amyloid peptide (Aβ) and alpha-synuclein in the brain has been implicated as a critical step in the development of AD and LBD, respectively. Thus, in addition to antioxidant strategies, increasing evidence points to the possibility of achieving neuroprotection by DA agonists through inhibition of fibril formation [42]. DAergic agents were indeed found to dose-dependently inhibit generation of, as well as destabilize preformed, β-amyloid fibrils [43]. Using fluorescence spectroscopy with thioflavin S, electron microscopy, and atomic force microscopy, the effects of selegiline, DA, pergolide, bromocriptine, and trihexyphenidyl on alpha-synuclein fibrils formation have been recently studied. All molecules except for trihexyphenidyl, dose-dependently inhibited the formation of alpha-synuclein fibrils. Moreover, these molecules dose-dependently destabilized preformed alpha-synuclein fibrils [44]. The anti-fibrillary actions elicited by the DA agonists appear to be independent of DA receptor stimulation and detectable in cell-free systems.

2. EXPERIMENTAL EVIDENCE FOR NEURORESCUE ACTIVITY

Apomorphine

Apomorphine, a non-ergoline DA receptor agonist, is a short-acting and non-selective DA D1/D2 receptor agonist used for treating PD since many decades. Subcutaneous intermittent injections or continuous infusions of apomorphine have been proposed for the management of sudden, unexpected and refractory levodopa-induced "off" states in fluctuating PD [45]. The original evidence for neuroprotective and antioxidant effects of apomorphine, and other DA receptor agonists, is from Youdim's group [46]. Rational for their pivotal studies was the relevance of ROS generation in the neurotoxicity associated with PD [1, 4]. Since compounds with a catechol structure have metal chelating properties and can act as reducing agents [47], it was attractive the idea that apomorphine may inhibit metal-catalyzed free radical processes and act as a radical scavenger. Studies on the effect of apomorphine on lipid peroxidation and protein carbonyl formation after ascorbate/iron-induced free radical formation in rat brain mitochondrial fractions clearly demonstrated the antioxidant properties of apomorphine in brain tissue [46, 48]. It has later been shown that apomorphine exhibited neuroprotection against DA depletion in 6-OH-DA lesioned-rats [49] and MPTP-treated mice [50]. Furthermore, continuous subcutaneous infusion of apomorphine was found to rescue striatal DAergic terminals and increase the tyrosine hydroxylase and DA-transporter immunoreactivity against toxicity induced by MPTP in mice and enhanced the number of tyrosine hydroxylase-positive cells in the ventral tegmental area in partially 6-OH-DA-lesioned rats [51, 52]. It has also been observed that apomorphine increased the survival of cultured mesencephalic DAergic cells [53] suggesting trophic effects of apomorphine either *in vivo* or *in vitro*. The neuroprotective effect of apomorphine was further supported by the result of different pharmacological properties including antioxidant activity, potent iron chelating action, inhibition of lipid peroxidation, induction of neurotrophic factors and anti-inflammatory effects [46, 50, 54-56]. Specific brain gene expression changes have been reported in the chronic MPTP model in the late stage of degeneration, employing cDNA expression array, which indicate a domino cascade of events involved in in oxidative-stress, inflammatory processes AND signal transduction and glutamate toxicity that finally lead to neuronal cell death [57].

Bromocryptine

Bromocryptine is the first DA receptor agonist that has been approved for anti-parkinsonian therapy since 1974 [58]. It was first used as adjunctive therapy to levodopa in patients experiencing motor fluctuations and later it was recommended as monotherapy in the early stage of the disease [59]. From the pharmacological point of view, bromocryptine was the first DA receptor agonist endowed with D2 receptor specificity to be described. Bromocryptine has been shown to protect mice and DAergic cells against 6-OH-DA and MPTP, and levodopa-induced cell loss, respectively; it also attenuated DA depletion in mouse striatum in response to methamphetamine [60-62]. The neuroprotective effect of bromocryptine was dependent on both its action as a D2 receptor agonist and its antioxidant capacity. In this context, it has been reported that bromocryptine is able to scavenge hydroxyl and superoxide radicals *in vitro* [61, 63] and to inhibit hydroxyl radical formation and lipid peroxidation *in vivo* [61]. There is evidence that excitotoxic mechanisms contribute to the pathogenesis of PD and that glutamate signaling could be an important mechanism for the death of dopaminergic neurons and trigger the induction of programmed cell death [64]. Bromocryptine has been shown to exert a protective effect against glutamate-induced cytotoxicity in primary cultures of rat cortical or mesencephalic neurons [65] The neuroprotective effect was mediated *via* D2 receptors, because it was attenuated by domperidone, a D2 DAergic receptor antagonist.

Pergolide

Pergolide is a synthetic ergoline derivate that has been indicated in monotherapy as an efficacious treatment in patients with early stage PD. Several studies demonstated the neuroprotective effect of pergolide, observed either *in vivo* or *in vitro* [66-68]. The neuroprotective effect of pergolide has been shown to be mediated by free radical scavenging activity particularly hydroxyl radicals and nitric oxide and by decreasing phospholipid peroxidation [69, 70] suppressing apoptotic pathways through inhibiting of NF-κB nuclear translocation [71] and stabilizing the mitochondrial function [68]. Data obtained in our laboratory demonstrated that pergolide protected SH-SY5Y neuroblastoma cells from cell death induced by H_2O_2 [72]. Unfortunately, peroglide therapy was associated with valvular heart fibrosis; for this reason it was withdrawn from sale in the United States and the European Medicines Agency has added new warnings and controindications to the product information [73-75].

Ropinirole

Ropinirole is a non-ergoline DA receptor agonist that exhibits a high affinity for the D2 and D3 receptor subtypes but little or no affinity for the D1 receptor [76, 77]. Symptomatically, it was reported that ropinirole was as effective as bromocriptine in reducing motor complications and decreasing levodopa dose without increasing adverse events including dyskinesia [78]. Also, ropinirole monotherapy was effective in treating resting tremor in early PD [79], in reducing periodic leg movements and in improving sleep efficiency in patients with restless legs syndrome [76, 80]. These positive effects of ropinirole in PD are believed to be due to stimulation of the post-synaptic DA D2-type receptor [77]. In experimental models for PD, it has been found that ropinirole reversed the motor and behavioral deficits induced by MPTP in marmosets [77] and showed neuroprotective effect against 6-OH-DA in mice [78]. Activation of glutathion and glutathion-regulating enzymes such as glutathione peroxidase, glutathione reductase and glutathione transferase as well as activation of catalase and superoxide dismutase were principal neuroprotective mechanisms mediated by ropinirole [81, 82].

In clinical trials ropinirole reduced the long-term decline of striatal fluorodopa uptake compared to levodopa therapy indicating a preserving effect on terminal function of DAergic neurons [83, 90]. Previous *in vitro* study showed that ropinirole can promote the differentiation and survival of DAergic neurons, and it can upregulate the expression and secretion of GDNF and BDNF [84]. Considering the potential effects of ropinirole in neuroprotection, it has been suggested that ropinirole may have an anti-apoptotic effect through interfering with MAP kinase pathway and caspase-dependent pathway. Recent studies showed that ropinirole has neuroprotective effect against rotenone-induced apoptosis in both SH-SY5Y cells and primary mesencephalic cultures [85]. Exposure to rotenone significantly activated p-JNK, p-P38 and p-c-Jun in SH-SY5Y cells. Activation of JNK and P38 was responsible for the inhibition of the anti-apoptotic protein Bcl-2, and the induction of phosphorylation of c-Jun, a nuclear transcription factor and a known target of JNK, which further promotes the release of cytochrome c from the mitochondria to the cytoplasm and it leads to caspase-9 activation [86-89]. Pretreatment with ropinirole inhibits p-JNK, p-P38 and p-c-Jun expression, indicating that ropinirole may act at early stage of apoptosis. These effects appear to be mediated by neurotrophic factors GDNF and BDNF, since both of them were found to increase in the primary mesencephalic cultures after treatment with ropinirole and pramipexole [84, 91]. Ropinirole also increased GDNF in mouse astrocytes taken from whole brain [92] The antiapoptotic effect of ropinirole can be suppressed by D2 and D3 receptor antagonists sulpiride and nafadotride. Nafadotride exhibited greater effect in blocking ropinirole mediated anti-apoptosis, suggesting that D3 receptor may play a significant role.

Pramipexole

Pramipexole (PPX), a non-ergot DA agonist, has been successfully applied to the treatment of early and severe PD. It appears efficacy in the reduction of motor fluctuations and, in association with levodopa, it allows a reduction of the dose of the latter. Furthermore, pramipexole is able to improve the affective symptomatology. Pramipexole exhibits an 8-fold higher affinity for D3 than D2 little or no affinity for the D1 receptor. The neuroprotective effects elicited by this drug have been associated with different mechanism, such as antioxidant effects, mitochondrial stabilization or induction of the antiapoptotic Bcl-2 family. In particular, Le et al. [93] reported that pramipexole protected DAergic MES 23.5 cell line against DA, 6-OH-DA and hydrogen peroxide-induced cytotoxicity possibly through antioxidant effects, and such neuroprotection was independent from DA receptor stimulation not being prevented by selective D2 or D3 antagonists. Similar results were obtained by Fujta et al. [94] and Uberti *et al.* [95], who demonstrated that pramipexole inhibited generation of H_2O_2-induced reactive oxygen species in PC12 cells and SH-SY5Y neuroblastoma cells, respectively. In a recent study we showed that, in retinoic acid differentiated SH-SY5Y cells, PPX preventing cell death inhibiting mithocondrial reactive oxygen and that these ability are held by both S (-) and R (+) enantiomers [96].

In a search for an appropriate cell model for studying neuroprotection, Presgraves *et al.* [97] characterized differentiation conditions of the SH-SY5Y neuroblastoma cell line for phenotypic markers of DAergic cells, Cells were differentiated with retinoic acid (RA), 12-O-tetradecanoyl-phorbol-13-acetate (TPA), and RA followed by TPA (RA/TPA). Interestingly, RA/TPA differentiated cells exhibited 3-fold and 6-fold higher levels, respectively, of DA D2 and D3 receptors than undifferentiated or RA-differentiated cells. Pretreatment with pramipexole was protective against MPP+ in the RA/TPA differentiated cells but not in undifferentiated or RA differentiated cells. The neuroprotective effect of pramipexole was concentration-dependent and DA D2/D3 receptor dependent. In contrast, protection by pramipexole against DA was not DA receptor dependent.

An additional mechanisms underlying protection by D2/D3 receptor agonist has been suggested to involve neurotrophic factors. To verify this hypothesis, the D3 receptor preferring agonists S32504 [(+)-*trans*-3,4,4*a*,5,6,10*b*-hexahydro-9-carbamoyl-4-propyl-2*H*-naphth[1,2-*b*]-1,4-oxazine] and pramipexole were utilized in a terminally differentiated neuroblastoma SH-SY5Y cell line exhibiting a DAergic phenotype [98]. The cytotoxic effects of MPP^+ were stereospecifically antagonized by S32504 and by pramipexole, but not by their inactive stereoisomers, R(+) pramipexole and S32601, respectively. Neuroprotective effects afforded by S32504 and pramipexole were specifically antagonized by the selective D3 antagonists S33084, U99194A, and SB269652, and by the D2/D3 antagonist raclopride. The preferential D2 receptor antagonist LY741626 was ineffective as the D1 antagonist SCH23390.

BDNF potently protected against MPP^+-induced neurotoxicity. Antibody directed against BDNF concentration-dependently blocked both the neuroprotective effects of BDNF and those of pramipexole and S32504 against MPP^+. The protection afforded by BDNF was blocked by the P3K-AKT pathway inhibitor LY249002. Neuroprotective effects of pramipexole and S32504 against MPP(+) toxicity appear to be mediated by D3 receptors stimulation. Their actions also reflect downstream recruitment of BDNF and *via* a PK3-AKT pathway.

PPX has been recently shown to promote adult neurogenesis in SVZ in an acute rat Parkinson's disease model [99] and to increase the levels of brain-derived neurotrophic factor and glial cell line-derived neurotrophic factor in a mouse model of PD induced by ubiquitin-proteasome system impairment [100] furthermore, the PPX-induced neurogenesis was better characterized by Merlo *et al.* [101]; the authors showed that PPX is able to favor expansion and neuronal differentiation of neural progenitors from the adult SVZ, by both D2 and D3 receptor-mediated mechanisms, which involves involves enhanced release of BDNF and subsequent activation of AKT. Finally, PPXwas demonstrated to inhibit the phosphorylation of α-synuclein with a mechanism independent of dopamine receptor activation, by inhibition of the ubiquitin proteasomal system. The phosphorylation of α-synuclein occurs in part at least through casein kinase 2, and PPX in turn reduces the phosphorylation of this enzyme, thereby inhibiting its activity [102].

Rotigotine

Rotigotine is a non-ergoline dopamine agonist developed for the once daily treatment of Parkinson's disease (PD) which provides a 24-hour continuous treatment through a transdermal delivery system (patch), to obtain the reduction of motor fluctuation. Rotigotine acts as a full agonist at dopamine receptors, with the highest affinity for the D_3 receptor [103]. Rotigotine exerted its protective effects *via* dopamine receptor significantly reducing the production of superoxide radicals. In the acute MPTP mouse or progressive MPTP macaque model the agonist was shown to exert partial protective effects on dopaminergic nerve endings [104, 105]. Rotigotine treatment appears to protect dopaminergic neurons from cell death mediated by glutamate. In a very recent study, Oster *et al.* revealed the signalling pathways that are associated with rotigotine-induced neuroprotection against glutamate excitotoxicity, likely through the activation of the PI3K/Akt pathway followed by GSK-3 β inactivation which was finally associated with an increased content of GSH [106].

3. PERSPECTIVES

Dopamine Receptor Agonists in Clinical Trials

The neuroprotective action of dopamine agonists has been demonstrated and repeatedly confirmed over the years by numerous studies *in vitro* and *in vivo*. However, some difficulties were encountered in translating these experimental results in a clinical benefit for the patients.

Two clinical trails investigated the neuroprotective potential of DA using imaging end-points; the CALM-PD was a multicenter randomized double-blind controlled clinical trial enrolling subjects with early PD, randomly assigned to receive pramipexole or levodopa. Using β-CIT SPECT scanning techniques, the study aimed to assess the uptake of radiolabeled β-CIT [107, 108]. In pramipexole-treated patients a reduction in β-CIT uptake was observed, suggesting a neuroportective role of the DA. Likely, the REAL-PET trial, which compared ropirinole with levodopa for the ^{18}F-dopa uptake in the putamen, showed less decrease in dopamine uptake in patient treated with ropinirole [109]. However, these studies are burdened by some important limitations, the most important the lack of a strong linkage to the clinical outcome, which did not differ significantly in DA treatment group compared with leovodopa treated patients [110, 111]. Confirming this limitation the PROUD study, recently designed to identify whether early *versus* delayed pramipexole initiation has clinical and neuroimaging benefits, has not reached the clinical endpoint [112].

Dopamine Agonists in Neurodegenerative Disease

Because their antioxidant properties, DA has been proposed to be effective against other pathologies, beside PD, where oxidative stress is the main mechanism implicated in pathophysiology of the diseases. In fact, oxidative abnormalities have been identified in several neurodegenerative diseases such as amyotrophic lateral sclerosis (ALS) and Alzheimer's disease (AD). The treatment of patients affected by ALS with pramipexole reduced the systemic production of oxygen radicals, as demonstrated by measuring the levels of 2,3-dihydroxybenzoic acid (2,3-DHBA) in the blood, suggesting that pramipexole may interrupt free radical production in ALS [113]. However, in the phase 3 trial EMPOWER, dexpramipexole (the R (+) enantiomer of pramipexole) did not differ from placebo on any prespecified efficacy endpoint measurement [114]. An increasing amount of evidences suggest a central role of oxidative stress also in AD pathogenesis [115, 116]. Furthermore, many findings link free radical formation with excessive deposition of Aβ derived from the amyloidogenic processing of the larger amyloid precursor protein [117-122]. DA agonists were found to be active in preventing Aβ-induced both aggregation and ROS formation [22, 42-44, 95]. We studied the role of free radical in the neurotoxic events caused by different Aβ aggregation states, and investigated the neuroprotective effects of pramipexole in neuronal death induced by unaggregated, oligomeric and fibrils Aβ species [95]. Increasing evidences suggest that protofibrillar aggregates of Aβ recognized as diffuse plaques by neuropathological examination, are indeed the most toxic Aβ species. Oligomers of Aβ, rather than monomers or large fibrils, may form pores in the cell membrane, allowing influx of ions, that disrupt neuronal signaling and initiate cell death cascade [123, 124]. These data strongly support the hypothesis that each of the Aβ aggregation state possesses different biological and pathological functions. We challenged the neuronal SK-SH5Y cell line with $A\beta_{1-42}$ peptide in different states of aggregation. Contrary to unaggregated peptide, oligomers and fibrils caused generation of ROS and this effect was inhibited by pramipexole in a DA receptors independent manner. The action of pramipexole on Aβ activity is further supported by the data showing that pramipexole prevented the induction of caspase 3 activated by $A\beta_{25-35}$ [125]. These experimental data could acquire clinical significance in AD therapy; in fact, in a recent study, Koch *et al.* [126] demonstrated that dopamine agonists may restore the long term potentiation that is impaired or abolished in AD patients, thus providing novel implications for therapies based on dopaminergic stimulation.

CONCLUSION

The neuroprotective and neurorescue properties of dopamine agonists have been reconfirmed over the years in several *in vitro* and *in vivo* experimental models. These drugs are endowed with intrinsic and peculiar antioxidant, antiapoptotic, neurotrophic and anti-fibril formation effects. These activities are molecule-specific and may represent additional pharmacological properties contributing to the clinical efficacy in PD and in the other neurodegenerative disease, in which oxidative stress and protein aggregation play a decisive role. However, an unequivocal disease-modifying effect of these molecules is difficult to prove, because some obstacles including the incomplete understanding of the mechanisms underlying neuronal death and the imprecise definition of doses and endpoints in clinical trials.

In conclusion, further studies to elucidate the PD pathogenesis and a better clinical trials design are needed to bridge the gap between the promising experimental results and the not yet optimal clinical efficacy of dopamine agonists.

ACKNOWLEDGEMENTS

Declared None.

CONFLICT OF INTEREST

The authors confirm that this chapter contents have no conflict of interest.

DISCLOSURE

This is an updated report of the earlier review article by G. Ferrari Toninelli *et al.* “Dopamine receptor agonists for protection and repair in Parkinson’s disease”. Curr Top Med Chem 2008; 8(12):1089-99.

ABBREVIATIONS

6-OH-DA	=	6-hydroxydopamine
AAV2	=	Adeno-associated virus type 2
Aβ	=	β-amyloid peptide
BDNF	=	Brain derived neurotrophic factor
COMT	=	Cathecol-O-methyl transferase
DA	=	Dopamine
DOPA	=	3,4-diidrossi-l-fenilalanina

GDNF = Glial derived neurothophic factor

LBD = Lewy bodies dementia

MPP^+ = 1-metyl-4-phenylpyridinium

MPTP = 1-methyl-4-phenyl-1,2,3,6-tetrahydropyridine

PD = Parkinson's disease

PI3K = Phosphatidylinositol 3 kinase

RA = Retinoic acid

ROS = Radical oxygen species

SNpc = Substantia nigra *pars compacta*

SVZ = Subventricular zone

TPA = 12-O-tetradecanoyl-phorbol-13-acetate

β-CIT = $[^{123}I]$2β-Carbomethoxy-3β–(4–iodophenil)tropane

REFERENCES

[1] Adams JD Jr, Odunze IN. Oxygen free radicals and Parkinson's disease. Free Radic Biol Med 1991; 10: 161– 69.

[2] Schapira AH. Neuroprotection in PD – A role for dopamine agonists? Neurology 2003; 61: S34–S42.

[3] Jenner P. Dopamine agonists, receptor selectivity and dyskinesia induction in Parkinson's disease. Curr Opin Neurol 2003; 16: S3–S7.

[4] Olanow CW A radical hypothesis for neurodegeneration. Trends Neurosci 1993; 16: 439–44.

[5] Ossowska K, Wardas J, Kuter K, Nowak P, Dabrowska J, Bortel A, *et al.* Influence of paraquat on dopaminergic transporter in the rat brain. Pharmacol Rep 2005; 57: 330–35.

[6] Miller GW. Paraquat: the red herring of Parkinson's disease research. Toxicol Sci 2007; 100: 1-2.

[7] Shepherd KR, Lee ES, Schmued L, Jiao Y, Ali SF, Oriaku ET, *et al.* The potentiating effects of 1-methyl-4-phenyl-1,2,3,6-tetrahydropyridine (MPTP) on paraquat-induced neurochemical and behavioral changes in mice. Pharmacol Biochem Behav. 2006; 83: 349-59.

[8] Lipton SA, Rosenberg PA. Excitatory amino acids as a final common pathway for neurologic disorders. N Engl J Med 1994; 330: 613-22.

[9] Howells DW, Porritt MJ, Wong JY, Batchelor PE, Kalnins R, Hughes AJ *et al.* Reduced BDNF mRNA expression in the Parkinson's disease substantia nigra. Exp Neurol 2000; 166(1):127-35.

[10] Mogi M, Togari A, Kondo T, Mizuno Y, Kogure O, Kuno S *et al.* Glial cell line-derived neurotrophic factor in the substantia nigra from control and parkinsonian brains. Neurosci Lett 2001; 300(3):179-8.

[11] Radad K, Gille G, Rausch WD. Short review on dopamine agonists: insight into clinical and research studies relevant to Parkinson's disease. Pharmacol Rep 2005; 57(6): 701-12.

[12] Gould E, Gross CG. Neurogenesis in adult mammals: some progress and problems. Neurosci 2002; 22(3): 619-23.

[13] Lidow MS, Rakic P. Neurotransmitter receptor in the proliferative one of the developing primate occipitale lobe. J Comp Neurol 1995; 360: 393-402.

[14] Hoglinger GU, Rizk P, Muriel MP, Duvckaerts C, Oertel WH, Caille I. Dopamine depletion impairs precursor cell proliferation in Parkinson disease. Nature Neurosci 2004; 7: 726-35.

[15] Freundlieb N, François C, Tandé D, Oertel WH, Hirsch EC, Höglinger GU. Dopaminergic substantia nigra neurons project topographically organized to the subventricular zone and stimulate precursor cell proliferation in aged primates. J Neurosci 2006; 26: 2321-25.

[16] Joyce JN, Millan MJ. Dopamine D3 receptor agonists for protection and repair in Parkinson's disease. Cur Op Pharmacol 2007; 7:100-105.

[17] Van Kampen JM, Hagg T, Robertson HA. Induction of neurogenesis in the adult rat subventricular zone and neostriatum following dopamine D3 receptor stimulation. Eur J Neurosci 2004; 19: 2377-87.

[18] Van Kampen JM, Eckman CB. Dopamine D3 receptor agonist delivery to a model of Parkinson's disease restores the nigrostriatal pathway and improves locomotor behaviour. Neurobiology of disease 2006; 26: 7272-80.

[19] Sayles M, Jain M, Barker RA. The cellular repair of the brain in Parkinson's disease--past, present and future. Transpl Immunol 2004; 12(3-4): 321-42.

[20] Baker SA, Baker KA, Hagg T. D3 dopamine receptors do not regulate neurogenesis in the subventricular zone of adult mice. Neurobiol Dis. 2005; 18: 523-27.

[21] Milosevic J, Schwarz SC, Maisel M, Poppe-Wagner M, Dieterlen MT, Storch A, Schwarz, J. Dopamine D2/D3 receptor stimulation fails to promote dopaminergic neurogenesis of murine and human midbrain-derived neural precursor cells *in vitro*. Stem Cells Dev 2007; 16(4): 625-35.

[22] Kippin TE, Kapur S, van der Kooy D. Dopamine specifically inhibits forebrain neural stem cell proliferation, suggesting a novel effect of antipsychotic drugs. J Neurosci 2005; 25: 5815-23.

[23] Elsworth JD, Redmond DE Jr, Leranth C, Bjugstad KB, Sladek JR Jr, Collier TJ *et al.* AAV2-mediated gene transfer of GDNF to the striatum of MPTP monkeys enhances the survival and outgrowth of co-implanted fetal dopamine neurons. Exp Neurol 2008; 211(1): 252-8.

[24] Airaksinen MS, Saarma M. The GDNF family: signalling, biological functions and therapeutic value. Nat Rev Neurosci 2002; 3: 383–94.

[25] Lin LF, Doherty DH, Lile JD, Bektesh S, Collins,F. GDNF: a glial cell line-derived neurotrophic factor for midbrain dopaminergic neurons. Science 1993; 260:1130–2.

[26] Choi-Lundberg DL, Bohn MC. Ontogeny and distribution of glial cell line-derived neurotrophic factor (GDNF) mRNA in rat. Brain Res Dev. 1995; 85: 80–8.

[27] Schaar DG, Sieber BA, Dreyfus CF, Black IB. Regional and cell-specific expression of GDNF in rat brain. Exp Neurol 1993; 124: 368–71.

[28] Stromberg I, Bjorklund L, Johansson M, Tomac A, Collins F, Olson L *et al.* Glial cell line-derived neurotrophic factor is expressed in the developing but not adult striatum and stimulates developing dopamine neurons *in vivo*. Exp Neurol 1993; 124: 401–12.

[29] Ahn YH, Bensadoun JC, Aebischer P, Zurn AD, Seiger A, Bjorklund A *et al.* Increased fiber outgrowth from xeno-transplanted human embryonic dopaminergic neurons with co-implants of polymer-encapsulated genetically modified cells releasing glial cell line-derived neurotrophic factor. *Brain Res Bull* 2005; 66: 135–42.

[30] Espejo M, Cutillas B, Arenas TE, Ambrosio S. Increased survival of dopaminergic neurons in striatal grafts of fetal ventral mesencephalic cells exposed to neurotrophin-3 or glial cell line-derived neurotrophic factor. Cell Transplant 2000; 9: 45–53.

[31] Rosenblad C, Martinez-Serrano A, Bjorklund A. Glial cell line-derived neurotrophic factor increases survival, growth and function of intrastriatal fetal nigral dopaminergic grafts, Neuroscience 1996; 75: 979–85.

[32] Sautter J, Tseng JL, Braguglia D, Aebischer P, Spenger C, Seiler RW *et al.* Implants of polymer-encapsulated genetically modified cells releasing glial cell line-derived neurotrophic factor improve survival, growth, and function of fetal dopaminergic grafts. Exp Neurol 1998; 149: 230–6.

[33] Sinclair SR, Svendsen CN, Torres E.M., Martin, D., Fawcett, J.W., Dunnett, S.B. GDNF enhances dopaminergic cell survival and fibre outgrowth in embryonic nigral grafts, NeuroReport 1996; 7: 2547–52.

[34] Sullivan AM, Pohl J, Blunt SB. Growth/differentiation factor 5 and glial cell line-derived neurotrophic factor enhance survival and function of dopaminergic grafts in a rat model of Parkinson's disease. Eur J Neurosci 1998; 10: 3681–88.

[35] Wilby MJ, Sinclair SR, Muir EM, Zietlow R, Adcock KH, Horellou P *et al.* A glial cell line-derived neurotrophic factor-secreting clone of the Schwann cell line SCTM41 enhances survival and fiber outgrowth from embryonic nigral neurons grafted to the striatum and to the lesioned substantia nigra. J Neurosci 1999; 19: 2301–12.

[36] Altar CA, Boylan CB, Jackson C, Hershenson S, Miller J, Wiegand SJ *et al.* Brain-derived neurotrophic factor augments rotational behavior and nigrostriatal dopamine turnover *in vivo*. PNAS 1992; 89: 11347–51.

[37] Altar A, Boylan CB, Fritsche M, Jones BE, Jackson C, Wiegand SJ *et al.* Efficacy of brain-derived neurotrophic factor and neurotrophin-3 on neurochemical and behavioral deficits associated with partial nigrostriatal dopamine lesions. J Neurochem 1994; 63: 1021–32.

[38] Hyman C, Hofer M, Barde YA, Juhasz M, Yancopoulos GD, Squinto SP, Lindsay RM. BDNF is a neurotrophic factor for dopaminergic neurons of the substantia nigra. Nature 1991; 350: 230–32.

[39] Levivier M, Przedborski S., Bencsics C, Kang UJ. Intrastriatal implantation of fibroblasts genetically engineered to produce brain-derived neurotrophic factor prevents degeneration of dopaminergic neurons in a rat model of Parkinson's disease. J Neurosci 1995; 15:7810-20.

[40] Yoshimoto Y, Lin Q., Collier TJ, Frim DM, Breakefield XO, Bohn MC. Astrocytes retrovirally transduced with BDNF elicit behavioral improvement in a rat model of Parkinson's disease. Brain Res. 1995; 691:25–36.

[41] Klein RL, Lewis MH, Muzyczka N, Meyer EM. Prevention of 6-hydroxydopamine-induced rotational behavior by BDNF somatic gene transfer. Brain Res. 1999 847:314–320.

[42] Batelli S, Albani D, Rametta R, Polito L, Prato F, Pesaresi M *et al.* DJ-1 modulates alpha-synuclein aggregation state in a cellular model of oxidative stress: relevance for Parkinson's disease and involvement of HSP70. PLoS ONE 2008; 3(4):e1884.

[43] Li J, Zhu M, Manning-Bog AB, Di Monte DA, Fink AL. Dopamine and L-dopa disaggregate amyloid fibrils: implications for Parkinson's and Alzheimer's disease. FASEB J 2004; 18(9): 962-4.

[44] Ono K, Hirohata M, Yamada M. Anti-fibrillogenic and fibril-destabilizing activities of anti-Parkinsonian agents for alpha-synuclein fibrils *in vitro*. J Neurosci Res 2007; 85(7):1547-57.

[45] Deleu D, Hanssens Y, Northway MG. Subcutaneous apomorphine: an evidence-based review of its use in Parkinson's disease. Drugs Aging 2004; 21(11): 687-709.

[46] Gassen M, Glinka Y, Pinchasi B, Youdim MB. Apomorphine is a highly potent free radical scavenger in rat brain mitochondrial fraction. Eur J Pharmacol 1996; 308(2): 219-25.

[47] Liu J, Mori A, Ogata K. Antioxidant and pro-oxidant assay for a new drug GEPC: detected by ESR spectrometry and by protective effects on lipid peroxidation and biomolecule degradation. Res Commun Chem Pathol Pharmacol 1993; 82(2):151-66.

[48] Gassen M, Gross A, Youdim MB. Apomorphine enantiomers protect cultured pheochromocytoma (PC12) cells from oxidative stress induced by H2O2 and 6-hydroxydopamine. Mov Disord. 1998; 13: 661-67.

[49] Youdim MB, Grünblatt E, Mandel S. The pivotal role of iron in NF-kappa B activation and nigrostriatal dopaminergic neurodegeneration. Prospects for neuroprotection in Parkinson's disease with iron chelators. *Ann N Y Acad Sci* 1999; 890:7-25.

[50] Grünblatt E, Mandel S, Berkuzki T, Youdim MB. Apomorphine protects against MPTP-induced neurotoxicity in mice. Mov Disord. 1999; 14(4):612-18.

[51] Battaglia G, Busceti CL, Cuomo L, Giorgi FS, Orzi F, De Blasi A. Continuous subcutaneous infusion of apomorphine rescues nigro-striatal dopaminergic terminals following MPTP injection in mice. Neuropharmacology 2002; 42(3): 367-73.

[52] Yuan H, Sarre S, Ebinger G, Michotte Y. Histological, behavioural and neurochemical evaluation of medial forebrain bundle and striatal 6-OHDA lesions as rat models of Parkinson's disease. J Neurosci Methods 2005; 144(1):35-45.

[53] Guo H, Tang Z, Yu Y, Xu L, Jin G, Zhou J. Apomorphine induces trophic factors that support fetal rat mesencephalic dopaminergic neurons in cultures. Eur J Neurosci. 2002; 16(10):1861-70.

[54] Sam EE, Verbeke N. Free radical scavenging properties of apomorphine enantiomers and dopamine: possible implication in their mechanism of action in parkinsonism. J Neural Transm Park Dis Dement Sect. 1995; 10(2-3):115-27.

[55] Ubeda A, Montesinos C, Payá M, Alcaraz MJ. Iron-reducing and free-radical-scavenging properties of apomorphine and some related benzylisoquinolines. Free Radic Biol Med. 1993; 15(2):159-67.

[56] Roceri M, Molteni R, Fumagalli F, Racagni G, Gennarelli M, Corsini G *et al.* Stimulatory role of dopamine on fibroblast growth factor-2 expression in rat striatum. J Neurochem 2001; 76(4): 990-7.

[57] Weinreb O, Mandel S, Youdim MB. Gene and protein expression profiles of anti- and pro-apoptotic actions of dopamine, R-apomorphine, green tea polyphenol (-)-epigallocatechine-3-gallate, and melatonin. Ann N Y Acad Sci 2003, 993:351-61.

[58] Foley P, Gerlach M, Double KL, Riederer P. Dopamine receptor agonists in the therapy of Parkinson's disease. J Neural Transm 2004; 111:1375–46.

[59] Watts RL. The role of dopamine agonists in early Parkinson's disease. Neurology 1997; 49: S34–S48.

[60] Kondo T, Ito T, Sugita Y. Bromocriptine scavenges methamphetamine-induced hydroxyl radicals and attenuates dopamine depletion in mouse striatum. Ann NY Acad Sci 1994; 738: 222–9.

[61] Ogawa N, Tanaka K, Asanuma M, Kawai M, Masumizu T, Kohno M *et al.* Bromocriptine protects mice against 6-hydroxydopamine and scavenges hydroxyl free radicals *in vitro.* Brain Res 1994; 657:207–13.

[62] Takashima H, Tsujihata M, Kishikawa M, Freed WJ. Bromocriptine protects dopaminergic neurons from levodopa-induced toxicity by stimulating D(2)receptors. Exp Neurol 1999; 159: 98–104.

[63] Yoshikawa T, Minamiyama Y, Naito Y, Kondo M. Antioxidant properties of bromocriptine, a dopamine agonist. J Neurochem 1994; 62: 1034–38.

[64] Meredith GE, Totterdel LS, Beales M, Meshul CK. Impaired glutamate homoeostasis and programmed cell death in a chronic MPTP mouse model of Parkinson's disease. Exp. Neurol 2009; 219: 34-340.

[65] Kihara T, Shimohama S, Sawada H, Honda K, Nakamizo T, Kanki R *et al.* Protective effect of dopamine D2 agonists in cortical neurons *via* the phosphatidylinositol 3 kinase cascade. J Neurosci Res 2002; 70(3):274-82.

[66] Felten DL, Felten SY, Fuller RW, Romano TD, Smalstig EB, Wong DT *et al.* Chronic dietary pergolide preserves nigrostriatal neuronal integrity in aged-Fischer-344 rats. Neurobiol Aging 1992; 13:339–51.

[67] Asanuma M, Ogawa N, Nishibayashi S, Kawai M, Kondo Y, Iwata E. Protective effects of pergolide on dopamine levels in the 6-hydroxydopamine-lesioned mouse brain. Arch Int Pharmacodyn Ther 1995; 329:221–30.

[68] Gill G, Rausch WD, Hung ST, Moldzio R, Janetzky B, Hundemer HP *et al.* Pergolide protects dopaminergic neurons in primary culture under stress conditions. J Neural Transm 2002; 109:633–43.

[69] Gomez-Vargas M, Nishibayashi-Asanuma S, Asanuma M, Kondo Y, Iwata E, Ogawa N. Pergolide scavenges both hydroxyl and nitric oxide free radicals *in vitro* and inhibits lipid peroxidation in different regions of the rat brain. Brain Res 1998; 790:202–8.

[70] Opacka-Juffry J, Wilson AW, Blunt SB. Effects of pergolide treatment on *in vivo* hydroxyl free radical formation during infusion of 6-hydroxydopamine in rat striatum. Brain Res 1998; 810:27–33.

[71] Uberti D, Carsana T, Francisconi S, Ferrari-Toninelli G, Canonico PL, Memo M. A novel mechanism for pergolide-induced neuroprotection: inhibition of NFkappaB nuclear translocation. Biochem Pharmacol 2004; 67:1743–50.

[72] Uberti D, Piccioni L, Colzi A, Bravi D, Canonico PL, Memo M. Pergolide protects SH-SY5Y cells against neurodegeneration induced by H(2)O(2). Eur J Pharmacol 2002; 434:17–20.

[73] Baseman DG, O'Suilleabhain, PE, Reimold SC, Laskar SR, Baseman JG, Dewey RB Jr. Pergolide use in Parkinson's disease is associated with cardiac valve regurgitation. Neurology 2004, 63:301–304.

[74] Horvath J, Fross RD, Kleiner-Fisman G, Lerch R, Stalder H, Liaudat S *et al.* Severe multivalvular heart disease: a new complication of the ergot derivative dopamine agonists. Mov Disord 2004, 19:656–62.

[75] Van Camp G, Flamez A, Cosyns B, Weytjens C, Muyldermans L, Van Zandijcke M *et al.* Treatment of Parkinson's disease with pergolide and relation to restrictive valvular heart disease. Lancet 2004; 363:1179–83.

[76] Cheer SM, Bang LM, Keating GM. Ropinirole: for the treatment of restless legs syndrome. CNS Drugs 2004; 18:747–54.

[77] Eden RJ, Costall B, Domeney AM, Gerrard PA, Harvey CA, Kelly ME *et al.* Preclinical pharmacology of ropinirole (SK&F 101468-A) a novel dopamine D2 agonist. Pharmacol Biochem Behav 1991; 38:147–54.

[78] Clarke CE, Deane KH. Ropinirole *versus* bromocriptine for levodopa-induced complications in Parkinson's disease. Cochrane Database Syst Rev 2001; CD001517.

[79] Schrag A, Keens J, Warner J. Ropinirole for the treatment of tremor in early Parkinson's disease. Eur J Neurol 2002; 9:253–57.

[80] Allen R, Becker PM, Bogan R, Schmidt M, Kushida CA, Fry JM *et al.* Ropinirole decreases periodic leg movements and improves sleep parameters in patients with restless legs syndrome. Sleep 2004; 27:907–14.

[81] Iida M, Miyazaki I, Tanaka K, Kabuto H, Iwata-Ichikawa E, Ogawa N. Dopamine D2 receptor-mediated antioxidant and neuroprotective effects of ropinirole, a dopamine agonist. Brain Res 1999; 838:51–9.

[82] Tanaka K, Miyazaki I, Fujita N, Haque ME, Asanuma M, Ogawa N. Molecular mechanism in activation of glutathione system by ropinirole, a selective dopamine D2 agonist. Neurochem Res 2001; 26: 31–36.

[83] Whone AL, Watts RL, Stoessl AJ, Davis M, Reske S, Nahmias C *et al.* Slower progression of Parkinson's disease with ropinirole *versus* levodopa: the REAL-PET study. Ann Neurol 2003; 54:93-101.

[84] Du F, Li R, Huang Y, Li X, Le W. Dopamine D3 receptor-preferring agonists induce neurotrophic effects on mesencephalic dopamine neurons. Eur J Neurosci 2005; 22:2422–30.

[85] Sheng C, Xiaojie Z, Dehua Y, Yunlan D, Liang L, Xuping L *et al.* D3 receptor agonist ropinirole protects dopaminergic cell line against rotenone-induced apoptosis through inhibition of caspase- and JNK-dependent pathways. FEBS Letters 2008; 582:603–10.

[86] Junn E, Mouradian MM. Apoptotic signaling in dopamine induced cell death: the role of oxidative stress, P38 mitogen-activated protein kinase, cytochrome c and caspases. J. Neurochem 2001; 78:374–83.

[87] Davis RJ. Signal transduction by the JNK group of MAP kinases. Cell 2000; 103, 239–52.

[88] Xia Z, Dickens M., Raingeaud J, Davis RJ, Greenberg ME. Opposing effects of ERK and JNK-P38 MAP kinases on apoptosis. Science 1995; 270, 1326–31.

[89] Li P, Nijhawan D, Budihardjo I, Srinivasula SM, Ahmad M, Alnemri ES *et al.* Cytochrome c and dATPdependent formation of Apaf-1/caspase-9 complex initiates an apoptotic protease cascade. Cell 2004; 91, 479–89.

[90] Schwarz, J. Rationale for dopamine agonist use as monotherapy in Parkinson's disease. Curr Opin Neurol 2003; 16 (S1): S27–S33.

[91] Consales C, Volpicelli F, Greco D, Leone L, Colucci-D'Amato L, Perrone-Capano C *et al.* GDNF signaling in embryonic midbrain neurons *in vitro*. Brain Res 2007 ; 1159 : 28-39.

[92] Ohta K, Kuno S, Inoue S, Ikeda E, Fujinami A, Ohta M. The effect of dopamine agonists: the expression of GDNF, NGF, and BDNF in cultured mouse astrocytes. J Neurol Sci 2010; 291(1-2):12–16.

[93] Le WD, Jankovic J, Xie W, Appel SH. Antioxidant property of pramipexole independent of dopamine receptor activation in neuroprotection. J Neural Transm 2000; 107:1165-73.

[94] Fujita Y, Izawa Y, Ali N, Kanematsu Y, Tsuchiya K, Hamano S *et al.* Pramipexole protects against H2O2-induced PC12 cell death. Naunyn Schmiedebergs Arch Pharmacol. 2006; 372: 257-66.

[95] Uberti D, Bianchi I, Olivari L, Ferrari-Toninelli G, Canonico P, Memo M. Neuroprotective effect of pramipexole against beta amyloid toxicity depends on peptide aggregated states. Eur J Pharmacol 2007, 569(3): 194-6.

[96] Ferrari-Toninelli G, Maccarinelli G, Uberti D, Buerger E, Memo M. Mitochondria-targeted antioxidant effects of S(-) and R(+) pramipexole. BMC Pharmacol 2010;10-12.

[97] Presgraves SP, Ahmed T, Borwege S, Joyce JN. Terminally differentiated SH-SY5Y cells provide a model system for studying neuroprotective effects of dopamine agonists. Neurotox Res 2004; 5(8), 579-98.

[98] Presgraves SP, Borwege S, Millan MJ, Joyce JN. Involvement of dopamine D(2)/D(3) receptors and BDNF in the neuroprotective effects of S32504 and pramipexole against 1-methyl-4-phenylpyridinium in terminally differentiated SH-SY5Y cells. Exp Neurol 2004; 190(1): 157-70.

[99] Winner B, Desplats P, Hagl C, Klucken J, Aigner R, Ploetz S *et al.* Dopamine receptor activation promotes adult neurogenesis in an acute Parkinson model. Exp Neurol 2009; 219(2): 543-52.

[100] Li C, Guo Y, Xie W, Li X, Janokovic J, Le W. Neuroprotection of pramipexole in UPS impairment induced animal model of Parkinson's disease. Neurochem Res 2010; 35(10):1546-56.
[101] Merlo S, Canonico PL, Sortino MA. Distinct effects of pramipexole on the proliferation of adult mouse sub-ventricular zone-derived cells and the appearance of a neuronal phenotype. Neuropharmacology 2011; 60(6): 892-900.
[102] Chau KY, Cooper JM, Schapira AHV. Pramipexole Reduces Phosphorylation of α-Synuclein at Serine-129. J Mol Neurosci 2013; 51: 573–80.
[103] Scheller D, Ullmer C, Berkels R, Gwarek M, Lubbert H. The *in vitro* receptor profile of rotigotine: a new agent for the treatment of Parkinson's disease. Naunyn Schmiedebergs Arch. Pharmacol 2009; 379: 73-86.
[104] Scheller D, Stichel-Gunkel C, Lubbert H, Porras G, Ravenscroft P, Hill M *et al.* Neuroprotective effects of rotigotine in the acute MPTP-lesioned mouse model of Parkinson's disease. Neurosci Lett 2008; 432: 30-4.
[105] Scheller D, Chan P, Li Q, Wu T, Zhang R, Guan L *et al.* Rotigotine treatment partially protects from MPTP toxicity in a progressive macaque model of Parkinson's disease. Exp. Neurol 2007; 203: 415-22.
[106] Oster S, Radad K, Scheller D, Hesse M, Balanzew W. Rotigotine protects against glutamate toxicity in primary dopaminergic cell culture. Eur J Pharmacol 2014; 724:31-42.
[107] No authors listed. A randomized controlled trial comparing pramipexole with levodopa in early Parkinson's disease: design and methods of the CALM-PD Study. Clin Neuropharmacol 2000; 23(1): 34-44.
[108] Parkinson Study Group. Dopamine transporter brain imaging to assess the effects of pramipexole *vs* levodopa on Parkinson disease progression. JAMA 2002; 287(13): 1653-61.
[109] Whone AL, Watts RL, Stoessl AJ, Davis M, Reske S, Nahmias C *et al.* REAL-PET Study Group. Slower progression of Parkinson's disease with ropinirole *versus* levodopa: The REAL-PET study. Ann Neurolm 2003;54(1): 93-101.
[110] Morrish PK. How valid is dopamine transporter imaging as a surrogate marker in research trials in Parkinson's disease? Mov Disord 2003; Suppl 7: S63-S70.
[111] Marek K, Jennings D, Seibyl J. Do dopamine agonists or levodopa modify Parkinson's disease progression? Eur J Neurol 2002; (Suppl 3): 15-22.
[112] Schapira AH, McDermott MP, Barone P, Comella CL, Albrecht S, Hsu HH *et al.* Pramipexole in patients with early Parkinson's disease (PROUD): a randomised delayed-start trial. Lancet Neurol 2013;12(8):747-55.
[113] Pattee GL, Post GR, Gerber RE, Bennett JP Jr. Reduction of oxidative stress in amyotrophic lateral sclerosis following pramipexole treatment. Amyotroph Lateral Scler Other Motor Neuron Disord 2003; 4: 90-5.
[114] Cudkowicz ME, van den Berg LH, Shefner JM, Mitsumoto H, Mora JS, Ludolph A *et al.* Dexpramipexole *versus* placebo for patients with amyotrophic lateral sclerosis (EMPOWER): a randomised, double-blind, phase 3 trial. Lancet Neurol 2013;12(11):1059-67.
[115] Moreira PI, Smith MA, Zhu X, Honda K, Lee HG, Aliev G. Oxidative damage and Alzheimer's disease: are antioxidant therapies useful? Drug News Perspect 2005; 18:13-19.
[116] Behl C. Oxidative stress in Alzheimer's disease: implications for prevention and therapy. Subcell Biochem 2005; 38:65-78.
[117] Aksenov MY, Aksenova MV, Carney JM, Butterfield DA. Oxidative modification of glutamine synthetase by amyloid beta peptide. Free Rad. Res 1997; 27: 267-81.
[118] Butterfield DA. Amyloid beta-peptide [1-42]-associated free radical-induced oxidative stress and neurodegeneration in Alzheimer's disease brain: mechanisms and consequences. Curr Med Chem 2003; 10: 2651-8.
[119] Lin MT, Beal F. Mitochondrial dysfunction and oxidative stress in neurodegenerative diseases. Nature 2006; 443: 787-95.
[120] Barnham KJ, Masters CL, Bush AI. Neurodegenerative diseases and oxidative stress. Nat Rev Drug Discov 2004; 3: 205-14.
[121] Zhang YW, Xu H. Molecular and cellular mechanisms for Alzheimer's disease: understanding APP metabolism. Curr Mol Med 2007; 7(7): 687-96.

[122] Bredesen DE, Rabizadeh S. APP-based neuroprotective strategies. Curr Alzheimer Res. 2007; 4(5): 541-43.

[123] Walsh DM, Klyubin I, Shankar GM, Townsend M, Fadeeva JV, Betts V *et al.* The role of cell-derived oligomers of Abeta in Alzheimer's disease and avenues for therapeutic intervention. Biochem Soc Trans 2005; 33(Pt 5):1087-90.

[124] Walsh DM, Selkoe DJ. A beta oligomers - a decade of discovery. J Neurochem 2007; 101(5): 1172-84.

[125] Abramova NA, Cassarino DS, Khan SM, Painter TW, Bennett JP Jr. Inhibition by R(+) or S(-) pramipexole of caspase activation and cell death induced by methylpyridinium ion or beta amyloid peptide in SH-SY5Y neuroblastoma. J Neurosci Res 2002; 67: 494-500.

[126] Koch G, Di Lorenzo F, Bonnì S, Giacobbe V, Bozzali M, Caltagirone C, Martorana A. Dopaminergic Modulation of Cortical Plasticity in Alzheimer's Disease Patients. Neuropsychopharmacology 2014; 39: 2654–61.

CHAPTER 4

The Role of Oxytocin in Neuropsychiatric Disorders

Donatella Marazziti*, Valentina Falaschi, Stefano Baroni, Mario Catena Dell'Osso and Liliana Dell'Osso

Dipartimento di Medicina Clinica e Sperimentale, Section of Psychiatry, University of Pisa, Pisa, Italy

Abstract: Oxytocin (OT) is a neurohypophysial hormone synthesized in the paraventricular and supraoptic nuclei of the hypothalamus. Although OT-like substances have been identified in all vertebrates, OT has been found only in mammals where it plays a major role in the onset and maintaining of behaviors which are typical of these animals, such as labour and lactation. Several data have suggested the involvement of OT in the formation of infant attachment, maternal behavior, pair bonding and, more generally, in linking social signals with cognition, behaviors and reward.

The aim of this paper was to review critically the role of OT in the regulation of different physiological functions and complex behaviors, as well as its possible involvement in the pathophysiology of some neuropsychiatric disorders. MEDLINE and PubMed (1972-2014) databases were searched for English language articles by using the following keywords: oxytocin, physiology, cognitive functions, attachment, psychopathology, psychiatric disorders. Papers were examined that addressed the following aspects of the OT system: synthesis and localization, receptors, physiology. In addition, latest findings showing abnormalities of OT and OT system in several neuropsychiatric disorders, including autism, obsessive-compulsive disorder, eating disorders, addiction, schizophrenia, post-traumatic stress disorder and Prader-Willy syndrome, will be also discussed together with the possible clinical use of OT or its analogues and/or antagonists.

Keywords: Addiction, anxiety, attachment, autism, depression, eating disorders, lactation and parturition, neuropeptides, neuropsychiatric disorders, obsessive-compulsive disorder, oxytocin, physiology, post-traumatic stress disorder, receptors, schizophrenia, sexual behavior, social behaviors, stress-related behavior, synthesis, vasopressin.

INTRODUCTION

The dawn of the neuropeptide research dates back to the end of the 19th century,

***Corresponding author Donatella Marazziti:** Dipartimento di Medicina Clinica e Sperimentale, Section of Psychiatry, University of Pisa, Via Roma, 67 56100 Pisa, Italy; E-mail: dmarazzi@psico.med.unipi.it

Atta-ur-Rahman, Allen B. Reitz, M. Iqbal Choudhary and Jizhou Wang (Eds.)

when it was reported the existence of a neural pathway from the supraoptic (SON) nucleus of the hypothalamus to the posterior pituitary [1]. In 1928 Scharrer described the so-called "glandular cells" in the fish hypothalamus, to indicate some hypothalamic neurons which could secrete substances *via* exocytosis of cytoplasmic vesicles, according to the concept of neurosecretion [2]. About two decades later, Du Vigneaud [3] isolated oxytocin (OT) in pituitary extracts that became the first peptide hormone to have its amino acid sequence identified and to be synthesized in its active form. This paper aimed to provide a comprehensive review of the OT system and of its physiological role in the regulation of different functions and behaviors, as well as of its possible involvement in some neuropsychiatric disorders.

SYNTHESIS AND LOCALIZATION

The neurohypophysial hormones are classified into the vasopressin (AVP) and OT families on the basis of the presence, at the position 8 of the chain, of a basic (AVP family) or a neutral (OT family) amino acid, which enables AVP and OT peptides to bind to their respective receptors [4]. OT, a very abundant nonapeptide, is constituted by a cyclic part of six-amino acid with a disulfide bridge between Cys 1 and 6 and a three-residue tail alfa-amidated at the COOH-terminal. The evidence that all vertebrates possess at least a OT-like and a AVP-like peptide suggested the existence of two evolutionary molecular lineages: the isotocin-mesotocin-OT line, implicated in reproductive functions, and the vasotocin-vasopressin line, involved in the water homeostasis. On the contrary, OT and AVP, that differ from each other in terms of two amino acids (Ile vs Phe at position 3 and Leu vs Arg at position 8, respectively), have been found only in mammals and probably have developed in parallel with typical mammalian behaviors, such as uterine contraction during labour and milk ejection essential for lactation. OT and AVP are mainly synthesized in the magnocellular neurons of the SON and paraventricular (PVN) nuclei of the hypothalamus, the most important nuclei of the hypothalamicneurohypophysial system [5]. OT and AVP are assembled on ribosomes at the level of the soma of the neurons as precursors which are subsequently processed in the neurosecretory vesicles. OT is synthesized as preprooxytocin, its largest precursor that comprises three components: a signal sequence of about 16-30 amino acid residues at the neuropeptide terminal, the neuropeptide sequence and the space parts. The signal and the space parts allow the precursor to deal with the endoplasmatic reticulum and, after the storage into the vesicles, it is rapidly cleaved and splitted [6]. During the intravescicular post-translational processing, OT precursor undergoes sequential proteolytic cleavage and other enzymatic modifications, such as glycosylation, phosphorylation, acetylation and amidation that lead to the three final products: OT, neurophysin and a carboxy-terminal glycoprotein. During this complex maturation process, OT is targeted along the axon to the posterior pituitary [7]. Within the neurohypophysis each axon produces several nerve terminals that constitute about 50% of the total volume of the neural lobe. Once activated, magnocellular neurons release into the blood OT and its transporting proteins, so that they can produce their effects on receptors located in distant target organs, such as mammary gland and kidney. Several other biologically active substances, including neuropeptide Y, tyrosine hydroxylase, dynorphin, thyrotropin-releasing hormone, atrial natriuretic factor, galanin and nitric oxide (NO) synthase, are co-released with OT and AVP from magnocellular neurons, however, the reciprocal effects between them and OT are still unknown [8, 9]. Oxytonergic magnocellular axons do not reach only the posterior pituitary, but also terminate in the arcuate nucleus, the lateral septum, the medial amygdaloid nucleus and the median eminence [10]. In the magnocellular SON and PVN nuclei, OT is also locally released from dendrites acting as self-neuromodulators: the intranuclear release is fundamental for the synchronization of the depolarization of OT neurons during lactation, or for the positive feed-back on OT dendritic release in the SON during parturition [11]. In fact, the somatodendritic release occurs in response to a variety of stimuli, including suckling, parturition, dehydration, hemorrhage, fever, physical restraint, pain, mating and territorial marking behaviors, administration of hypertonic solutions or pharmacological challenges. Hypothalamic OT can also reach the anterior pituitary through the hypothalamic-pituitary portal vascular system; at this level, OT might influence anterior pituitary hormones, while acting as a regulating factor. In rats, OT is released into the portal vessels and specific OT receptors are present in the adenohypophysis. OT seems to be involved in the regulation of the release of different adenohypophysial hormones, in particular prolactin, adrenocorticotropic hormone (ACTH) and gonadotropins. It is unclear, however, whether OT released during lactation, is responsible for the concomitant secretion of prolactin from the adenohypophysis, since it is supposed to act as a prolactin-releasing factor only when the dopamine levels are low, like during periods of dopamine withdrawal that characterize the onset of prolactin secretion. Moreover, pituitary OT receptor gene expression is restricted to lactotrophs and increases at the end of gestation [12]; these data suggest that OT might function as a prolactin-releasing factor only around the end of gestation. The endocrine response to stress is mediated by the activation of the hypothalamic-pituitary-adrenal axis (HPA), in particular, corticotropin- releasing hormone (CRH) and AVP stimulate ACTH secretion from the anterior pituitary. In rats, OT has been demonstrated to potentiate the release of ACTH induced by CRH: in fact, if CRH is responsible for the immediate secretion of ACTH following an acute stress, when CRH levels begin to decrease during prolonged stress, the persistent level of OT in the median eminence seems to be related to the delayed ACTH response and the generation of ACTH pulsatile secretory bursts [13]. On the contrary, in humans, OT infusion inhibited the plasma ACTH responses to CRH, and suckling and breast stimulation increased and decreased, respectively, plasma OT and ACTH levels; these evidences would indicate an inhibitory influence of OT on ACTH secretion. Luteinizing hormone (LH) secretion from the adenohypophysis is primarily regulated by the gonadotropin-releasing hormone (GnRH). OT has been demonstrated to stimulate LH release: its administration to proestrous rats can advance the LH surge with an earlier ovulation. Moreover, OT seems to sensitize the pituitary before full GnRH stimulation: in women, pre-ovulatory OT administration leads to the onset of the mid-cycle LH surge [14]. Overall, the physiological relationships between OT and LH have yet to be clarified [15]. OT neurons are also localized in the dorsal-caudal part of PVN and are called parvicellular given their smaller size, as compared with that of the magnocellular neurons. Their axons are part of the descending tract directed to the sympathetic centers of the spinal cord and to the parasympathetic caudal autonomic centers, including the dorsal motor nucleus of the nervus vagus and the nucleus of tractus solitarii [16, 17]. A peripheral synthesis of OT has also been demonstrated in placenta, uterus, corpus luteum, amnion, testis and heart.

OXYTOCIN RECEPTORS

A single population of OT receptors has been identified in both the brain and peripheral organs [18]. They belong to the class I of G protein-coupled receptor family and are primarily coupled to phospholipase C-beta which, once activated, leads to the generation of 1,2-diacyl-glycerol and inositol trisphosphate. These latter, in turn, promote the Ca^{2+} release from the intracellular storages and the activation of the protein kinase of type C with a final increase of intracellular Ca^{2+} that may trigger several cellular events, such as smooth cell contraction, changes of cellular excitability, modifications of gene transcription and protein synthesis. Progesterone has been demonstrated to bind and inhibit OT receptors in the rat, and in humans, too, a progesterone metabolite, the 5 beta-pregnane- 3,20-dione, seems to act as a negative allosteric modulator, while cholesterol and Mg^{2+} or Mn^{2+} would function as positive ones. The brain distribution of OT receptors shows a wide interspecies variability. In the rat, they are present in the olfactory system, basal ganglia, thalamus, lymbic system (bed nucleus of the stria terminalis, central amygdaloid nucleus, ventral subiculum), hypothalamus (ventromedial nucleus), brain stem and spinal cord with age-related changes in their density. In the rabbit, no receptors have been found in the ventral subiculum of the hippocampus or in the hypothalamic ventromedial nucleus. In humans, OT binding sites have been mainly found in the pars compacta of substantia nigra and globus pallidus, as well as in the anterior cingulate and medial insula [19], but not in the hippocampus, amygdala, entorinal cortex and olfactory bulb. Theoretically, the different distribution of OT and OT receptors throughout the brain within each species might be related to the variety of functions potentially regulated by them, but attempts to relate anatomy to function are still inconclusive.

PHYSIOLOGICAL EFFECTS OF OXYTOCIN

Lactation and Parturition

The only irreplaceable role of OT is represented by the induction of milk production given its powerful galactokinetic effect. The cascade of events during parturition which leads to the activation of OT cells is not well studied in humans; in rats, at the time of delivery, the first step is the mechanical stimulation of the uterus that provokes, *via* the vagal and pelvic nerves, the activation of the nucleus tractus solitarii and the ventromedial medulla which, in turn, project to magnocellular neurons. Several projecting cells are noradrenergic [20] and depolarise OT neurons mainly *via* receptors [21] which seem to be fundamental in the bursting activity of OT cells during suckling [22]. During this activity, a dual mode of OT secretion has been identified: OT neurons secrete a large amount of neuropeptide into the blood and, in parallel, they secrete it in a paracrine manner from their dendrites of the SON to synchronize the bursts of the release [23]. Milk ejection would occur through the contraction of myoepithelial cells as a reflex response to the stimulation of the nipple by suckling, although other important factors, such as the sight, smell and sound of the baby, would facilitate it [24]. The properties of OT cells change also during parturition: electrophysiological studies showed the presence of a positive feedback mechanism during delivery similar to that occurring during suckling [25].

Social Behavior

OT seems to play an important role in the initiation and maintenance of complex social behaviors [26-28]: the most intriguing data will be reviewed, according to Harlow's classification of social bonds (pair attachment, parental attachment, infant attachment) [29].

Pair Attachment

Little is known on the neurobiological mechanisms regulating the development of adult-adult pair. A natural animal model for studying the neural substrates of pair bonding is provided by prairie (Microtus Ochrogaster) and montane voles (Microtus Montanus) which shares several characteristics, but differs significantly in social behaviors [30]. The prairie vole, a mousesized rodent, is usually found in multigenerational family and shows the classic features of monogamy: a breeding pair lives in the same nest and territory, males, too, take care of infants and intruders of either sex are refused [31]. On the contrary, montane voles live in isolated conditions showing little interest in social contact, and are not monogamous. The most salient difference between the two species is represented by the neural distribution of OT receptors: the prairie voles show OT receptors in those brain regions associated with the reward system, such as the nucleus accumbens and prelimbic cortex, while suggesting that OT might have reinforcing properties selectively in these species. On the other hand, in the montane voles the OT receptors are mainly distributed in the lateral septum and are supposed to be responsible for the effects of OT on self-grooming behavior observed only in this species. Similar variability in receptor distribution have also been observed in other voles (pine and meadow ones) selected for differences in social organization (*i.e.*, monogamous *versus* non-monogamous) [32]. Moreover, after parturition, the pattern of OT receptor binding observed in the female montane voles changes and becomes similar to that observed in the parental prairie voles. Recently, elevated levels of the c-fos protein immunoreactivity have been detected in the medial and cortical amygdala, medial preoptic area, and bed nucleus of the stria terminalis in female voles exposed to mating. In parallel, studies in OT knock-out mice showed that the wild-type and OT knock-out animals had a similar pattern of neuronal activation, except that in the medial amygdala where only the wild-type exhibited an induction of c-fos [33]. The OT knock-out mice showed, however, a dramatic increase of c-fos in the somatosensory cortex and the hippocampus which suggests alternative processing of social cues. In any case, OT seems to produce different effects in male and female prairie voles: central OT administration in females, but not in males, facilitates the development of a partner preference in the absence of mating [34], while OT antagonists given before mating block the formation of a partner preference [35]. In humans, OT administration has been demonstrated to increase the trust towards the others, possibly through the involvement of the amygdala, the main component of the neurocircuit of fear and social cognition which highly expresses OT receptors [36]. Some functions of human amygdala seem to be modulated by OT: a functional magnetic resonance imaging study showed that OT reduced significantly the activation of amygdala and its coupling to brain regions implicated in autonomic and behavioral responses to fear [37]. The property of OT to facilitate the formation of social bonds has been related to the improvement of the inference of the affective mental state of the other subjects, which, in turn, would lead to a reduction of the ambiguity experienced during a social interaction with subsequent decreases of anxiety levels [38]. This is in line with previous findings showing a reduction of autonomic response to aversive pictures under OT [39]. It is worth noticing that OT administration did not affect self-report scales of psychological state regarding anxiety and mood [36]; however, it seems that the presence of a social interaction is necessary to elicit the OT effect, since it would become evident only in the social context, but not when subjects rate themselves in isolation. Moreover, the decrease of amigdala activation after OT administration was more significant for socially relevant stimuli (faces) than for socially less relevant ones; this suggests that social and non-social fear may depend on different neural systems [40]. In addition, it has been observed that a critical requirement for the formation of pair bonds is the ability of animals to identify conspecifics [41]. Both AVP and OT seem to play a critical role in social learning and recognition, the so-called social memory, with opposite effects. OT seems to be mainly involved in acquisition rather than in consolidation of social bonds: in rats, on the basis of olfactory stimulus, OT can lead to the onset of a partner preference [42]. OT knock-out mices are not able to recognize previously encountered conspecifics and do not show any attachment behavior, while central OT administration, before the first contact, can restore normal attachment behaviors [43]. The regulation of social bonding also requires the modification of feelings and behaviors that may influence the likelihood of an encounter, for example, anxiety and novelty avoidance may reduce the likelihood of approaching a conspecific; while OT decreases anxiety-like behaviors in males, AVP increases them [44]. The opposite effects produced by OT and AVP may be explained by the need of a gender-specific modulation of different behaviors. Although OT influences sexual and social interactions in both males and females, the onset of maternal behaviors is fundamental in females [18] and requires the inhibition of novelty avoidance, the suppression of prior social avoidance learning and a decrease of aggression. On the contrary, AVP promotes behavioral modifications leading to the establishment of territories and dominant hierarchy characteristics of males.

To date only one study investigated the role of OT in behavior and physiology in human couple interaction. In this double-blind placebo-controlled study 47 heterosexual couples received OT or placebo intranasally before a standard instructed couple conflict discussion. During the couple conflict OT administration was significantly associated to an increase of positive communication behavior in relation to negative behavior. After the conflict, subjects treated with OT showed a reduction of salivary cortisol levels compared to those treated with placebo. These results would support the hypothesis that also in humans OT would play a critical role in couple interaction and in the development of adult-adult pair bonds [45].

Sex related difference have been reported in the OT system [46]. In humans intranasally applied OT improved kinship recognition only in women, and competition recognition [46]. Moreover, intranasally applied OT reduces amygdala activity in men, but increases it in women when viewing emotional facial expressions [38, 47]. These findings suggest that the OT system functions differently in the two sexes.

Parental Attachment

The onset of maternal behaviors is critical for the survival of mammals. Rats represent an ideal model for studying the maternal care: nulliparous female rats do not show any interest in infants until the parturition, when a dramatic change in motivation occurs and typical maternal features, such as nest building and retrieval of pups, became established [48]. It has been demonstrated that OT injection in the lateral ventricles of nulliparous ovariectomized rats may induce maternal behaviors [49]. Moreover, the central injection of OT antagonists or lesions of OT-producing magnocellular neurons in the hypothalamus resulting in the blockade of OT neurotransmission, inhibit their onset. These evidences, coupled with the fact that, once a female has become maternal, OT antagonists have no effect, may indicate that OT is critical for the onset, but not for the maintenance of these behaviors [50]. On the contrary, the cutaneous injection of the nitric oxide (NO) sodium nitroprusside has been shown to prolong parturition and inhibit maternal behavior in rats. Given the demonstration that OT may restore intrapartum maternal behavior, NO was suggested to interfere with its development by inhibiting the OT release. The neurobiological mechanisms underlying OT-related onset of maternal behaviors are still unclear, but it seems crucial in this process the increase of OT receptors in the bed nucleus of the stria terminalis and the ventromedial nucleus of the hypothalamus that occurs just before parturition [51]. In humans, little is known on the influence of OT in maternal behaviors: in 1974 it was reported that breast-feeding within 1 h of birth, when OT levels are high, might contribute to a long-lasting mother-infant bond with beneficial effects on the development of the child [52] but this finding was never replicated.

Infant Attachment

Since the birth, infants begin to bond to their caregivers and this relationship influences emotional functions throughout the whole life span. In mammals, the mother-infant interaction and the early social experiences occurring during the post-natal period have profound behavioral effects which, in turn, produce long-lasting changes in the brain. OT is considered to be one of the potential candidates involved in the transduction of early experiences (birth process, breast-feeding and other aspects of parent-infant interactions) into physio- (patho-) logical changes, including brain growth, later stress reactivity and ovarian disorders [53]. In humans, the deprivation of the normal parental cares has been recently shown to produce long-lasting changes in the sensitivity to OT during adulthood [54] and to alter the development of children's OT and AVP systems, which interfere with the comforting effects that typically emerge between children and familial adults who take care of them; in fact, OT and AVP levels are increased by socially pleasant experiences, including comforting touches and smells [55]. In animals, the increased blood levels of these hormones may induce positive social interactions, including the formation of social bonds and of their memories, as well as of selective infant-parent attachments. OT seems to be critical in the infant's attachment response, as this peptide was demonstrated to facilitate a rapid conditioned association to maternal odor cues [56]. Moreover, OT administration reduces the separation response of the rat pups, consistently with the role of this peptide in either attachment or separation response [34, 57]. Interestingly, OT receptors have been found in the developing brain with a transient but marked "overproduction" (relative to the adult) in the limbic areas in the first two postnatal weeks [58, 59]. In addition, as already discussed above, OT receptors are present in the reward circuit that includes the nucleus accumbens, the cortex, the thalamus and the pallidus, and which, during infant development, has been implicated in the regulation of that sense of safety and protection which makes social and parental interactions highly rewarding.

Stress-Related Behavior

OT is released from the pituitary gland in response to different stressful stimuli, including pain, conditioned fear and exposure to novel environments; noradrenergic neurons containing prolactin-releasing peptide seem to be, at least in part, responsible for the stress-related OT secretion into the circulation [60]. In rats, the acute exposure to immobilization stress resulted in increased OT-mRNA levels, while both forced swimming and shaker stress provoked a raise of central and plasma OT concentrations. OT would facilitate the activation of the HPA axis by increasing the glucocorticoid release. Along this line, it is supposed that the stress-induced central release of OT can ameliorate some stress-related disorders, such as depression and anxiety: in mice and in rats, OT showed anxiolytic properties in estrogen-treated females possibly mediated by its influence on dopaminergic neurotransmission in the limbic areas. Given that stress and anxiety impair maternal caretaking, a reduced stress responsiveness during lactation appear to be adaptive for both mother and infant. Consistently with these observations, lactating women showed reduced hormonal responses to exercise stress, as compared with postpartum women who bottle-feed their infants [61]. Furthermore, some women with panic disorder have been demonstrated to experience a reduction of their anxiety symptoms during lactation [62].

Memory and Learning

OT and AVP are involved in several cognitive functions. AVP seems to reinforce memory, whereas OT has the opposite effect, as it reduces learning processes and memory. In animals, OT was demonstrated to facilitate the extinction of avoidance reaction and to attenuate the storage of verbal memory. In cultured neurons, OT is able to reduce the activity of N-methyl-D-aspartic acid receptors, which are one of the most important substrates for the induction of learning and memory [63].

Sexual Behavior

OT has been strongly related to sexual behavior and functions in mammals, including humans. Plasma OT levels begin to increase during sexual arousal and are significantly higher during orgasm than at baseline in both men and women [27, 28]. In men, AVP concentrations increased significantly during arousal and returned to basal levels at the time of ejaculation, while mean plasma OT rose about five-fold during ejaculation and returned to basal concentrations within 30 minutes [64]. In male rats OT can stimulate penile erection [65], while in females it is secreted during vaginal stimulation [66]. In both men and women, the level of muscular contractions during orgasm was highly correlated with OT plasma levels [67], while suggesting that some of OT effects may depend on its ability to stimulate smooth muscles contraction in the genital-pelvic area. In addition, intranasal OT administration seems to enhance the sexual arousal and orgasm intensity: interestingly, a woman who had used a synthetic OT spray, reported to experience a marked increase of sexual desire associated with intense vaginal transudate [68, 69]. Overall, during sexual arousal, OT seems to act peripherally on reproductive organs and activates the sexual functions in both women and men. Beyond its peripheral effects on reproductive organs, OT might also sensitize the neurons responsible for the cognitive feelings of orgasm, while representing a physiological substrate for both sexual behavior and performances. OT appears to fluctuate throughout the menstrual cycle in normally cycling fertile women with a regular sexual activity and not taking any oral contraceptives; plasma OT levels are significantly lower during the luteal than during the ovulatory or follicular phases [70, 71]; in addition, a suppression of the endogenous OT activity has been described during the ovulation [72].

OXYTOCIN AND NEUROPSYCHIATRIC DISORDERS

A few data suggest a link between OT and neuropsychiatric disorders, in particular obsessive-compulsive disorder, eating disorders, addiction, post-traumatic stress disorder, anxiety, depression, schizophernia, autism and Prader-Willy syndrome. Although most of them should be considered at a level of suggestions, nevertheless they are intriguing and would indicate the need of further research in this promising area.

Obsessive-Compulsive Disorder

Obsessive-compulsive disorder (OCD) is a common condition characterized by typical symptoms represented by obsessions and/or compulsions that bear some relationship with OT. Scattered findings suggest that also the neuropeptide oxytocin (OT) might be involved in the pathophysiology of OCD [73, 74, 114] both *per se* and given its interactions with the 5-HT system [75].

As already shown, OT may influence physiological activities, including memory acquisition, maintenance and retrieval, as well as grooming, maternal and sexual behaviors, which may be related to some OCD features. Furthermore, OT receptors have been identified in some brain areas, which have also been implicated in the pathophysiology of OCD [76-79]. In animals, the central OT microinjection produces a significant increase of grooming behavior [80, 81] which is considered a model of compulsions, as cleaning behaviors are prototypical symptoms in OCD patients [82-84]. The fact that violent, horrific and sexual thoughts, images, and impulses are common types of obsessions [28, 85, 86] and that central OT injections in animals is associated also with sexual behaviors and increased aggression [87] may be considered another cue of OT involvement in OCD. However, the most consistent data derive mainly from the evidence that pregnancy and the postpartum period show an increased risk for the onset of a subtype of OCD [87-91] characterized by contamination obsessions, as compared with the disorder prior to pregnancy. It was, therefore, suggested that a subgroup of women (between 11% and 47%) are vulnerable to the induction or exacerbation of OCD after the exposure to the elevated levels of OT, such as those occurring during the pregnancy [92, 93]. The direct measurement of OT parameters showed increased OT levels in the cerebrospinal fluid (CSF) of adults with OCD and Tourette'syndrome, as compared with healthy controls [94], and correlating with the current severity of OCD. However, a negative report has been also published [95]. Controversial results came from the attempts to administer OT to OCD patients in order to improve avoidance behaviors associated with the disorder [96-100], so that further data are necessary to draw firm conclusions on the potentiality of OT or its analogues as antiobsessional treatment. Ansseau *et al.* [90] reported the case of a patient who seemed to improve with 8.4-16.8 IU of OT intranasally administered over a 4-week period. On the contrary, others did not replicate this finding, as perhaps approximately 0.002% of the peripherally applied amount of OT has been demonstrate to reach the CNS in animals.

A positive correlation between OCD and OT was noted in a group of OCD patients irrespective of the subtype [101] or only in non-tic related OCD [88]. A study of our group that is then submitted showed an opposite find (Marazziti *et al.* submitted)

To summarize, evidence of OT involvement in OCD is unclear and further studies are needed to explore its role.

Eating Disorders

Eating disorders are characterized by abnormal eating behaviors, body image alterations, impulse and mood disturbances, as well as by several neuroendocrine dysfunctions, including those of the HPA axis [102]. As discussed, OT and AVP have been demonstrated to influence feeding behavior [103] and a few data show preliminary alterations of OT and AVP in eating disorders [104-107]. The serum activity of the prolyl endopeptidase (PEP), an enzyme that cleavages many active behaviorally active neuropeptides, such as AVP, thyrotropin releasing hormone, substance P, OT, bradykinin, neurotensin and angiotensin [108] resulted decreased in both bulimic and anorectic patients [109], while the CSF OT levels were found to be lowered in restricting anorectic patients only [107]. In any case, as these alterations tend to normalize after weight restoration, they are interpreted as secondary to malnutrition and abnormalities in fluid balance [109, 110]. However, elevated CSF AVP levels with normal OT concentrations were found in recovered anorectic patients, while suggesting that this persistent abnormality might be trait-related and potentially contribute to the pathophysiology of the disorder [111]. Moreover, autoantibodies against OT were described in both anorexia and bulimia nervosa and would indicate that immune dysfunctions may also be involved [112].

Addiction

Abuse drugs act by enhancing the brain reward mechanisms: a crucial drug-sensitive component of this circuit is represented by the mesolimbic dopaminergic system which is under the modulatory control of several neurotransmitters and hormones [113]. OT may act at this level and could be involved in the development of tolerance and dependence towards abuse substances including opiates and cocaine [114]. Given that adaptation and learning are likely to be implicated in the neural events leading to drug tolerance and dependence [115], OT is supposed to modulate dopamine in the reward circuit. In mice, OT has been demonstrated to inhibit the onset of tolerance to morphine [116] and to attenuate the severity of symptoms of morphine withdrawal [114]: moreover, after OT administration, rats showed a reduction of intravenous self-administration of heroin [117]. OT attenuated also the cocaine-induced hyperactivity and inhibited the behavioral tolerance to the effect of this drug, while facilitating the development of behavioral sensitization [118-120]. In humans, cocaine abuse during pregnancy seems to be associated with lower OT levels, depression and hostility [121]. As far as ethanol is concerned, OT was shown to inhibit the development of tolerance to ethanol in mice [122]; acute alcohol administration inhibits OT secretions [123], while its chronic use stimulates it [124]. It has also been hypothesized that OT might be involved in the cognitive dysfunctions observed in alcoholics [124, 125].

Post-Traumatic Stress Disorder

OT attenuates memory consolidation and retrieval, facilitates the extinctions of an activated avoidance response and decreases passive avoidance behavior [126, 127]. The intranasal OT administration was shown to reduce the memory retrieval and conditioned response in patients with post-traumatic stress disorder (PTSD) [39]. It seems that alterations of the OT system following severe early stress and abuse experiences may interfere with the brain development and increase the subsequent risk of developing PTSD and, more in general, psychiatric disorders [128]. Moreover, the activity of PEP (prolyl endopeptidase) was found to be increased in PTSD patients and particularly in those with a concomitant major depression: it was, therefore, hypothesized that increased PEP activity might play a role in the pathophysiology of behavioral and affective symptoms of PTSD through an increased degradation of different neuropeptides [129].

Other Anxiety Disorders

In humans, OT generally facilitates social interactions and feelings of attachment and is becoming increasingly established as a "prosocial neuropeptide", with therapeutic potential in treatment of social, cognitive, and mood disorders [130-136]. A growing body of evidence implicates OT in mediation of complex social behaviors. In fact, in highly social species, OT has been shown to increase social approach behavior and pair bonding by reducing behavioral and neuroendocrine responses to social stress, and it is also suggested to mediate the rewarding aspects of attachment [137, 138].

Given the established effects of OT on affiliative behaviors in non-human mammals, several speculations on its possible role in human emotions and relationships were proposed [41]. In mothers, OT levels have been demonstrated to positively relate with sociality, calm and tolerance [139] and with a reduction of the incidence of stress and anxiety disorders [58]; in addition, pregnancy, a period characterized by increased OT levels, is considered to protect from some anxiety disorders, including panic disorder. OT, which is released during stress, seems to be an important modulator of anxiety and fear response, mainly with anxiolytic effect [140-142]. Dysfunctions of the amygdala, which is implicated in the biological response to danger signals in social interaction, have been detected in depression and anxiety disorders: interestingly, its function is strongly modulated by OT, as its intranasal administration may reduce amygdala activation and its coupling to the brain regions involved in the autonomic and behavioral response to fear [37]. Recently, a downregulation of OT receptors has been related to the pathophysiology of social anxiety disorder that might explain the cognitive misappraisals typical of the patients affected by this condition [37].

However, no significant differences in OT level were found in patients with SAD, as compared with healthy control subjects; in the SAD patients, however, higher OT levels were associated with higher social anxiety symptom severity and dissatisfaction with social relationships [143].

These findings support the notion that OT is a prosocial hormone and, as such, primarily involved in SAD. However, research in this field in humans are just at its dawn and there is a great need for data in clinical samples.

In a recent study, OT cerebrospinal fluid (CSF) concentrations were measured in 22 healthy women. Exposure to maltreatment, in particular emotional abuse, was associated with decreased CSF OT concentrations. The number, severity and duration of the abuse, as well as current anxiety ratings, were inversely related to CSF OT concentrations, while suggesting that alterations in the OT system may be involved in the adverse outcomes of childhood trauma [144].

Depression

Since OT has been shown to decrease stress response and anxiety levels, as well as to modulate cognitive functions and promote positive social relationships, some symptoms of depression, in particular, social withdrawal, cognitive impairment, appetite modifications and stress reactivity [145], have been related to OT [146]. The first studies carried out in CSF showed no difference in OT levels between depressed patients and control subjects [147, 148], while an increased density of AVP- and OT-expressing neurons was detected in the PVN nucleus postmortem of patients with both major unipolar and bipolar depression [130]. The few data concerning OT in plasma have shown decreased levels of the neuropeptide in depressed patients [149], but also no difference in a larger group of patients, as compared with healthy subjects [150]. More recently, a significant and negative correlation was found between plasma OT and scored symptoms of depression and anxiety in 25 patients affected by major depression [151]. The OT abnormalities in depression, although requiring further support, would suggest that they may be related also to dysregulation of the HPA axis reported in this condition, together with that of multiple neurotransmitters and modulators acting at this level.

Animal studies supported a role of OT in depressive-like behaviors. Long-term peripheral OT administration among female praire voles reversed anhedonia elicited by persistent isolation [152]. Circulating OT levels were reduced in a sample of patients with major depression and during depressive episode amongst women diagnosed with bipolar affective disorder [153]. In addition CSF OT concentrations did not differ from those in healthy controls [154]. Moreover, several reports indicated that depressed patients of both sexes were characterized by elevtaed baseline OT levels and greater OT variability than healthy control subjects [132, 155, 156].

It is uncertain why plasma OT levels in depressed individuals are reduced in some studies, but elevated in others. A recent study reported that depressed women displayed lower plasma OT concentrations compared to controls, while depressed men showed a trend in the opposite direction [157].

Lower plasma OT concentrations were associated with increased risk for later post partum depression and women experiencing post partum depression displayed lower levels of this hormone [158]. Depression is often associated with cognitive processing biases towards negative stimuli and OT treatment could potentially influence this disorder by affecting sensitivity to social cues [159]. Amongst individuals with high depression scores, oxytocin attenuated the attentional bias that otherwise existed in relation to masked angry faces [160]. In fact it is thought that the inability to inhibit the influence of negative stimuli on cognitive and emotional responses contributes to major depression, which may be modulated by OT. Further other behavioral features associated with depressive disorders, such as anxiety and a tendency towards greater emotion-focused coping, are also influenced by OT [161]. Despite the evidence that OT might have the effect of diminuishing depressive mood, there are data suggesting that this hormone could actually exacerbate depressed affect. Depressed patients, for example, when presented with emotional faces after intranasal oxytocin administration, showed enhanced neuronal activaction within the superior frontal gyrus and insula, suggesting that this hormone can enhance neural representation of affective states [162]. When individuals with high depression scores received OT treatment, they were less able to ignore emotionally salient (sad) facies relative to placebo-treated individuals [163].

A study investigated the impact of antidepressant or electroconvulsive therapy in 40 depressed patients and in 32 healthy control subjects. Serum OT levels were decreased in the patients before and after treatment. In particular, women had significantly lower OT levels than the control subjects, whereas no differences were detected in man or between unipolar and bipolar depressed patients [153]. Taken all together these findings suggest that OT may play a role in depression, although it is still unclear whether it is at primary or secondary level.

Schizophrenia

Little is known on the relationship between the OT system and schizophrenia or, more generally, psychoses, because only a few data are available. One study showed that OT levels were increased in schizophrenic patients, as compared with healthy controls, particularly in those taking neuroleptics and that, in drug-free patients, they were significantly higher after three weeks of treatment [164]. In addition, a morphometric evaluation of neurophysin-immunoreactivity in the brain of untreated schizophrenic patients suggested the presence of an altered OT function [165].

Plasma OT levels were measured in patients with schizophrenia and in healthy control subjects after neutral and trust-related interpersonal interactions. Trust-related interactions were associated with increased OT levels in controls whereas this effect was absent in patients with schizophrenia. Moreover, the presence of low OT levels measured after trust-related interactions significantly predicted the negative symptoms of schizophrenia, but were not related to positive symptoms, depression, anxiety, and neuropsychological functions. This may suggest that decreased trust-related OT release is related to the negative symptoms, social withdrawal, isolation and flattened typical of schizophrenic patients [166].

Although a role of OT in schizophrenia cannot be excluded, further data are necessary to draw firm conclusions.

Autism and Related Disorders

Autism is a complex developmental disorder characterized by impairment in three behavioral domains including social interactions, language, communication and imaginative play, and reduced range of interests and activities [167]. As already shown, OT and AVP seem to be implicated in social skills [26-28, 168, 169] and abnormalities of their neural pathways have been proposed to underlie several aspects of autism including repetitive behaviors, cognitive and social deficits, early onset, and genetic loading [170, 171]. It has also been proposed that the central regulation and expression of OT and AVP may contribute to the higher prevalence of the disorder in male subjects. In fact, centrally active AVP has been related to increased vigilance, anxiety, arousal and activation, while OT seems to have opposite effects including reduced anxiety, relaxation, growth and restoration. On one hand, higher activity of AVP, due to an increased exposure to androgens, could contribute to the male vulnerability to autism, on the other hand, OT, which is estrogen-dependent and is higher in female subjects, especially during early development, may be protective [172]. There are only a few evidences of a direct involvement of OT in autism: decreased plasma OT levels have been reported in 29 autistic children, as compared with age-matched healthy control individuals and lower OT levels were associated to lower scores on social and developmental measures [173]. Moreover, an altered, extended form of OT, which is normally detected only during the fetal life, has been found in the blood of autistic children at higher levels than in normal subjects [174]; this fetal form of OT is considered to be less active than the adult OT and may interfere with the functioning of the OT system. It is possible that different genetic expression of OT or the enzyme implicated in its synthesis may be involved in the pathophysiology of autism and related disorders. In addition, animal models and linkage data from the genome screen in humans indicate that the OT receptor gene may be an excellent candidate for the susceptibility to autism [170, 175, 176, 177]. Two specific nucleotide polymorphisms of OT receptors, rs2254298 and rs53576, have been found in a significantly higher rate of autistic subjects than healthy individuals in a Chinese Han population [178]; the association has been replicated in a Caucasian sample of United States, but only for the rs2254298 polymorphism [179]. Another association study has recently confirmed that specific haplotypes of the OT receptor gene may confer the risk to develop autism, and a link with communication, daily living skills and socialization has also been reported [180].

In one study, family-based single-marker and haplotype association analyses with 22 SNPs in the OT receptor and its 5' region were performed in 100 families with autistic disorders on high-functioning level, including Asperger syndrome (ASD), high-functioning autism, and atypical autism. Significant associations with autism were found for one single SNP and one haplotype. Furthermore, ASD patients carrying the haplotype associated with autism showed significant impairments in comparison to noncarriers of the haplotype in social interaction and communication [181]. In a more recent association study carried out in a Japanese population, the associations between OT receptor gene and ASD was performed by analyzing 11 SNPs using both family-based association test and population-based case-control test. Although no significant signal was found in the family-based association test, significant differences were observed between patients and control subjects in allelic frequencies of four SNPs, including rs2254298. In addition, haplotype analysis exhibits a significant association between a five-SNP haplotype and ASD, including rs2254289 [182]. All these data seem to support that OT receptor gene may play a significant role in conferring the risk of ASD. Interestingly, in a recent study the epigenetic regulation of OT receptor gene has been implicated in the development of the disorder for the first time [183]. In fact, a genomic deletion containing the OT receptor gene previously implicated in autism was present in an autism proband and his mother. However, the proband's affected sibling did not harbor this deletion, but exhibited epigenetic misregulation of this gene through aberrant gene silencing by DNA methylation.

Another evidence of the possible involvement of OT in autism is that OT infusion has been recently shown to reduce repetitive behaviors in patients with autism spectrum disorders, as compared with placebo infusion; however, these preliminary results need to be confirmed by controlled trial [184].

In a double-blind, randomized, placebo-controlled, crossover design, OT nasal spray or a placebo were administered to 16 male ASD youth. In comparison with placebo, OT administration improved performance on the Reading the Mind in the Eyes Task. This is the first evidence that OT nasal spray improves emotion recognition in ASD young people, while suggesting its potential role in the early treatment of ASD patients [185].

Prader-Willy Syndrome

The Prader Willy syndrome (PWS) is a genetic disorder characterized by mental retardation, hypogonadism, short stature and distinctive dysmorphic features. In a postmortem study, a 42% reduction of OT-expressing neurons was described in the PVN nucleus of PWS subjects, as compared with healthy controls [186]. Increased OT CSF levels have also been found in PWS subjects, similarly to what described in OCD patients [82]: since the prevalence of OCD in PWS is quite high (45-50%) [187], taken together, these findings suggest that the two conditions might share a common dysfunction at the level of the OT system [188].

CONCLUSION

Oxytocin and the OT system are currently attracting an increasing interest and have become one of the main topics of several research lines [189]. Different data in both animals and humans suggest that OT plays a major role in the modulation of a broad range of functions, such as delivery and lactation, and of complex behaviors including memory, positive social bondings and stress decrease. Not surprisingly, OT has been implicated in the pathophysiology of different neuropsychiatric disorders, however data are scattered and the abnormalities described in the patients are quite meagre and, therefore, should be considered preliminary. This is also due to the fact that research in this area is strongly limited by the lack of reliable peripheral models of central OT to be used routinely in clinical samples. If substantiated by further data, however, the demonstration that OT abnormalities might contribute to the development of psychopathology, might open new horizons to novel and more “natural” treatment interventions.

ACKNOWLEDGEMENTS

Declared None.

CONFLICT OF INTEREST

The authors confirm that this chapter contents have no conflict of interest.

DISCLOSURE

This is an updated report of the earlier review article by D. Marazziti and M. Catena-Dell'Osso "The role of oxytocin in neuropsychiatric disorders". Curr Med Chem. 2008; 15(7): 698-704.

REFERENCES

[1] Ramon y Cajal S. The croonian lecture: La fine structure des centres nerveux. Proc R Soc 1894; 55: 444-468.
[2] Scharrer E, Scharrer B. Secretory cells within the hypothalamus. Res Publ Assoc Res Nerv Ment Dis 1940; 20: 170-94.
[3] Du Vigneaud V, Ressler C, Trippett S. The sequence of amino acids in oxytocin, with a proposal for the structure of oxytocin. J Biol Chem 1953; 205: 949-57.
[4] Barberis C, Movillac B, Durroux T. Structural bases of vasopressin/oxytocin receptor function. J Endocrinol 1998; 156: 223-9.
[5] Swaab DF, Pool CW, Nijveldt F. Immunofluorescence of vasopressin and oxytocin in the rat hypothalamo-neurohypophyseal system. J Neural Transm 1975; 36: 195-215.
[6] Holmgren S, Jensen J. Evolution of vertebrate neuropeptides. Brain Res Bull 2001; 55: 723-35.
[7] Arvan P, Castle D. Sorting and storage during secretory granule biogenesis: looking backward and looking forward. Biochem J 1998; 332: 593-610.
[8] Pretel S, Piekut D. Coexistence of corticotropin-releasing factor and enkephalin in the paraventricular nucleus of the rat. J Comp Neurol 1990; 294: 192-201.
[9] Xiao M, Ding J, Wu L, *et al.* The distribution of neural nitric oxide synthase-positive cerebrospinal fluid-contacting neurons in the third ventricular wall of male rats and coexistence with vasopressin or oxytocin. Brain Res 2005; 1038: 150-62.
[10] Pittman QJ, Blume HW, Renaud LP. Connections of the hypothalamic paraventricular nucleus with the neurohypophysis, median eminence, amygdala, lateral septum, and median periaqueductal gray: an electrophysiological study in the rat. Brain Res 1981; 215: 15-28.
[11] Neumann I, Douglas AJ, Pittman QJ, *et al.* Oxytocin released within the supraoptic nucleus of the rat brain by positive feedback action is involved in parturition-related events. J Neuroendocrinol 1996; 8: 227-33.
[12] Breton C, Pechoux C, Morel G, *et al.* Oxytocin receptor messenger ribonucleic acid: characterization, regulation, and cellular localization in the rat pituitary gland. Endocrinology 1995; 136: 2928-36.
[13] Boyle LL, Brownfield MS, Lent SJ, *et al.* Intensive venous sampling of adrenocorticotropic hormone in rats with sham or paraventricular nucleus lesions. J Endocrinol 1997; 153: 159-67.
[14] Hull ML, Reid RA, Evans JJ, *et al.* Pre-ovulatory oxytocin administration promotes the onset of the luteinizing hormone surge in human females. Hum Reprod 1995; 10: 2266-9.
[15] Evans JJ. Oxytocin and the control of LH. J Endocrinol 1996; 151: 169-74.
[16] Amico JA, Challinor SM, Comeron JL. Pattern of oxytocin concentrations in the plasma and cerebrospinal fluid of lactating rhesus monkeys (Macaca mulatta): evidence for functionally independent oxytocinergic pathways in primates. J Clin Endocrinol Metab 1990; 71: 1531-5.
[17] Palkovits M, Toth ZE, Gallatz K, *et al.* Decussations of the descending paraventricular pathways to the brainstem and spinal cord autonomic centers. Comp Neurol 1999; 414: 255-66.
[18] Gimpl G, Fahrenholz F. The oxytocin receptor system: structure, function, and regulation. Physiol Rev 2001; 81: 629-83.
[19] Brown LL, Veliskova J, Miller AM, *et al.* Regional neural activity within the substantia nigra during peri-ictal flurothyl generalized seizure stages. Neurobiol Dis 2005; 20: 752-9.
[20] Day TA, Sibbald JR. Direct catecholaminergic projection from nucleus tractus solitarii to supraoptic nucleus. Brain Res 1988; 454: 387-92.
[21] Yamashita H, Inenaga K, Kannan H. Depolarizing effect of noradrenaline on neurons of the rat supraoptic nucleus *in vitro*. Brain Res 1987; 405: 348-52.

[22] Crowley WR, Armstrong WE. Neurochemical regulation of oxytocin secretion in lactation. Endocr Rev 1992; 13: 33-65.
[23] Ludwig M. Dendritic release of vasopressin and oxytocin. J Neuroendocrinol 1998; 10: 881-95.
[24] Leng G, Caquineau C, Sabatier N. Regulation of oxytocin secretion. Vitam Horm 2005; 71: 27-58.
[25] Fuchs AR, Romero R, Keefe D, *et al.* Oxytocin secretion and human parturition: pulse frequency and duration increase during spontaneous labor in women. Am J. Obst Gynecol 1991; 165: 1515-23.
[26] Wakerley JB, Lincoln DW. The milk-ejection reflex of the rat: a 20- to 40-fold acceleration in the firing of paraventricular neurones during oxytocin release. J Endocrinology 1973; 57: 477-93.
[27] Carmichael MS, Humbert R, Dixen J, *et al.* Plasma oxytocin increases in the human sexual response. J Clin Endocrinol Met 1987; 64: 27-31.
[28] Carter CS. Oxytocin and sexual behavior. Neurosci Biobehav Rev 1992; 16: 131-44.
[29] Harlow HF, Mears C, Eds. The Human Model: Primate Perspectives. New York: John Wiley & Sons 1979.
[30] Insel TR. A neurobiological basis of social attachment. Am J Psychiatry 1997; 154: 726-35.
[31] Carter CS, DeVries AC, Getz LL. Physiological substrates of mammalian monogamy: the prairie vole model. Neurosci Biobehav Rev 1995; 19: 303-14.
[32] Insel TR, Wang ZX, Ferris CF. Patterns of brain vasopressin receptor distribution associated with social organization in microtine rodents. J Neurosci 1994; 14: 5381-92.
[33] Curtis JT, Wang Z. Ventral tegmental area involvement in pair bonding in male prairie voles. Physiol Behav 2005; 86: 338-46.
[34] Winslow JT, Hastings N, Carter CS, *et al.* A role for central vasopressin in pair bonding in monogamous prairie voles. Nature 1993; 365: 545-8.
[35] Insel TR, Winslow JT, Wang ZX, *et al.* Oxytocin and the molecular basis of monogamy. Adv Exp Med Biol 1995; 395: 227-34.
[36] Kosfeld M, Heinrichs M, Zack PJ, *et al.* Oxytocin increases trust in humans. Nature 2005; 435: 673-6.
[37] Kirsch P, Esslinger C, Chen Q, *et al.* Oxytocin modulates neural circuitry for social cognition and fear in humans. J Neurosci 2005; 25: 11489-93.
[38] Domes G, Heinrichs M, Michel A, *et al.* Oxytocin improves "mind-reading" in humans. Biol Psychiatry 2007; 61: 731-3.
[39] Pittman RK, Orr SP, Lasko NB. Effects of intranasal vasopressin and oxytocin on physiologic responding during personal combat imagery in Vietnam veterans with posttraumatic stress disorder. Psychiatry Res 1993; 48: 107-17.
[40] Meyer-Lindenberg A, Hariri AR, Munoz KE, *et al.* Neural correlates of genetically abnormal social cognition in Williams syndrome. Nat Neurosci 2005; 8: 991-3.
[41] Insel TR, Frenald RD. How the brain processes social information: searching for the social brain. Rev Neurosci Annu 2004; 27: 697-722.
[42] Popik P, Van Ree JM. Social transmission of flavored tea preferences: facilitation by a vasopressin analog and oxytocin. Behav Neural Biol 1993; 59: 63-8.
[43] Dantzer R, Bluthe RM, Koob GF, *et al.* Modulation of social memory in male rats by neurohypophyseal peptides. Psychopharmacology 1987; 91: 363-8.
[44] Bielsky IF, Young LJ. Oxytocin, vasopressin, and social recognition in mammals. Peptides 2004; 25: 1565-74.
[45] Ditzen B, Schaer M, Gabriel B, *et al.* Intranasal oxytocin increases positive communication and reduces cortisol levels during couple conflict. Biol Psychiatry 2009; 65: 728-31.
[46] Fischer-Shofty M, Levkovitz Y, Shamay-Tsoory SG. Oxytocin facilitates accurate perception of competition in men and kinship in women. Soc Cogn Affect Neurosci 2013; 8: 313-7.
[46] Fischer-Shofty M, Levkovitz Y, Shamay-Tsoory SG. Oxytocin facilitates accurate perception of competition in men and kinship in women. Soc Cogn Affect Neurosci 2013; 8: 313-7.
[47] Domes G, Lischke A, Berger C, *et al.* Effects of intranasal oxytocin on emotional face processing in women. Psychoneuroendocrinology 2010; 35: 83-93.
[48] Numan M. A neural circuitry analysis of maternal behavior in the rat. Acta Paediatr Suppl 1994; 397: 19-28.
[49] Pedersen CA, Prange AJ. Induction of maternal behavior in virgin rats after intracerebroventricular administration of oxytocin. Proc Natl Acad Sci USA 1979; 76: 6661-5.

[50] Skutella T, Weber T, Jirkowski GF. Coexistence of oxytocin and tyrosine hydroxylase in the rat hypothalamus, an immunocytochemical study. J Neural Transm Gen Sect 1993; 94: 55-61.
[51] Caldwell JD, Walker CH, Pedersen CA, *et al.* Estrogen increases affinity of oxytocin receptors in the medial preoptic area-anterior hypothalamus. Peptides 1994, 15: 1079-84.
[52] Kennell JH, Jerauld R, Wolfe H, *et al.* Maternal behavior one year after early and extended post-partum contact. Develop Med Child Neurol 1974; 16: 172-9.
[53] Carter CS. Developmental consequences of oxytocin. Physiol Behav 2003; 79: 383-97.
[54] Meinlschmidt G, Heim C. Sensitivity to intranasal oxytocin in adult men with early parental separation. Biol Psychiatry 2007; 61: 1109-11.
[55] Fries AB, Ziegler TE, Kurian JR, *et al.* Early experience in humans is associated with changes in neuropeptides critical for regulating social behavior. Proc Natl Acad Sci USA 2005; 102: 17237-40.
[56] Nelson E, Panksepp J. Oxytocin mediates acquisition of maternally associated odor preferences in preweanling rat pups. Behav Neurosci 1996; 110: 583-92.
[57] Tribollet E, Goumaz M, Raggenbass M, *et al.* Appearance and transient expression of vasopressin and oxytocin receptors in the rat brain. J Recept Res 1991; 11: 333-46.
[58] Shapiro LE, Insel TR. Ontogeny of oxytocin receptors in rat forebrain: a quantitative study. Synapse 1989; 4: 259-66.
[59] Insel TR, Winslow JT. Central administration of oxytocin modulates the infant rat's response to social isolation. Eur J Pharmacol 1991; 203: 149-52.
[60] Onaka T. Neural pathways controlling central and peripheral oxytocin release during stress. J Neuroendocrinol 2004; 16: 308-12.
[61] Altemus M. Neuropeptides in anxiety disorders. Effects of lactation. Ann NY Acad Sci 1995; 771: 697-707.
[62] Klein DF, Skrobola AM, Garfinkel R. Preliminary look at the effects of pregnancy on the course of panic disorder. Anxiety 1995; 1: 227-32.
[63] Caruso S, Agnello C, Campo MG, *et al.* Oxytocin reduces the activity of N-methyl-D-aspartate receptors in cultured neurons. J Endocrinol Invest 1993; 16: 921-4.
[64] Murphy CR, Dwarte DM. Increase in cholesterol in the apical plasma membrane of uterine epithelial cells during early pregnancy in the rat. Acta Anatom 1987; 128: 76-9.
[65] Argiolas A, Gessa GL. Oxytocin: a powerful stimulant of penile erection and yawning in male rats. Adv Biochem Psychopharmacol 1987; 43: 153-63.
[66] Keverne EB, Levy F, Poindron, P, *et al.* Vaginal stimulation: an important determinant of maternal bonding in sheep. Science 1983; 219: 81-3.
[67] Carmichael MS, Warburton VL, Dixen J, *et al.* Relationships among cardiovascular, muscular, and oxytocin responses during human sexual activity. Arch Sex Behav 1994; 23: 59-79.
[68] Anderson HM, Dennerstein L. Increased female sexual response after oxytocin. BMJ 1994; 309: 929.
[69] Anderson HM, Dennerstein L. Oxytocin and female sexuality. Gynecol Obs Investig 1995; 40: 217-21.
[70] Altemus M, Redine LS, Leong Y, *et al.* Responses to laboratory psychosocial stress in postpartum women. Psychosom Med 2001; 63: 814-21.
[71] Salonia A, Nappi RE, Pontilloc M, *et al.* Menstrual cycle-related changes in plasma oxytocin are relevant to normal sexual function in healthy women. Horm Behav 2005; 47: 164-9.
[72] Evans JJ, Reid RA, Wakeman SA, *et al.* Evidence that oxytocin is a physiological component of LH regulation in non-pregnant women. Hum Reprod 2003; 18: 1428-31.
[73] Marazziti D,Catena Dell'osso M. The role of oxytocin in neuropsychiatric disorders. Curr Med Chem 2008 15: 698-704.
[74] Scantamburlo G, Ansseau M, Geenen V, *et al.* Oxytocin: From milk ejection to maladaptation in stress response and psychiatric disorders. A psychoneuroendocrine perspective. Ann Endocrinol (Paris) 2009; 70: 449-54.
[75] Marazziti D, Baroni S, Giannaccini G, *et al.* A link between oxytocin and serotonin in humans: supporting evidence from peripheral markers. Eur Neuropsychopharmacol 2012, 22: 578-83.
[76] Rapoport JL, Wise SP. Obsessive-compulsive disorder: evidence for basal ganglia dysfunction. Psychopharmacol Bull 1988; 24: 380-4.
[77] Stahl SM. Basal ganglia neuropharmacology and obsessive-compulsive disorder: the obsessive-compulsive disorder hypothesis of basal ganglia dysfunction. Psychopharmacol Bull 1988; 24: 370-4.

[78] Modell JG, Mountz, JM, Curtis GC, *et al.* Neurophysiologic dysfunction in basal ganglia/limbic striatal and thalamocortical circuits as a pathogenetic mechanism of obsessive-compulsive disorder. J Neuropsychiatry 1989; 1: 27-36.

[79] Insel TR. Toward a neuroanatomy of obsessive-compulsive disorder. Arch Gen Psychiatry 1992; 49: 739-44.

[80] Van Wimersma Greidanus TB, Kroodsma JM, Po MLH, *et al.* Neurohypophyseal hormones and excessive grooming behaviour. Eur J Pharmacol 1990; 187: 1-8.

[81] Marroni SS, Nakano FN, Gati CD, *et al.* Neuroanatomical and cellular substrates of hypergrooming induced by microinjection of oxytocin in central nucleus of amygdala, an experimental model of compulsive behavior. Mol Psychiatry 2007; 12: 1103-17.

[82] Holzer JC, Goodman WK, McDougle CJ, *et al.* Obsessive-compulsive disorder with and without a chronic tic disorder. A comparison of symptoms in 70 patients. Br J Psychiatry 1994; 164: 469-73.

[83] Leckman JF, Grice DE, Barr LC, *et al.* Tic-related vs. non-tic-related obsessive compulsive disorder. Anxiety 1995; 1: 208-15.

[84] McDougle CJ. Possible role of neuropeptides in obsessive compulsive disorder. Psychoneuroendocrinology 1999; 24: 1-24.

[85] Jenike MA, Baer L, Minichiello WE. Obsessive-Compulsive Disorders. Littleton: Year Book Medical 1990.

[86] Winslow JT, Insel TR. Social status in pairs of male squirrel monkeys determines the behavioral response to central oxytocin administration. J Neurosci 1991; 11: 2032-8.

[87] Jenike MA. Obsessive-compulsive disorders in pregnancy and childbirth. Littleton: Year Book Medical 1990.

[88] Neziroglu F, Anemone R, Yaryura-Tobias JA. Onset of obsessive-compulsive disorder in pregnancy. Am J Psych 1992; 149: 947-50.

[89] Sichel DA, Cohen LS, Dimmock JA, *et al.* Postpartum obsessive compulsive disorder: a case series. J Clin Psychiatry 1993; 54: 156-9.

[90] Epperson CN, McDougle CJ, Brown RM, *et al.* Obsessive compulsive disorder during pregnancy and the puerperium. American Psychiatric Association New Research Abstract 1995; 112: 84

[91] Williams KE, Koran LM. Obsessive-compulsive disorder in pregnancy, the puerperium, and the premenstruum. J Clin Psychiatry 1997; 58: 330-4.

[92] Leckman JF, Mayes LC. Preoccupations and behaviors associated with romantic and parental love. Perspectives on the origin of obsessive-compulsive disorder. Child Adolesc Psych Clin North Am 1999; 8: 635-65.

[93] Albert U, Maina G, Bogetto F, Ravizza L. The role of recent life events in the onset of obsessive-compulsive disorder. CNS Spectr 2000; 5: 44-50.

[94] Leckman JF, Goodman WK, North WG, *et al.* Elevated cerebrospinal fluid levels of oxytocin in obsessive-compulsive disorder. Comparison with Tourette's syndrome and healthy controls. C J Arch Gen Psychiatry 1994; 51: 782-92.

[95] Altemus M, Jacobson KR, Debellis M, *et al.* Normal CSF oxytocin and NPY levels in OCD. Biol Psychiatry 1999; 45: 931-3.

[96] Ansseau M, Legros JJ, Mormont C, *et al.* Intranasal oxytocin in obsessive-compulsive disorder. Psychoneuroendocrinology 1987; 12: 231-6.

[97] Charles G, Guillaume R, Schittecatte M, *et al. Oxytocin* in the t*reatment* of *obsessive-compulsive disorder*: a *report* on t*wo cases*. Psychiatry Psychobiol 1989; 4: 111-15.

[98] Den Boer JA, Westenberg HG. Oxytocin in obsessive compulsive disorder. Peptides 1992; 13: 1083-5.

[99] Salzberg AD; Swedo SE. Oxytocin and vasopressin in obsessive-compulsive disorder. Am J Psychiatry 1992; 149: 713-4.

[100] Epperson CN, McDougle CJ, Price LH. Intranasal oxytocin in obsessive-compulsive disorder. Biol Psychiatry 1996; 40: 547-9.

[101] Humble MB, Uvnäs-Moberg K, Engström I, *et al.* Plasma oxytocin changes and anti-obsessive response during serotonin reuptake inhibitor treatment: a placebo controlled study. BMC Psychiatry 2013; 13: 344.

[102] Kaye WH, Gendall K, Kye C. The role of the central nervous system in the psychoneuroendocrine disturbances of anorexia and bulimia nervosa. Psychiatr Clin North Am 1998; 21: 381-96.

[103] Olson BR, Drutarosky MD, Stricker EM, *et al.* Brain oxytocin receptor antagonism blunts the effects of anorexigenic treatments in rats: evidence for central oxytocin inhibition of food intake. Endocrinology 1991; 129: 785-91.
[104] Baranowska B. Are disturbances in opioid and adrenergic systems involved in the hormonal dysfunction of anorexia nervosa? Psychoneuroendocrinology 1990; 15: 371-9.
[105] Gold PW, Gwirtsman H, Avgerinos PC, *et al.* Abnormal hypothalamic-pituitary-adrenal function in anorexia nervosa. Pathophysiologic mechanisms in underweight and weight-corrected patients. N Engl J Med 1986; 314: 1335-42.
[106] Kaye WH. Neuropeptide abnormalities in anorexia nervosa. Psychiatry Res 1996; 62: 65-74.
[107] Demitrack MA, Lesem MD, Listwak SJ, *et al.* CSF oxytocin in anorexia nervosa and bulimia nervosa: clinical and pa- thophysiologic considerations. Am J Psychiatry 1990; 147: 882-6.
[108] Welches WR, Brosnihan KB, Ferrario CM. A comparison of the properties and enzymatic activities of three angiotensin processing enzymes: angiotensin converting enzyme, prolyl endopeptidase and neutral endopeptidase 24.11. Life Sci 1993; 52: 1461-80.
[109] Maes M, Monteleone P, Bencivenga R, *et al.* Lower serum activity of prolyl endopeptidase in anorexia and bulimia nervosa. Psychoneuroendocrinology 2001; 26: 17-26.
[110] Chiodera P, Volpi R, Capretti L, *et al.* Effect of estrogen or insulin-induced hypoglycemia on plasma oxytocin levels in bulimia and anorexia nervosa. Metabolism 1991; 40: 1226-30.
[111] Frank GK, Kaye WH, Altemus M, *et al.* CSF oxytocin and vasopressin levels after recovery from bulimia nervosa and anorexia nervosa, bulimic subtype. Biol Psychiatry 2000; 48: 315-8.
[112] Fetissov SO, Harro J, Jaanisk M, *et al.* Autoantibodies against neuropeptides are associated with psychological traits in eating disorders. Proc Natl Acad Sci U S A 2005; 102: 14865-70.
[113] Koob GF. Neural mechanisms of drug reinforcement. Ann New York Acad Sci 1992; 654: 171-91.
[114] Kovacs GL, Sarnyai Z, Szabo G. Oxytocin and addiction: a review. Psychoneuroendocrinology 1998; 23: 945-632.
[115] De Wied D, Gispen WH, Van Wiersma Greidanus TJB. Neuropeptides and Behavior. Oxford: Pergamon Press1986.
[116] Kovacs GL, Telegdy G. Beta-endorphin tolerance is inhibited by oxytocin. Pharmacol Biochem Behav 1987; 26: 57-60.
[117] Kovacs CL, Van Ree JM. Behaviorally active oxytocin fragments simultaneously attenuate heroin self-administration and tolerance in rats. Life Sci 1985; 37: 1895-900.
[118] Kovacs GL, Sarnyai Z, Babarczy E, *et al.* The role of oxytocin-dopamine interactions in cocaine-induced locomotor hyperactivity. Neuropharmacology 1990; 29: 365-8.
[119] Sarnay Z, Birò E, Babarczy E, *et al.* Oxytocin modulates behavioural adaptation to repeated treatment with cocaine in rats. Neuropharmacology 1992; 31: 593-8.
[120] Sarnay Z, Szabò G, Kovacs GL, *et al.* Opposite actions of oxytocin and vasopressin in the development of cocaine-induced behavioral sensitization in mice. Pharmacol Biochem Behav 1992; 43: 491-4.
[121] Light KC, Grewen KM, Amico JA, *et al.* Deficits in plasma oxytocin responses and increased negative affect, stress, and blood pressure in mothers with cocaine exposure during pregnancy. Addic Behav 2004, 29: 1541-69.
[122] Szabò G, Kovacs GL, Telegdy G. Intraventricular administration of neurohypophyseal hormones interferes with the development of tolerance to ethanol. Acta Physiol Hungar 1989; 73: 97-103.
[123] Nemeroff CB, Loosen PT. Handbook of Clinical Endocrinology. New York: Guilford Press 1987.
[124] Marchesi C, Chiodera P, Brusamonti E, *et al.* Abnormal plasma oxytocin and beta-endorphin levels in alcoholics after short and long term abstinence. Prog Neuropsychopharmacol Biol Psychiatry 1997; 21: 797-807.
[125] Holden JJ, McLaughlin EJ, Reilly EL, *et al.* Accelerated mental aging in alcoholic patients. J Clin Psychiatry 1988; 44: 286-92.
[126] Bohus B, Kovacs GL, de Wied D. Oxytocin, vasopressin and memory: opposite effects on consolidation and retrieval processes. Brain Res 1978, 157: 414-7.
[127] Amico JA, Robinson G. Oxytocin: Clinical and Laboratory Studios. New York: Elsevier 1985.
[128] Teicher MH, Andersen SL, Polcari A, *et al.* Developmental neurobiology of childhood stress and trauma. Psychiatr Clin North Am 2002; 25: 397-426.

[129] Maes M, Lin A, Bonaccorso S, *et al.* Higher serum prolyl endopeptidase activity in patients with post-traumatic stress disorder. J Affect Dis 1999; 53: 27-34.

[130] Guastella AJ, Einfeld SL, Gray KM, *et al.* Intranasal oxytocin improves emotion recognition for youth with autism spectrum disorders. Biol Psychiatry 2010; 67: 692-4.

[131] Hollander E, Bartz J, Chaplin W, *et al.* Oxytocin increases retention in social cognition in autism. Biol Psychiatry 2007; 61: 498-503.

[132] Bakermans-Kranenburg MJ, van I Jzendoorn MH. Sniffing around oxytocin: review and meta-analyses of trials in healthy and clinical groups with implications for pharmacotherapy. Transl Psychiatry 2013; 3: 258.

[133] Domes G, Heinrichs M, Kumbier E, *et al.* Effects of intranasal oxytocin on the neural basis of face processing in autism spectrum disorder. Biol Psychiatry 2013; 74: 164-171.

[134] Hall SS, Lightbody AA, McCarthy BE, *et al.* Effects of intranasal oxytocin on social anxiety in males with fragile X syndrome. Psychoneuroendocrinology 2012; 37: 509-518.

[135] Eisenstein M. Treatments: In the waiting room. Nature 2012; 491: S14-16.

[136] Stevens FL, Wiesman O, Feldman R, *et al.* Oxytocin and behavior: evidence for effects in the brain. J Neuropsychiatry Clin Neurosci 2013; 25: 96-102.

[137] Bartz JA, Zaki J, Bolger N, *et al.* Oxytocin selectively improves empathic accuracy. Psychol Sci 2010; 21: 1426-1428.

[138] Meyer-Lindenberg A, Domes G, Kirsch P, *et al.* Oxytocin and vasopressin in the human brain: social neuropeptides for translational medicine. Nat Rev Neurosc 2011; 12: 524-538.

[139] Nissen E, Gustavsson P, Widstrom AM, *et al.* Oxytocin, prolactin, milk production and their relationship with personality traits in women after vaginal delivery or Cesarean section. Psychosom Obstet Gynaecol 1998; 19: 49-58.

[140] Jezova D, Skultetyova I, Tokarev DI, *et al.* Vasopressin and oxytocin in stress. Ann N Y Acad Sci 1995; 771: 192-203.

[141] McCarthy MM, McDonald CH, Brooks PJ, *et al.* An anxiolytic action of oxytocin is enhanced by estrogen in the mouse. Physiol Behav 1996; 60: 1209-15.

[142] Marazziti D, Dell'Osso B, Baroni S, *et al.* A relationship between oxytocin and anxiety of romantic attachment. Clin Pract Epidemol Ment Health 2006; 2: 28.

[143] Hoge EA, Pollack MH, Kaufman RE, *et al.* Oxytocin levels in social anxiety disorder. CNS Neurosci Ther 2008; 14:165-70.

[144] Heim C, Young LJ, Newport DJ, *et al.* Lower CSF oxytocin concentrations in women with a history of childhood abuse. Mol. Psychiatry 2009; 14: 954-8.

[145] Pariante CM, Miller AH. Glucocorticoid receptors in major depression: relevance to pathophysiology and treatment. Biol Psychiatry 2001; 49: 391-404.

[146] Bell CJ, Nicholson H, Mulder RT, *et al.* Plasma oxytocin levels in depression and their correlation with the temperament dimension of reward dependence. J Psychopharmacol 2006; 20: 656-60.

[147] Newport DJ, Heim C, Owens MJ, *et al.* Cerebrospinal fluid corticotropin-releasing factor (CRF) and vasopressin concentrations predict pituitary response in the CRF stimulation test: a multiple regression analysis. Neuropsychopharmacology 2003; 28: 569-76.

[148] Purba JS, Hoogendijk WJ, Hofman MA, *et al.* Increased number of vasopressin- and oxytocin-expressing neurons in the paraventricular nucleus of the hypothalamus in depression. Arch Gen Psychiatry 1996; 53: 137-43.

[149] Frasch A, Zetzsche T, Steiger A, *et al.* Reduction of plasma oxytocin levels in patients suffering from major depression. Adv Exp Med Biol 1995; 395: 257-8.

[150] van Londen L, Goekoop JG, van Kempen GM, *et al.* Plasma levels of arginine vasopressin elevated in patients with major depression. Neuropsychopharmacology 1997; 17: 284-92.

[151] Scantamburlo G, Hansenne M, Fuchs S, *et al.* Plasma oxytocin levels and anxiety in patients with major depression. Psychoneuroendocrinology 2007; 32: 407-10.

[152] Grippo AJ, Trahanas DM, Zimmerman RR, *et al.* Oxytocin protects against negative behavioral and autonomic consequences of long-term social isolation. Psychoneuroendocrinology 2009; 34: 1542-53.

[153] Ozsoy S, Esel E, Kula M. Serum oxytocin levels in patients with depression and the effects of gender and antidepressant treatment. Psychiatry Res 2009; 169: 249-52.

[154] Sasayama D, Hattori K, Teraishi T, *et al.* Negative correlation between cerebrospinal fluid oxytocin levels and negative symptoms of male patients with schizophrenia. Schizophr Res 2012; 139: 201-6.

[155] Holt-Lunstad J, Birmingham W, Light KC. The influence of depressive symptomatology and perceived stress on plasma and salivary oxytocin before, during and after a support enhancement intervention. Psychoneuroendocrinology 2011; 36: 1249-56.

[156] Parker KJ, Kenna HA, Zeitzer JM, *et al.* Preliminary evidence that plasma oxytocin levels are elevated in major depression. Psychiatry Res 2010; 178: 359-62.

[157] Yuen KW, Garner JP, Carson DS, *et al.* Plasma oxytocin concentrations are lower in depressed vs. healthy control women and are independent of cortisol. J Psychiatr Res 2014; 51: 30-6.

[158] Skrundz M, Bolten M, Nast I, *et al.* Plasma oxytocin concentration during pregnancy is associated with development of postpartum depression. Neuropsychopharmacology 2011; 36: 1886-93.

[159] Bartz JA, Zaki J, Bolger N, *et al.* Social effects of oxytocin in humans: context and person matter. Trends Cogn Sci 2011; 15: 301-9.

[160] Ellenbogen MA, Linnen AM, Grumet R, *et al.* The acute effects of intranasal oxytocin on automatic and effortful attentional shifting to emotional faces. Psychophysiology 2012; 49: 128-37.

[161]Cardoso C, Linnen AM, Joober R, *et al.* Coping style moderates the effect of intranasal oxytocin on the mood response to interpersonal stress. Exp Clin Psychopharmacol 2012; 20: 84-91.

[162] Pincus D, Kose S, Arana A, *et al.* Inverse effects of oxytocin on attributing mental activity to others in depressed and healthy subjects: a double-blind placebo controlled FMRI study. Front Psychiatry 2010; 1: 134.

[163] Ellenbogen MA, Linnen AM, Cardoso C, *et al.* Intranasal oxytocin impedes the ability to ignore task-irrelevant facial expressions of sadness in students with depressive symptoms. Psychoneuroendocrinology 2013; 38: 387-98.

[164] Beckmann H, Lang RE, Gattaz WF. Vasopressin--oxytocin in cerebrospinal fluid of schizophrenic patients and normal controls. Psychoneuroendocrinology 1985; 10: 187-91.

[165] Mai JK, Berger K, Sofroniew MV. Morphometric evaluation of neurophysin-immunoreactivity in the human brain: pronounced inter-individual variability and evidence for altered staining patterns in schizophrenia. J Hirnforsh 1993; 34: 133-54.

[166] Kéri S, Kiss I, Kelemen O. Sharing secrets: oxytocin and trust in schizophrenia. Soc Neurosci 2009; 4: 287-293.

[167] Muhle R, Trentacoste SV, Rapin I. The genetics of autism. Pediatrics 2004; 113: 472-86.

[168] Insel TR. Oxytocin--a neuropeptide for affiliation: evidence from behavioral, receptor autoradiographic, and comparative studies. Psychoneuroendocrinology 1992; 17: 3-35.

[169] Panskepp J. Oxytocin effects on emotional processes: separation distress, social bonding, and relationships to psychiatric disorders. Ann N Y Acad Sci 1992; 652: 243-52.

[170] Insel TR, O'Brien DJ, Leckman JF. Oxytocin, vasopressin, and autism: is there a connection? Biol Psychiatry 1999; 45: 145-57.

[171] Insel TR, Young LJ. The neurobiology of attachment. Nat Rev Neurosci 2001; 2: 129-36.

[172] Carter CS. Sex differences in oxytocin and vasopressin: implications for autism spectrum disorders? Behav Brain Res 2007; 176: 170-86.

[173] Modahl C, Green L, Fein D, *et al.* Plasma oxytocin levels in autistic children. Biol Psychiatry 1998; 43: 270-7.

[174] Green L, Fein D, Modhal C, *et al.* Oxytocin and autistic disorder: alterations in peptide forms. Biol Psychiatry 2001; 50: 609-13.

[175] Young LJ. Oxytocin and vasopressin as candidate genes for psychiatric disorders: lessons from animal models. Am J Med Genet 2001; 105: 53-4.

[176] Young LJ, Pitkow LJ, Ferguson JN. Neuropeptides and social behavior: animal models relevant to autism. Mol Psychiatry 2002; 7: 538-9.

[177] Ylisaukko-oja T, Alarcon M, Cantor RM, *et al.* Search for autism loci by combined analysis of Autism Genetic Resource Exchange and Finnish families. Ann Neurol 2006; 59: 145-55.

[178] Wu S, Jia M, Ruan Y, *et al.* Positive association of the oxytocin receptor gene (OXTR) with autism in the Chinese Han population. Biol Psychiatry 2005; 58: 74-7.

[179] Jacob S, Brune CW, Carter CS, *et al.* Association of the oxytocin receptor gene (OXTR) in Caucasian children and adolescents with autism. Neurosci Lett 2007; 417: 6-9.

[180] Lerer E, Levi S, Salomon S, *et al.* Association between the oxytocin receptor (OXTR) gene and autism: relationship to Vineland Adaptive Behavior Scales and cognition. Mol Psychiatry 2008; 13: 980-8.

[181] Wermter AK, Kamp-Becker I, Hesse P, *et al.* Evidence for the involvement of genetic variation in the oxytocin receptor gene (OXTR) in the etiology of autistic disorders on high-functioning level. Am J Med Genet B Neuropsychiatr Genet 2010; 153(B): 629-39.

[182] Liu X, Kawamura Y, Shimada T, *et al.* Association of the oxytocin receptor (OXTR) gene polymorphisms with autism spectrum disorder (ASD) in the Japanese population. J Hum Genet 2010; 55: 137-41.

[183] Gregory SG, Connelly JJ, Towers AJ, *et al.* Genomic and epigenetic evidence for oxytocin receptor deficiency in autism. BMC Med 2009; 7: 62.

[184] Hollander E, Bartz J, Chaplin W, *et al.* Oxytocin increases retention of social cognition in autism. Biol Psychiatry 2007; 61: 498-503.

[185] Guastella AJ, Einfeld SL, Gray KM, *et al.* Intranasal Oxytocin Improves Emotion Recognition for Youth with Autism Spectrum Disorders. Biol Psychiatry 2010; 67: 692-4.

[186] Swaab DF, Purba JS, Hofman MA. Alterations in the hypothalamic paraventricular nucleus and its oxytocin neurons (putative satiety cells) in Prader-Willi syndrome: a study of five cases. J Clin Endocrinol Metab 1995; 80: 573-9.

[187] Dykens EM, Leckman JF, Cassidy SB. J Child Psychol. Obsessions and compulsions in Prader-Willi syndrome. Psychiatry 1996; 37: 995-1002.

[188] Martin A, State M, Anderson MA, *et al.* Cerebrospinal fluid levels of oxytocin in Prader-Willi syndrome: a preliminary report. Biol Psychiatry 1998; 44: 1349-52.

[189] Marazziti D, Bani A, Casamassima F, *et al.* Clin Neuropsychiatry 2006; 5: 301-321.

CHAPTER 5

Structural and Computational Insights into Conformational Diseases: A Review

Xavier Fernàndez-Busquets[1,2], Natalia S. de Groot[3] and Salvador Ventura[4,5,*]

[1]Nanomalaria Group, Institute for Bioengineering of Catalonia (IBEC), Baldiri Reixac 10-12, ES-08028 Barcelona, Spain; [2]Barcelona Centre for International Health Research (CRESIB, Hospital Clínic-Universitat de Barcelona), Rosselló 149-153, ES-08036 Barcelona, Spain; [3]Medical Research Council Laboratory of Molecular Biology, Francis Crick Avenue, Cambridge, CB2 0QH United Kingdom; [4]Departament de Bioquímica i Biologia Molecular, Facultat de Biociències and [5]Institut de Biotecnologia i Biomedicina, Universitat Autònoma de Barcelona, E-08193 Bellaterra, Spain

Abstract: Protein aggregation correlates with the development of several deleterious human disorders such as Alzheimer's disease, Parkinson's disease, prion-associated transmissible spongiform encephalopathies, type II diabetes and several types of cancers. The polypeptides involved in these disorders may be globular proteins with a defined 3D-structure or natively unfolded proteins in their soluble conformations. In either case, proteins associated with these pathogenesis all aggregate into amyloid fibrils sharing a common structure, in which β-strands of polypeptide chains are perpendicular to the fibril axis. Because of the prominence of amyloid deposits in many of these diseases, much effort has gone into elucidating the structural basis of protein aggregation. A number of recent experimental and theoretical studies have significantly increased our understanding of the process. On the one hand, solid-state NMR, X-ray crystallography and single molecule methods have provided us with the first high-resolution 3D structures of amyloids, showing that they exhibit conformational plasticity and are able to adopt different stable tertiary folds, with impact both their transmissibility and neurotoxicity. On the other hand, several computational approaches have identified regions prone to aggregation in disease-linked polypeptides, predicted the differential aggregation propensities of their genetic variants and simulated the early, crucial steps of the oligomerization reaction. This review summarizes these findings and their therapeutic relevance, as by uncovering specific structural or sequential targets they may provide us with a means to tackle the debilitating diseases linked to protein aggregation.

Keywords: Aggregation prediction, Alzheimer's disease, amyloid fibrils, atomic force microscopy, computational biology, conformational diseases, electron microscopy, fluorescence spectroscopy, hydrophobicity, neurodegenerative diseases, oligomerization, Parkinson's disease, prion, protein aggregation, protein

***Corresponding author Salvador Ventura:** Departament de Bioquímica i Biologia Molecular, Facultat de Biocències, Universitat Autònoma de Barcelona, E-08193 Bellaterra, Spain; Tel: +34 93 581 41 47; Fax: +33 34 93 581 12 64; E-mail: salvador.ventura@uab.es

Atta-ur-Rahman, Allen B. Reitz, M. Iqbal Choudhary and Jizhou Wang (Eds.)

folding, protein structure, scanning microscopy, single-molecule, solid-state nuclear magnetic resonance, X-ray crystallography.

INTRODUCTION

Polypeptide chains are built up of repetitive amino acid units that differ in the chemical nature of their substituents at the α-carbon. Although inside the dense cellular environment amino acid side chains are exposed to manifold interactions, they usually manage to form the specific intramolecular contacts that direct the folding of polypeptides towards the stable, native structure. Such a conformation is necessary for the proteins to perform their biological functions. However, under certain conditions proteins misfold, lose their native structure, and adopt non-native conformations, leading to self-assembly and formation of amyloid protein deposits. The ability to adopt an amyloid-like structure or to undergo fibrillogenesis has emerged as a common, and perhaps fundamental, property of polypeptide chains [1-4]. Although protein aggregates may constitute a stable and secure way to accumulate unwanted proteinaceous material in the cytosol, they can also trigger a cascade of pathological events. Misfolding and protein aggregation are related to more than 40 distinct human conformational disorders [5, 6]. Some examples include the following: Alzheimer's disease, which is possibly linked to the formation of extracellular plaques and intraneuronal tangles by amyloid β (Aβ) protein and hyperphosphorylated tau protein, respectively; Huntington's disease, in which long glutamine stretches in huntingtin make it more prone to deposit into intranuclear inclusions and cytoplasmic aggregates; Parkinson's disease, which is related to α-synuclein aggregation and Lewy body formation; and the spongiform encephalopathies, including Creutzfeldt-Jakob disease, which is associated with human prion protein [7-10]. Also, some types of cancer are being found to be associated with protein misfolding and aggregation [11].

In many cases, conformational diseases exhibit degenerative pathologies with a major impact on the elderly. In the rapidly ageing developed world, such diseases threaten to lead to a collapse of public health services in the near future. Because of the urgent need to develop measures to prevent and treat conformational diseases, recent years have seen an overwhelmingly vast amount of research dedicated to in depth investigation of events that lead to protein misfolding and to the formation of harmful protein deposits. Structure-focused studies aimed at elucidating the conformation and deposition dynamics of polypeptide molecules in aggregates are likely to yield valuable rewards in the search for specific targets for therapeutic intervention. At the same time, bioinformatic approaches to protein aggregation may reveal detailed information on the mechanisms of aggregation at a molecular level, a challenge for wet lab studies. The methods employed to deduce structural models of amyloid aggregate architecture and the computational tools available to study these protein assemblies are the focus of this review.

NEW STRUCTURAL INSIGHTS INTO CONFORMATIONAL DISEASES

Until recently, one of the main obstacles to characterizing the structure of amyloid fibrils at the molecular level was that they are neither crystalline nor small enough to be studied by solution NMR spectroscopy [2]. However, in the last few years, high-resolution structures of amyloid fibrils have been obtained through major advances in solid-state nuclear magnetic resonance [12-14] and through the use of nano- and microcrystals of small amyloidogenic peptides that can be subjected to single crystal X-ray diffraction analysis [15-17]. A third major technique by which the structure of amyloid fibrils not in solution might be elucidated is transmission electron microscopy (TEM) [18]. Finally, single-molecule methods have recently emerged as a promising new approach for studying amyloid structure [19].

Solid-State Nuclear Magnetic Resonance (SSNMR)

SSNMR is a spectroscopic method ideally suited to study the structure and dynamics of densely packed molecular assemblies at the atomic level [20]. This technique was developed specifically for structural studies of molecular systems in noncrystalline solid and solid-like states such as phospholipid bilayers or precipitated protein aggregates, and it has consolidated as a principal methodology for amyloid fibril analysis (as reviews see [21-23]). SSNMR allows for high-resolution calculation of the distances between ^{13}C labels placed selectively in amyloid-forming peptides [18]. These data are accurate to ca. 0.2 Å for distances less than about 6 Å, although this precision is not maintained for greater distances. Current studies improve the data obtained with SSNMR by upgrading the technique using isotopic labeling [24], low temperatures [25] or magic-angle spinning (MAS) [24, 26-29]. Additionally, the combination of SSNMR with other methods such as dynamic nuclear polarization [24], fluorescence recovery after photobleaching (FRAP) [28], X-ray fiber diffraction [30] or electron microscopy [31, 32] also permits the refinement of the structures studied. Overall, amyloid fibrils are a good system for SSNMR investigations because (i) although they are noncrystalline, amyloid fibrils do have well-ordered molecular structures and therefore yield SSNMR data that are of high quality and easy to interpret; (ii) they can be prepared with selective or uniform isotopic labeling in the milligram amounts required for most SSNMR measurements; (iii) the fibrils can be obtained in high concentrations, leading to good signal-to-noise ratios; and (iv) the structural information resulting from SSNMR measurements is arguably more direct and specific than information derived from other techniques [22].

The first SSNMR studies of amyloid fibrils were focused on the structure of Aβ fibrils associated with Alzheimer's disease (AD) pathology. In a pioneering study of $A\beta_{34-42}$ fibrils, a structural model of antiparallel β-sheets with an alternating hydrogen bond registry was proposed [33]. Later, analysis of $A\beta_{10-35}$ fibrils [34] revealed for the first time the parallel, in-register β-sheet organization that has since been found in other systems and is now widely believed to be the most common (but not universal) β-sheet organization in amyloid fibrils [35-39], especially those formed by relatively long polypeptide chains. Later, Tycko and coworkers have undertaken SSNMR studies of fibrils formed by the full-length β-amyloid peptides $A\beta_{1-40}$ [12, 40-44] and $A\beta_{1-42}$ [35], and by the fragments $A\beta_{16-22}$ [45, 46] and $A\beta_{11-25}$ [46, 47]. A model has been put forward for the structure of $A\beta_{1-40}$ protofilaments [22] (Fig. **1**), where each $A\beta_{1-40}$ molecule contributes a pair of β-strands, spanning approximately residues 12-24 and 30-40, to the core region of the fibrils. These strands, connected by the loop residues 25-29, are not part of the same β-sheet, but participate in the formation of two distinct parallel and in-register β-sheets within the same protofilament. In $A\beta_{1-42}$, residues 1-17 are disordered, whereas residues 18-42 form a β-strand-turn-β-strand motif that contains two intermolecular β-strands formed by residues 18-26 (β_1) and 31-42 (β_2) [48]. More recently, Tycko and coworkers studied the molecular structure of β-Amyloid Fibrils obtained from brain tissue of two AD patients [32]. They combine SSNMR and electron microscopy measurements showing a single predominant 40 residue Aβ ($A\beta_{40}$) fibril structure in each patient. This data suggests that fibrils in the brain may spread from a single nucleation site. However, the structures observed were different between patients, which may be correlated with variations in AD neuropathology [32].

In AD, Aβ may exert its neurotoxic activity through interactions with the cell membrane [49-51]. Therefore, the study of Aβ-membrane interactions at the molecular level is a key approach towards understanding the pathology of amyloidogenic proteins. Steady-state interactions of $A\beta_{29-42}$ with neutral lipid bilayers have been investigated by SSNMR [52]. Results from these experiments and molecular modeling support the conclusion that when Aβ is inserted in the membrane, its steady-state structure is an oligomeric association of β-sheet peptides located at the bilayer interface. Furthermore, these peptides preferentially recruit phosphatidylethanolamine, suggesting that the inaccessibility of this crucial lipid could be a feature of the cellular disorders induced in AD [23].

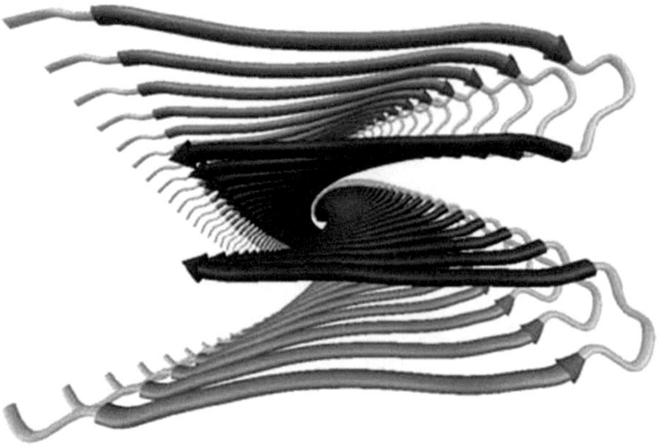

Figure 1: Ribbon representation of a structural model for the Aβ1-40 protofilament, based on SSNMR and TEM data. In the protofilament, viewed along its long axis, the β-sheets in the cross-β motif are parallel and in-register and are formed by two β-strand segments from each Aβ1-40 molecule (medium and darkest grey segments) that are separated by a loop (lightest grey segment). Two molecular layers form a four-layered structure with a predominantly hydrophobic core. (Ref. [22], with permission).

The structural diversity characteristic of Aβ fibrils has been also observed in other amyloidogenic polypeptides [24, 28, 53]. This conformational variability is important for yeast prions to develop and regulate their biological roles and generate phenotype diversity [28]. Hence, the distinct aggregated conformations of the yeast prion Sup35 result in different PSI+ phenotypes [28]. Frederick and co-workers studied the dynamic properties of the Sup35 conformation at different times using MAS NMR. They complemented this data with (FRAP) to investigate their interaction with the Hsp104 chaperone *in vivo*. Their results suggest that the prion conformation and dynamics will regulate the interaction with chaperones [28]. SSNMR studies have also contributed to the development of a model for the fibrils formed by the prion protein HET-s that could explains many of the properties of this functional amyloid [14]. In the proposed structure, HET-s forms a left-handed β solenoid, with each molecule forming two helical windings, a compact hydrophobic core, all charged residues face outside and two asparagine ladders. Specifically, each molecule contributes to four β-strands, with strands one and three forming the same parallel β-sheet and strands two and four forming another parallel β-sheet. Interestingly, the constrains and charge compensation involved in this structure ensures high order and regulation supporting the absence of polymorphism characteristic of the HET-s fibrils [54].

Other interesting SSNMR studies analyze the cross-β-sheet structure of α-Synuclein [26], peptides derived from the transthyretin (TTR) [13, 55], the *de novo* designed peptide ccβ [56] bacterial curli amyloid [57] and synthetic peptides representing residues 89-143 of mammalian prion protein [58].

X-Ray Crystallography

Recently, short amyloid-forming peptides, which exhibit key characteristics of amyloid fibrils, have been successfully induced to form 3D crystals, allowing for the high-resolution elucidation of the packing structure of the peptides [16, 17]. The seven-residue peptide GNNQQNY and its relative NNQQNY, both derived from the Sup35 prion protein of *Saccharomyces cerevisiae*, form elongated microcrystals that have made X-ray diffraction studies possible [16]. Atomic structure determination of the cross-β spine revealed that it is a double β-sheet, with each sheet formed from parallel segments stacked in-register. In these crystals, side chains protruding from the two sheets form a tightly self-complementing steric zipper, stabilizing bonding between the sheets. For the longer peptide, the two sheets interact with each other through the side chains of Asn2, Gln4, and Asn6, and these interactions are so tight that water is excluded from the region between them. Within each sheet every segment is bound to its two neighboring segments through stacks of both backbone and side-chain hydrogen bonds. Similarly, fibrous crystals were grown from a 12-mer peptide containing two KFFE motifs separated by an AAAK motif, and they yielded high-resolution X-ray and electron diffraction data [17]. In this case, the peptide associated to form antiparallel β-sheets. However, as in the case of GNNQQNY, these sheets were found to be zipped together *via* a staggered arrangement of contacts between the side-chains that also excluded water. More recently, Wiltzius and coworkers analyzed nine new structures derived from fragments and full-length sequences of amyloid proteins [53]. The data obtained from these structures provide clues about how prion conformations are inheritance and transmitted. The authors deduced two distinct structural mechanisms: (i) in packing polymorphism, prion strains are encoded by alternative packing arrangements of β-sheets formed by the same segment of a protein and (ii) in segmental polymorphism, prion strains are encoded by distinct β-sheets built from different segments of a protein [53].

Success in the formation of crystals from full-length amyloidogenic proteins is allowing the field of X-ray crystallography to solve a wide variety of structures [53, 59-62]. X-ray crystallography was used to elucidate the 3D structures of two important TTR variants: TTR Y78F, an amyloidogenic protein, and TTR R104H, a protein associated with a protective effect over the amyloidogenic V30M mutation [59]. While the former structure strongly suggests a relevant role for an α-helix in the overall stability of TTR, the latter suggests that N-terminal stabilization might be the key determinant of its protective effects. Inspection of a limited number of crystal structures of β2 microglobulin (β2m) as an isolated chain separated from the major histocompatibility complex I heavy chain revealed that the corresponding 3D structure is based on an antiparallel β-barrel fold, with an immunoglobulin (Ig) domain topology [60]. The structural bases of amyloidogenic potential in β2m can be related to local unfolding, to the tendency to aggregate laterally through non-compensated β-strands, and partially to its trend towards N-terminal proteolytic degradation [60].

Electron Microscopy

Electrons interact with atoms much more strongly than X-rays or neutrons, thus allowing the observation of individual molecules. The wavelength of the electron beam is much smaller than that of visible light, allowing magnifications ca. 500 times higher than with light microscopes. TEM examination of a substantial number of amyloidogenic proteins has provided images of amyloid fibrils as long, unbranching filaments 6-12 nm across, characterized by the lateral association of protofilaments, and exhibiting a clear helical twist [63-66]. As a drawback of TEM, the intense flow of electrons required to obtain good contrast leads to substantial radiation and sample damage. Staining improves the contrast but it often leads to loss in resolution of internal fibril structure, which adds to the possible distortion induced by dehydration. Cryo-TEM represents an alternative strategy in which the sample is frozen in liquid nitrogen or helium in order to reduce the magnitude of damage from ionization. The proteins are maintained in a hydrated state and low-dose TEM is used to minimize radiation damage. Cryo-TEM does not necessarily give higher resolution than standard TEM, but it provides more reliable structural information, which can be extracted from the micrographs through either direct visualization or 3D reconstruction of the fibril [67]. The protofilament structure of an SH3 peptide was successfully studied by cryo-TEM [68], revealing an elliptical cross-section formed by four protofilaments. Studies performed on lysozyme and apolipoprotein A fibrils indicated varying protofilament arrangements within the same sample [69]. High-resolution cryo-TEM images of $A\beta_{11-25}$ protofilaments revealed striations running across the filament [18], that likely corresponded to individual β strands within a single β-sheet/protofilament, in agreement with Fourier-transform infrared spectroscopy data showing specific bonding patterns consistent with β-sheet structure. A strong reflection at 4.7 Å, which is the hydrogen bond spacing of β strands, was also observed in Fourier transforms of cryo-TEM images of full-length islet amyloid polypeptide (IAPP) fibrils [70], supporting the view that this is a common feature of amyloid fibrils.

Cryo-TEM data has been complemented by information derived from scanning TEM (STEM), which allows the quantitative determination of the mass-per-unit-length of a fibril by comparison with a standard, such as the tobacco mosaic virus [71]. If the specimen is thin, the image intensity is directly proportional to the mass of the irradiated region. STEM measurements of $A\beta_{1-40}$ protofilaments were consistent with the peptide being folded on itself and stacked to form two β-sheets in contact with each other through their side-chains [12]. In contrast, other studies observed a three-fold symmetry in certain $A\beta_{1-40}$ fibrils [32, 72]. The cross-β structure deduced for IAPP protofilaments from X-ray diffraction data [73] was also confirmed by diameter and mass-per-unit-length measurements determined using TEM and STEM [70, 74]. Recently, the quantification of the mass-per-unit-length of individual fibrils was achieved using dark-field TEM images of unstained samples [75]. The tilted-beam dark-field images can be obtained on common transmission electron microscopes, providing a technique to determine the mass-per-length accessible to many research groups [75, 76].

Image reconstruction from electron micrographs has been used to examine the structure of TTR amyloid fibrils [77]. Averaged cross-sections of 200 different fibrils produced a detailed view of the substructure, revealing a fibril diameter of ca. 130 Å. Cross-sections of the fibrils exhibit 4-fold symmetry with four proto-filaments, each measuring 40 to 50 Å across, arranged around a central hollow core. 3D maps of SH3 fibrils based on cryo-TEM data have also revealed a hollow core with a maximum diameter of 5 nm [68]. Meinhardt and coworkers combined negative stain TEM, cryo-EM and 3D reconstruction to analyze 12 samples of Aβ(1-40), all formed under the same solution conditions and all acquiring a different amyloid conformation [78]. Interestingly, the comparison of these morphologies shows an almost continuous range of altered structural properties and supports that the same polypeptide sequence can lead to many different inter- or intraresidue interactions. Recently, Fitzpatrick and coworkers developed a 4D cryo-electron microscopy by integrating time to the 3D reconstruction [30]. The technique consists in a time-resolved diffraction of the amyloid fibrils trapped in a thin layer of vitrified water at cryogenic temperatures [30]. With this system the authors are able to detect picometer movements of protein molecules on a nanosecond time scale, which permits the visualization of the ultrafast dynamics of biomolecules in a fully hydrated state [30].

Whereas SSNMR, X-ray crystallography, and TEM have provided valuable high-resolution data on the structure of amyloid fibrils, they have important limitations. Some of the most fundamental characteristics of amyloid fibrils that are still largely unknown include their dynamics and kinetics of growth and disassembly, their stiffness, their mechanical response to compression, tension, or pulling and their resistance to breakage, among others. These properties are being increasingly acknowledged to be important to understanding the *in vivo* roles of amyloid deposits [79-81]. Study of the conformational plasticity of amyloids has required new experimental approaches that materialized with the development of nanometric methods [19]. The advantages of most nanometric techniques arise from their ability to observe and/or manipulate single molecules in a liquid environment.

Amyloid Fibril Structure Deduced from SSNMR, X-Ray, and TEM Data

The similarities between amyloid fibrils derived from very different peptides and proteins suggest that their core structures have similar features primarily dictated by the intrinsic conformational preferences of polypeptide chains [2]. However, the specific nature of the side-chain packing, including such characteristics as the alignment of adjacent strands and the separation of the sheets, provides an explanation for the occurrence of variations in the structural details of different types of fibrils. Putting together the above data derived from SSNMR, X-ray crystallography, and TEM, a picture of the common properties of amyloid fibrils emerges that includes, as a universal element, the cross-β X-ray diffraction pattern. This pattern consists of an X-ray reflection at ~4.8 Å resolution along the fibril axis and another X-ray reflection at ~10-12 Å resolution perpendicular to the fibril [82], indicating that the fibrils contain β-sheets parallel to the fibril axis with their extended protein strands perpendicular to the axis. Amyloid structures can be built of parallel or antiparallel sheets, that in turn may or may not be in register [15]. Unfortunately, if the strands are out of register rather than being antiparallel or if the sheet is composed of a mixture of parallel and antiparallel strands, misleading results may be generated [18]. This might explain the wide variety in reported results. To date, antiparallel β-sheet structures have been identified only in fibrils formed by relatively short peptides containing only one β-strand segment, indicating that studies of model peptides can not generally be used to infer structural properties of full-length amyloid-forming sequences [22]. The more fundamental feature, which is common to all amyloid and amyloid-like fibrils, appears to be the dry steric zipper motif present in the structure of the cross-β spine [83].

Single-Molecule Studies

Single-molecule approaches have the capacity to provide previously unattainable data on elementary biological processes. Until recently, most experimental techniques derived data from molecular ensembles: examples include bands in electrophoresis and in Western, Southern, and Northern blots; ordered structures in NMR and crystallographic studies; absorbance, fluorescence, or diffraction in solution studies; and optical microscopy. Although these methods have obviously generated and will continue providing essential structural and functional information, the data obtained only represent the mean values of large numbers of molecules. This "molecular sociology" is tremendously informative for the understanding of biomolecular processes. However, just as human sociology is adequate to study groups of people, other disciplines like anatomy are much more appropriate for the study of the individual. Thus, single-molecule techniques can be considered to be the study of "molecular anatomy" in the sense that they explore and manipulate individual molecules. Clearly, single-molecule techniques can tap an unfathomable ocean of new information that can not be obtained with "multiple-molecule" approaches. In a recent survey of literature employing single-molecule studies [84], a search on PubMed (www.pubmed.gov) revealed exponential growth over the last two decades. Conventional multi-molecule tools can only provide an averaged picture of a system under study, where much of the subtle or short-lived information is lost. Emerging nanotools might complement this limitation by opening novel paths for the development of early diagnostic and therapeutic approaches [85-87].

Application of Nanometric Methods to the Study of Self-Aggregating Proteins

In order to fully understand the process leading to fibril deposition it is paramount to have access to methods that allow the study of the growth and/or disassembly of individual amyloid fibrils. Such knowledge will be essential to the design of novel therapeutic methods for the removal of amyloid aggregates. Among the plethora of an ever-growing array of single-molecule approaches, we will focus our attention on those that are being currently employed to explore amyloid structure and dynamics. These include (i) scanning probe microscopies such as scanning tunneling microscopy (STM) and atomic force microscopy (AFM), the latter of which is used both for imaging and for single molecule force spectroscopy (SMFS); (ii) optical methods based on the properties of the evanescent wave, such as total internal reflection fluorescence microscopy

(TIRFM) and near-field scanning optical microscopy (NSOM); and (iii) fluorescence correlation spectroscopy (FCS).

Scanning Probe Microscopy (SPM) Methods

SPM methods are a group of techniques where a surface is imaged at high (and in some cases atomic) resolution by rastering an atomically sharp tip in close, but not direct, contact with the surface [88]. The strength of the interaction between the tip and the surface and the relative position of the tip is measured to produce an image of interaction strength as a function of position, which, depending on the particular technique, represents surface topography or chemistry. SPM allows investigation of structural as well as functional properties of native biomolecules in liquid and physiological environments by a unique combination of subnanometer spatial resolution, millisecond temporal definition and piconewton force sensitivity [89-91]. This method relies on the highest possible precision for the movement of the cantilever holding the tip, which is achieved by the use of piezoelectric transducer elements and whose accuracy is well below 1 Å (smaller than one single atom). Most commonly, the cantilever deflections are monitored *via* a laser beam, where the reflected laser spot is converted on a position sensitive photodetector into an electric signal.

In scanning tunnelling microscopy (STM) a voltage is applied between the cantilever tip and a conductive or semiconductive sample. Under these conditions, electrons can flow between the probe and the surface, generating an electric "tunnelling" current whose intensity is inversely proportional to the distance that it has to span. A 3D map is generated by fixing a constant value for the current between the probe and the specimen and simultaneously recording the vertical displacement of the cantilever during the scan. Despite the subnanometer resolution that can be obtained by STM, the application of this technique to biological molecules is limited by the requirement that the imaged sample be conductive, although significant progress has been made in imaging Aβ fibrils by STM [92-95]. STM has permitted not only to monitor the assembly of Aβ peptides to but also to locate the arrangement of Congo read and ThT dyes a long the amyloid fibril [96].

In atomic force microscopy (AFM), forces between the atoms of the scanning tip and those of the surface-immobilized molecules under examination induce the vertical displacement of the cantilever. The topographic image of the scanned sample is generated by monitoring these signals with a spatial resolution that is equal to or even higher than that obtained for the same sample with the electron microscope [97], but with the benefit that SPM images can be obtained with staining-free protocols on functional biomolecules in physiological solutions. The relative large radius of the AFM tip apex respective to the size of the molecules being scanned requires mathematical deconvolution of raw data in order to calculate the real horizontal dimensions. On the other hand, although STM can provide highly precise distance measurements in the *x-y* plane, it is much less accurate than AFM for height measurements in the *z* axis. AFM has been widely used to monitor the assembly of numerous amyloid-forming peptides and proteins [98], including Aβ, Ig light chain, α-synuclein, β2m, IAPP, insulin, the B1 domain of protein G, TTR, lysozyme, and the 90-residue Src homology (SH3) domain of the α subunit of bovine phosphatidylinositol-3'-kinase [64, 65, 99-103]. Interestingly, the combination of AFM with topography and recognition (TREC) imaging permits to analyze protein aggregates in a physiological environment. This modification consist in a receptor-functionalized tip on a magnetic-coated cantilever driven by a magnetic field that records simultaneously topography and binding site recognition [104]. Creasey and coworkers achieve for the first time to use TREC and immunofluorescence on untreated tissue human samples. By comparing samples from patients and healthy subjects, they detected differences on the aggregation pattern of the proteins implicated in the Pseudoexfoliation syndrome.

AFM imaging in real time (also termed on-line or time-lapse AFM) allows for the identification and tracking of prefibrillar structures and the continuous monitoring of individual molecular assemblies in a liquid environment with temperature and buffer control. Such studies have provided valuable data on elongation rates, directionality of growth, changes in morphology for individual fibrils, and the assembly process of higher order polymorphic species. One application of time-lapse AFM in this area has been the direct visualization of amyloid fibril growth *in vitro* [105]. This experimental approach holds promise for the future testing of potential therapeutic drugs, for example, by directly visualizing at which level of fibril assembly (nucleation, elongation, branching, or lateral association of protofibrils) a given active compound will interfere. With conventional AFMs, it takes more than a minute to capture an image, while biomolecular processes generally occur on a millisecond timescale or less. High-speed atomic force microscopy (HS-AFM) has been developed as a nano-dynamics visualization technique [106]. This microscopy permits direct observation of structure dynamics and dynamic processes of biological molecules in physiological solutions, at a subsecond to sub-100 ms temporal resolution and a ~2 nm lateral and a 0.1 nm vertical resolution. Importantly, tip-sample interactions do not disturb the biomolecules' functions. Lithostathine, a protein overexpressed in the pre-clinical stages of Alzheimer's disease and present in the pathognomonic lesions associated with this disease, forms fibrillar aggregates after its N-terminal truncation. HS-AFM has allowed the visualization of the growth and assembly of lithostathine protofibrils under physiological conditions with a time resolution of one image/s [107]. Real-time imaging highlighted a very high velocity of elongation, and the formation of fibrils *via* protofibril lateral association and stacking was also monitored revealing a zipper-like mechanism of association.

Perhaps the main potential of AFM is its ability to be used as a tool for the manipulation of biomolecules. Direct manipulation of individual amyloid fibrils or monomers can provide essential information on aspects such as fibril brittleness or elasticity, or on the adhesion forces that govern the self-aggregation process of individual polypeptide molecules.

Single Molecule Force Spectroscopy

During the last two decades SMFS has developed into a highly sensitive tool for investigation of the interaction of single biomolecules. Most SMFS experiments use either optical tweezers [108] or AFM [109] to measure piconewton dissociation forces of single ligand-receptor complexes. In the AFM configuration, the molecular binding partners are attached to the micro- or nanoscale force sensor and a sample holder, respectively, by covalent chemistry. When both moieties are brought into close contact, a specific bond between the individual molecules can form. By increasing the distance between the two surfaces again, the molecular bond is loaded under an external force until it finally breaks, yielding the molecular dissociation force. By systematically varying the externally applied load and monitoring the mechanistic elasticity of the complex, information about the kinetic reaction rates, the mean lifetime, the equilibrium rate of dissociation, dissociation length and the energy landscape of the interaction can be derived [110, 111]. AFM has the sensitivity to measure forces comparable with that of a single hydrogen bond [112], a magnitude far below the force exerted in most enzyme-substrate interactions.

The combination of high sensitivity and the possibility to operate in liquid environments has made of the AFM a very useful tool for the characterization of biological samples, including amyloid fibrils [64, 65, 98, 103]. SMFS has been applied to dissecting the binding forces responsible for the formation and stability of amyloid structures, *e.g.* to detect unzipping events in Aβ fibrils [113], to measure the strength and mechanical stiffness of fibrillar structures of insulin

[114], to determine the force required for the extraction of a prion protein from its amyloid fibrils [115], and to characterize the self-binding between Aβ, Tau, α-synuclein, lysozyme, or amylin molecules [116-122] (Fig. **2**). Quantitative analysis of SMFS data demonstrated that Aβ interaction leads to the formation of dimers with a lifetime in the range of a second [123]. In mechanically manipulated individual amyloid fibrils formed from either $A\beta_{1-40}$ or $A\beta_{25-35}$, β-sheets behave as elastic structures that can be "unzipped" from the fibril with different amounts of constant force [113]. Unzipping was fully reversible across a wide range of stretch rates provided that coupling, *via* the β-sheet, between bound and dissociated states was maintained. These data suggested that the rapid, cooperative zipping together of β-sheets could be an important mechanism behind the self-assembly of amyloid fibrils. The use of appropriate surface chemistry enabled the anchoring of Aβ through the N-terminal ends, allowing the measurement of the rupture of Aβ-Aβ contacts at single molecule level [119]. The rupture of these interactions was accompanied by the extension of the peptide chain detected by a characteristic elasto-mechanical component of the force-distance curves. SMFS has also been used to study the effect of pH on the interactions and misfolding of α-synuclein, Aβ, and lysozyme [118], showing that the attractive force between homologous protein molecules is minimal at physiological pH and increases dramatically in an acidic milieu.

The severing of amyloid fibrils generates seeds for new fibril formation, and thus represents a key determinant of their physiological impact [81]. In an elegant approach, a detailed mechanical characterization of individual insulin amyloid fibrils revealed that they have a strength close to that of steel and a mechanical stiffness comparable to that of silk [114], indicating that amyloid fibrils possess properties that make them potentially useful materials for biotechnological applications. Several studies have demonstrated the use of amyloids as functional templates for the synthesis of conductive nanowires [124-127], and the structural characteristics of some amyloid fibrils suggest that they can be adequate scaffolds for the fabrication of coaxial nanocables [128, 129]. The design of new nanowire fabrication strategies requires a detailed knowledge of the forces governing the growth of monomeric peptides into amyloid protofibrils and fibrils, and methods for the oriented deposition of fibrillar peptide assemblies. The spontaneous self-assembly of amylin-derived peptides into thin, straight and very long protofibrils [120] suggests that these molecules can be interesting candidates as templates for nanowire fabrication. Substrates like graphite or DMSO-coated graphite are able to orientate those thin fibrillar assemblies along well defined parallel or 60/120-

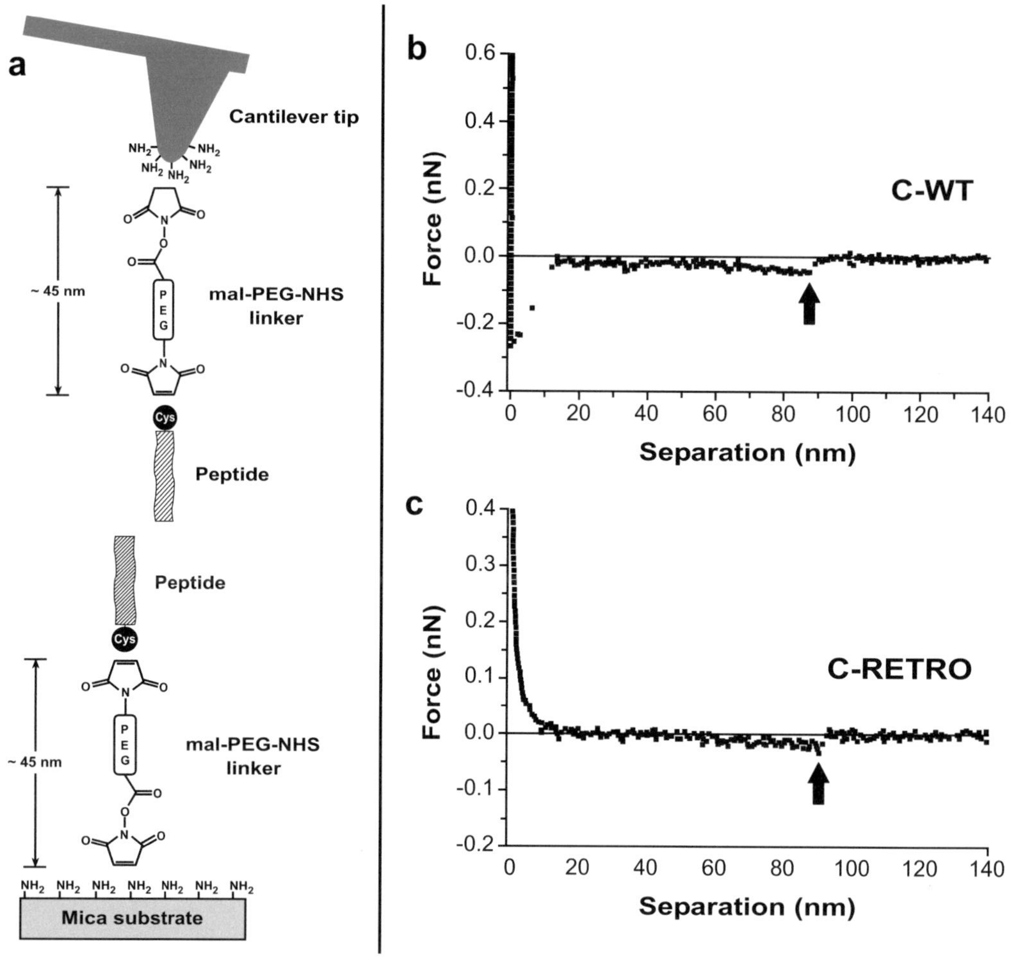

Figure 2: SMFS analysis of the self-adhesion forces for the highly amyloidogenic peptide comprising residues 10-29 of human amylin (WT) and for its reverse sequence (RETRO); to both peptides a terminal cysteine was added (C-WT and C-RETRO). **a)** Scheme of the immobilization of the peptides on a cantilever tip and a mica substrate (Cys = cysteine). **b)** and **c)** Two examples of retraction force curves for C-WT and C-RETRO showing unbinding events (arrows) at the expected separation distances of about twice the length of a linker molecule. (Ref. [120], with permission).

degree directions. Easy peptide assembly and effective control of the directions of the deposited fibrillar structures are a very attractive combination for the development of new amyloid-based nanotechnological systems. The growth of amylin-derived peptide protofibrils at physiological ionic strength and neutral pH suggests that they can be used *in vivo* as template structures with potential biomedical applications ranging from nerve cell growth guidance to bone healing and reconstruction. Ironically, naturally occurring amyloid fibrils associated with some of the most devastating diseases, could provide new materials for technologies that can be applied to the improvement of human health.

The forces required to mechanically fracture amyloid fibrils range from 300 to 500 piconewtons (pN) [114], which are values on the same order of magnitude as the forces required to unfold individual protein domains [130] or to break carbohydrate-carbohydrate interactions in cell-to-cell proteoglycan-mediated adhesion [131]. This suggests that the interactions within the amyloid core, which include hydrogen bonding, van der Waals forces, and electrostatic forces, are of similar origin and nature as those responsible for the folding of native structural motifs in proteins. The high level of stability of amyloid aggregates is an essential factor underlying their involvement in a range of clinical disorders [132]. The high strength and mechanical stiffness of amyloid fibrils makes them very resistant to degradation by the endogenous mechanisms of living organisms, thus leading to an accumulation of protein aggregates.

Total Internal Reflection Fluorescence Microscopy (TIRFM)

The ~250 nm resolution limit of conventional optical microscopy was the primary factor behind the development of higher-resolution electron microscopy and scanning probe techniques. These and related microscopic methods have enabled phenomenal gains in resolution, up to the level of visualizing individual atoms [133]. However, the requirement for highly purified molecular preparations limits the application of these high-resolution methods in many biological investigations, especially those measuring *in vivo* dynamics. A first approximation to obtaining nanometer resolution in live-cell optical fluorescence microscopy was the development of TIRFM [134]. In this method, when a beam of light hits the interface between two media of different refractive index at above a critical angle, all of the light is reflected but some of the incident energy enters the second medium forming an electromagnetic field that oscillates with the same frequency as the incident light, generating the so-called evanescent wave. For a laser light at 455 nm, this evanescent field penetrates only for ca. 150 nm into the sample [135], although this distance is also a function of the refraction indexes and the angle of incidence beyond the critical angle. As a result, fluorescent molecules are excited within a very small volume near the interface and there is a low background of out-of-focus fluorescence compared to epifluorescence, from which TIRFM derives its potential for detecting single molecules. Finally, the excitation light, which can also contribute to background noise as it is usually much more intense than the emitted fluorescence, is cleanly removed from the TIRFM image, as any that is not absorbed gets carried away in the reflected beam.

Individual amyloid fibril growth can be visualized in real time by following the incorporation of protein monomers or oligomers into preformed seeding fibrils or protofibrils. Polymerization of the amyloidogenic yeast prion protein Sup35 has been studied using this approach [80, 136]. Sup35 fibers labeled with a red fluorescent dye were deposited on a slide and treated with a solution of monomeric Sup35 labeled with a green fluorescent dye. Fibril growth observed both by epifluorescence [136] and by TIRFM [80]. occurred mainly unidirectionally through direct monomer addition, in the absence of observable intermediates. By monitoring Thioflavin T (ThT) fluorescence with TIRFM, the growth of amyloid fibrils formed by β2m could be followed without the need for covalent fluorescent labeling [137]. The results obtained showed that the extension of β2m fibrils was mostly unidirectional. Since ThT binding is common to all amyloid fibrils, this method will likely have general applicability. ThT fluorescence measurements have also been used to study the dynamics of fibrillogenesis for an octapeptide derived from the C terminus of human medin and for $A\beta_{1-40}$ [137].

For live cell imaging, TIRFM illuminates only the basal membrane proximal to the microscope slide, imposing a constriction on the experimental design. Yet, TIRFM can provide valuable dynamic information on the interaction of living cell membranes with different oligomeric or fibrillar amyloid species with which the slides can be previously functionalized. There are evidences indicating that the appearance of insoluble fibrillar structures enriched in β-sheets is facilitated by diverse environmental factors [138], biological membranes among them. Aβ is cleaved from its precursor protein in the membrane interface, and its cytotoxic effect is likely related to the amyloid-lipid interaction [23]. TEM and AFM studies performed with artificial membranes have shown that, upon interaction with the membrane lipids, Aβ in fibrillar form reverts to soluble globular peptide oligomers that associate into disordered domains [139]. TIRFM often requires to stably immobilize integral membrane proteins in a bilayer without covalent modification, enabling a range of surface-based experiments where accurate localization of the protein of interest is needed. Nanodiscs are a new class of model membranes that are being used to solubilize and study a range of integral membrane proteins and membrane-associated proteins [140]. Unlike other model membranes, the Nanodisc bilayer is bounded by a scaffold protein coat that confers enhanced stability and a narrow particle size distribution. The bilayer diameter can be precisely controlled by changing the diameter of the protein coat. All these properties make Nanodiscs excellent model membranes for single-molecule fluorescence applications such as TIRFM.

Near-Field Scanning Optical Microscopy

In NSOM, as in the case of AFM, a sharp probe scans the sample surface, but in addition to topography, NSOM also generates optical images (for a review see [141]). The most generally applied near-field optical probe consists of a small aperture, typically 20-120 nm in diameter (*i.e.*, much smaller than the wavelength of the excitation light), at the end of a metal-coated tapered optical fiber. In fluorescence mode, it serves as a constriction that funnels an incident light wave to dimensions that are substantially below the diffraction limit, resulting in a light source that has the size of the aperture. However, in contrast to common light sources such as lightbulbs and lasers, the light emitted by the probe is predominantly composed of evanescent waves rather than propagating waves. As described above, the intensity of the evanescent light decays exponentially and to insignificant levels ~100 nm from the aperture, and thus the probe can only excite fluorophores that reside within a layer of <100 nm from the probe, in what is termed the near-field region. Sample fluorescence can subsequently be collected by conventional optics and transformed into an optical image of the sample surface in which the resolution is now primarily dictated by the aperture dimensions rather than by the wavelength of the light. An electronic feedback system keeping the probe-sample distance during scanning at less than 10 nm is used, as in AFM, to generate a topographic map of the sample surface. Unique to NSOM is the fact that a corresponding fluorescence map is simultaneously generated. Although NSOM has not been widely applied yet to the study of amyloidogenesis, it is foreseeable that in the near future this technique will provide valuable data on the dynamics of the interaction of different amyloid species with membranes.

In a pioneering study, the binding of two amyloid fibril stain dyes, ThT and its neutral analog 2-[4'-(dimethylamino)phenyl]-benzothiazole (BTA-2), have been measured using NSOM [142]. Polarized NSOM images reveal that both dyes bind to individual insulin amyloid fibrils with the long axis of the molecule aligned parallel to the long axis of the fibrils. This indicates that the dyes bind along the surface of the beta-sheet within the grooves of the fibril that run parallel to the fibril axis. The similarity in the binding motifs of the two dyes showed that electrostatic interaction of the charged amine group on the ThT dye plays a minimal role in the affinity of the dyes for the amyloids. The polarized NSOM images confirmed that the enhanced fluorescence of the ThT and BTA-2 result from binding of the monomeric dye rather than micelles or excimer species.

Fluorescence Correlation Spectroscopy

FCS is based on the fluctuation of light emitted by fluorescently labelled molecules crossing a small laser spot and detected with confocal optics [143, 144]. Fluorescence fluctuations are the result of molecular diffusion, chemical reactions, and physical processes of a few fluorescent molecules in an optically restricted sub-micron observation volume (~1 fL = 10^{-15} L), that can be studied with a temporal resolution in the range of 1 ms to >10 s. The elegance of FCS lies in its ability to extract a wealth of molecular and environmental data from a weak signal that is comparable with background noise. This information is obtained by using correlation analysis of the fluorescence fluctuations of very small samples of molecules at nanomolar concentrations. Recent technological advances have enhanced the number of biological and chemical applications of FCS, such as study of binding interactions between biomolecules, sparse molecule detection, intramolecular protein dynamics, and diffusion in the membranes of living cells.

The first FCS data on the kinetics of amyloid formation reported on the cooperativity of Aβ polymerization [145], showing that the formation of very large aggregates preceded the formation of fibrils. FCS has been also applied to the study of the interaction of Aβ with cell membranes. Rhodamine-labelled Aβ showed different diffusion times that likely corresponded to Aβ either unbound or complexed to a target molecule in the cell membrane [146], thus adding to the hypothesis that Aβ affects neurons through its binding to a receptor [147]. Aβ binding to certain plasma membrane receptors can damage the cell by promotion of tau protein phosphorylation [148], generation of oxidative stress and stimulation of macrophages [149-152], blocking of protein function [153, 154], or induction of apoptosis [155]. In a recent study, FCS imaging revealed that monomeric human islet amyloid polypeptide (hIAPP) significantly destabilized both model and live cell membranes by increasing membrane fluidity [155]. This interaction with membranes happened *via* carpet formation [156] followed by lipid extraction in a concentration dependent manner, which led to propose that hIAPP aggregation prior to membrane interaction may not be necessary for its cytotoxicity.

Single Aβ aggregates could also be detected in the cerebrospinal fluid of AD patients by FCS [157, 158], suggesting that this technique could enable easy *in vivo* detection of cerebral amyloid and might hold potential value for enhancing routine diagnosis. FCS has also been applied to study the oligomerization of amyloidogenic polypeptides such as Aβ [159] and the polyQ stretch within disease-causing proteins [160]. In both cases, FCS-derived data on the dynamics of individual fluorescent molecules in solution provided valuable insights into the respective polymerization processes. In sodium phosphate buffer, several different sized $A\beta_{1-42}$ components (monomer, oligomers, protofibrils and fibrils) could be identified simultaneously, whereas in artificial cerebrospinal fluid, the faster kinetics of fibrillogenesis led to the formation of fibrils on very short timescales [161].

Different Models for the Amyloid Fibril Structure

The similarities between amyloid fibrils derived from very different peptides and proteins suggest that their core structures have similar features primarily dictated by the intrinsic conformational preferences of polypeptide chains [2]. However, the specific nature of the side-chain packing, including such characteristics as the alignment of adjacent strands and the separation of the sheets, provides an explanation for the occurrence of variations in the structural details of different types of fibrils. Putting together the above data derived from SSNMR, X-ray crystallography, and TEM, a picture of the common properties of amyloid fibrils emerges that includes, as a universal element, the cross-β X-ray diffraction pattern. This pattern consists of an X-ray reflection at ~4.8 Å resolution along the fibril axis and another X-ray reflection at ~10-12 Å resolution perpendicular to the fibril [82], indicating that the fibrils contain β-sheets parallel to the fibril axis with their extended protein strands perpendicular to the axis. Amyloid structures can be built of parallel or antiparallel sheets, that in turn may or may not be in register [18, 162]. Unfortunately, if the strands are out of register rather than being antiparallel or if the sheet is composed of a mixture of parallel and antiparallel strands, misleading results may be generated [18], which might explain the wide variety in reported results. To date, antiparallel β-sheet structures have been identified only in fibrils formed by relatively short peptides containing only one β-strand segment, indicating that studies of model peptides can not generally be used to infer structural properties of full-length amyloid-forming sequences [22]. The more fundamental feature, which is common to all amyloid and amyloid-like fibrils, appears to be the dry steric zipper motif present in the structure of the cross-β spine [83].

Other alternative models center around a β-helical or nanotube structure that has been suggested as a possible generic architecture for the amyloid fibril (reviewed in [17]). In these models, one or more extended β-sheets wrap around a hollow core in a helical manner. Modeling studies suggested that tilted arrays of β-strands constitute tubular, 30 Å-diameter protofilaments, and that three to five of these protofilaments constitute the Aβ fibril. This type of multimeric array of protofilaments organized as a tubular fibril resembles that formed by the shorter $A\beta_{6-25}$, $A\beta_{11-28}$, and $A\beta_{1-28}$ [163, 164], which hints at a common structural motif in Aβ fibril organization. AFM data of $A\beta_{1-42}$ fibrillogenesis [65], suggested that the process starts with the association of peptide monomers or oligomers to form ribbon-like β-sheets 1.5 nm high and 5.5 nm across, termed protofibrils. Protofibrils have been observed to grow into long protofilaments that intertwine into helical structures [23, 48, 65], but also to form short 100-nm bundles [65]. Helices and bundles are present in immature nodular fibrils, but not in mature smooth fibrils, 11.5 nm in diameter and up to several microns long, suggested to be formed by a cylindrical array of tightly packed parallel protofilaments [129] (Fig. **3**).

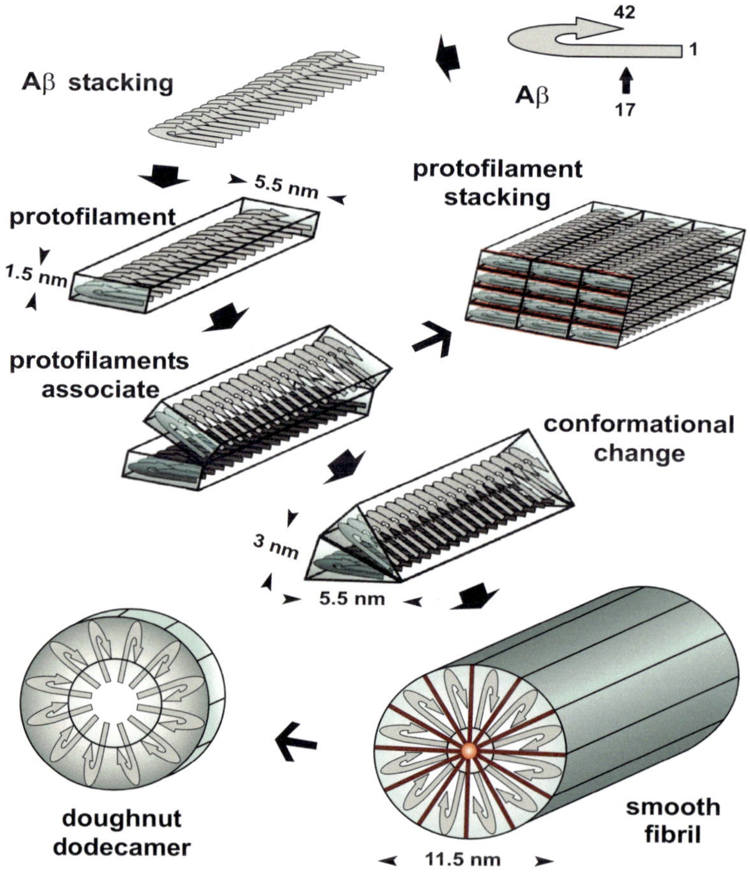

Figure 3: Proposed model for $A\beta_{1-42}$ smooth fibrils. Wedged cylinder *vs*. protofilament stacking models. Hydrophobic boundaries are drawn as red lines. (Ref. [129], with permission).

Glycosaminoglycans as Amyloid Fibrillogenesis Enhancers

Since amyloid fibril deposition is an early process in the development of AD, inhibition of plaque formation is usually proposed as one of the main therapeutic strategies [165, 166]. However, if the pathogenic species is a fibril precursor, blocking fibrillogenesis could accelerate the disease by facilitating the accumulation of the toxic component [167]. Indeed, current data have shown that oligomers of Aβ that are intermediate species in the assembly of fibrils are powerful neurotoxins, and can therefore be the key effectors of cytotoxicity [168-171]. In this context, fibril formation could be a defense mechanism that would act by hiding and immobilizing the neurotoxic soluble forms [171, 172]. Other evidences, nevertheless, indicate that the neurotoxic activity of Aβ requires its aggregation [173]. These apparently contradictory observations can be conciliated by the existence of different fibril forms: intermediate nodular fibrils are unstable and might be reservoirs leaking toxic soluble Aβ species, whereas smooth fibrils are the highly stable dead end of fibrillogenesis from where peptide molecules can hardly escape [29].

There are evidences indicating that the appearance of insoluble structures enriched in β-sheets is facilitated by diverse environmental factors [173]. Proteoglycans are highly glycosylated proteins carrying multiple negatively charged glycosaminoglycans (GAGs), which figure among the most abundant components in the extracellular matrix of many tissues including the brain [174]. From a chemical point of view, GAGs are unbranched polymers of repeated disaccharide units, usually composed by an amino sugar and an uronic acid. Except in the case of hyaluronan (HA), GAGs have varying degrees of sulfation. Proteoglycans and GAGs are found associated with all types of amyloid deposits [165, 175, 176], including Aβ plaques formed in several regions of AD brains, like the hippocampus and the cerebral cortex [177-181]. They can also stabilize mature fibrils against proteolytic degradation [182], and facilitate the formation of fibrils of amylin [183, 184], apo-serum amyloid A [185], α-synuclein [186], prion protein [187], β_2-microglobulin [188], Tau [189], and Aβ [190, 191]. Heparan and chondroitin sulfate GAGs have been described to attenuate the neurotoxic effect of Aβ in primary neuronal cultures [192] and in neuron-like cell lines [129, 193, 194]. A recent study has also shown that heparin strongly induces the aggregation of Dutch-mutated $A\beta_{40}$ and significantly reduces the toxicity of this peptide on cerebrovascular cells [195]. It is conceivable then that GAG-mediated neuroprotection is due to the sequestration of Aβ, in agreement with the hypothesis that the generation of senile plaques in AD would be a protective response aimed at reducing soluble Aβ neurotoxicity [129]. Supportive of this view is the finding that low susceptibility of neurons and cortical areas to neurofibrillar deposition corresponds with high proportions of aggregating chondroitin sulfate proteoglycans in the neuronal microenvironment [196]. $A\beta_{1-42}$ and other self-aggregating peptides share cationic motifs which may be involved in binding to the negative charges of sulfated GAGs [39, 197]. It had been proposed that GAGs may have a scaffolding role, promoting fibrillogenesis-prone conformations of the amyloid precursor proteins [175]. To explore this hypothesis, the association of $A\beta_{1-42}$ with sulfated GAGs, hyaluronan, chitosan, and poly(vinyl sulfate) (PVS) was studied by circular dichroism, Fourier transform infrared spectroscopy, fluorescence microscopy and spectroscopy, protease digestion, AFM, and molecular dynamics simulations [198]. The results indicated that the formation of stable $A\beta_{1-42}$ fibrils was promoted by polymeric GAGs with negative charges placed in-frame with the 4.8-Å separating $A\beta_{1-42}$ monomers within protofibrillar β-sheets (Fig. **4**).

NEW COMPUTATIONAL INSIGHTS INTO CONFORMATIONAL DISEASES

The increasing knowledge on the sequential and structural constraints in protein aggregation, together with a solid classical background on the determinants of protein folding and stability, has allowed the recent development of a number of theoretical methodologies to predict and model protein aggregation. In this second part of the review we shall try to describe some of the new *in silico* approaches to studying protein deposition. First, we will discuss predictive methods relying on the analysis of the physico-chemical and/or structural properties of amino acids in amyloidogenic protein sequences. Then, we will address approaches based on the experimental determination of protein aggregation propensities within amyloidogenic sequence stretches. For the description of already classical programs such as those developed by Serrano's, Caflisch's and Dobson's groups, the reader is directed to excellent reviews published elsewhere [199-201]. We will close this section with a review of recent insights on the mechanistic details of protein self-assembly and deposition, as determined by *in silico* simulation methods.

Prediction Methods Based on Sequence-Structure Relationships in Local Protein Regions

The prediction of aggregation-prone regions in polypeptides does not necessarily need to rely on complex assumptions. A recent example is the program SALSA [202].

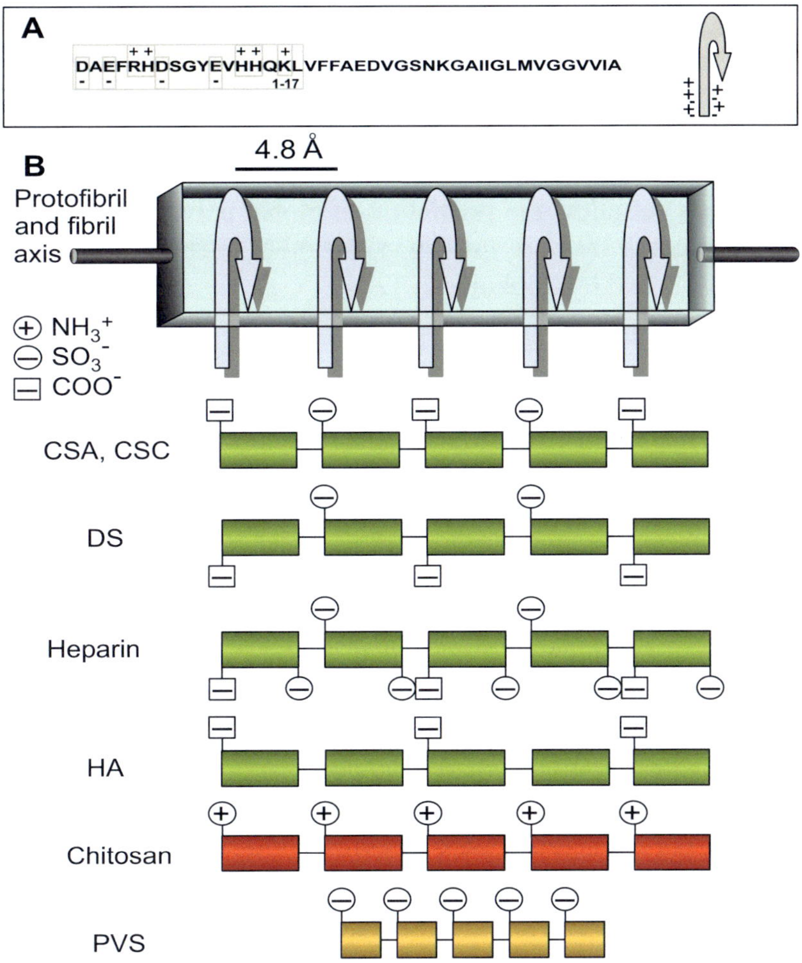

Figure 4: Schematic model of the proposed association of $A\beta_{1-42}$ peptides with polysaccharides and PVS. *A*) Amino acid sequence and representation (arrow folded in a hairpin) of an $A\beta_{1-42}$ peptide. *B*) Illustration of the interaction between $A\beta_{1-42}$ monomers inside a protofibril and different polysaccharides and the synthetic polymer PVS. $A\beta_{1-42}$ self-aggregates into long β-sheets (protofibrils, blue box) where individual β-strands run perpendicular to the axis of amyloid protofibrils and fibrils. The fibrillation of $A\beta_{1-42}$ is enhanced by GAG chains (green models), where the spatial negative charge distribution is in-frame with the distance between $A\beta_{1-42}$ peptides inside protofibrils. This allows an optimal electrostatic attraction between the sulfate and/or carboxyl groups in GAGs and the basic residues in the HHQK cluster at the N-terminus of consecutive or alternate $A\beta_{1-42}$. Chitosan (red model), with in-frame positive charges, hinders $A\beta_{1-42}$ fibrillation, whereas PVS (amber model), a negatively charged polymer with a charge distribution out of frame with protofibrillar $A\beta_{1-42}$ units, facilitates the formation of fibrils of different structure from those obtained in the presence of GAGs. (Ref. [198], with permission).

It calculates the average β-strand propensity score of a peptide window, which the authors name "β-strand contiguity", by a very simple treatment of Chou and

Fasman's secondary structure preference numbers [203]. Despite employing just a single physicochemical property of amino acids, the authors demonstrate that peaks in the SALSA plots correlate well with the location of amyloid fibril cores in three pathogenic proteins: α-synuclein, $A\beta_{1\text{-}40}$ and tau protein. The simplicity of this approach permits fast identification of protein regions with latent β-strand propensity but does not allow the prediction of global polypeptide aggregation or deposition rates because they are influenced by other intrinsic protein properties, such as global charge and hydrophobicity [204].

In contrast to SALSA, Yoon and Welsh [205, 206] based their approach on the hypothesis that the propensity of individual amino acids to adopt a particular secondary structure is influenced by their overall tertiary environment, in addition to their simple physicochemical properties. Therefore, it was argued that secondary structure propensities cannot be determined directly from the sequence, and structural information is also necessary. They quantified the influence of tertiary effects on secondary structural preferences by using a simplified approach that counts the number of atom-to-atom tertiary contacts between nonadjacent residues that are spatially close to one another in the native conformation of a given protein. The benchmark used in the study was SCOP20, a collection of protein domains that exhibit <20% sequence identity between any two members. The sequence-structure relationship of any query sequence was systematically evaluated in terms of tertiary contacts by analyzing the secondary structure preferences of similar template sequences in the database. They use this conformation-based approach to detect nonnative (hidden) sequence propensity for amyloid fibril formation. It assumes that segments with high amyloidogenic propensity should display elevated tendency towards β-sheet formation in tightly packed environments (*i.e.*, those with a high number of tertiary contacts). The method correctly assigns high scores to minimal peptide fragments shown experimentally to mediate amyloid fibril formation in Aβ peptide, IAPP, α-synuclein, and human acetylcholinesterase.

Protein structural information allowed Galzitskaya and coworkers [207] to derive a new parameter, mean packing density (number of residues within a given distance from the considered residue), to be incorporated in the prediction of amyloidogenic and intrinsically disordered regions in protein sequences. The mean packing density of each amino acid was derived from the analysis of a database of protein structures, looking at the number of residues close to any given non-covalently bound neighbour. The observed mean packing density was found to be maximal for the three aromatic residues, Tyr, Phe, and Trp, which have been shown to be relevant for aggregation in different polypeptide systems. The authors realized that protein regions possessing strong packing density correlate with aggregating sequences, which presumably intersect with amyloid promoting regions in proteins. In contrast, regions with weak packing density correspond in many cases to disordered regions of proteins. For any query sequence, the expected packing densities are averaged over a sliding window, and a packing density profile is produced. A region is predicted as amyloidogenic if the expected packing density is above a certain, calibrated, threshold. For eight out of twelve examined disease-related proteins and peptides, the predictions were consistent with experimentally tested amyloidogenic regions.

As discussed above, a breakthrough advance in the amyloid field is the resolution of the atomic structure of a common amyloid cross-β spine involving the formation of steric zippers [16]. This information has opened new avenues of research and paved the way to novel predictions that make use of precise structural information. The central idea in these new structure-based methods is that the formation of the cross-β spine in a fibril might be achieved by a segment of the protein, independent of its location within a particular secondary-structure element [208].

Eisenberg, Baker and co-workers used a structure-based approach for the prediction of fibril formation starting from the crystal structure of the fibril-forming peptide NNQQNY from Sup35 [209]. They assumed that a six-residue sequence stretch was sufficent to drive polypeptide amyloid formation. To identify those segments that might be capable of nucleating fibrillogenesis they used 3D profiling. The side chains in the cross-β spine of NNQQNY were mutated *in silico* to those of the sequence of interest and the energetic fit of these variants to the template ensemble was evaluated. Hexapeptide segments that fit well into the template were selected using the ROSETTADESIGN algorithm energy function [210]. This function incorporates contributions from apolar interactions, hydrogen bonds, and steric overlaps. Because it includes these various factors, it can identify fibril-forming sequences that would not have been selected on the basis of simple intrinsic properties, such as hydrophobicity or β-strand propensity. Fibril-forming segments that have been experimentally observed for lysozyme, muscle myoglobin, $A\beta_{1-42}$, and tau were correctly predicted by this method. Interestingly enough, the method identified fibril-forming regions irrespective of their secondary structure context in the native polypeptide. Thus, segments other than native β-strands of lysozyme and muscle myoglobin, which is mainly an α-helical protein, were identified.

Saiki *et al.* [211] described a method for the evaluation of the propensity to amyloidogenicity using a simplified structural model based in two essential assumptions: The first one is that hydrophobic and hydrogen-bonding interactions occur between residues on neighbouring β-strands aligned in an antiparallel orientation. The second assumption is that interacting hydrophobic residues are present at both faces of the protofibril. The latter condition assumes that the line-matching interactions at both surfaces of a β-sheet protofibril are essential for repetitive stacking. To construct a model for the sheet-to-sheet lamination, the authors looked at interactions between β-sheets in globular proteins and selected the antiparallel system because it is found at a relatively high frequency (*e.g.* in the Greek-key motif). To model the interactions among paired β-strands, they developed a multi-term mathematical expression, which takes into account the hydrophobic interactions of coupled residues and hydrogen bonding interactions, and introduces one term that checks the required coupling of hydrophobic interacting residues at each face of a β-sheet. Twelve polypeptides or protein fragments known to form amyloids were evaluated using a sliding-window technique. The reliability of the method was assessed by experimentally characterizing peptides that according to the predictions have some degree of amyloid formation propensity. Most of these peptides showed fibrillogenic properties. Conversely, peptides predicted to have low amyloid formation propensity were shown to be unable to form amyloids.

Along this line of thought, Trovato and co-workers [212] derived the PASTA algorithm. In this approach, the propensities of two residues to be found within a β-sheet facing one another on neighbouring strands were determined from the analysis of a dataset of globular protein structures. The model assumes that distinct protein molecules involved in fibril formation will adopt the minimum-energy β-pairings in order to better stabilise the cross-β core of the fibril. Two identical protein chains are assumed to associate by means of an ordered pairing of two hydrogen-bonded β-strands of the same length, whereas the remaining parts of the polypeptide chains remain unstructured. All possible pairings are studied by sliding the two strand-forming regions along the corresponding sequences and varying their length and their relative orientations. The two possible orientations, parallel and antiparallel, are considered. In summary, PASTA does a comparison of the energy score of the parallel and antiparallel β-pairings of a sequence stretch with itself. The authors realize that the parallel in-register arrangement provides a natural way of maximizing the number of favourable stacking interactions, lining up hydrophobic and hydrophilic residues in long rows along the fibril axis. PASTA was employed to analyze the sequence of five natively unfolded systems, namely $A\beta_{1-40}$, α-synuclein, IAPP, the PHF43 fragment from tau, and the HET-s prion domain. Good agreement between the predictions and the experimental information available on these amyloid structures was found for most systems.

The formation of short hydrogen bonds between adjacent and parallel β-strands is the main stabilizing force of amyloid structures. Many polypeptide sequences might be accommodated to some extent into a β-sheet conformation. However, the relative β-sheet propensity of the different regions in a sequence and their tendency to join into parallel pairs of β-strands is strongly dependent on the sequence. The regions of the sequence with the best local likelihood to adopt those local conformations will nucleate the aggregation reaction if they become exposed to solvent. This is the principle exploited by BETASCAN [213] (http://betascan.csail.mit.edu/), by assuming that the sequence not only determines the secondary structure, but also the supersecondary structural assembly of individual β-strands into an ordered β-sheet. BETASCAN analyze sequences using pairwise probability tables to capture the preference of each amino acid pair to form hydrogen bonds in a β-sheet. These probabilities were derived by analysing the structures of amphipathic β-sheets in the protein data bank, since β-sheets in amyloid fibrils exhibit differences in solubility between the two faces, one face at the hydrophobic core of the fibril and the other being exposed solvent. BETASCAN is able to identify β-sheet forming regions both in amyloidogenic and repetitive native assemblies, like β-helices. It is a useful algorithm to define the edges of β-strands in regions with an overall high β-sheet propensity and for the identification of protein regions with very low propensity to form pairs of β-strands.

Prediction Methods Based on Experimental Aggregation Propensity Measurements

Although, as described above, theoretical approaches based on structural or sequential intrinsic protein properties have achieved a high accuracy in their predictions, extrinsic factors can dramatically modulate polypeptide aggregation propensities *in vitro* and especially *in vivo*. Hence, several groups have focused their efforts on developing algorithms using *in vitro* or cell-based experimentally determined aggregation propensities, employing different amyloidogenic protein systems. Rojas Quijano and co-workers [214] have used the tau protein as a model. Tau is a largely unstructured protein, but the aggregates it forms display all the physical characteristics of amyloids. Several short nucleating regions have been identified experimentally in tau, among them PHF6 ($_{306}$VQIVYK$_{311}$). The goal of the study was to formulate a model in which the chemical properties of amino acid residues substituted at a single site in the PHF6 structure (VQIVXK, where X was individually substituted by the 20 standard amino acids, except cysteine) could be correlated to the propensity of each peptide mutant to form amyloid fibrils. The kinetics, conformation and morphology of the aggregates formed by the 20 different hexapeptides were analyzed *in vitro*. This allowed for the extrapolation of a scale of amino acid aggregation propensities, which was used to successfully estimate the amyloidogenic propensities of sequences within the tau protein capable of amyloid formation.

The *in vivo* aggregation of polypeptides does not necessarily have to correlate with their *in vitro* properties, since polypeptides within the cell encounter a highly complex and crowded intracellular environment where the protein quality control machinery modulates the accumulation of aggregation-prone polypeptide chains by facilitating their folding, masking hydrophobic regions and targeting improperly folded proteins towards degradation pathways [215, 216].

Measuring *in vivo* polypeptide folding and aggregation has traditionally been a challenging task. Recently, several molecular probes capable of indicating the solubility of target proteins within living cells have been engineered. Using these probes, Waldo and co-workers [217] showed that the folding trajectory of a protein of interest fused upstream to the green fluorescent protein (GFP) dictates the fluorescent behavior of the reporter protein in such a way that the emitted fluorescent signal is directly proportional to the amount of correctly folded target protein. We sought to exploit this approach to study a complete set of mutations in one of the best characterized "hot spots" of aggregation in a disease-linked polypeptide: the central hydrophobic cluster (CHC) of Aβ. We fused the $A\beta_{1-42}$ protein upstream of the GFP and expressed it, as well as 19 other mutants differing only in a single side chain at the central CHC residue, inside prokaryotic cells. We confirmed that the cellular levels of GFP fluorescence depended exclusively on the *in vivo* aggregation propensity of the $A\beta_{42}$ variant. The study allowed us to obtain for the first time an individual intrinsic aggregation propensity scale of the 20 common amino acids in an *in vivo* context [218]. These values were further used to develop a simple approach able to accurately predict protein fragments involved in the aggregation of disease-related proteins and the effect of genetic mutations on their deposition propensities [219]. The AGGRESCAN algorithm was implemented into a web server [220] that allows the simultaneous analysis of the aggregation properties of large sets of protein sequences. This analytical ability might be important for protein production in large-scale structural initiatives, for the analysis of the distribution of aggregation-prone regions in complete genomes or for evolution studies, since it is likely that natural protein sequences have evolved in part to code for avoidance of aggregation. The application of AGGRESCAN to the prediction of aggregation-prone regions in globular proteins is illustrated in (Fig. **5**).

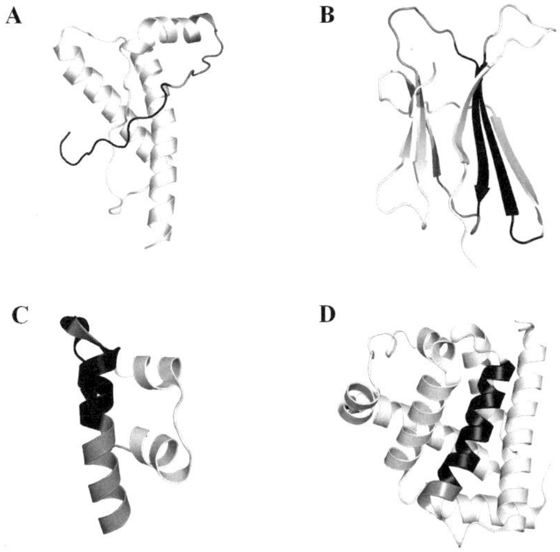

Figure 5: AGGRESCAN predicted aggregation-prone sequences in amyloidogenic proteins. Structural representation of different proteins showing in black the regions where the AGGRESCAN predictions coincide with the experimental data and in grey the regions where experimentally derived data were not predicted by the algorithm. A) Human prion protein (PDB 1QLX), B) Human β2m (1LDS), C) Human insulin (2OMI), and D) Horse heart myoglobin (1AZI).

Remarkably, a comparative analysis of the aggregation propensities obtained by theoretical calculations with those obtained experimentally [220] shows that, despite the diversity of approaches, there is a striking correlation between *in vitro*, *in vivo* and *in silico* data [220]. Likely, this is because aggregation depends on a combination of characteristics, such as hydrophobicity, charge, secondary structure propensity and packing density, all of which are included in either an implicit or explicit way in most algorithms (Table **1**). As a consequence, the performance of programs like AGGRESCAN, not aimed at the specific identification of short amyloidogenic peptides, but rather of aggregation prone sequences within large natural proteins, compares well with those of pure structural algorithms, when analyzing databases of very short peptides containing both amyloid formers and nonformers [220]. Accordingly, as shown in Table **1**, most reviewed algorithms provide overlapping predictions on the determinants of aggregation in disease-linked polypeptides. In (Table **2**) the reader can find a list of World Wide Web available algorithms to predict relative aggregation rates and/or amyloidogenic segments of protein sequences.

Table 1: Comparison between the amyloid forming regions experimentally discovered in disease-linked polypeptides and the stretches predicted by different programs

Protein	Experimental Regions	Predicted Regions						
		SALSA [202]	Yoon and Welsh [205]	Galzitskaya *et al.* [207]	Nelson *et al.* [208]	Saiki *et al.* [211]	PASTA [212]	AGGRESCAN [220]
Aβ protein	17-21 31-36 38-42	**8-40***	11-13; **16-20**	**15-22**	8-15; **18-37**	13; 15; **17-19; 21; 27; 31; 33; 35; 37-38**	**12-20; 31-40**	**17-22; 30-42**
α-synuclein	31-109	**32-89**	no data available	no predicted regions	no data available	4-5; 13; 15-16; 26; **38-39; 46; 48; 50-53; 55; 61; 64; 66; 68; 70; 72-75; 77; 79; 81; 85; 91; 95;** 121	**48-55; 70-77**	**36-42; 49-55; 87-94**
IAPP	8-20	no data available	**6-9; 14-20;** 25-30	**12-18**	no data available	**14; 16-17**	**12-32**	**13-18**
tau	301-320	120-130; 220-230; **296-326;** 390-410; 412-430	no data available	no predicted regions	11-17; 27-31; 61-64; 100-106; 116-150; 162-167; 254-267; 273-287; 294-299 **305-310; 316-319;** 326-331; 337-346; 353-362; 369-377; 385-396; 411-422; 429-441	no data available	**306-310**	**304-311**

* The predicted regions that match with the experimental results are indicated in bold.

Table 2: WWW available algorithms to predict relative aggregation rates and/or amyloidogenic segments of protein sequences

Name	Availability	Reference
AGGRESCAN	http://bioinf.uab.es/aggrescan/	[220]
BETASCAN	http://betascan.csail.mit.edu/	[211]
PAGE	caflisch@bioc.unizh.ch	[200]
PASTA	http://protein.cribi.unipd.it/pasta	[212]
SALSA	l.c.serpell@sussex.ac.uk	[202]
TANGO	http://tango.embl.de/	[201]
ZYGGREGATOR	http://www-vendruscolo.ch.cam.ac.uk/zyggregator.php	[199]

In Silico Simulation of Protein Aggregation

In conformational diseases the appearance of pathological forms of proteins involves a complex aggregation pathway in which the protein can coexist in a number of different monomeric or oligomeric conformational ensembles, each with characteristic kinetic, thermodynamic and binding properties. Thus, the pathway can be viewed as a ragged free-energy landscape whose shape is delineated by the collection of protein conformations. Although the complete characterization of the different conformational states of a protein is a highly desirable goal and would provide the most accurate description of the pathway, the experimental means to achieve this objective are not obvious. Theoretically, one can obtain a picture of one of these particular states by applying techniques in the solid state (*i.e.*, X-ray diffraction or SSNMR) or by NMR in solution. However, folding intermediates and oligomers are generally difficult to isolate due to their short half-life and highly flexible structures, and thus they are usually not amenable to these experimental approaches. Computational approaches and specifically molecular dynamics (MD) simulations might aid to fill this gap by providing insights, mechanistic explanations, and guiding experimental studies to tackle the problem of conformational diversity in pathological proteins. Importantly, the effects of specific changes in the polypeptide sequence or in the simulated environment can be also evaluated *in silico*. In addition, computational simulations might be of help in targeted drug design, *e.g.* by identifying chemical compounds aimed to interfere selective steps in the aggregation pathway. The areas of methodology development, simulations of aggregation mechanisms, and computational evaluation of amyloid structures have been the subject of excellent reviews [221, 222]. Here, we will shortly review recent results on MD simulations of pathological protein systems and their possible use towards the development of therapeutic agents with the potential to treat conformational diseases.

All-atom MD simulations with explicit solvent are the most precise *in silico* approximations to model protein folding and assembly processes. These approaches employ high-resolution protein models based on a realistic representation of protein geometry and monitor continuously the positions and forces among all protein atoms along with their surrounding water molecules. They have been used mainly to explore conformational fluctuations in native or intermediate species and to analyze the stability and dynamics of particular proposals for the oligomeric or aggregated amyloid β-sheet structure. The potential of this approach is illustrated by a recent study by Thirumalai and co-workers in which extensive all-atom MD simulations were performed to understand how a preformed highly dynamic oligomeric assembly interacts with a nascent monomer. This is of great interest because metastable oligomers are being increasingly identified as the cytotoxic forms related to onset of disease. The authors studied the incorporation of a monomer of the CHC region of Aβ to well-defined preformed oligomers of different sizes formed by the same Aβ region [223]. The authors observed that when disordered monomer was added to an ordered oligomer, growth occurs largely by a two-phase dock-lock mechanism. The maximum change in the conformation of the monomer occurs during the rapid dock phase whereas in the much slower lock phase, the monomer forms a β-strand that is in register with the rest of the oligomer. An *a priori* unexpected observation was that, unlike during fibril growth, the initially ordered oligomer also partially disorders before forming a stable ordered structure. This dock and lock behavior could be a generic path for formation of toxic oligomers of amyloidogenic peptides and thus a target for therapeutic intervention.

Pure all-atom simulations are extremely demanding in computational terms because the length of a simulation is approximately proportional to the number of considered atoms to the third power. This complexity can be reduced by eliminating the solvent through its implicit consideration within the potential function governing the interactions between the different species in the system. This strategy was employed by Cheon and co-workers [224] to investigate the early stages of the oligomerization process for two fragments of the Aβ peptide: $A\beta_{16-22}$ (KLVFFAE) and $A\beta_{25-35}$ (GSNKGAIIGLM) respectively. The MD simulations were carried out with PROFASI [225], a Monte Carlo algorithm developed by Irback and co-workers, which implements an implicit water all-atom model, and included systems of 20 peptides for each Aβ fragment. The conversion of the monomeric peptides into oligomers could be rationalized into a process occurring through a generic two-step mechanism modulated by the competition between hydrogen bonding and peptide hydrophobicity (Fig. **6**). Depending on the balance of forces, the first step may be the coalescing of the peptides into more or less ordered oligomers. This step could be fast or even nonexistent, and, since hydrophobic forces drive it, it is nonspecific. The slower, second step comprises the reorganization of the peptides to form ordered oligomers, a process dependent on the formation of specific interchain hydrogen bonds by atoms in the peptide backbone, which results in the generation of β-sheets within the oligomeric species.

The kinetics of spontaneous amyloid fibril formation might take from seconds to days. The access to those time scales in the simulations can be only gained by

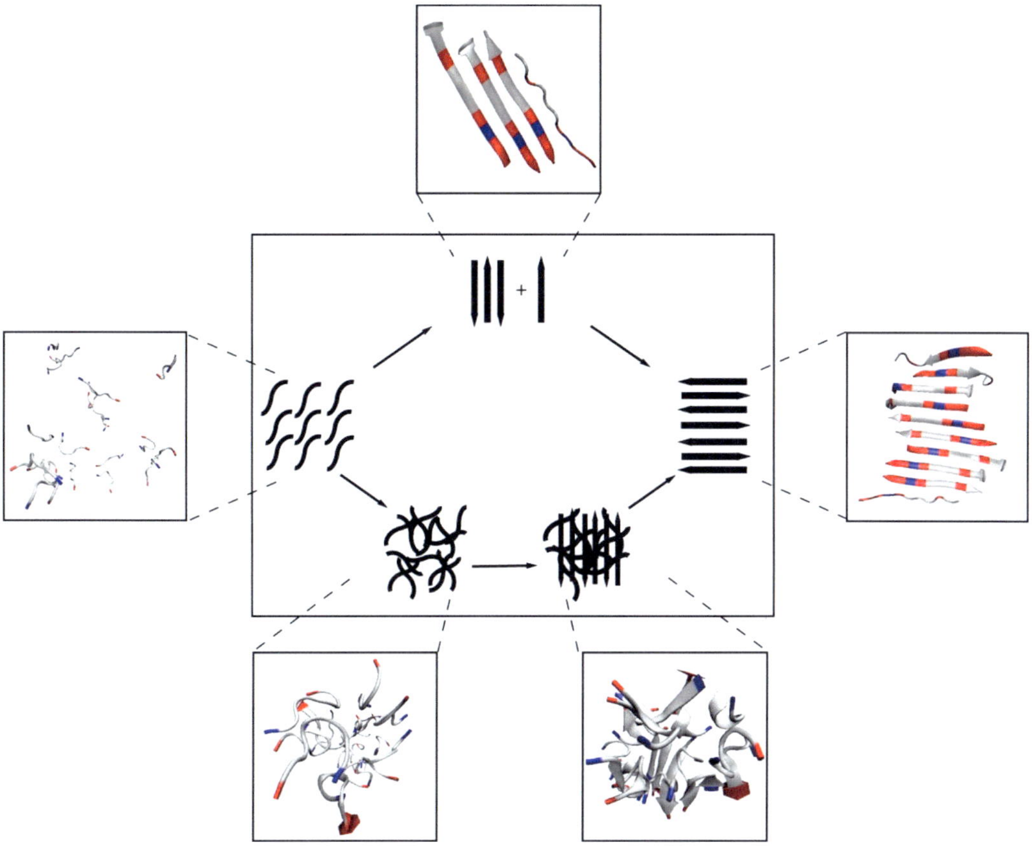

Figure 6: General scheme of amyloid fibril formation pathways simulated by MD. The figure ilustrates the possible pathways that may follow peptide monomers to form myloid fibrils (far right). Representative structures of the simulated Aβ_{25-35} (GSNKGAIIGLM) and Aβ_{16-22} (KLVFFAE) fragments are shown. For Aβ_{25-35} the formation of an ordered oligomer proceeds directly *via* selective backbone interchain hydrogen bonding. Representative structures of are indicated by an arrow. In contrast, for Aβ_{16-22} the first assembly step is nonspecific, sustained by hydrophobic interactions, and leads to the formation of weakly ordered oligomers. A second slower step results in the reorganization of the oligomers through backbone hydrogen bonding to gain the β-sheet structure. (Ref. [224], with permission).

sacrificing some detail in the models. In these simplified systems, beads with associated specific physicochemical properties represent the amino acids on the poylypeptide chain and, usually, accelerated methods are used to sample the resulting vast conformational space. This way, Caflisch and co-workers have reduced computational demand by using a coarse-grained model polypeptide composed of ten spherical beads disposed to have an overall amphipathic character and simulating 125 monomers in a cubic box [226]. They provide the monomer with internal flexibility and consider simplified free energy profiles with only two minima at the β-amyloid-competent state and the amyloid-protected state. The latter corresponds to the ensemble of conformers not compatible with the cross-β structure in a fibril. This strategy allows kinetic and thermodynamic analysis of the system. Interestingly, the results provide insights into the experimentally observed diversity of fibril formation mechanisms, suggesting that they depend on the stability of the amyloid-competent state of the monomer. It appears that in spite of the common final structure of the fibrils, high and low β-prone sequences encode for totally different aggregation kinetics. According to the model, fibrillogenesis of β-stable polypeptides would follow a downhill pathway without significant accumulation of intermediates, whereas aggregation of β-unstable sequences requires travelling through a series of on-pathway oligomeric states in the lag phase, before fibril elongation occurs. It follows that, in this latter case, the presence of metastable off-pathway assemblies can compete with and decrease the likelihood of on-pathway interactions that would finally promote fibril formation. Still, even in this situation, fibrillar structures would represent a lower free energy state than metastable off-pathway oligomers, thus accounting for the thermodynamic bias towards the population of amyloidogenic pathways. Overall, these results imply that different strategies should be used to block the growth of fibrils depending on the nature of the polypeptide.

In general, despite their simplicity, coarse-grained approaches using minimalist models have allowed the construction of free-energy landscapes for protein aggregation processes. Nevertheless, although they are well suited to delineate general rules, they can hardly be used to address specific problems where selective backbone and side chains interactions modulate fibrillogenesis. This task can be undertaken with intermediate-resolution models containing explicit representations of the protein backbone and side chains. This way, Hall *et al.* developed the PRIME model [227, 228], which represents each amino acid with four beads, three for the backbone and one for the side chain. This allows the treatment of large multichain systems while maintaining a fairly realistic description of protein dynamics. They combined the model with discontinuous molecular dynamics, a fast sampling procedure that is applicable to systems of molecules interacting *via* discontinuous potentials. Solvent was modeled implicitly and backbone hydrogen bonding in explicit detail. Overall, the approach allowed sampling much wider regions of conformational space, longer time scales, and larger systems than in traditional molecular dynamics. They used polyalanine peptides to benchmark the approach and found that small amorphous aggregates populated the pathway before the formation of the critical nucleus [229]. After nucleation, fibril growth depends on β-sheet elongation by the addition of monomers to the end of each β-sheet and on lateral extension by incorporation of preformed β-sheets.

One of the most promising applications of MD to the protein aggregation field is the possibility to design chemical compounds against specific intermediate conformers in the fibrillogenesis pathway. In the case of AD, the structural complexity of Aβ is a major problem for the design of chemical compounds that bind to it specifically. Importantly, for Aβ, the group of Nussinov has shown that computational simulations might become a powerful discovery tool, since their approach [230] was able to foresee with great detail the solid-NMR structure of the fibrillar state of the peptide [230, 231]. The formation of Aβ fibrils *in vivo* is thought to be mediated by the conformational transition of Aβ from α-helix to β-sheet or from random coil to β-sheet. According to various experimental assays, Aβ may adopt multiple conformations *in vitro*, such as α-helices, β-sheets, or random coils. These conformations depend on the buffer conditions of pH, ionic strength, and solvent properties. Although some chemical compounds have been demonstrated to reduce the cytotoxicity of Aβ peptides, in most cases it is not clear to which Aβ conformation or assembly they bind. The development of toxic conformation-directed drugs to treat AD is especially attractive because of their predicted specificity and low toxicity. MD simulations on the full $A\beta_{1-40}$ peptide by Jiang and co-workers showed that the conformational transition from α-helix to random coil passes through an α-helix/β-sheet intermediate structure. Furthermore, it was observed that the α-helix/β-sheet intermediate structure possesses a core domain, within which four glycine residues (G25, G29, G33, and G37) are essential to β-sheet formation and amyloid fibrillogenesis. The authors proposed that chemical compounds that lock the structure of Aβ into the α-helix/β-sheet intermediate state could potentially inhibit Aβ fibrillogenesis [232]. Under this hypothesis, virtual screening based on molecular docking was performed, targeting an Aβ peptide α-helix/β-sheet intermediate structure extracted from the trajectory of a 50-ns MD simulation on $A\beta_{1-40}$. A commercial database of small-molecule compounds was queried using DOCK4.0 (http://dock.compbio.ucsf.edu/) as the search engine. The top 1,000 ranked entities according to the energy score were re-evaluated, and the interaction with $A\beta_{1-40}$ was carefully inspected. More than hundred candidate compounds were used in biological assays. One of them inhibited $A\beta_{1-42}$ fibrillogenesis in a concentration-dependent manner. Results from several assays led to the proposal that the inhibitor stabilized the β-sheet conformation of Aβ. Modelling suggested that the inhibitor binds to the target mainly through hydrophobic interactions [233]. This work offers a proof of principle that computational simulations might contribute to render valuable chemicals against conformational diseases.

A promising computational technique for the study of amyloid formation is the use of metadynamics (MetaD). MetaD approaches have proven to be useful for study of the conformational landscape of different protein models. However, its use in the study of protein aggregation reactions is still limited [234, 235]. MetaD can overcome the sampling limitations inherent to classical MD simulations [236], since it is a technique relying on the introduction of a bias on a small number of selected parameters, which are able to catch remarkable features of the biological problem under investigation, named “collective variables”. The quantitative comparison of MetaD with experimental biophysical data obtained from *in vitro* aggregation assays might turn a successful to address the mechanistic underlying this complex process [234, 237].

FUTURE DIRECTIONS

The information contained in the present review clearly demonstrates that both structural and computational approaches to the investigation of protein aggregation associated with conformational diseases are undergoing a rapid and very productive phase of growth. Even if in the past, both kind of analysis worked largely independently, we are witnessing the construction of a highly synergic environment in which both kinds of data are integrated to attain an unambiguous and accurate description of the mechanism underlying protein aggregation, the conformers eliciting the cytotoxic effect and the way these processes are regulated within the cell. The knowledge gained at this basic research level its already beginning to be translated into novel and effective alternative therapeutics addressed to the treatment of Alzheimer’s, Parkinson’s and other debilitating diseases caused by protein misfolding and deposition.

ACKNOWLEDGEMENTS

Space limitation precludes citation of a large body of excellent work by colleagues in the fields discussed. We acknowledge these efforts here.

CONFLICT OF INTEREST

The authors confirm that this chapter contents have no conflicts of interest.

DISCLOSURE

This is an updated report of the earlier review article by Fernàndez-Busquets X, de Groot NS, Fernandez D, Ventura S “Recent structural and computational insights into conformational diseases (review)" Curr Med Chem, 2008, 15(13): 1336-49.

ABBREVIATIONS

3D	= Three-dimensional
Aβ	= Amyloid-β protein
AD	= Alzheimer's disease
AFM	= Atomic force microscopy
β2m	= β2 microglobulin
CHC	= Central hydrophobic cluster
Cryo-TEM	= Cryogenic transmission electron microscopy
FRAP	= Fluorescence recovery after photobleaching
FCS	= Fluorescence correlation spectroscopy
GFP	= Green fluorescent protein
IAPP	= Islet amyloid polypeptide
Ig	= Immunoglobulin
MAS	= Magic-angle spinning
MD	= Molecular dynamics
NMR	= Nuclear magnetic resonance
NSOM	= Near-field scanning optical microscopy
PASTA	= Prediction of amyloid structure aggregation
PDB	= Brookhaven protein data bank
polyQ	= Polyglutamine

RNAse A = Ribonuclease A

SALSA = Single algorithm for sliding averages

SH3 = Src homology 3

SMFS = Single molecule force spectroscopy

SPM = Scanning probe microscopy

SSNMR = Solid-state nuclear magnetic resonance

STEM = Scanning transmission electron microscopy

STM = Scanning tunneling microscopy

TEM = Transmission electron microscopy

ThT = Thioflavin T

TIRFM = Total internal reflection fluorescence microscopy

TTR = Transthyretin

REFERENCES

[1] Luheshi LM, Tartaglia GG, Brorsson AC, *et al.* Systematic *In Vivo* Analysis of the Intrinsic Determinants of Amyloid beta Pathogenicity. PLoS Biol 2007; 5(11): e290.

[2] Chiti F, Dobson CM. Protein misfolding, functional amyloid, and human disease. Annu Rev Biochem 2006; 75: 333-66.

[3] Dobson CM. Protein misfolding, evolution and disease. Trends Biochem Sci 1999; 24: 329-32.

[4] Fandrich M, Dobson CM. The behaviour of polyamino acids reveals an inverse side chain effect in amyloid structure formation. EMBO J 2002; 21(21): 5682-90.

[5] Huff ME, Balch WE, Kelly JW. Pathological and functional amyloid formation orchestrated by the secretory pathway. Curr Opin Struct Biol 2003; 13(6): 674-82.

[6] Invernizzi G, Papaleo E, Sabate R, Ventura S. Protein aggregation: mechanisms and functional consequences. Int J Biochem Cell Biol 2012; 44(9): 1541-54.

[7] Lomas DA, Carrell RW. Serpinopathies and the conformational dementias. Nat Rev Genet 2002; 3(10): 759-68.

[8] Selkoe DJ. Folding proteins in fatal ways. Nature 2003; 426(6968): 900-4.

[9] Bossy-Wetzel E, Schwarzenbacher R, Lipton SA. Molecular pathways to neurodegeneration. Nat Med 2004; 10 Suppl: S2-9.

[10] Stefani M. Protein misfolding and aggregation: new examples in medicine and biology of the dark side of the protein world. Biochim Biophys Acta 2004; 1739: 5-25.

[11] Xu J, Reumers J, Couceiro JR, *et al.* Gain of function of mutant p53 by coaggregation with multiple tumor suppressors. Nat Chem Biol 2011; 7(5): 285-95.

[12] Petkova AT, Ishii Y, Balbach JJ, *et al.* A structural model for Alzheimer's beta-amyloid fibrils based on experimental constraints from solid state NMR. ProcNatlAcadSciUSA 2002; 99(26): 16742.
[13] Jaroniec CP, MacPhee CE, Astrof NS, Dobson CM, Griffin RG. Molecular conformation of a peptide fragment of transthyretin in an amyloid fibril. ProcNatlAcadSciUSA 2002; 99(26): 16748.
[14] Ritter C, Maddelein ML, Siemer AB, *et al.* Correlation of structural elements and infectivity of the HET-s prion. Nature 2005; 435(7043): 844.
[15] Makin OS, Serpell LC. Structures for amyloid fibrils. FEBS J 2005; 272(23): 5950.
[16] Nelson R, Sawaya MR, Balbirnie M, *et al.* Structure of the cross-beta spine of amyloid-like fibrils. Nature 2005; 435(7043): 773-8.
[17] Makin OS, Atkins E, Sikorski P, Johansson J, Serpell LC. Molecular basis for amyloid fibril formation and stability. ProcNatlAcadSciUSA 2005; 102(2): 315.
[18] Makin OS, Serpell LC. Examining the structure of the mature amyloid fibril. BiochemSocTrans 2002; 30(4): 521.
[19] Uversky VN, Kabanov AV, Lyubchenko YL. Nanotools for megaproblems: probing protein misfolding diseases using nanomedicine modus operandi. J ProteomeRes 2006; 5(10): 2505.
[20] Baldus M. Solid-state NMR spectroscopy: molecular structure and organization at the atomic level. AngewChemIntEd Engl 2006; 45(8): 1186.
[21] Tycko R. Characterization of amyloid structures at the molecular level by solid state nuclear magnetic resonance spectroscopy. Methods Enzymol 2006; 413: 103.
[22] Tycko R. Solid-state NMR as a probe of amyloid structure. Protein PeptLett 2006; 13(3): 229.
[23] Naito A, Kawamura I. Solid-state NMR as a method to reveal structure and membrane-interaction of amyloidogenic proteins and peptides. Biochim Biophys Acta 2007; 1768(8): 1900-12.
[24] Debelouchina GT, Bayro MJ, Fitzpatrick AW, *et al.* Higher order amyloid fibril structure by MAS NMR and DNP spectroscopy. Journal of the American Chemical Society 2013; 135(51): 19237-47.
[25] Tycko R. NMR at low and ultralow temperatures. Accounts of chemical research 2013; 46(9): 1923-32.
[26] Comellas G, Lemkau LR, Zhou DH, George JM, Rienstra CM. Structural intermediates during alpha-synuclein fibrillogenesis on phospholipid vesicles. Journal of the American Chemical Society 2012; 134(11): 5090-9.
[27] Comellas G, Rienstra CM. Protein structure determination by magic-angle spinning solid-state NMR, and insights into the formation, structure, and stability of amyloid fibrils. Annual review of biophysics 2013; 42: 515-36.
[28] Frederick KK, Debelouchina GT, Kayatekin C, *et al.* Distinct prion strains are defined by amyloid core structure and chaperone binding site dynamics. Chemistry & biology 2014; 21(2): 295-305.
[29] Tycko R. Solid-state NMR studies of amyloid fibril structure. Annual review of physical chemistry 2011; 62: 279-99.
[30] Fitzpatrick AW, Debelouchina GT, Bayro MJ, *et al.* Atomic structure and hierarchical assembly of a cross-beta amyloid fibril. Proceedings of the National Academy of Sciences of the United States of America 2013; 110(14): 5468-73.
[31] Fitzpatrick AW, Lorenz UJ, Vanacore GM, Zewail AH. 4D cryo-electron microscopy of proteins. Journal of the American Chemical Society 2013; 135(51): 19123-6.
[32] Lu JX, Qiang W, Yau WM, *et al.* Molecular structure of beta-amyloid fibrils in Alzheimer's disease brain tissue. Cell 2013; 154(6): 1257-68.
[33] Lansbury PT, Jr., Costa PR, Griffiths JM, *et al.* Structural model for the beta-amyloid fibril based on interstrand alignment of an antiparallel-sheet comprising a C-terminal peptide. NatStructBiol 1995; 2(11): 990.
[34] Benzinger TL, Gregory DM, Burkoth TS, *et al.* Two-dimensional structure of beta-amyloid(10-35) fibrils. Biochemistry 2000; 39(12): 3491.
[35] Antzutkin ON, Leapman RD, Balbach JJ, Tycko R. Supramolecular structural constraints on Alzheimer's beta-amyloid fibrils from electron microscopy and solid-state nuclear magnetic resonance. Biochemistry 2002; 41(51): 15436.
[36] Kajava AV, Baxa U, Wickner RB, Steven AC. A model for Ure2p prion filaments and other amyloids: the parallel superpleated beta-structure. ProcNatlAcadSciUSA 2004; 101(21): 7885.
[37] Wille H, Michelitsch MD, Guenebaut V, *et al.* Structural studies of the scrapie prion protein by electron crystallography. ProcNatlAcadSciUSA 2002; 99(6): 3563.

[38] Guo JT, Wetzel R, Xu Y. Molecular modeling of the core of Abeta amyloid fibrils. Proteins 2004; 57(2): 357.

[39] Diaz-Avalos R, Long C, Fontano E, *et al.* Cross-beta order and diversity in nanocrystals of an amyloid-forming peptide. J MolBiol 2003; 330(5): 1165.

[40] Antzutkin ON, Balbach JJ, Leapman RD, *et al.* Multiple quantum solid-state NMR indicates a parallel, not antiparallel, organization of beta-sheets in Alzheimer's beta-amyloid fibrils. ProcNatlAcadSciUSA 2000; 97(24): 13045.

[41] Balbach JJ, Petkova AT, Oyler NA, *et al.* Supramolecular structure in full-length Alzheimer's beta-amyloid fibrils: evidence for a parallel beta-sheet organization from solid-state nuclear magnetic resonance. BiophysJ 2002; 83(2): 1205.

[42] Antzutkin ON, Balbach JJ, Tycko R. Site-specific identification of non-beta-strand conformations in Alzheimer's beta-amyloid fibrils by solid-state NMR. BiophysJ 2003; 84(5): 3326.

[43] Petkova AT, Leapman RD, Guo Z, *et al.* Self-propagating, molecular-level polymorphism in Alzheimer's beta-amyloid fibrils. Science 2005; 307(5707): 262-5.

[44] Oyler NA, Tycko R. Absolute structural constraints on amyloid fibrils from solid-state NMR spectroscopy of partially oriented samples. J AmChemSoc 2004; 126(14): 4478.

[45] Balbach JJ, Ishii Y, Antzutkin ON, *et al.* Amyloid fibril formation by A beta 16-22, a seven-residue fragment of the Alzheimer's beta-amyloid peptide, and structural characterization by solid state NMR. Biochemistry 2000; 39(45): 13748.

[46] Tycko R, Ishii Y. Constraints on supramolecular structure in amyloid fibrils from two-dimensional solid-state NMR spectroscopy with uniform isotopic labeling. J AmChemSoc 2003; 125(22): 6606.

[47] Petkova AT, Buntkowsky G, Dyda F, *et al.* Solid state NMR reveals a pH-dependent antiparallel beta-sheet registry in fibrils formed by a beta-amyloid peptide. J MolBiol 2004; 335(1): 247.

[48] Luhrs T, Ritter C, Adrian M, *et al.* 3D structure of Alzheimer's amyloid-beta(1-42) fibrils. ProcNatlAcadSciUSA 2005; 102(48): 17342.

[49] Kawahara M, Kuroda Y. Molecular mechanism of neurodegeneration induced by Alzheimer's beta-amyloid protein: channel formation and disruption of calcium homeostasis. Brain ResBull 2000; 53(4): 389.

[50] Kayed R, Sokolov Y, Edmonds B, *et al.* Permeabilization of lipid bilayers is a common conformation-dependent activity of soluble amyloid oligomers in protein misfolding diseases. JBiol Chem 2004; 279(45): 46363.

[51] Lin MC, Kagan BL. Electrophysiologic properties of channels induced by Abeta25-35 in planar lipid bilayers. Peptides 2002; 23(7): 1215.

[52] Ravault S, Soubias O, Saurel O, *et al.* Fusogenic Alzheimer's peptide fragment Abeta (29-42) in interaction with lipid bilayers: secondary structure, dynamics, and specific interaction with phosphatidyl ethanolamine polar heads as revealed by solid-state NMR. Protein Sci 2005; 14(5): 1181.

[53] Wiltzius JJ, Landau M, Nelson R, *et al.* Molecular mechanisms for protein-encoded inheritance. Nature structural & molecular biology 2009; 16(9): 973-8.

[54] Wasmer C, Lange A, Van Melckebeke H, *et al.* Amyloid fibrils of the HET-s(218-289) prion form a beta solenoid with a triangular hydrophobic core. Science 2008; 319(5869): 1523-6.

[55] Jaroniec CP, MacPhee CE, Bajaj VS, *et al.* High-resolution molecular structure of a peptide in an amyloid fibril determined by magic angle spinning NMR spectroscopy. ProcNatlAcadSciUSA 2004; 101(3): 711.

[56] Kammerer RA, Kostrewa D, Zurdo J, *et al.* Exploring amyloid formation by a *de novo* design. ProcNatlAcadSciUSA 2004; 101(13): 4435.

[57] Shewmaker F, McGlinchey RP, Thurber KR, *et al.* The functional curli amyloid is not based on in-register parallel beta-sheet structure. The Journal of biological chemistry 2009; 284(37): 25065-76.

[58] Laws DD, Bitter HM, Liu K, *et al.* Solid-state NMR studies of the secondary structure of a mutant prion protein fragment of 55 residues that induces neurodegeneration. ProcNatlAcadSciUSA 2001; 98(20): 11686.

[59] Neto-Silva RM, Macedo-Ribeiro S, Pereira PJ, *et al.* X-ray crystallographic studies of two transthyretin variants: further insights into amyloidogenesis. Acta CrystallogrD Biol Crystallogr 2005; 61(Pt 3): 333.

[60] Rosano C, Zuccotti S, Bolognesi M. The three-dimensional structure of beta2 microglobulin: results from X-ray crystallography. BiochimBiophysActa 2005; 1753(1): 85.
[61] Wiltzius JJ, Sievers SA, Sawaya MR, *et al.* Atomic structure of the cross-beta spine of islet amyloid polypeptide (amylin). Protein science: a publication of the Protein Society 2008; 17(9): 1467-74.
[62] McDonald M, Box H, Bian W, *et al.* Fiber diffraction data indicate a hollow core for the Alzheimer's abeta 3-fold symmetric fibril. Journal of molecular biology 2012; 423(3): 454-61.
[63] Goldsbury CS, Wirtz S, Muller SA, *et al.* Studies on the *in vitro* assembly of a beta 1-40: implications for the search for a beta fibril formation inhibitors. JStructBiol 2000; 130(2-3): 217.
[64] Chamberlain AK, MacPhee CE, Zurdo J, *et al.* Ultrastructural organization of amyloid fibrils by atomic force microscopy. BiophysJ 2000; 79(6): 3282.
[65] Arimon M, Diez-Perez I, Kogan MJ, *et al.* Fine structure study of Abeta1-42 fibrillogenesis with atomic force microscopy. FASEB J 2005; 19(10): 1344.
[66] Ruben GC, Wang JZ, Iqbal K, Grundke-Iqbal I. Paired helical filaments (PHFs) are a family of single filament structures with a common helical turn period: negatively stained PHF imaged by TEM and measured before and after sonication, deglycosylation, and dephosphorylation. MicroscResTech 2005; 67(3-4): 175.
[67] Serpell LC, Smith JM. Direct visualisation of the beta-sheet structure of synthetic Alzheimer's amyloid. J MolBiol 2000; 299(1): 225.
[68] Jimenez JL, Guijarro JI, Orlova E, *et al.* Cryo-electron microscopy structure of an SH3 amyloid fibril and model of the molecular packing. EMBO J 1999; 18(4): 815-21.
[69] Jimenez JL, Tennent G, Pepys M, Saibil HR. Structural diversity of *ex vivo* amyloid fibrils studied by cryo-electron microscopy. J MolBiol 2001; 311(2): 241.
[70] Sumner MO, Serpell LC. Structural characterisation of islet amyloid polypeptide fibrils. J MolBiol 2004; 335(5): 1279.
[71] Tycko R. Progress towards a molecular-level structural understanding of amyloid fibrils. CurrOpinStructBiol 2004; 14(1): 96.
[72] Paravastu AK, Leapman RD, Yau WM, Tycko R. Molecular structural basis for polymorphism in Alzheimer's beta-amyloid fibrils. Proceedings of the National Academy of Sciences of the United States of America 2008; 105(47): 18349-54.
[73] Kajava AV, Aebi U, Steven AC. The parallel superpleated beta-structure as a model for amyloid fibrils of human amylin. J MolBiol 2005; 348(2): 247.
[74] Goldsbury CS, Cooper GJ, Goldie KN, *et al.* Polymorphic fibrillar assembly of human amylin. J StructBiol 1997; 119(1): 17.
[75] Chen B, Thurber KR, Shewmaker F, Wickner RB, Tycko R. Measurement of amyloid fibril mass-per-length by tilted-beam transmission electron microscopy. Proceedings of the National Academy of Sciences of the United States of America 2009; 106(34): 14339-44.
[76] Goldsbury C, Baxa U, Simon MN, *et al.* Amyloid structure and assembly: insights from scanning transmission electron microscopy. Journal of structural biology 2011; 173(1): 1-13.
[77] Serpell LC, Sunde M, Fraser PE, *et al.* Examination of the structure of the transthyretin amyloid fibril by image reconstruction from electron micrographs. J MolBiol 1995; 254(2): 113.
[78] Meinhardt J, Sachse C, Hortschansky P, Grigorieff N, Fandrich M. Abeta(1-40) fibril polymorphism implies diverse interaction patterns in amyloid fibrils. Journal of molecular biology 2009; 386(3): 869-77.
[79] Dobson CM. Protein folding and misfolding. Nature 2003; 426(6968): 884-90.
[80] Collins SR, Douglass A, Vale RD, Weissman JS. Mechanism of prion propagation: amyloid growth occurs by monomer addition. PLoSBiol 2004; 2(10): e321.
[81] Tanaka M, Collins SR, Toyama BH, Weissman JS. The physical basis of how prion conformations determine strain phenotypes. Nature 2006; 442(7102): 585-9.
[82] Sunde M, Serpell LC, Bartlam M, *et al.* Common core structure of amyloid fibrils by synchrotron X-ray diffraction. JMolBiol 1997; 273(3): 729.
[83] Eisenberg D, Nelson R, Sawaya MR, *et al.* The structural biology of protein aggregation diseases: Fundamental questions and some answers. AccChemRes 2006; 39(9): 568.
[84] Cornish PV, Ha T. A survey of single-molecule techniques in chemical biology. ACS ChemBiol 2007; 2(1): 53.

[85] Lyubchenko YL, Sherman S, Shlyakhtenko LS, Uversky VN. Nanoimaging for protein misfolding and related diseases. J Cell Biochem 2006; 99(1): 52.
[86] Ban T, Hoshino M, Takahashi S, *et al.* Direct observation of Abeta amyloid fibril growth and inhibition. Journal of molecular biology 2004; 344(3): 757-67.
[87] Krasnoslobodtsev AV, Portillo AM, Deckert-Gaudig T, Deckert V, Lyubchenko YL. Nanoimaging for prion related diseases. Prion 2010; 4(4): 265-74.
[88] Binnig G, Quate CF, Gerber C. Atomic force microscope. Phys Rev Lett 1986; 56(9): 930-3.
[89] Poggi MA, Gadsby ED, Bottomley LA, *et al.* Scanning probe microscopy. Anal Chem 2004; 76(12): 3429-43.
[90] Engel A, Muller DJ. Observing single biomolecules at work with the atomic force microscope. NatStructBiol 2000; 7(9): 715.
[91] Horber JK, Miles MJ. Scanning probe evolution in biology. Science 2003; 302(5647): 1002.
[92] Shivji AP, Brown F, Davies MC, *et al.* Scanning tunnelling microscopy studies of beta-amyloid fibril structure and assembly. FEBS Lett 1995; 371(1): 25.
[93] Wang Z, Zhou C, Wang C, *et al.* AFM and STM study of beta-amyloid aggregation on graphite. Ultramicroscopy 2003; 97(1-4): 73.
[94] Losic D, Martin LL, Mechler A, Aguilar MI, Small DH. High resolution scanning tunnelling microscopy of the beta-amyloid protein (Abeta1-40) of Alzheimer's disease suggests a novel mechanism of oligomer assembly. J StructBiol 2006; 155(1): 104.
[95] Yang Y, Wang C. Single-molecule studies on individual peptides and peptide assemblies on surfaces. Philosophical transactions Series A, Mathematical, physical, and engineering sciences 2013; 371(2000): 20120311.
[96] Ma X, Liu L, Mao X, *et al.* Amyloid beta (1-42) folding multiplicity and single-molecule binding behavior studied with STM. Journal of molecular biology 2009; 388(4): 894-901.
[97] Fritz J, Anselmetti D, Jarchow J, Fernandez-Busquets X. Probing single biomolecules with atomic force microscopy. JStructBiol 1997; 119(2): 165.
[98] Gosal WS, Myers SL, Radford SE, Thomson NH. Amyloid under the atomic force microscope. Protein PeptLett 2006; 13(3): 261.
[99] Kowalewski T, Holtzman DM. *In situ* atomic force microscopy study of Alzheimer's beta-amyloid peptide on different substrates: new insights into mechanism of beta-sheet formation. ProcNatlAcadSciUSA 1999; 96(7): 3688.
[100] Khurana R, Ionescu-Zanetti C, Pope M, *et al.* A general model for amyloid fibril assembly based on morphological studies using atomic force microscopy. BiophysJ 2003; 85(2): 1135.
[101] Hoyer W, Cherny D, Subramaniam V, Jovin TM. Rapid self-assembly of alpha-synuclein observed by *in situ* atomic force microscopy. J MolBiol 2004; 340(1): 127.
[102] Kad NM, Myers SL, Smith DP, *et al.* Hierarchical assembly of beta2-microglobulin amyloid *in vitro* revealed by atomic force microscopy. J MolBiol 2003; 330(4): 785.
[103] Goldsbury C, Kistler J, Aebi U, Arvinte T, Cooper GJ. Watching amyloid fibrils grow by time-lapse atomic force microscopy. J MolBiol 1999; 285(1): 33.
[104] Creasey R, Sharma S, Craig JE, *et al.* Detecting protein aggregates on untreated human tissue samples by atomic force microscopy recognition imaging. Biophysical journal 2010; 99(5): 1660-7.
[105] Stolz M, Stoffler D, Aebi U, Goldsbury C. Monitoring biomolecular interactions by time-lapse atomic force microscopy. Journal of structural biology 2000; 131(3): 171-80.
[106] Ando T. High-speed atomic force microscopy. Microscopy (Oxf) 2013; 62(1): 81-93.
[107] Milhiet PE, Yamamoto D, Berthoumieu O, *et al.* Deciphering the structure, growth and assembly of amyloid-like fibrils using high-speed atomic force microscopy. PloS one 2010; 5(10): e13240.
[108] Capitanio M, Pavone FS. Interrogating biology with force: single molecule high-resolution measurements with optical tweezers. Biophysical journal 2013; 105(6): 1293-303.
[109] Zlatanova J, Lindsay SM, Leuba SH. Single molecule force spectroscopy in biology using the atomic force microscope. ProgBiophysMolBiol 2000; 74(1-2): 37.
[110] Merkel R, Nassoy P, Leung A, Ritchie K, Evans E. Energy landscapes of receptor-ligand bonds explored with dynamic force spectroscopy. Nature 1999; 397(6714): 50.
[111] Evans E, Ritchie K. Dynamic strength of molecular adhesion bonds. BiophysJ 1997; 72(4): 1541.
[112] Hoh JH, Cleveland JP, Prater CB, Revel JP, Hansma PK. Quantized Adhesion Detected with the Atomic Force Microscope. J Am Chem Soc 1992; 114(12): 4917.

[113] Kellermayer MS, Grama L, Karsai A, *et al.* Reversible mechanical unzipping of amyloid beta-fibrils. JBiol Chem 2005; 280(9): 8464.
[114] Smith JF, Knowles TP, Dobson CM, MacPhee CE, Welland ME. Characterization of the nanoscale properties of individual amyloid fibrils. ProcNatlAcadSciUSA 2006; 103(43): 15806.
[115] Ganchev DN, Cobb NJ, Surewicz K, Surewicz WK. Nanomechanical properties of human prion protein amyloid as probed by force spectroscopy. Biophysical journal 2008; 95(6): 2909-15.
[116] Wegmann S, Scholer J, Bippes CA, Mandelkow E, Muller DJ. Competing interactions stabilize pro- and anti-aggregant conformations of human Tau. The Journal of biological chemistry 2011; 286(23): 20512-24.
[117] Hane FT, Attwood SJ, Leonenko Z. Comparison of three competing dynamic force spectroscopy models to study binding forces of amyloid-á (1-42). Soft Matter 2014; 10(12): 1924-30.
[118] McAllister C, Karymov MA, Kawano Y, *et al.* Protein interactions and misfolding analyzed by AFM force spectroscopy. J MolBiol 2005; 354(5): 1028.
[119] Kransnoslobodtsev AV, Shlyakhtenko LS, Ukraintsev E, *et al.* Nanomedicine and protein misfolding diseases. Nanomedicine 2005; 1(4): 300.
[120] Valle-Delgado JJ, Liepina I, Lapidus D, *et al.* Self-assembly of human amylin-derived peptides studied by atomic force microscopy and single molecule force spectroscopy. Soft Matter 2012; 8: 1234-42.
[121] Hoffmann A, Neupane K, Woodside MT. Single-molecule assays for investigating protein misfolding and aggregation. PhysChemChemPhys 2013; 15(21): 7934-48.
[122] Krasnoslobodtsev AV, Volkov IL, Asiago JM, *et al.* alpha-Synuclein misfolding assessed with single molecule AFM force spectroscopy: effect of pathogenic mutations. Biochemistry 2013; 52(42): 7377-86.
[123] Kim BH, Lyubchenko YL. Nanoprobing of misfolding and interactions of amyloid á 42 protein. Nanomedicine: nanotechnology, biology, and medicine 2014; 10(4): 871-8.
[124] Scheibel T, Parthasarathy R, Sawicki G, *et al.* Conducting nanowires built by controlled self-assembly of amyloid fibers and selective metal deposition. Proceedings of the National Academy of Sciences 2003; 100(8): 4527-32.
[125] Reches M, Gazit E. Casting metal nanowires within discrete self-assembled peptide nanotubes. Science 2003; 300(5619): 625-7.
[126] Carny O, Shalev DE, Gazit E. Fabrication of coaxial metal nanocables using a self-assembled peptide nanotube scaffold. Nano Lett 2006; 6(8): 1594-7.
[127] Malisauskas M, Meskys R, Morozova-Roche LA. Ultrathin silver nanowires produced by amyloid biotemplating. Biotechnol Progress 2008; 24(5): 1166-70.
[128] Perutz MF, Finch JT, Berriman J, Lesk A. Amyloid fibers are water-filled nanotubes. ProcNatlAcadSciUSA 2002; 99(8): 5591.
[129] Bravo R, Arimon M, Valle-Delgado JJ, *et al.* Sulfated polysaccharides promote the assembly of amyloid á 1-42 peptide into stable fibrils of reduced cytotoxicity. JBiolChem 2008; 283(47): 32471-83.
[130] Kessler M, Gaub HE. Unfolding barriers in bacteriorhodopsin probed from the cytoplasmic and the extracellular side by AFM. Structure 2006; 14(3): 521.
[131] Garcia-Manyes S, Bucior I, Ros R, *et al.* Proteoglycan mechanics studied by single-molecule force spectroscopy of allotypic cell adhesion glycans. JBiolChem 2006; 281(9): 5992.
[132] Pepys MB. Pathogenesis, diagnosis and treatment of systemic amyloidosis. Philos Trans R Soc Lond B Biol Sci 2001; 356(1406): 203-10; discussion 10-1.
[133] Leapman RD, Rizzo NW. Towards single atom analysis of biological structures. Ultramicroscopy 1999; 78(1-4): 251.
[134] Wazawa T, Ueda M. Total internal reflection fluorescence microscopy in single molecule nanobioscience. AdvBiochemEng Biotechnol 2005; 95: 77.
[135] Ban T, Goto Y. Direct observation of amyloid growth monitored by total internal reflection fluorescence microscopy. Methods Enzymol 2006; 413: 91-102.
[136] Inoue Y, Kishimoto A, Hirao J, Yoshida M, Taguchi H. Strong growth polarity of yeast prion fiber revealed by single fiber imaging. J Biol Chem 2001; 276(38): 35227.
[137] Ban T, Hamada D, Hasegawa K, Naiki H, Goto Y. Direct observation of amyloid fibril growth monitored by thioflavin T fluorescence. The Journal of biological chemistry 2003; 278(19): 16462-5.

[138] Alexandrescu AT. Amyloid accomplices and enforcers. Protein science: a publication of the Protein Society 2005; 14(1): 1-12.
[139] Widenbrant MJ, Rajadas J, Sutardja C, Fuller GG. Lipid-induced beta-amyloid peptide assemblage fragmentation. Biophysical journal 2006; 91(11): 4071-80.
[140] Nath A, Trexler AJ, Koo P, *et al.* Single-molecule fluorescence spectroscopy using phospholipid bilayer nanodiscs. Methods in enzymology 2010; 472: 89-117.
[141] de Lange F, Cambi A, Huijbens R, *et al.* Cell biology beyond the diffraction limit: near-field scanning optical microscopy. J Cell Sci 2001; 114(Pt 23): 4153.
[142] Kitts CC, Bout DAV. Near-field scanning optical microscopy measurements of fluorescent molecular probes binding to insulin amyloid fibrils. JPhysChemB 2009; 113(35): 12090-5.
[143] Hess ST, Huang S, Heikal AA, Webb WW. Biological and chemical applications of fluorescence correlation spectroscopy: a review. Biochemistry 2002; 41(3): 697.
[144] Levin MK, Carson JH. Fluorescence correlation spectroscopy and quantitative cell biology. Differentiation 2004; 72(1): 1.
[145] Tjernberg LO, Pramanik A, Bjorling S, *et al.* Amyloid beta-peptide polymerization studied using fluorescence correlation spectroscopy. ChemBiol 1999; 6(1): 53.
[146] Hossain S, Grande M, Ahmadkhanov G, Pramanik A. Binding of the Alzheimer amyloid beta-peptide to neuronal cell membranes by fluorescence correlation spectroscopy. ExpMolPathol 2007; 82(2): 169.
[147] Verdier Y, Zarandi M, Penke B. Amyloid beta-peptide interactions with neuronal and glial cell plasma membrane: binding sites and implications for Alzheimer's disease. J Pept Sci 2004; 10(5): 229-48.
[148] Wang HY, Li W, Benedetti NJ, Lee DH. Alpha 7 nicotinic acetylcholine receptors mediate beta-amyloid peptide-induced tau protein phosphorylation. The Journal of biological chemistry 2003; 278(34): 31547-53.
[149] Le Y, Gong W, Tiffany HL, *et al.* Amyloid (beta)42 activates a G-protein-coupled chemoattractant receptor, FPR-like-1. J Neurosci 2001; 21(2): RC123.
[150] Coraci IS, Husemann J, Berman JW, *et al.* CD36, a class B scavenger receptor, is expressed on microglia in Alzheimer's disease brains and can mediate production of reactive oxygen species in response to beta-amyloid fibrils. Am J Pathol 2002; 160(1): 101-12.
[151] Husemann J, Loike JD, Anankov R, Febbraio M, Silverstein SC. Scavenger receptors in neurobiology and neuropathology: their role on microglia and other cells of the nervous system. Glia 2002; 40(2): 195-205.
[152] Du Yan S, Zhu H, Fu J, *et al.* Amyloid-beta peptide-receptor for advanced glycation endproduct interaction elicits neuronal expression of macrophage-colony stimulating factor: a proinflammatory pathway in Alzheimer disease. Proceedings of the National Academy of Sciences of the United States of America 1997; 94(10): 5296-301.
[153] Xie L, Helmerhorst E, Taddei K, *et al.* Alzheimer's beta-amyloid peptides compete for insulin binding to the insulin receptor. J Neurosci 2002; 22(10): RC221.
[154] Dineley KT, Bell KA, Bui D, Sweatt JD. beta-Amyloid peptide activates alpha 7 nicotinic acetylcholine receptors expressed in Xenopus oocytes. The Journal of biological chemistry 2002; 277(28): 25056-61.
[155] Yaar M, Zhai S, Fine RE, *et al.* Amyloid beta binds trimers as well as monomers of the 75-kDa neurotrophin receptor and activates receptor signaling. The Journal of biological chemistry 2002; 277(10): 7720-5.
[156] Guo SM, Bag N, Mishra A, Wohland T, Bathe M. Bayesian total internal reflection fluorescence correlation spectroscopy reveals hIAPP-induced plasma membrane domain organization in live cells. Biophysical journal 2014; 106(1): 190-200.
[157] Funke SA, Birkmann E, Henke F, *et al.* Single particle detection of Abeta aggregates associated with Alzheimer's disease. Biochem Biophys Res Commun 2007; 364(4): 902-7.
[158] Pitschke M, Prior R, Haupt M, Riesner D. Detection of single amyloid beta-protein aggregates in the cerebrospinal fluid of Alzheimer's patients by fluorescence correlation spectroscopy. NatMed 1998; 4(7): 832.
[159] Matsumura S, Shinoda K, Yamada M, *et al.* Two distinct amyloid á-Protein (Aá) assembly pathways leading to oligomers and fibrils identified by combined fluorescence correlation spectroscopy, morphology, and toxicity analyses. Journal of Biological Chemistry 2011; 286(13): 11555-62.

[160] Takahashi Y, Okamoto Y, Popiel HA, *et al.* Detection of polyglutamine protein oligomers in cells by fluorescence correlation spectroscopy. The Journal of biological chemistry 2007; 282(33): 24039-48.
[161] Mittag JJ, Milani S, Walsh DM, R„dler JO, McManus JJ. Simultaneous measurement of a range of particle sizes during Aá 1-42 fibrillogenesis quantified using fluorescence correlation spectroscopy. Biochemical and Biophysical Research Communications 2014; 448(2): 195-9.
[162] Tycko R, Wickner RB. Molecular structures of amyloid and prion fibrils: consensus *versus* controversy. Accounts of chemical research 2013; 46(7): 1487-96.
[163] Inouye H, Fraser PE, Kirschner DA. Structure of beta-crystallite assemblies formed by Alzheimer beta-amyloid protein analogues: analysis by x-ray diffraction. BiophysJ 1993; 64(2): 502.
[164] Kirschner DA, Inouye H, Duffy LK, *et al.* Synthetic peptide homologous to beta protein from Alzheimer disease forms amyloid-like fibrils *in vitro*. ProcNatlAcadSciUSA 1987; 84(19): 6953.
[165] Hirschfield GM, Hawkins PN. Amyloidosis: new strategies for treatment. IntJBiochemCell Biol 2003; 35(12): 1608-13.
[166] Selkoe DJ. Alzheimer's disease: genotypes, phenotypes, and treatments. Science 1997; 275(5300): 630-1.
[167] Lambert MP, Barlow AK, Chromy BA, *et al.* Diffusible, nonfibrillar ligands derived from Aá 1-42 are potent central nervous system neurotoxins. ProcNatlAcadSciUSA 1998; 95(11): 6448-53.
[168] Kayed R, Head E, Thompson JL, *et al.* Common structure of soluble amyloid oligomers implies common mechanism of pathogenesis. Science 2003; 300(5618): 486-9.
[169] Demuro A, Mina E, Kayed R, *et al.* Calcium dysregulation and membrane disruption as a ubiquitous neurotoxic mechanism of soluble amyloid oligomers. JBiolChem 2005; 280(17): 17294-300.
[170] Chromy BA, Nowak RJ, Lambert MP, *et al.* Self-assembly of Aá1-42 into globular neurotoxins. Biochemistry 2003; 42(44): 12749-60.
[171] Roher AE, Baudry J, Chaney MO, *et al.* Oligomerization and fibril asssembly of the amyloid-á protein. BiochimBiophysActa 2000; 1502(1): 31-43.
[172] Fernandez-Busquets X. Amyloid fibrils in neurodegenerative diseases: villains or heroes? Future Med Chem 2013; 5(16): 1903-6.
[173] Serpell LC. Alzheimer's amyloid fibrils: structure and assembly. BiochimBiophysActa 2000; 1502(1): 16-30.
[174] Ruoslahti E. Brain extracellular matrix. Glycobiology 1996; 6(5): 489-92.
[175] Ancsin JB. Amyloidogenesis: historical and modern observations point to heparan sulfate proteoglycans as a major culprit. Amyloid 2003; 10(2): 67-79.
[176] van Horssen J, Wesseling P, van den Heuvel LP, de Waal RM, Verbeek MM. Heparan sulphate proteoglycans in Alzheimer's disease and amyloid-related disorders. Lancet Neurol 2003; 2(8): 482-92.
[177] Snow AD, Mar H, Nochlin D, *et al.* The presence of heparan sulfate proteoglycans in the neuritic plaques and congophilic angiopathy in Alzheimer's disease. AmJ Pathol 1988; 133(3): 456-63.
[178] Snow AD, Mar H, Nochlin D, Kresse H, Wight TN. Peripheral distribution of dermatan sulfate proteoglycans (decorin) in amyloid-containing plaques and their presence in neurofibrillary tangles of Alzheimer's disease. J HistochemCytochem 1992; 40(1): 105-13.
[179] Snow AD, Nochlin D, Sekiguichi R, Carlson SS. Identification in immunolocalization of a new class of proteoglycan (keratan sulfate) to the neuritic plaques of Alzheimer's disease. ExpNeurol 1996; 138(2): 305-17.
[180] Su JH, Cummings BJ, Cotman CW. Localization of heparan sulfate glycosaminoglycan and proteoglycan core protein in aged brain and Alzheimer's disease. Neuroscience 1992; 51(4): 801-13.
[181] DeWitt DA, Silver J, Canning DR, Perry G. Chondroitin sulfate proteoglycans are associated with the lesions of Alzheimer's disease. ExpNeurol 1993; 121(2): 149-52.
[182] Gupta-Bansal R, Frederickson RC, Brunden KR. Proteoglycan-mediated inhibition of Aá proteolysis. A potential cause of senile plaque accumulation. JBiolChem 1995; 270(31): 18666-71.
[183] Castillo GM, Cummings JA, Yang W, *et al.* Sulfate content and specific glycosaminoglycan backbone of perlecan are critical for perlecan's enhancement of islet amyloid polypeptide (amylin) fibril formation. Diabetes 1998; 47(4): 612-20.
[184] Potter-Perigo S, Hull RL, Tsoi C, *et al.* Proteoglycans synthesized and secreted by pancreatic islet á-cells bind amylin. ArchBiochemBiophys 2003; 413(2): 182-90.

[185] Ancsin JB, Kisilevsky R. The heparin/heparan sulfate-binding site on apo-serum amyloid A. Implications for the therapeutic intervention of amyloidosis. JBiolChem 1999; 274(11): 7172-81.
[186] Cohlberg JA, Li J, Uversky VN, Fink AL. Heparin and other glycosaminoglycans stimulate the formation of amyloid fibrils from à-synuclein *in vitro*. Biochemistry 2002; 41(5): 1502-11.
[187] Supattapone S. Prion protein conversion *in vitro*. JMolMed 2004; 82(6): 348-56.
[188] Borysik AJ, Morten IJ, Radford SE, Hewitt EW. Specific glycosaminoglycans promote unseeded amyloid formation from á2-microglobulin under physiological conditions. Kidney Int 2007; 72(2): 174-81.
[189] Goedert M, Jakes R, Spillantini MG, *et al.* Assembly of microtubule-associated protein tau into Alzheimer-like filaments induced by sulphated glycosaminoglycans. Nature 1996; 383(6600): 550-3.
[190] McLaurin J, Franklin T, Zhang X, Deng J, Fraser PE. Interactions of Alzheimer amyloid-á peptides with glycosaminoglycans. Effects on fibril nucleation and growth. EurJBiochem 1999; 266(3): 1101-10.
[191] Castillo GM, Lukito W, Wight TN, Snow AD. The sulfate moieties of glycosaminoglycans are critical for the enhancement of á-amyloid protein fibril formation. JNeurochem 1999; 72(4): 1681-7.
[192] Woods AG, Cribbs DH, Whittemore ER, Cotman CW. Heparan sulfate and chondroitin sulfate glycosaminoglycan attenuate á-amyloid(25-35) induced neurodegeneration in cultured hippocampal neurons. Brain Res 1995; 697(1-2): 53-62.
[193] Fernandez-Busquets X, Ponce J, Bravo R, *et al.* Modulation of amyloid beta peptide(1-42) cytotoxicity and aggregation *in vitro* by glucose and chondroitin sulfate. Current Alzheimer research 2010; 7(5): 428-38.
[194] Pollack SJ, Sadler II, Hawtin SR, Tailor VJ, Shearman MS. Sulfated glycosaminoglycans and dyes attenuate the neurotoxic effects of á-amyloid in rat PC12 cells. NeurosciLett 1995; 184(2): 113-6.
[195] Timmer NM, Schirris TJ, Bruinsma IB, *et al.* Aggregation and cytotoxic properties towards cultured cerebrovascular cells of Dutch-mutated Abeta40 (DAbeta(1-40)) are modulated by sulfate moieties of heparin. NeurosciRes 2010; 66(4): 380-9.
[196] Bruckner G, Hausen D, Hartig W, *et al.* Cortical areas abundant in extracellular matrix chondroitin sulphate proteoglycans are less affected by cytoskeletal changes in Alzheimer's disease. Neuroscience 1999; 92(3): 791-805.
[197] McLaurin J, Fraser PE. Effect of amino-acid substitutions on Alzheimer's amyloid-á peptide-glycosaminoglycan interactions. EurJBiochem 2000; 267(21): 6353-61.
[198] Valle-Delgado JJ, Alfonso-Prieto M, de Groot NS, *et al.* Modulation of Abeta42 fibrillogenesis by glycosaminoglycan structure. FASEB journal: official publication of the Federation of American Societies for Experimental Biology 2010; 24(11): 4250-61.
[199] Bemporad F, Calloni G, Campioni S, *et al.* Sequence and structural determinants of amyloid fibril formation. Acc Chem Res 2006; 39(9): 620-7.
[200] Caflisch A. Computational models for the prediction of polypeptide aggregation propensity. Curr Opin Chem Biol 2006; 10(5): 437-44.
[201] Rousseau F, Schymkowitz J, Serrano L. Protein aggregation and amyloidosis: confusion of the kinds? Curr Opin Struct Biol 2006; 16(1): 118-26.
[202] Zibaee S, Makin OS, Goedert M, Serpell LC. A simple algorithm locates beta-strands in the amyloid fibril core of alpha-synuclein, Abeta, and tau using the amino acid sequence alone. Protein Sci 2007; 16(5): 906-18.
[203] Chou PY, Fasman GD. Prediction of the secondary structure of proteins from their amino acid sequence. Adv Enzymol Relat Areas Mol Biol 1978; 47: 45-148.
[204] Chiti F, Stefani M, Taddei N, Ramponi G, Dobson CM. Rationalization of the effects of mutations on peptide and protein aggregation rates. Nature 2003; 424(6950): 805-8.
[205] Yoon S, Welsh WJ. Detecting hidden sequence propensity for amyloid fibril formation. Protein science: a publication of the Protein Society 2004; 13(8): 2149-60.
[206] Yoon S, Welsh WJ. Rapid assessment of contact-dependent secondary structure propensity: relevance to amyloidogenic sequences. Proteins 2005; 60(1): 110-7.
[207] Galzitskaya OV, Garbuzynskiy SO, Lobanov MY. Prediction of amyloidogenic and disordered regions in protein chains. PLoS Comput Biol 2006; 2(12): e177.
[208] Nelson R, Eisenberg D. Recent atomic models of amyloid fibril structure. CurrOpinStructBiol 2006; 16(2): 260.

[209] Thompson MJ, Sievers SA, Karanicolas J, *et al.* The 3D profile method for identifying fibril-forming segments of proteins. Proc Natl Acad Sci U S A 2006; 103(11): 4074-8.
[210] Kuhlman B, Baker D. Native protein sequences are close to optimal for their structures. Proceedings of the National Academy of Sciences of the United States of America 2000; 97(19): 10383-8.
[211] Saiki M, Konakahara T, Morii H. Interaction-based evaluation of the propensity for amyloid formation with cross-beta structure. Biochem Biophys Res Commun 2006; 343(4): 1262-71.
[212] Trovato A, Chiti F, Maritan A, Seno F. Insight into the structure of amyloid fibrils from the analysis of globular proteins. PLoS Comput Biol 2006; 2(12): e170.
[213] Bryan AW, Jr., Menke M, Cowen LJ, Lindquist SL, Berger B. BETASCAN: probable beta-amyloids identified by pairwise probabilistic analysis. PLoS Comput Biol 2009; 5(3): e1000333.
[214] Rojas Quijano FA, Morrow D, Wise BM, Brancia FL, Goux WJ. Prediction of nucleating sequences from amyloidogenic propensities of tau-related peptides. Biochemistry 2006; 45(14): 4638-52.
[215] Ma Y, Hendershot LM. The unfolding tale of the unfolded protein response. Cell 2001; 107(7): 827-30.
[216] Castillo V, Grana-Montes R, Sabate R, Ventura S. Prediction of the aggregation propensity of proteins from the primary sequence: aggregation properties of proteomes. Biotechnol J 2011; 6(6): 674-85.
[217] Waldo GS, Standish BM, Berendzen J, Terwilliger TC. Rapid protein-folding assay using green fluorescent protein. Nat Biotechnol 1999; 17: 691-5.
[218] de Groot NS, Aviles FX, Vendrell J, Ventura S. Mutagenesis of the central hydrophobic cluster in Abeta42 Alzheimer's peptide. Side-chain properties correlate with aggregation propensities. FEBS J 2006; 273(3): 658-68.
[219] de Groot N, Pallares I, Aviles F, Vendrell J, Ventura S. Prediction of "hot spots" of aggregation in disease-linked polypeptides. BMC Struct Biol 2005; 5(1): 18.
[220] Conchillo-Sole O, de Groot NS, Aviles FX, *et al.* AGGRESCAN: a server for the prediction and evaluation of "hot spots" of aggregation in polypeptides. BMC Bioinformatics 2007; 8: 65.
[221] Cellmer T, Bratko D, Prausnitz JM, Blanch HW. Protein aggregation *in silico*. Trends Biotechnol 2007; 25(6): 254-61.
[222] Ma B, Nussinov R. Simulations as analytical tools to understand protein aggregation and predict amyloid conformation. Curr Opin Chem Biol 2006; 10(5): 445-52.
[223] Nguyen PH, Li MS, Stock G, Straub JE, Thirumalai D. Monomer adds to preformed structured oligomers of Abeta-peptides by a two-stage dock-lock mechanism. Proceedings of the National Academy of Sciences of the United States of America 2007; 104(1): 111-6.
[224] Cheon M, Chang I, Mohanty S, *et al.* Structural reorganisation and potential toxicity of oligomeric species formed during the assembly of amyloid fibrils. PLoS Comput Biol 2007; 3(9): 1727-38.
[225] Irback A, Mohanty S. PROFASI: A Monte Carlo simulation package for protein folding and aggregation. J Comput Chem 2006; 27(13): 1548-55.
[226] Pellarin R, Caflisch A. Interpreting the aggregation kinetics of amyloid peptides. Journal of molecular biology 2006; 360(4): 882-92.
[227] Smith AV, Hall CK. Protein refolding *versus* aggregation: computer simulations on an intermediate-resolution protein model. J Mol Biol 2001; 312(1): 187-202.
[228] Nguyen HD, Marchut AJ, Hall CK. Solvent effects on the conformational transition of a model polyalanine peptide. Protein Sci 2004; 13(11): 2909-24.
[229] Nguyen HD, Hall CK. Molecular dynamics simulations of spontaneous fibril formation by random-coil peptides. Proceedings of the National Academy of Sciences of the United States of America 2004; 101(46): 16180-5.
[230] Ma B, Nussinov R. Stabilities and conformations of Alzheimer's beta-amyloid peptide oligomers (Abeta 16-22, Abeta 16-35, and Abeta 10-35): Sequence effects. Proceedings of the National Academy of Sciences of the United States of America 2002; 99(22): 14126-31.
[231] Petkova AT, Yau WM, Tycko R. Experimental constraints on quaternary structure in Alzheimer's beta-amyloid fibrils. Biochemistry 2006; 45(2): 498-512.
[232] Xu Y, Shen J, Luo X, *et al.* Conformational transition of amyloid beta-peptide. Proceedings of the National Academy of Sciences of the United States of America 2005; 102(15): 5403-7.
[233] Liu D, Xu Y, Feng Y, *et al.* Inhibitor discovery targeting the intermediate structure of beta-amyloid peptide on the conformational transition pathway: implications in the aggregation mechanism of beta-amyloid peptide. Biochemistry 2006; 45(36): 10963-72.

[234] Barducci A, Chelli R, Procacci P, *et al.* Metadynamics simulation of prion protein: beta-structure stability and the early stages of misfolding. J Am Chem Soc 2006; 128(8): 2705-10.
[235] Camilloni C, Schaal D, Schweimer K, Schwarzinger S, De Simone A. Energy landscape of the prion protein helix 1 probed by metadynamics and NMR. Biophys J 2012; 102(1): 158-67.
[236] Bonomi M, Parrinello M. Enhanced sampling in the well-tempered ensemble. Phys Rev Lett 2010; 104(19): 190601.
[237] Barducci A, Bonomi M, Parrinello M. Linking well-tempered metadynamics simulations with experiments. Biophys J 2010; 98(9): L44-6.

CHAPTER 6

Cholinesterase Inhibitors Useful for the Treatment of Neurodegenerative Disorders from Natural Sources: Past and Present

Monica R. Loizzo[1,*], Rosa Tundis[1], Federica Menichini[2] and Francesco Menichini[1]

[1]Department of Pharmacy Health and Nutritional Sciences, University of Calabria, I-87036 Rende (CS), Italy and [2]AIFA Agenzia Italiana Del Farmaco, via del Tritone 181, 00187 Roma, Italy

Abstract: Alzheimer's disease (AD) is the most common form of dementia affecting older people (60-80% cases). Besides the neuropathologic hallmark abnormalities that are deposits of the protein fragment β-amyloid (plaques) and twisted strands of the protein tau (tangles), AD is characterized by a consistent deficit in cholinergic neurotransmission, consequently leading to destruction of cholinergic neurons in basal forebrain. Cholinesterases (acetylcholinesterase - AChE, and butyrylcholinesterase - BChE) are serine hydrolases that split acetylcholine (ACh) and terminate its action. Cholinesterase inhibitors increase the levels of ACh and prolong its action. For this reason, they have relevance to the treatment of neurodegenerative disorders. Moreover, AChE inhibition holds a key role to reduce the aggregation of β amyloid and the formation of the neurotoxic fibrils. Recently, researchers move their interest in finding drugs that are able to inhibit both of these events. This chapter covers the current knowledge of natural products as cholinesterase inhibitors and how these compounds could serve as lead compounds for semi-synthetic analogs with improved properties. In particular, studies in the period 2000-2007 and from 2013 until now (2014 first semester) are reported.

Keywords: Acetylcholinesterase, alkaloids, alzheimer's disease, butyrylcholinesterase, derivatives, neurodegenerative disorders, polyphenols, terpenoids.

INTRODUCTION

Neurodegenerative diseases result in progressive degeneration and/or death of nerve cells. This causes problems with movement (called ataxias) or mental functioning (called dementias). AD represents approximately 60-70% of cases and

***Corresponding author Monica R. Loizzo:** Department of Pharmacy Health and Nutritional Sciences, University of Calabria, I-87036 Rende (CS), Italy; Tel: +39 984 493073; Fax: +39 984 493107; E-mail: monica_rosa.loizzo@unical.it

Atta-ur-Rahman, Allen B. Reitz, M. Iqbal Choudhary and Jizhou Wang (Eds.)

affects more than 20 million people worldwide. Between now and 2050, over 680 million new cases of dementia will be diagnosed worldwide. Moreover, if more effective therapies than the ones currently available are not developed that either prevent AD or other forms of dementia or block progression of the diseases in its very early stages, the economic and societal cost of caring for AD patients will be devastating. In fact, it is expected that 680 million people living with dementia in the next 40 years worldwide [1-3]. Difficulty in remembering names and recent events are often early clinical symptoms; Similarly, apathy and depression are also considered as early symptoms. Later symptoms include, impaired judgment, disorientation, confusion, behavior changes and difficulty in speaking, swallowing and walking. Four relevant pathogenic events are involved in AD, which are primary events including genetic alterations, neuronal apoptosis-like processes leading to premature neuronal death and brain dysfunction; secondary events including β-amyloid deposition in senile plaques and brain vessels, neurofibrillary tangles due to hyperphosphorilation of tau proteins, synaptic loss; tertiary events including neurotransmitter deficits, neurotrophic alterations, neuroinmune dysfunction, neuroinflammatory processes; and quaternary events including accelerated neuronal death due to excitotoxic reactions, alterations in calcium homeostasis, free radical generation, cerebrovascular dysfunction. Actually, no treatment is available to slow or stop the deterioration of brain cells in AD so the approach to treat these patients is symptomatic. This strategy is based on the so-called cholinergic hypothesis of cognitive dysfunction [4]. This hypothesis postulates that at least some of the cognitive decline experienced by patients of AD results from a deficiency in neurotransmitter acetylcholine (ACh) and thus in cholinergic neurotransmission in brain cortical or hippocampal regions, which seems to play a fundamental role in memory [5]. Moreover, in late AD stage, levels of AChE declined by up to 85% and BChE represents the predominant cholinesterase in the brain [6]. The enzyme BChE is associated with glial and cleaves ACh in a manner similar to AChE [7, 8]. Such studies, together with a statistically slower decline in the cognitive performance of dementia patients possessing specific BChE polymorphisms that naturally lower BChE activity, have targeted BChE as a new approach to intercede in the progression of AD [9, 10]. To restore the level of ACh through inhibition of both two major forms of cholinesterase AChE and BChE continues to be a useful therapeutic approach to treat not only AD but also other forms of dementia, such as vascular dementia, Parkinson's dementia, dementia/Lewy body and cognitive symptoms associated with multiple sclerosis and Down syndrome.

AChE and BChE differ genetically, structurally and for their kinetics [11]. Structural characteristics are well described by Houghton *et al.* [12].

Cholinesterase inhibitors could perhaps help prevent the formation of β-amyloid plaques. In fact, recent investigations demonstrated that the peripheral site of AChE promotes aggregation of β-amyloid peptides [13], which causes plaque formation known to play an essential role in the development of AD,while BChE did not affect amyloid formation [14, 15]. Moreover, a recent study demonstrated that cholinergic up-regulation obtained with the use of acetylcholinesterase inhibitors (AChEI) was associated with an anti-inflammatory effect through the suppression of lymphocyte proliferation and pro-inflammatory cytokine production. This activity was mediated by the α7 nicotinic acetylcholine receptor. In AD, there is a well recognized inflammatory component, in which microglia were found to be the major cellular component in AD plaques. This new role of AChEI emphasizes the importance of cholinergic balance in this neurological disorder [16]. In medicinal chemistry the discovery of new lead compounds is based on the modification and improvement of known active compounds such as a naturally derived product. The possible approach is a disjunctive approach (simplification of the leadership structure), analogical approach (keeping the same level of structural complexity) and conjunctive approach (enrichment of structural complexity). Several review articles have covered the discovery of natural products as cholinesterase inhibitors [12, 17-35].

This chapter describes the cholinesterase inhibitory activity of natural products and their derivatives covering studies in the period 2000-2007 and from 2013 until now (2014 first semester). Moreover, the drug design strategy and the structure activity relationship (SAR) of the most active compounds are also discussed.

ALKALOIDS

(-)-Galanthamine (**1**), isolated from *Galanthus woronowii* (Amaryllidaceae), is a centrally acting, competitive and reversible AChEI that enhances cognitive functions in Alzheimer's patients [36]. Galanthamine (**1**) has been shown to be much less potent against BChE than AChE activity [37]. Moreover, it has been reported that it appears to act directly on nicotinic receptors, thus enhancing the release of each [38]. A variety of synthetic galanthamine derivatives has been chemically synthesized including rings C and D derivatives, quaternary ammonium derivatives, esters and carbamates of 6-*O*-demethylgalanthamine and bis-interacting ligands [39-43]. In particular, open ring D of galanthamine analogs as in **2** and **3**, was used as starting compound for the synthesis of new bis-interacting ligand. All synthesized compounds **4-8** were characterized by the replacement of the *N*-methyl group of galanthamine (**1**) with a CO_2Et group. Compounds **4-8** exhibited IC_{50} values ranging from 60 to 1000 μM for

Electrophorus electricus AChE (*Ee*AChE) and from 155 to 960 μM for Horse serum BChE (*Hs*BChE), respectively. All compounds revealed a less potency compared with **1** (IC$_{50}$ 0.35 and 18.6 μM for *Ee*AChE and *Hs*BChE, respectively). This could be explained by the loss of a non-conventional hydrogen bond between the *N*-methyl group of galanthamine (**1**) and Asp-72 residue of *Ee*AChE. Moreover, the potency of AChE inhibition decreased when the size of the ring D increased as well as when the same ring is opened (*i.e.*, **5**, **6** and **8**). The presence of an electron withdrawing group on the ring A of galanthamine (**1**) enhanced the *Hs*BChE inhibition potency (*i.e.*, **8**) compared to the other analogs. Analogs **6**, **7** and **8** exerted a marked selectivity on *Hs*BChE probably due to the ability of this enzyme to accommodate steric bulk around the enzyme catalytic site [44]. Mary *et al.* [45] synthesized a series of *O*- and *N*-substituted galanthamine derivatives, as bis-interacting ligands. Among them compounds **9** and **10** exhibited the highest activity with IC$_{50}$ values of 0.1 x $10^{-7}$and 0.2 x 10^{-7} M, respectively, while in *O*-substituted series compounds **11** and **12** (that contain iminium function at oxygen and present an alkyl chain with *n* =8 and 10, respectively) exhibited the highest activity with IC$_{50}$ values of 0.7 x $10^{-7}$and 0.5 x 10^{-7} M, respectively.

The same research group compared the homodimeric and heterodimeric alkylene linked bis-galanthamine derivatives [46]. In the homodimeric series, compound **13** showed the highest inhibitory activity against AChE (IC$_{50}$ of 230 nM) while in heterodimeric derivatives compounds **14** (IC$_{50}$ of 22 nM) and **15** (IC$_{50}$ of 16 nM) were more potent than galanthamine (**1**) and tacrine (**16**). SAR analysis revealed that the potency of AChE inhibition depends on the length of the alkylene tether as well as the presence of the iminium function. A variation in the length of the chain should affect the possibility of an efficient interaction with the peripheral site. A length of eight methylenes (*i.e.,* **13**) is favored in the biz-galanthamine series, whereas this increases to ten methylenes in the heterodimeric galanthamine-galan-thaminium series (*i.e.,* **15**).

From *Nerine bowdenii* (Amaryllidaceae) ungeremine (**17**) was isolated and tested for *Ee*AChE inhibition. This alkaloid exhibited an IC$_{50}$ value of 0.35 μM. This value is about 6-10-times stronger than **1** [47]. Lycorine (**18**), isolated from *Narcissus pseudonarcissus* (Amaryllidaceae), has been demonstrated to possess an IC$_{50}$ of 450 μM against AChE [48]. Its higher activity was explained by the presence of an aromatic ring C, providing a planar-like skeleton. From this evidence authors supposed that the intact ring D is essential for anti-AChE activity. Based upon these, a series of *N*-alkyl-5,6-secolycorine derivatives, possessing a planar-like dihydrophenanthridine moiety devoid of ring D, was synthesized and evaluated for their anti-AChE activity. The secolycorine derivatives **19-24** and **25-30** exhibited a strong inhibitory activity against AChE with IC_{50} values ranging from < 5 μM to 10.59 μM [49].

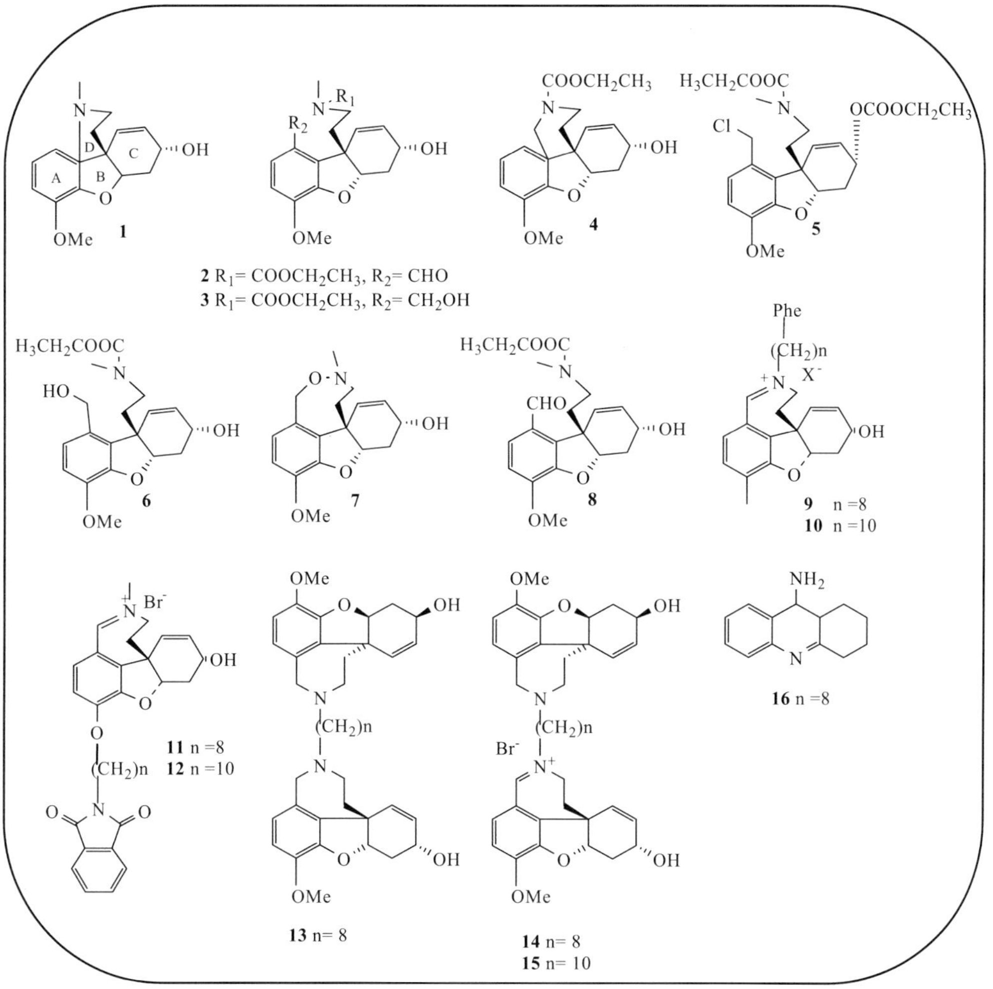

The cleavage of ring D does not decrease the anti-AChE activity. Instead, some advantages of such modification, including enhancing both potency and solubility relative to those of assoanine (**31**), are observed. For example 4-ethyl-*N*-methyl-dihydrophenanthridine (**19**) (IC_{50} of 2.89 μM) is more potent than **31** (IC_{50} of 3.87 μM) [50]. The length of *N*-substitution in dihydrophenanthridines series **19-24** and **25-30** is critical for the activity. *N*-Methyl or *N*-butyl substitution seems to be optimized for anti- AChE activity, which decreases 2 to 4-times for *N*-ethyl **20**, **26** or *N*-propyl derivatives **21** and **27**. In fact **21** exerted an interesting anti-AChE activity with an IC_{50} of 2.60 μM. This value is comparable to **19** and lesser than commercially available drug galanthamine (**1**, IC_{50} of 3.18 μM). Although the structures of 4-vinyl-5,6-dihydrophenanthridines **19-24** are more planar than those of the corresponding 4-ethyl derivatives **25-30**, their anti-AChE activities seem to show no significant difference. The increase of solubility with non planar ring C causes the dramatic loss of the bioactivity as observed for 5-hydroxy-5,6-secolycorine (**32**) and 5-hydroxy-5,6-seco-dihydrolycorine (**33**). The peracetyl derivatives **34** and **35** are much less polar but more bulky relative to the parent compounds. They did not show significant anti-AChE activity at 10 μM level similar to that observed in **32** and **33**. Docking analysis revealed that the *N*-alkyl chain needs are tailored to an optimal length. In fact, initially the binding affinity decreases with increasing the *N*-alkyl chain length, but then increases again with even longer alkyl chain, although the predicted alkyl chain length for the minimum of binding affinity has been offset from that of experimental results. The *N*-alkyl chain length also depends on the solubility of secolycorine derivatives. Regarding the interaction between derivatives and enzyme active site simulation analysis identifies in the numbers 11 and 12 hydrophobic residues for **22** and **19**, respectively, while **20** and **21** can interact maximum with 8 and 6 residues, and this reduction of binding affinities is evident in the obtained high values of IC_{50} of 8.35 and 10.59 μM, respectively.

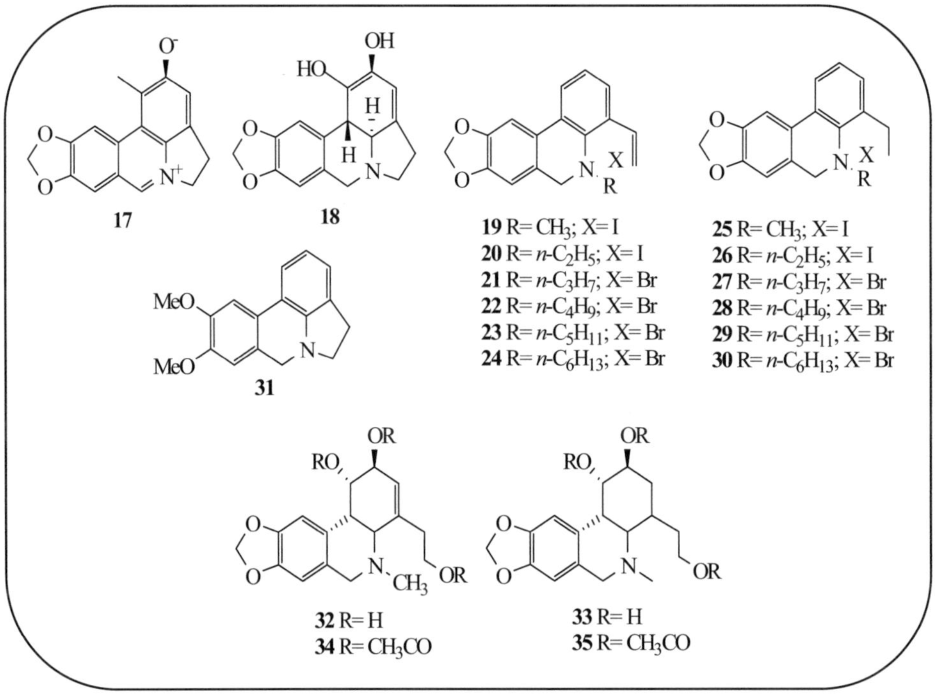

Wang *et al.* [51] synthesized a series of lycorine derivatives. Among tested compounds 2-*O*-*tert*-butyldimethylsilyl-1-*O*-(methylthio)-methyllycorine was a dual inhibitor of both human AChE and BChE with IC_{50} values of 11.40 and 4.17 μM, respectively. SAR analysis indicated that a) the 1-*O*-(methylthio)methyl substituent in lycorine was better than the 1-*O*-acetyl group for the inhibition of cholinesterase; b) acylated or etherified derivatives of lycorine and lycorin-2-one were more potent against hBChE than hAChE; and c) the oxidation of lycorine at C-2 decreases the activity. In order to evaluate if different ring type of Amaydillaceae alkaloids can influence the AChE activity, Elgorashi *et al.* [52] screened twenty three alkaloids for cholinesterase inhibition including crinine (**36**), epibuphasine (**37**), powelline (**38**), crinamidine (**39**), undulatine (**40**), hamayne (**41**), 3-*O*-acetylhamayne (**42**), epivittatine (**43**), crinamine (**44**), 6-hydroxycrinamine (**45**), 1-epideacetylbowdensine (**46**), papyramine (**47**), maritidine (**48**), and *O*-methylmaritidine (**49**) for the crinine-type group; tazettine (**50**), 8-α-ethoxyprecriwelline (**51**), *N*-desmethyl-8α-ethoxypretazzettine (**52**), and *N*-desmethyl-8β-ethoxypretazzettine (**53** for tazettine-type group; while for lycorine-type alkaloids, lycorine (**18**), *O*-acetyllycorine (**54**), 1,2-diacetyllycorine (**55**). Cherylline (**56**) and galanthamine (**1**) were also tested. Among tested alkaloids, lycorine-type exhibited the higher activity. In particular, *O*-acetyllycorine (**54**, IC_{50} 0.96 μM) exhibited a 2-fold more potency than galanthamine (**1**) against AChE. Lycorine (**18**) and 1,2-diacetyllycorine (**55**) exhibited a 100-time less activity that **54**.

SAR analysis revealed that the presence of an acetoxy group and hydroxyl group in C1 and C2, respectively, is required in **54** for the proper binding and the consequent inhibition of the enzyme. Crinine (**36**) and epivittatine (**43**), which represent the simplest forms of the crinine-type alkaloids exhibited the highest inhibitory activity of this ring-type group (IC_{50} 461 and 239 μM for **36** and **43**, respectively) but a less activity compared to lycorine-type group. The highest activity could be structurally related to the presence of ethano-bridge in β- and α-positions, respectively. The replacement of hydroxy group in crinine (**36**) with methoxy substituent as in epibuphasine (**37**) causes a reduction of the activity (IC_{50} for **37** 547 μM). Other structure modifications, such as methoxylation of **36** at the aromatic ring as in powelline (**38**), decreased the activity.

Another crinamine analog was identified by Adewusi *et al.* [53]. Researchers reported the AChE inhibitory activity of a mixture 3:1 of epimers with 6α-hydroxycrinamine as the major one with an IC_{50} value of 445 μM. Recently, Nair *et al.* [54] reported the AChE inhibitory activity of narciprimine from *Cyrtanthus contractus*. This Amaryllidaceae alkaloid represents the first example of a lycorine compound with an intact, untethered phenanthridone nucleus (IC_{50} value of 78.9 μM). The unsubstituted, ethyl-tethered alkaloids lycorine (**18**) exhibited relatively low activities with IC_{50} of 213 μM, respectively. However, mono-substitution in ring C as in the natural compound 1-*O*-acetyllycorine was seen to affect a dramatic spike in activity with an IC_{50} value of 0.96 μM that is almost twice as active as reference compound galanthamine. *Habranthus tubispathus* and *H. Jamesonii* also belongs to the Amaryllidaceae family. Both species screened for cholinesterase inhibitors lead to the isolation of hippeastidine that showed IC_{50} values of 104.6 and 200.8 μM against AChE and BChE, respectively [55].

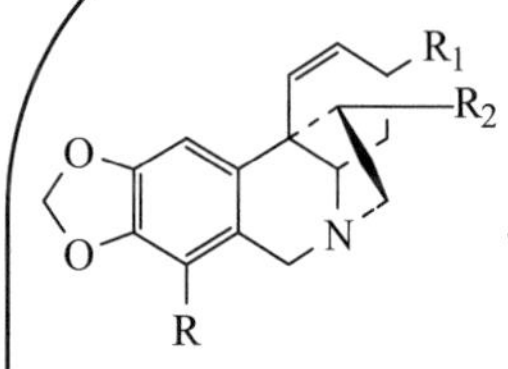

36 R= H; R_1= OH
37 R= H; R_1= OMe
38 R= OMe; R_1= OH

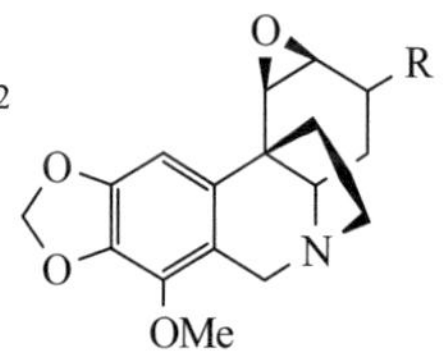

39 R=OH
40 R= OMe

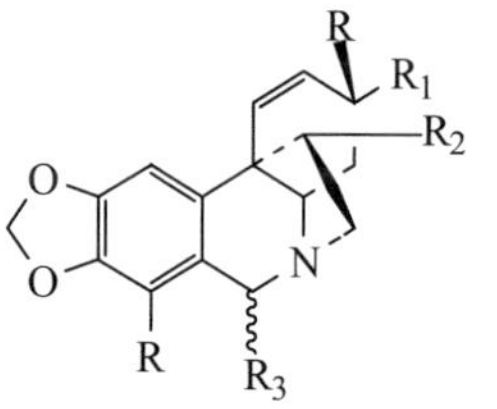

41 R= H; R_1= OH; R_2=OH; R_3= H
42 R= R_3=H; R_1= OAc; R_2=OH
43 R= H; R_1= OH; R_2=R_3=H
44 R= R_3=H; R_1= OMe; R_2= OH
45 R= H; R_1= OMe; R_2= OH; R_3= OH (a,b)

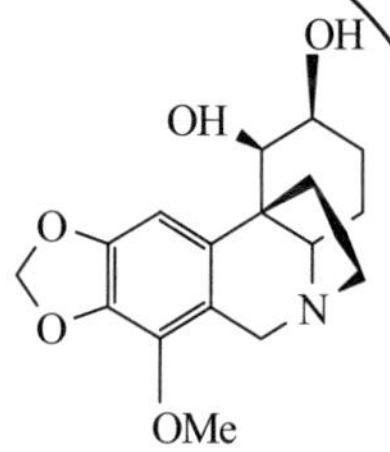

46

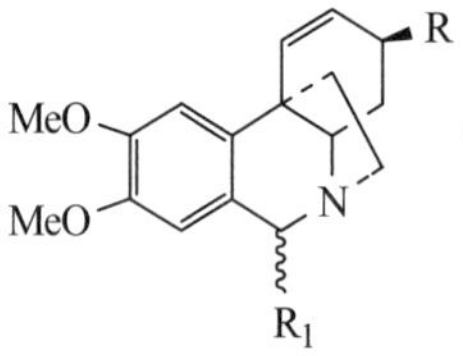

47 R= OMe; R_1= OH (a,b)
48 R= OH; R_1=H
49 R= OMe; R_1= H

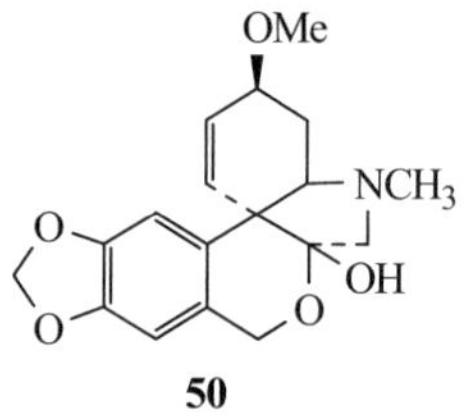

50

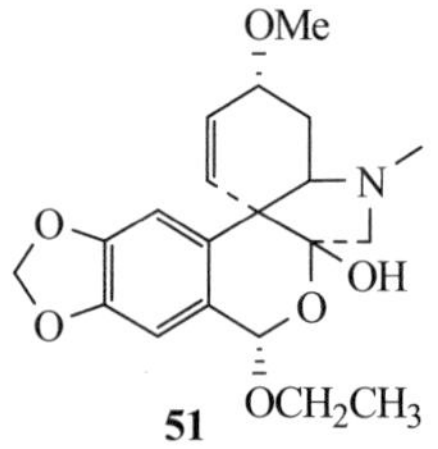

51

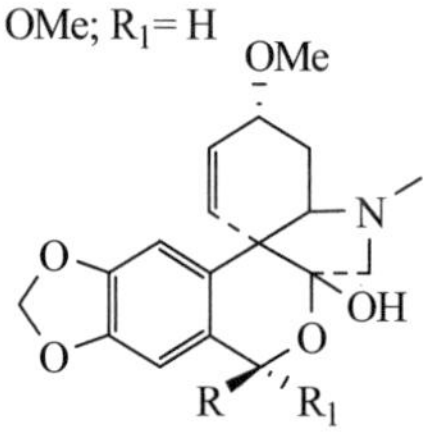

52 R= H; R_1= OCH_2CH_3
53 R= OCH_2CH_3; R_1= H

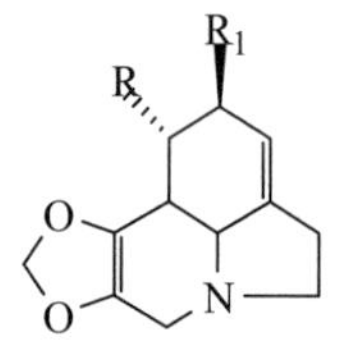

54 R= OAc; R_1= OH
55 R=R_1= OAc

Ingkanian *et al.* [56] through bioguided-fractionation of *Stephania venosa* (Menispermaceae) extract provided the isolation of stepharanine (**57**), cyclanoline (**58**), *N*-methylstepholidine (**59**), stepholidine (**60**) and coryldammine (**61**) that were compared with their ability to inhibit AChE with previously isolated palmatine (**62**), jatrorrhizine (**63**) and berberine (**64**). The AChE inhibitory activity of the tertiary protoberberine alkaloids stepholidine (**60**) and corydalmine (**61**) was lower than that of the quaternary protoberberine alkaloids tested with IC_{50} value about 100 μM. In particular, stepholidine (**60**) had significantly less potency compared with *N*-methyl stepholidine (**59**) and stepharanine (**57**) with IC_{50} values of 31.30 and 14.10 μM, respectively. Moreover, aromatization at ring C and the absence of a methyl group at the nitrogen improved the AChE activity of stepharanine (**57**) in comparison with **59**. This information suggests that a positive charge at the nitrogen of the tetrahydroisoquinoline portion, steric substitution at the nitrogen or the planarity of molecule influence the AChE inhibitory activity. These compounds have four substitutions (hydroxy or methoxy groups) at positions C2, C3, C9 and C10. These substitutions may be responsible for the binding pocket or they may be the parts that can interact with enzymes as proton donor or proton acceptor. Comparison of protoberberine (**57**) and palmatine (**62**) highlights the importance of substitutions at C2 and C10 in the structure of **62** for AChE inhibitory activity (**62**, IC_{50} of 0.26 μM). The substitution of the methoxy group at C3 in palmatine (**62**) with a hydroxyl group in jatrorrhizine (**63**) (IC_{50} of 0.93 μM) caused a reduction in the activity of about 3-fold. It should be emphasized that palmatine (**62**), jatrorrhizine (**63**) and berberine (**64**) (IC_{50} 0.58 μM) showed AChE inhibitory activity in the same range as that of galanthamine (**1**) (IC_{50} 0.59 μM). Other interesting isoquinoline alkaloids were obtained from *Chelidonium majus* (Papaveraceae), including 8-hydroxydihydrochelerythrine (**65**), 8-hydroxydihydrosanguinarine (**66**) and berberine (**64**). These alkaloids strongly inhibited AChE with IC_{50} values of 0.61, 1.37 and 1.85 μM for **65**, **66** and **64**, respectively. Moreover, compounds **64-66** exhibited an SI ranging from 9.3 to 56.7 μM [57].

Recently, Bonesi *et al.* [58] investigated the cholinesterase inhibitory activity of berberine (**64**) and palmatine (**62**) and obtained IC_{50} values of 2.2 and 7.4 μg/ml for AChE and 116.7 and 105.1 μg/ml for BChE, respectively.

The alkaloid stephaoxocanidine (**67**), isolated from *Stephania cepharantha* (Menspermaceae), was used by Bianchi *et al.* [59] as a lead compound for the development of new AChEIs. Among them, the oxazaphenalene location **68** exhibited the highest activity, compared to other synthesized analogs with an IC_{50} of

19.6 μM. However, the activity diminished to the half in 8-bromo derivative **69** (IC_{50} 49 μM) and it was less than a quarter of the original in quaternary isoquinolinium compound **70** (IC_{50} 105 μM). The simplest allyl-derivative **71** was also found to be the less active; however, a notorious improvement of enzyme inhibition was detected when its free hydroxyl was converted to the corresponding methyl acetal **72** (IC_{50} 100 μM). Glycols **73** and **74** were more active that their olefinic precursors and, as observed before, the acetal outperformed the hemiacetal (IC_{50} ranging from 96 to 137 μM). The lactone moiety does not seem to be the main responsible for the inhibition, but it may contribute to the effect found in **68**, since structural modifications of the latter with retention or the transformation of the lactone moiety furnished less active compounds but did not abolish the AChE inhibiting activity. Unexpectedly, however, the introduction of a functioned side chain, partially resembling ring D of the tetracycline natural products did not improve the activity.

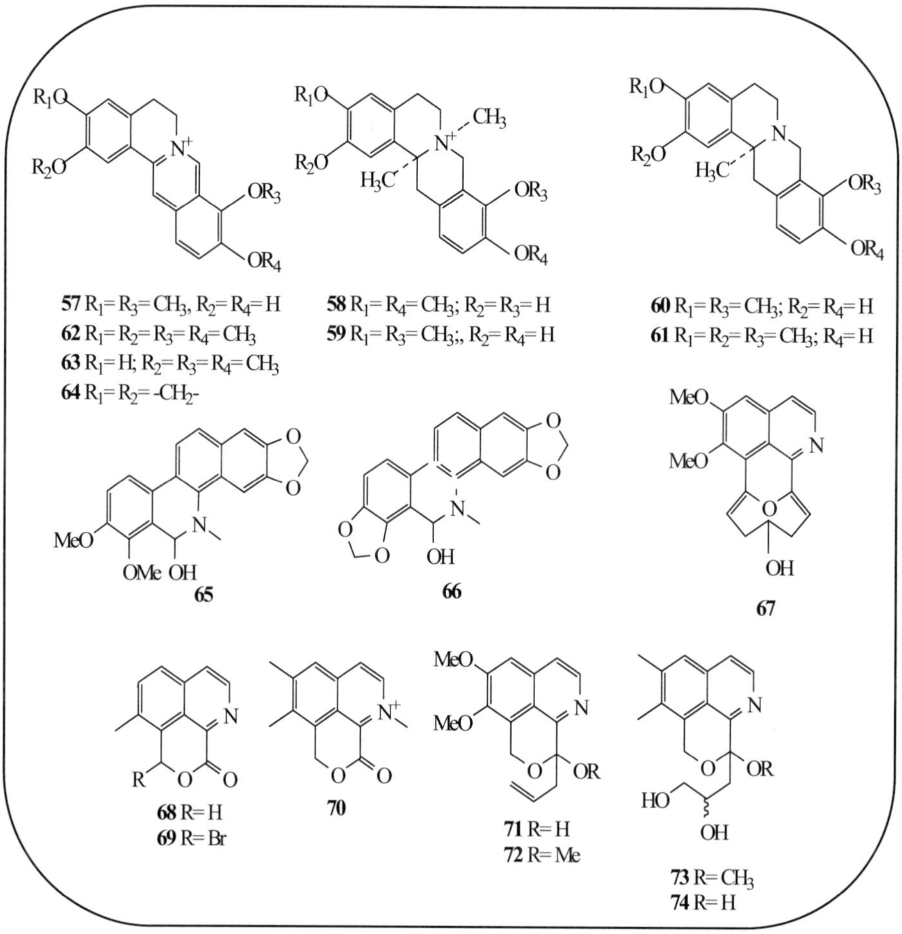

From the tubers of *Corydalis cava* (Papaveraceae) corylucinine and 8-trichloromethyl-7,8-dihydropalmatine were obtained. Both compounds were able to selectively inhibit human AChE with IC_{50} values of 127.6 and 82.9 μM respectively [60].

Another alkaloid with known neuroprotective properties was (-)-physostigmine (**75**). The major drawbacks of this drug are a low therapeutic window and short duration of action [61]. A phenylcarbamate derivative of physostigmine (**75**), and (-)-phenserine (**76**), is a new potent and highly selective AChE inhibitor with a > 50-fold activity *vs.* BChE in clinical trials for the treatment of AD. Compared to physostigmine (**75**) and tacrine (**16**), phenserine (**76**) was less toxic and robustly enhances cognition *in vivo*. Moreover, it reduced β-amyloid peptide formation *in vitro* and *in vivo* and inhibits the formation of the β-amyloid precursor protein [62]. On the basis of this interesting result, several phenserine (**76**) and physostigmine (**75**) analogs were prepared and tested for cholinesterase inhibition. These derivatives contain methyl, dimethyl and trimethyl substituents in each position of the phenyl group of the phenylcarbamoyl moieties. In comparison to unsubstituted phenserine (**76**), which has a selectivity for AChE inhibition of 70-fold (IC_{50} 22 nM), methyl substitution in the C2' position leads tolserine (**77**), provided a 2-fold improvement in AChE potency (IC_{50} 10 nM) and a marginal decline in BChE activity to afford an AChE selectivity of 195-fold [63]. The 4'-methyl substitution, as in **78**, reduced AChE and increased BChE potency compared to **76**, affording a compound that was unselective. In contrast, methyl substitution in the C3' position as in **79**, maintained AChE potency, compared to **76** and increased BChE potency to afford a 12-fold AChE selective compound (IC_{50} 28 nM). The 2'-ethyl substitution, as in **80**, rendered the compound less BChE active, with an AChE selectivity of 290-fold (IC_{50} 10 nM). Compounds with dimethyl substituents at the C2', C3' (**81**, IC_{50} 23 nM), C2', C4' (**82**, IC_{50} 14 nM) and C2', C5' positions (**83**, IC_{50} 26 nM) possessed a similar AChE inhibitory activity to that of the lead compound physostigmine (**75**, IC_{50} 28 nM) and unsubstituted phenserine (**76**, IC_{50} 22 nM). All these compounds were AChE selective. Dimethyl substitution in the C3', C4'-positions as in **84** likewise maintained AChE activity, but provided significant potency against BChE, rendering an unselective inhibitor (IC_{50} 31 and 66 nM for AChE and BChE, respectively). In contrast, C2', C6'-dimethyl substitutions as in **85** reduced inhibitory action against both enzyme subtypes, with the former primarily affecting AChE and the latter BChE activity. The C2', C4', C6'-trimethyl substitution as in **86** resulted in a lack of inhibitory action against both enzymes [64]. Cymserine (**87**) demonstrated a potent concentration-dependent binding with BChE (IC_{50} ranging from 63 to 100 nM). Both **76** and **87** interact similarly with the same two binding domains. However, nucleophilic attack of its carbonyl moiety results in a transient **87**-BChE intermediate, in a tetrahedral conformation, which then promptly collapses to a carbamylated drug-enzyme complex that is more stable than the acetyl enzyme one associated with each; enzyme inhibition occurs consequently due to the slow rate of enzyme decarbamylation. The acyl-binding pockets within BChE and AChE deviates from each other in shape and size at their deepest point and can be distinguished by two amino acid residues positioned at the bottom of the acyl-loop. These residues are aromatic in AChE (Phe-288 and Phe-290) but have smaller aliphatic ones, Leu-286 and Val-288, in BChE [65]. Hence, the acyl pocket with BChE is larger due to the smaller protruding amino acids. The C4'-isopropyl substitution of cymserine (**87**), which elongates the compound in comparison to the unsubstituted analog, phenserine (**76**), provides cymserine its enzyme subtype selectivity. In contrast, the phenylcarbamate moiety of unsubstituted **76** can form a π-π interaction with the phenyl groups of Phe-288 and Phe-290 that increases the stability of the transition state and, consequently, favors AChE selectivity [64]. The hydrophobic characteristics of cymserine (**87**), together with the basic nitrogens within its pyrroloindole backbone, allow for additional aromatic interactions and hydrogen bonding with the primarily aromatic amino acids [65] known to define the catalytic region of BChE. Bisnorcymserine (**88**), obtained by *N*-demethylation of **87**, exhibited an IC_{50} value of 0.228 nM against BChE. This modification leads to increase its potency and extends its pharmacokinetic half-life, and retains sufficient lipophilicity to readily penetrate the blood brain barrier [66]. The bioassay-guided fractionation of *Tabernaemontana divaricata* (Apocynaceae) led to the isolation of 19,20-dihydrotabernamine (**89**) and 19,20-dihydroervahanine A (**90**). Both compounds inhibited the *Ee*AChE activity with IC_{50} values of 227 and 71 nM, for **89** and **90**, respectively. The evidence that alkaloid **90** is more potent than **89** suggests that a carbomethoxy substitution at C16' increases the AChE inhibitory activity. Moreover, **90** acted as specific, reversible and competitive inhibitor [67]. From the same plant Chaiyana *et al.* [68] isolated 3'-R/S-hydroxyvoacamine that acted as a non-competitive inhibitor against AChE (IC_{50} of 7.0 μM).

Recently, Passos *et al.* [69] screened thirteen *Psychotria* (Rubiaceae) alkaloids regarding their interactions with cholinesterase. Two quaternary β-carboline alkaloids namely prunifoleine and 14-oxoprunifoleine inhibited cholinesterase with IC_{50} values corresponding to 10 and 3.39 μM for AChE, 100 and 11 μM for BChE, respectively. Both compounds seem to behave as non competitive AChE inhibitors. In addition, the monoterpene indole alkaloids angustine, vallesiachotamine lactone, *E*-vallesiachotamine and *Z*-vallesiachotamine inhibited BChE with IC_{50} values ranging from 3.47 to 14 μM.

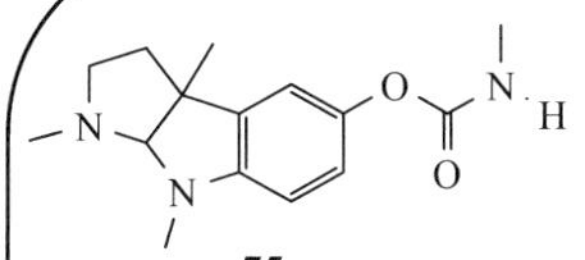

75

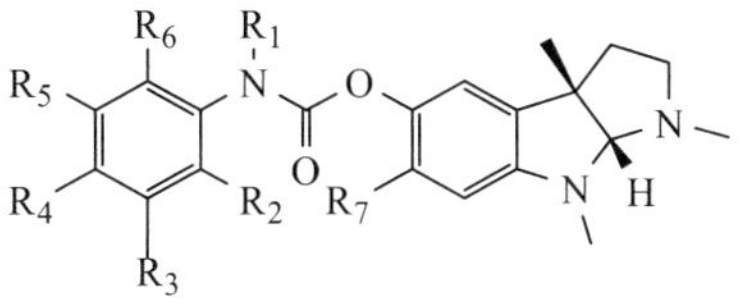

76 R_1- R_7= H
77 R_2= CH_3; R_1, R_3-R_7= H
78 R_4= CH_3; R_1-R_3, R_5-R_7= H
79 R_3= CH_3; R_1, R_2, R_4-R_7= H
80 R_2= CH_3CH_2; R_1, R_2-R_7= H
81 R_2, R_3= CH_3; R_1, R_4-R_7– H
82 R_2, R_4= CH_3; R_1, R_3, R_5-R_7= H
83 R_2, R_5= CH_3; R_1, R_3, R_4, R_6, R_7= H
84 R_3, R_4= CH_3; R_1, R_2, R_5-R_7= H
85 R_2, R_6= CH_3; R_1, R_3, R_5, R_7= H
86 R_2, R_4, R_6= CH_3; R_1, R_3, R_5, R_7= H
87 R_2= H; R_4= $CH(CH_3)_2$; R_1, R_3, R_5, R_7= H
88 R_4= $CH(CH_3)_2$; R_1-R_3, R_5,R_6, R_7= H

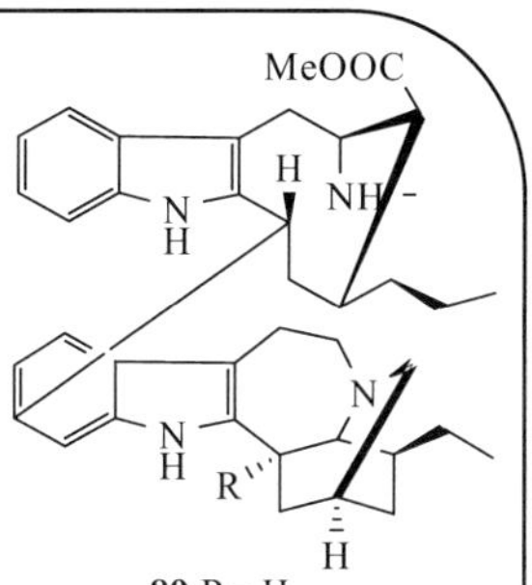

89 R= H
90 R= $COOH_3$

Carinatumins A (**91**) and B (**92**), isolated from *Lycopodium carinatum* (Lycopodiaceae), inhibited AChE with IC_{50} value of 4.6 and 7.0 μM, respectively [70]. Carinatumin A (**91**) was characterized by the presence of a hydroxyl group at C10. Alkaloid (**91**) showed less potent inhibition than huperzine A ((-)-Hup A) (**93**, IC_{50} 0.8 μM), while carinatumin B (**92**) showed AChE inhibition comparable to that of huperzine B ((-)-Hup B) (**94**, IC_{50} 8 μM). Annotine (**95**) isolated from *L. annotinum*, exhibited a promising AChE inhibitory activity [71]. The interesting activity of *Lycopodium* alkaloids, has drawn attention recently to the research on *L. serratum* from which 2α,11α-dihydroxyfawcettiine (**96**) and lycoposerramine (**97**) were isolated and tested for their AChE inhibitory property. Both compounds demonstrated an interesting activity with IC_{50} values of 27.9 and 16.7 μM, in **96** and **97**, respectively [72]. The most promising known alkaloid with lycodine skeleton was (-)-Hup A (**93**). Several studies described the synthesis and cholinesterase activity of (-)-Hup A analogs. Zhou *et al.* [73] demonstrated the importance of amino group in C5 in the AChE inhibition using 5-hydroxyl (**98**) and 5-fluoro (**99**) (-)-Hup A analogs. The choice of this substitute was based on the consideration that hydroxyl group and fluoro are isosteres of amino group and the hydroxyl group and amino group are also similar in polarity but the hydroxyl group has much lower basicity than the amino group. The substitution of amino group by hydroxy group makes the activity decrease sharply in the physiological condition, the amino group can form quaternary ammonium to imitate the function of acetylcholine and combine the active site of AChE, but the hydroxy group cannot act in this way (**98**, IC_{50} of 35 μM). On the other hand, both analogs exhibited comparable activity indicating that, except for the tendency to form the cation, the polarity has less influence on inhibitory activity. Another interesting structural modification of **93** led to the synthesis of 5-desamino analog **100** that showed about 100-fold less activity than the natural product **93** [74]. The substitution at C10 of **93** with methyl group leads to analog (-)-10,10-dimethylhuperzine A (**101**) that was found 8-fold more potent than **93** [75]. Several HupA-based bivalent ligands were described by Haviv *et al.* [76] and Du & Carlier [76]. The synthesis of tacrine-huperzine A-hybrids (huprines) opened the search of a new class of very potent and selective *Torpedo californica* AChE (*Tc*AChE) inhibitors. These compounds were designed by combining the carbobicyclic moiety of (-)-Hup A (**93**) with the 4-aminoquinoline subunit of tacrine (**16**). Until now, the most powerful huprines are the so called huprine X (**102**) and huprine Y (**103**). Huprine X (**102**) exhibits a 1200-fold higher than that of **16** and 180-fold higher than **93** because the introduction of methanesulfonamido group at C13 leads to an extended binding near the active site of *Tc*AChE, and consequently improve the enzyme inhibition [77]. Molecular modeling analysis revealed that huprine-X could complex the *Tc*AChE through three sites: the aromatic portion of **102** occupies the same binding site as **16**; the chlorine substituent lies in a hydrophobic pocket, establishing additional interactions with other residues of the enzyme; and, finally, the carbobicyclic unit of **102** occupies the same binding pocket as the corresponding subunit of **93**. From *H. serrata* (Lycopodiaceae) the congener of **93**, (-)-Hup B (**94**) was isolated. This alkaloid also acted as a reversible inhibitor of AChE with a considerable potency (K_i= 0.79 μM) [78]. Interestingly, when it is tested *in vivo*, **94** exhibited a higher therapeutic index due to its longer duration of action in comparison to **93** [79]. A series of parent molecules which incorporated structural features of both **93** and **94** was synthesized and tested for AChE inhibition. The analog **104** was less active than its parent molecules, probably because the cation-π interaction between the ammonium group and Trp-84, and Phe-330 is slightly altered by the presence of the piperidine ring methylene groups. Based on these findings, analogs **105** and **106** were designed to explore the effects of the size of the heterocyclic ring and its substituents on the AChE activity. Although **105** and **106** were racemic, they are 2- and 4-fold more potent than natural (-)-Hup B (**94**) with K_i values of 0.40 and 0.22 μM, respectively. Thus, if either the size of the substitute or the heterocyclic ring size is increased, the activity decreased, possibly indicating a steric interaction with the walls of the active site gorge [80]. Based on the hypothesis with respect to two binding sites on the active-site gorge of AChE and supporting numerous examples of bivalent ligands related to **93**, a series of new derivatives of (-)-HupB (**94**) with two pharmacophoric moieties, which may interact with both central and peripheral binding sites on the active-site gorge of AChE, respectively, was synthesized and tested for its AChE inhibitory activity. Ligands **107-110**, were much more potent and selective compared to the lead compound **95**. Probably, this activity is related to a double interaction with enzyme [81]. Another series of bis-Hup B analogs **112-119** showed more activity than **93** and **94**. The inhibitory potency and selectivity of bis-Hup B derivatives are dependent on the length of the chain, suggesting a dual-site binding of bis-Hup B compounds to AChE. The optimal compound is bis(18)-Hup B (**117**), which showed significantly higher potency (3915-fold) and selectivity (932-fold) for the inhibition of AChE in comparison to **94** [82]. Interestingly, the optimal length among bis-HupB derivatives was a chain of 18 atoms, which was longer than other bis-molecule AChE inhibitors. A length of 8 to 10 methylenes is favored in the bis-galanthamine series. Among the bis-Hup A derivatives, the most favorable length was 7 atoms, but the inhibitory potency of the (-)-Hup A dimers was less than that of (-)-Hup A (**93**) itself. An intuitive reason inferred that the endocyclic secondary amino group in the bridge-ring (-)-Hup B (**94**) molecule possessed the special steric hindrance and, therefore, the longer chain was needed for optimal binding to AChE. Bis-Hup B derivatives in which compounds with 12-20 atom tethered compounds **119-123**, displayed significantly high potent inhibition of AChE and the inhibitory activity enhanced about two to three orders of magnitude than parental natural product. Another interesting derivative was **124**, in which the length of the tether was 15 atoms with decamethylenediamine between (-)-Hup B (**94**) and the aromatic group. Compound **124** exhibited a 914-fold more potent activity and 34-fold more selectivity in the inhibition of AChE than parental **94** [15]. From *H. Serrata*, another *Lycopodium* alkaloid, Huperzine R (**125**) was isolated, but its ability to inhibit AChE was less pronounced than parent (-)-Hup A (**93**) [83].

From *Skimmia laureola* (Rutaceae), 3-hydroxy,2,2,6-trimethyl-3,4,5,6-tetrahydro-2*H*-pyrano[3,2c] quinoline 5-one (**126**), ribalinine (**127**) and methyl isoplatydesmine (**128**) were isolated and evaluated for cholinesterase inhibition. All compounds were found to be partially competitive and showed pure non competitive inhibitions of both cholinesterases except **127** which exhibited a pure non competitive inhibition [84]. Alkaloids **126** and **127** exhibited the same potency in the inhibition of AChE. Compound **128** exhibited the highest BChE inhibitory activity. On the other hand, **126** and **128** showed certain selectivity against BChE probably due the ability of these compounds to accommodate in the larger aromatic gorge of this enzyme. On the contrary **127** was an AChE selective inhibitor probably due to its reduced size.

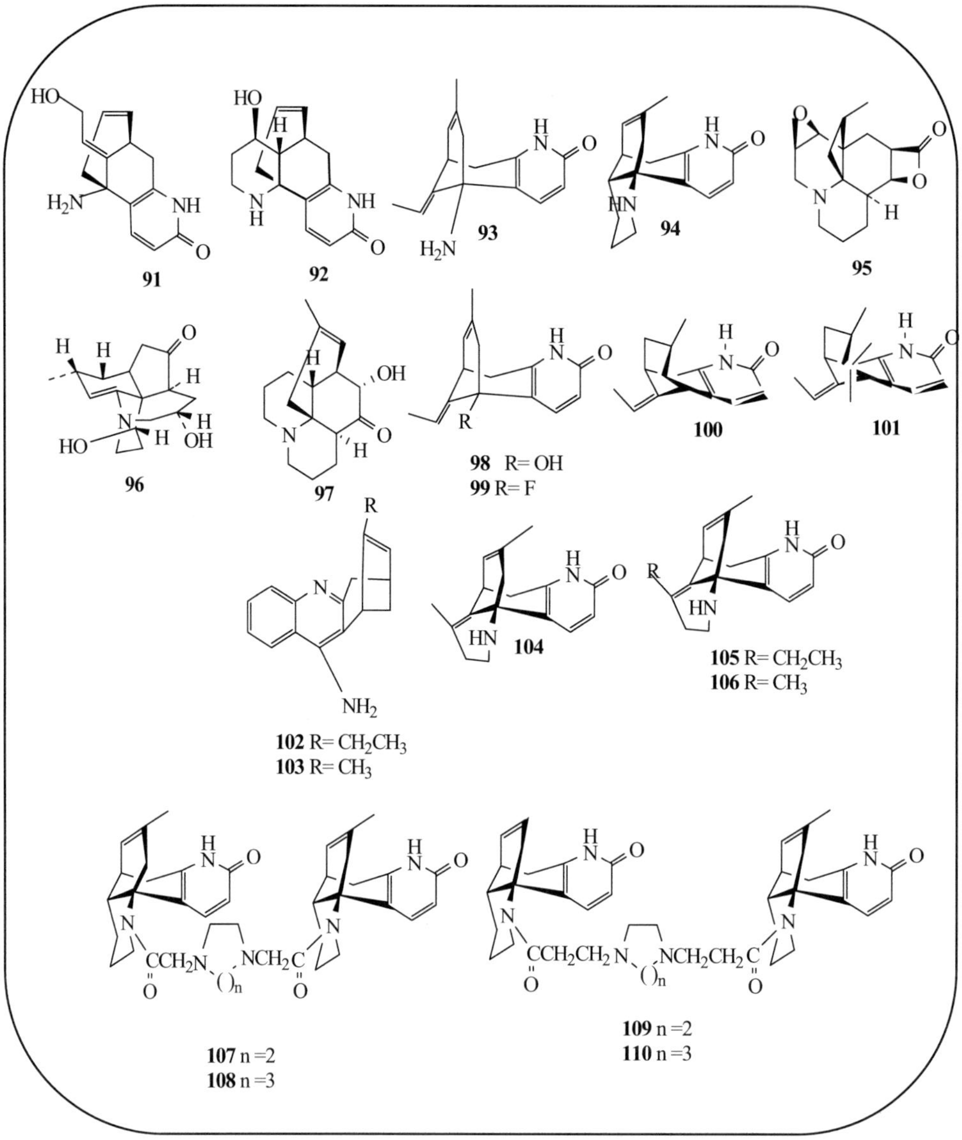
91
92
93
94
95
96
97
98 R= OH
99 R= F
100
101
102 R= CH_2CH_3
103 R= CH_3
104
105 R= CH_2CH_3
106 R= CH_3
107 n =2
108 n =3
109 n =2
110 n =3

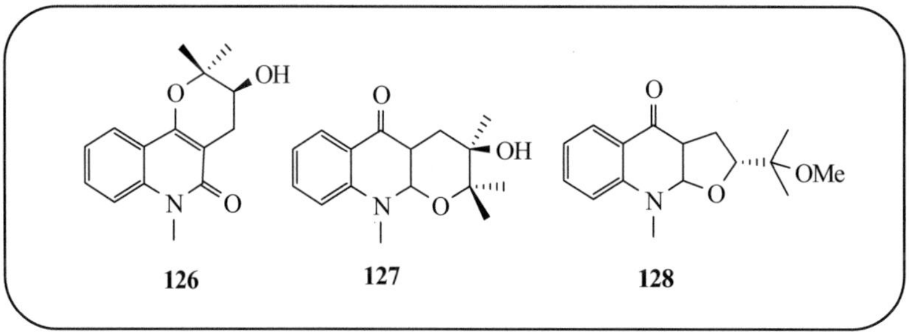
126
127
128

Homoproaporphine and homoaporphine alkaloids, isolated from *Colchicum speciosum* (Liliaceae), showed anticholinesterase activity [85]. The homoproaporphine derivatives include kesselridine (**129**), kesselringine (**130**), 12-*n*-butoxy-demethoxykesselringine (**131**), regelamine (**132**), regeline (**133**), kesselridine iodomethylate (**134**), kesselringine iodomethylate (**135**), regeline iodomethylate (**136**) and the quaternary base of regeline (**137**), dec-*N*-dimethylregeline (**138**), luteidine (**139**), luteinone (**140**), luteidine iodomethylate (**141**), and iolantamine iodomethylate (**142**). Merenderin (**143**) presented the homoaporphine base skeleton. Using kesselridine (**129**) as a reference compound for SAR analysis, it was found that the methylation of the hydroxy group of spirocyclohexanoic ring **130** nearly did not change its ability to inhibit AChE but significantly (10-fold) increased the activity against BChE. The substitution in **130** of the methoxy group with the *n*-butoxy group as in **131** resulted in an increased inhibitory activity against both enzymes in 5- and 10-times, respectively. The methylation in the kesselringine (**130**) of the hydroxy group in the benzene ring as in regeline (**133**) decreased AChE and increased BChE inhibitory activity. It was interesting to compare the effect of formal isomers, such as kesselringine (**130**) and regelamine (**132**) differing by the presence of the methoxy group in the spirocyclohexanoic and benzene rings. The presence of the methoxy group only in the spirocyclohexanoic ring as in **131** makes the alkaloid effect more active against both cholinesterases. From the comparison of the homoproaporphine alkaloid group containing the enonoic ring-luteidine (**139**), luteinone (**140**) and merenderine (**143**) emerges that all compounds exhibited nearly the same anticholinesterase effectiveness *since* that the homoproaporphine alkaloids contain the spirocyclohexanoic ring.

The analysis of the effect of quaternization revealed that onium compounds usually have a more pronounced anti-cholinesterase effect than their nonalkylated analogs as in derivatives of kesselridine (**129**, **134**) and of regeline (**133**, **137**). In all other cases, the effect of iodomethylation was either completely absent (**130** and **135**, **137** and **141** for both enzymes and compounds **129** and **134** for BChE) or opposite (**133** and **136** in the case of BChE). Some oxoisoaporphine alkaloids, isolated from *Menispermum dauricum* (Menispermaceae), were characterized by a 1-azabenzanthrone moiety in their structures. Based on the structural information of AChE and AChEI, the planar 1-azabenzanthrone moiety with an ammonium group might bind to peripheral anionic site (PAS) by π– π stacking and electronic interaction. Introduction of a side chain with terminal amines or ammonium groups in the 1-azabenzanthrone could greatly improve AChE/BChE selectivity and water-solubility of the alkaloids. In view of the above reasons, a series of oxoisoaporphine derivatives **144-152** and **153-159** with different side chain (*n* = 1, 2, and 3) at C9 were reported [86]. All the synthesized compounds presented IC_{50} values in the nM range and higher selectivity for AChE than the lead compound **160** (IC_{50} > 10^5 mM). These results indicated that the introduction of the amino group side chains could increase the inhibitory capacity and selectivity. Compounds **153** and **154** with quaternary nitrogen exhibited highest inhibitory activity on AChE (IC_{50} of 1.08 and 1.06 nM, respectively), compared with the corresponding non-quaternary nitrogen **144** and **145** (IC_{50} Values of 6.18 and 2.47 nM, respectively). A similar tendency was also found in **146-150** and their corresponding quaternary methiodide salts **155-159**. The structural resemblance of quaternary functionality among the synthesized cationic compounds and each would be responsible for the competition for the binding site of AChE. The structure of the terminal groups of side chain has also produced effects on their inhibitory activities. High inhibitory potency was found to be associated with pyrrolidine at the end of side chain **144-146** and **153-155**. Dimethylamine derivatives are characterized by less potency **147-149** and **156-158**. All the synthetic compounds showed high selectivity for AChE. The range of AChE/BChE selectivity ratios was from 45 to 1980-fold. Compound **154** was characterized by the highest inhibitory activity (IC_{50} value of 1.06 nM) with the highest selectivity ratios (1850-fold) [86].

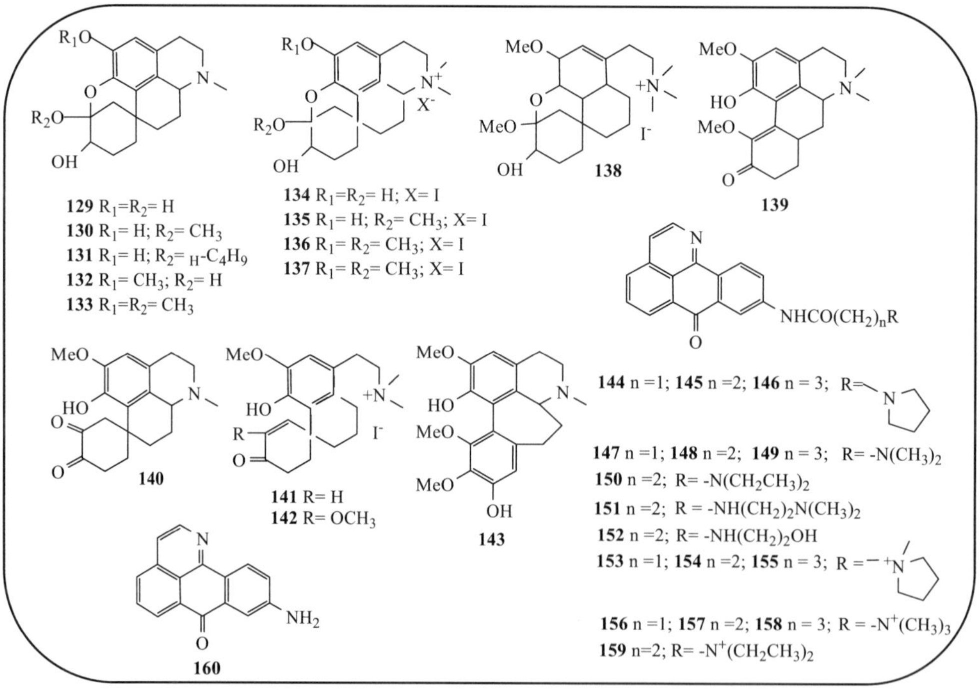

From *Corydalis cava* (Papaveraceae) bulbocapnine (**161**), corydaline (**162**) and corydine (**163**) were isolated and tested. Corydaline (**162**) was found to be the most active and selective AChEI with IC_{50} value of 15 μM. Bulbocapnine (**161**) inhibited AChE as well as BChE with IC_{50} values of 40 and 83 μM, respectively, while corydine (**163**) selectively inhibited BChE with an IC_{50} value of 52 μM [87].

Markmee *et al.* [88] compared a series of tetrahydroquinoline derivatives **164-167**, aporphine derivatives **168**, **169**, dihydroquinoline derivatives **170** and isoquinoline derivatives **171-175** for their ability to inhibit AChE. These compounds exhibited an IC_{50} value ranging from 5.50 to 24.33 μM. When an unsaturation was added to the piperidine ring or quaternarized at basic nitrogen of piperidine, the activity improved by about 10- to 100-fold as shown by compounds **170** and **175**. A similar result was also found in compound **164** and its aromatic analog **173**. Substituents on the benzene ring had quite an influence on the activity as seen by **171** and **172**. The type of substituents and the position of the substitution also affected the AChE inhibition. The amine derivative **166** lost activity when changed to its corresponding amide analog **167**. Rigid molecule as in **168** and **169** displayed slightly reduced activity. Hybridization at benzyl part seemed to play an important role in inhibition as shown by hydroxyl derivative of **173** and the activity decreased when the molecule was oxidized to its ketone derivatives **175**.

Epimediphine, isolated from the leaves of *Epimedium koreanum* (Berberidaceae), inhibited AChE in a dose-dependent manner with an IC_{50} value of 3.1 μM [89].

Hyrcanone (**176**), hyrcatrienine (**177**), N_b-dimethylcycloxobuxoviricine (**178**), hyrcamine (**179**), buxidin (**180**), buxandrine (**181**), buxabenzacinine (**182**) and buxippine-K (**183**), isolated from *Buxus hyrcana* (Buxaceae) exhibited IC_{50} values ranging from 83.0 to 468.0 μM against *Tc*AChE and from 1.12 to 350.0 μM against *Hs*BChE. From SAR studies emerged that the presence of Me_2N moieties at C3 and C20 may be the most important factor influencing the activity of the individual compounds of this class. On the other hand, the presence of oxygen functions at C3 and C20, as in **183**, causes a decrease in the inhibitory activity against the AChE. No clear difference in inhibition profiles, against both cholinesterases, between compounds containing a benzamido group at C3 (*i.e.*, **176**, **180**, **181**, and **182**) and compounds lacking a benzamido group at C3 (*i.e.*, **178**, **179**, and **183**) was observed. However, when *N*-benzoyl containing compounds **176**, **180**, **181**, and **182** were compared, **176** was found to be more active with IC_{50} values of 145.0 and 20.0 μM for AChE and BChE, respectively. This difference in activity seems to be due to the negative effect of other substituents, such as OH at C16 and AcO at C31 in **180-182**. This suggests that, in the absence of these substituents, **176** can approach the aromatic gorge of the AChE and BChE enzymes in a relatively more effective way. The negative impact of the OH substituent at C16 on AChE inhibition was also evident in the case of **178** (IC_{50} of 310.0 μM). Interestingly, the OH group at C16 had no adverse effect on BChE inhibition, as **178** exhibits pronounced activity against the BChE (IC_{50} of 1.12 μM).

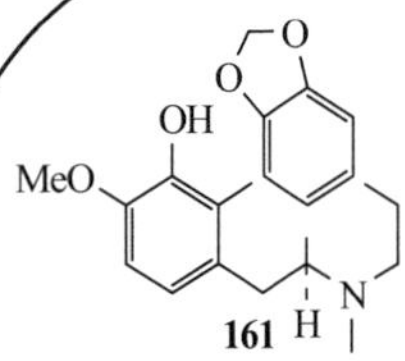

161

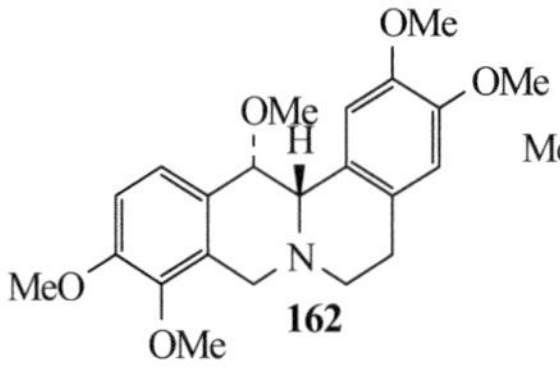

162

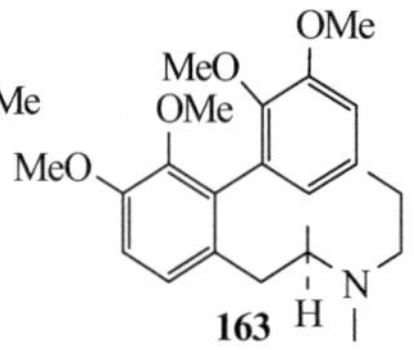

163

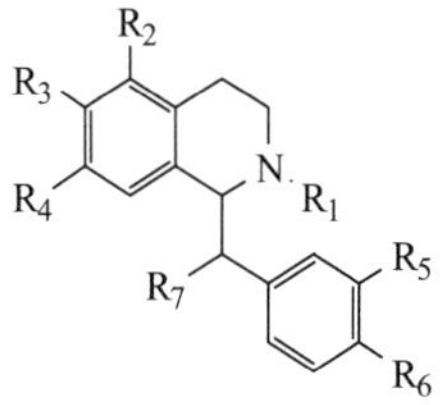

164 $R_1 = R_2 = H$; $R_3 = R_4 = R_5 = R_6 = OCH_3$; $R_7 = OH$
165 $R_1 = CH_3$; $R_2 = H$; $R_3 = R_4 = R_5 = R_6 = OCH_3$; $R_7 = H$
166 $R_1 = CH_3$; $R_2 = H$; $R_3 = R_4 = OCH_3$; $R_5 = R_6 = OCH_2O$; $R_7 = H$
167 $R_1 = CHO$; $R_2 = H$; $R_3 = R_4 = OCH_3$; $R_5 = R_6 = OCH_2O$; $R_7 = H$

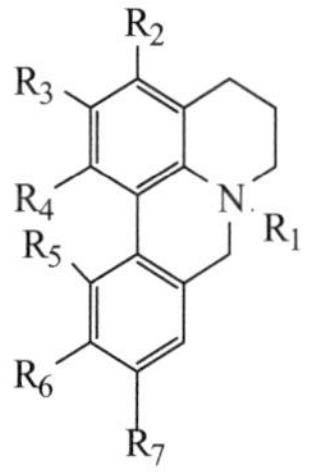

168 $R_1 = CH_3$; $R_2 = R_7 = H$; $R_3 = R_4 = R_5 = R_6 = OCH_3$
169 $R_1 = COOCH_3$; $R_2 = R_3 = R_4 = R_6 = R_7 = OCH_3$; $R_5 = H$

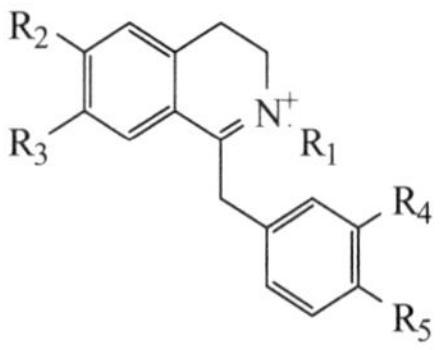

170 $R_1 = CH_3$; $R_2 = R_3 = R_4 = R_5 = OCH_3$

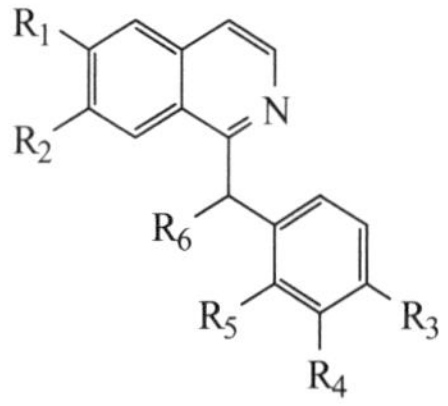

171 $R_1 = R_2 = R_3 = R_4 = H$; $R_5 = OCH_3$; $R_6 = OH$
172 $R_1 = R_2 = R_5 = OCH_3$; $R_3 = R_4 = H$; $R_6 = OH$
173 $R_1 = R_2 = R_3 = R_4 = OCH_3$; $R_5 = H$; $R_6 = OH$
174 $R_1 = R_2 = R_3 = R_4 = OCH_3$; $R_5 = H$; $R_6 = {=}O$
175 $R_1 = R_2 = R_3 = R_4 = OCH_3$; $R_5 = R_6 = H$

The presence of cyclopropane or other unsaturation sites apparently do not influence the activity of the molecule against both enzymes. This may be due to the inherent conformational flexibility of the aromatic gorge residues of the enzymes, which enables the gorge to accommodate equally effectively the different conformations of these compounds. On the other hand, if the methoxy group in C31 is substituted with AcO functionality, as in **182**, the steric bulk of the ring A increases, which, in turn, reduces the ability of the molecule to enter into the aromatic gorge of the enzyme. This further supports the preference of the hydrophobic interactions of the methyl group at this position [90].

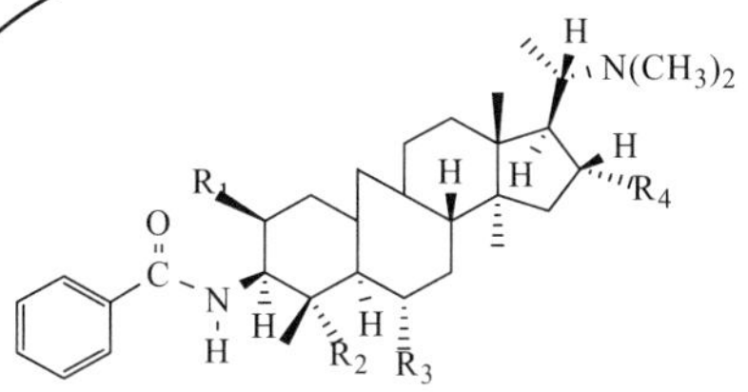

176 $R_1= R_3= R_4=$ H; $R_2=$ CH_3; $D^{1,10}$; 11-C=O

177 $R_1= R_3= R_4=$ H; $R_2=$ CH_3, $D^{1,10}$; $D^{9,11}$

182 $R_1= R_3=$ H; $R_2=$ CH_3COCH_2; $R_4=$ OH; $D^{1,10}$; $D^{9,11}$

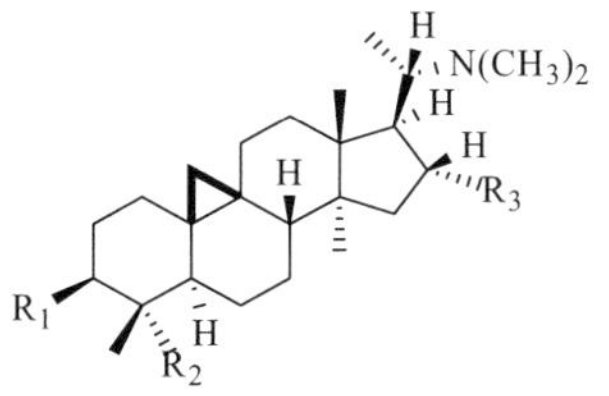

178 $R_1=$ C=O; $R_2=$ CH_3; $R_3=$ OH; $D^{1,2}$

179 $R_1=$ NH-tigoyl; $R_2=$ $HOCH_2$; $R_3=$ CH_3CO

180 $R_1=$ NH-benzoyl; $R_2=$ $HOCH_2$; $R_3=$ HO; $D^{6,7}$

181 $R_1=$ NH-benzoyl; $R_2=$ $HOCH_2$; $R_3=$ CH_3CO; $D^{6,7}$

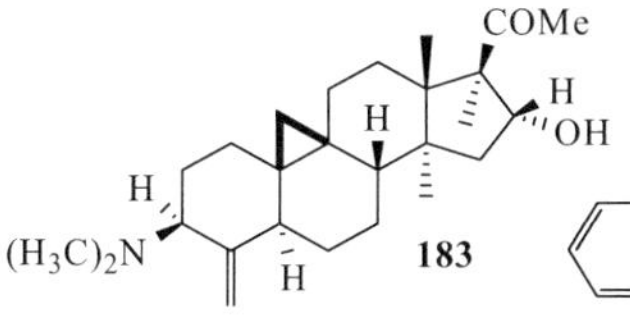

183

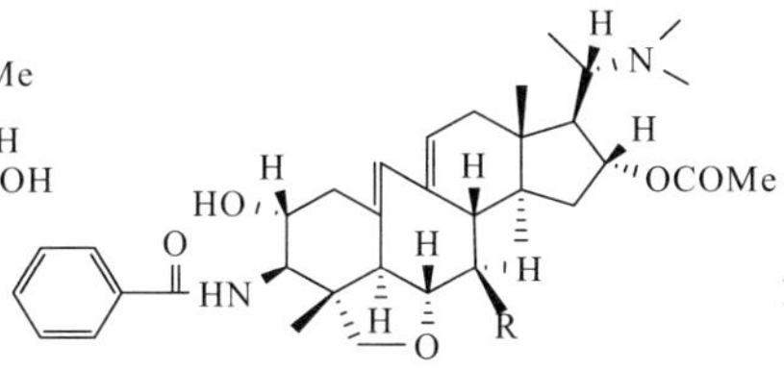

184 R= OH

185 R= H

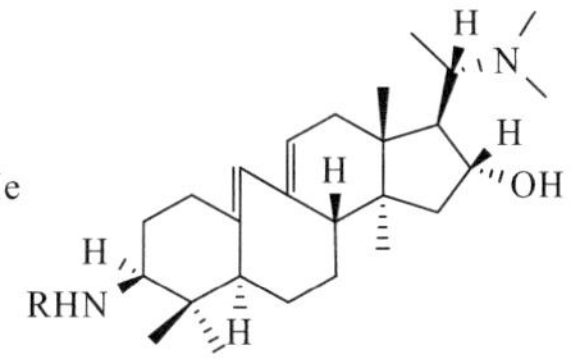

186 R= PhCO

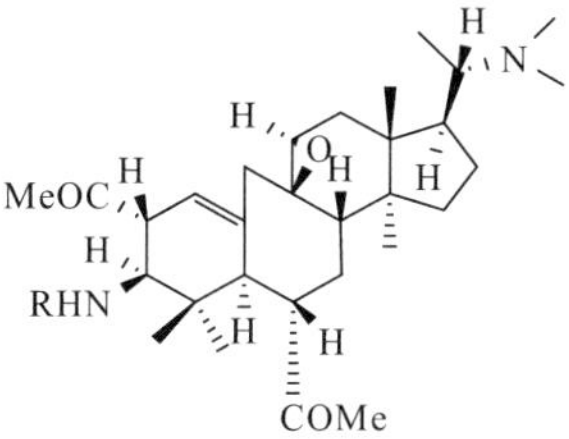

187 R= PhCO

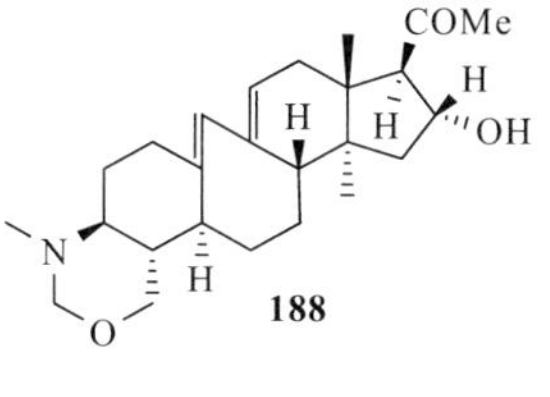

188

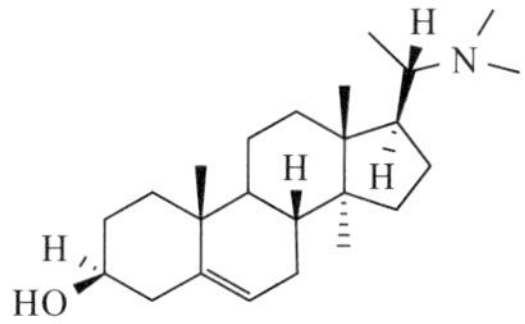

189

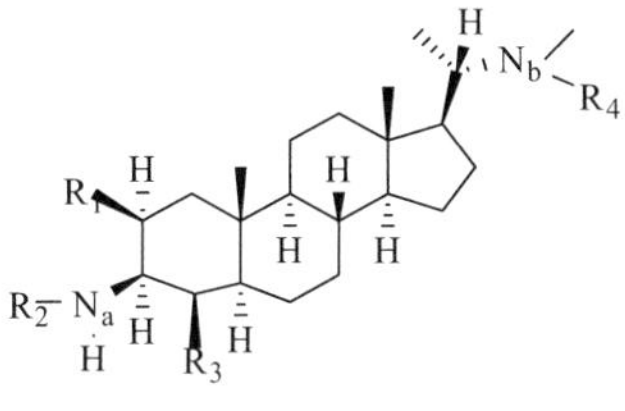

190 $R_1=$ H; $R_2=$ $COCCH_3=CHCH_3$; $R_3=$ CH_3CO; $R_4=$ CH_3

191 $R_1=$ OH; $R_2=$ $COCH=C(CH_3)_2$; $R_3=$ H; $R_4=$ CH_3; $D^{5,6}$

From *B. hyrcana* (Buxaceae), (+)-O^6-buxafurandiene (**184**) and (+)-7-deoxy- O^6-buxafurandiene (**185**) (+)-benzoylbuxidienine (**186**), (+)-buxapapillinine (**187**),

(+)-buxaquamarine (**188**) and (+)-irehine (**189**) were isolated [91]. Compounds **184** and **185**, characterized by the presence of a tetrahydrofuran ring incorporated in their structures, exhibited the most promising activity with IC_{50} values of 17 and 13 μM, respectively. Furthermore, **184** and **185** exhibited nearly the same potency and this indicated that C7 hydroxyl group does not play any role in enzyme inhibition activity. Sarcovagenine-C (**190**) and salignarine-C (**191**), isolated from *Sarcococca saligna* (Buxaceae), showed an interesting activity on *Hs*BChE with IC_{50} values of 1.5 and 1.25 μM, respectively. Moreover **191,** demonstrated a promising activity on *Ee*AChE (IC_{50} of 19.7 μM) [92].

Zaheer-ul-haq *et al.* [93] reported the binding models of several steroidal alkaloids isolated from *S. saligna* (Buxaceae) including salignenamide C (**192**), salignenamide D (**193**), 2β-hydroxyepipachysamine D (**194**), salignenamide E (**195**), and salignenamide F (**196**), axillarine C (**197**), axillarine F (**198**), sarcorine (**199**), N^3-demethylsaracodine (**200**), saligcinnamide (**201**), salignenamide A (**202**), vaganine A (**203**), axillaridine A (**204**), sarsalignone (**205**) and sarsalignenone (**206**) [94]. Among them axillaridine A (**204**), sarsalignone (**205**) and sarsalignenone (**206**) were the most potent with the IC_{50} values of 5.21, 7.02 and 5.83 μM, respectively, against AChE. Compounds **205** and **206** also showed an interesting activity against BChE with IC_{50} values of 2.49 and 2.18 μM, respectively. A comparable activity against BChE was found with vaganine A (**204**) (IC_{50} of 2.32 μM). The anti-cholinesterase activity may be associated with the presence of an α,β-unsaturated amide functionality at C3 of ring A. In addition, the presence of an α,β-unsaturated carbonyl moiety between rings A and B probably contribute to the enhancement of the activity. The presence of an electron-rich functionality in ring A apparently decreases the anticholinesterase activity of the concerned compounds, *e.g.* **192-194**, **196**, and **198** with high IC_{50} values as compared to those of **205** and **206**. Another factor that probably contributes to the anticholinesterase activity of these compounds is the presence of dimethylamino moiety at C20 [94]. Recently, haloxylines A (**207**) and B (**208**), isolated from *Haloxylon salicornicum* (Chenopodiaceae), were found to be able to inhibit cholinesterase (IC_{50} values of 25.3 and 20.2 μM against *Ee*AChE and 19.0 and 14.7 μM against *Hs*BChE for **207** and **208**, respectively) [95]. With piperidine skeleton, Viegas *et al.* [96] designed five analogs of natural (-)-3-*O*-acetyl-spectaline (**209**) and (-)-spectaline (**210**) that were previously obtained from the flowers of *Senna spectabilis* (Leguminosae) [97]. Compound **209** was selected as a lead compound because it is possible to recognize each moiety in its structure. Among the tested compounds **211** and **212** were found to be effective inhibitors of Rat brain AChE (*Rb*AChE) with IC_{50} values of 7.32 and 15.1 μM, respectively.

Thus, the *N*-methyl substitution in the piperidine ring as in **213** would be expected to increase potency, because of the greater similarity with the ammonium group in ACh. However, the inhibitory potency was markedly reduced in **213**, suggesting that the methyl group introduced steric hindrance to the ionic interaction of the charged nitrogen. The addition of the *tert*-Butoxycarbonyl (*t*-Boc) group to the acetyl moiety in **212** slightly reduced the potency as compared to **211**.

Casuarinine C and casuarinine I, isolated from *Lycopodiastrum casuarinoides* (Lycopodiaceae), showed moderate inhibitory activity against AChE with IC_{50} values of 20.9 and 12.1 μM, respectively [98].

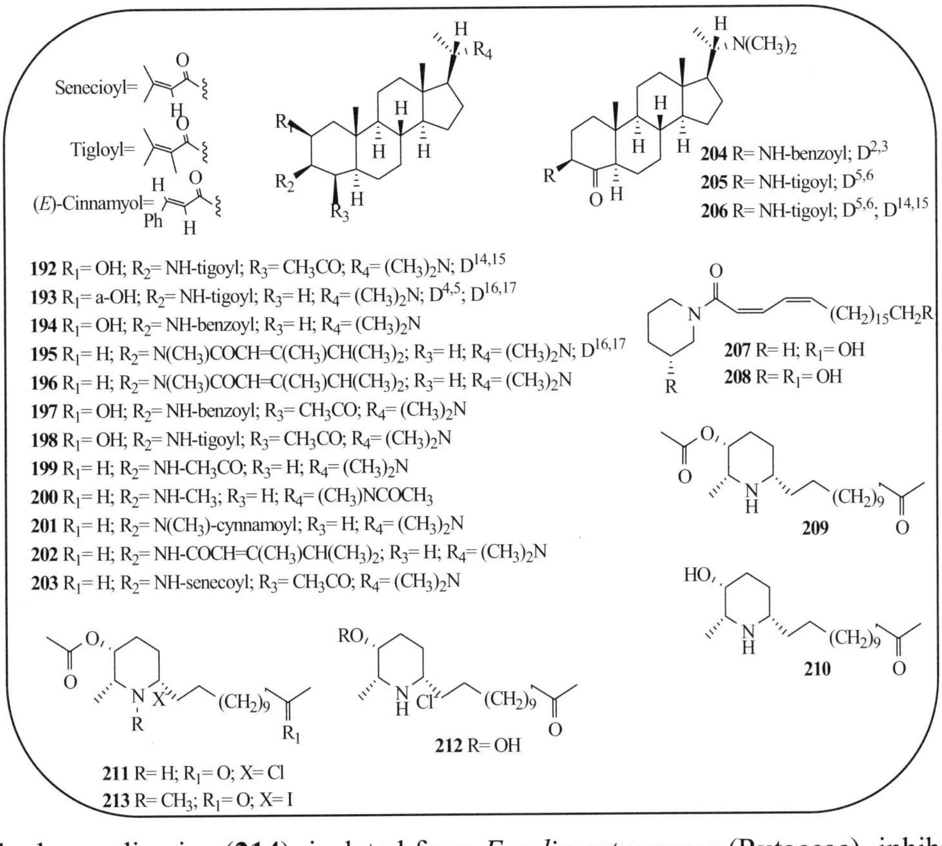

Dehydroevodiamine (**214**), isolated from *Evodia rutaecarpa* (Rutaceae), inhibited *eE*AChE activity in a non-competitive manner with an IC_{50} value of 37.8 μM [99]. A single administration of **214** (6.25 mg/kg) to rats significantly reversed the scopolamine-induced memory impairment in a passive avoidance test. This potent antiamnesic effect was due to the combined effects of AChE inhibition and the known cerebral blood flow enhancement. Compound **214** and related alkaloid rutaecaerpine (**215**) were lead compounds for the synthesis of 13-methyl-5,8-dihydro-6*H*-isoquino[1,2-*b*]quinazolin-13-ium chloride (**216**), (8*Z*)-5,6-dihydro-8*H*-isoquino[1,2-*b*]quinazolin-8-imine (**217**), 5,8-dihydro-6*H* isoquino[1,2*b*]quinazoline (**218**), 13-methyl-5,8-dihydro-6*H*-isoquino[1,2-*b*]quinazolin-13-ium chloride (**219**), 5,7,8,13-tetrahydroindolo [2′,3′:3,4]pyrido[2,1-*b*]quinazoline (**220**), and *N*-(2-phenylethyl)-*N*-[(12*Z*)-7,8,9,10-tetrahydroazepino [2,1-*b*]quinazolin-12(6*H*)-ylidene]amine (**221**) [100]. Alkaloid **220** was a strong cholinesterase inhibitor with IC_{50} of value 3.4 and 0.5 μM for AChE and BChE, respectively, therefore representing a kind of a reversed binding profile of physostigmine (**75**). Compounds **217**, **220** and **221**, possessed a 10-fold higher affinity to BChE with an IC_{50} values ranging from 0.5 to 2.2 μM [101]. The analog **218** (IC_{50} of 7.7 and 4.4 μM for AChE and BChE, respectively), was also quite a potent inhibitor, but with decreased selectivity. The synthesized compounds showed different selectivity profiles reaching from higher affinities to AChE (**216**) and to BChE (**220**). Noteworthy, the imines **217** and **221** were moderate cholinesterase inhibitors with a 10-fold higher affinity at BChE. Alkaloid **214** and its dibenzo-analog **216** are moderate inhibitors of AChE. Compound **219** did not show significant affinity towards AChE, and only a low affinity towards BChE. This is actually quite surprising, because this compound combines the structures of the **218** and **216**, respectively.

Tapsine (**222**), isolated from *Magnolia x soulangiana* (Magnoliaceae), showed long-lasting and a 10-time more potent effect on *Ee*AChE than galanthamine (**1**), with IC_{50} of 0.33 μM. Docking studies indicate that **222** interact with AChE through three major interaction patters: (i) **222** π-stacked on one side with Trp-84 and on the other side with Phe-330. (ii) The cationic side chain of **222** occupied by a positively ionized feature reaches into the esteratic site anchoring at His-440, Ser-200, and Glu-199. (iii) A small hydrogen-bonding network with active site water molecules and Ser-122 further stabilized ligand binding [102].

Sessilistemonamines A (**223**) and B (**224**), isolated from *Stemona sessilifolia* (Stemonaceae), were moderately active against AChE with IC_{50} values of 68.8 and 17.1 μM, respectively [103]. From the roots of this plant, stenine B and stenine were isolated. These compounds showed significant anti-AChE activities with IC_{50} values of 2.1 and 19.8 μM, respectively. Stenine B was reversible and competitive inhibitor [104].

From the bulbs of *Zephyranthes robusta* (Amaryllidaceae) 8-*O*-demethylmaritidine was isolated. This compound exhibited significant AChE inhibitory activity with IC_{50} value of 28.0 μM [105].

Phytochemical investigation of *Beilschmiedia alloiophylla* (Lauraceae) resulted in the isolation of several alkaloids namely 2-hydroxy-9-methoxyaporphine, laurotetanine, liriodenine, boldine, secoboldine, isoboldine, asimilobine, oreobeiline, 6-epioreobeiline, and (*S*)-3-methoxynordomesticine. All compounds inhibited AChE in the range 2.0-10.0 μM [106].

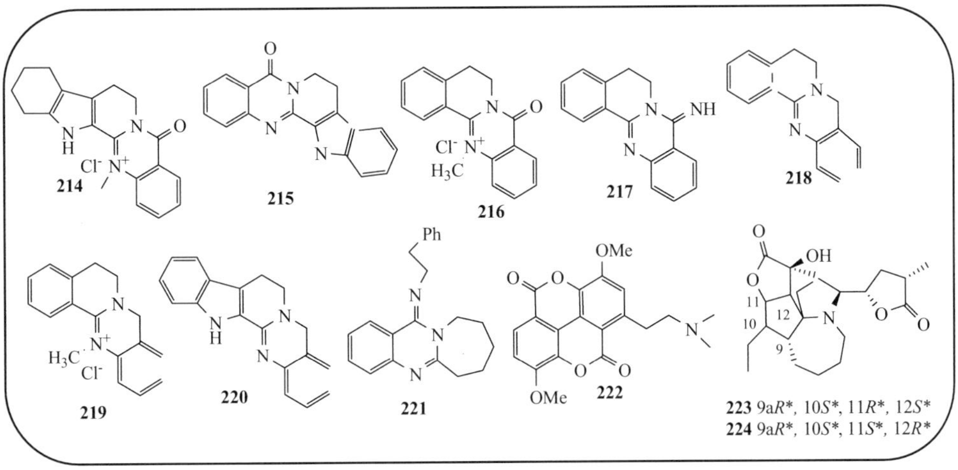

TERPENOIDS

The AChE inhibitory activity exerted by the main constituents of essential oil of *Thymus vulgaris* (Lamiaceae) was recently examined [107]. In particular, linalool (**225**), thymol (**226**), carvacrol (**227**) and their derivatives thymoquinone (**228**) and thymohydroquinone (**229**) were tested. This activity decreased in the following order: thymohydroquinone (**229**) > carvacrol (**227**) > thymoquinone (**228**) > essential oil > thymol (**226**) > linalool (**225**).

The AChE inhibitory effect exerted by carvacrol (**227**) was 10-times stronger than that exerted by its isomer thymol (**226**). The main components of *Melaleuca alternifolia* (Myrtaceae), such as terpinen-4-ol (**230**), γ-terpinene (**231**), α-terpinene (**232**), *p*-cymene (**233**) and 1,8-cineole (**234**) were also investigated [80]. The most potent monoterpenoid was 1,8-cineole (**234**) with an IC_{50} value of 49.0 μg/ml. The cholinesterase inhibitory activity of the main components of *Salvia fruticosa*, *S. lavandulaefolia*, *S. officinalis* and *S. officinalis* var. *Purpurea* (Lamiaceae) essential oils were carried out [108]. 3-Carene (**235**) was the only compound that reached 50% inhibition of the BChE at 2 mM during the 5 min pre-incubation period. Incubation time affected the inhibitory activity of **235** with the IC_{50} values being significantly decreased from 2 mM in 5 min to 0.7 mM in 60 min. The anti-AChE activity of β-caryophyllene (**236**), 3-carene (**235**), 1,8-

cineole (**234**), α-pinene (**237**) and β-pinene (**238**) was apparent. β-Caryophyllene (**237**) at 0.06 mM gave 32% of AChE inhibition, whereas 3-carene, at a final concentration of 0.08 mM gave 33% of AChE inhibition. β-Asarone (**239**), identified in *Acorus calamus* (Acoraceae), showed an interesting activity with an IC_{50} value of 3.33 μM, while the *trans* isomer, α-asarone (**240**) exhibited a lower activity with IC_{50} value of 46.38 μM [109]. Recently, Kundu [110] reported the ability of vanillin to inhibit AChE with a greater potency than 2-hydroxy-4-methoxybenzaldehyde with IC_{50} values of 0.037 and 0.047 mM, respectively. Three main identified monoterpenes of *Pimpinella anisoides* (Apiaceae) namely sabinene, limonene and *trans*-anethole were tested for their cholinesterase inhibitory activity. Among them *trans*-anethole and sabinene were the most active against AChE with IC_{50} values of 134.7 and 176.5 μg/ml, respectively [111].

Fractionation of *Detarium microcarpum* (Caesalpiniaceae) extract led to the isolation of several diterpenes. 5α,8α-(2-oxokolavenic acid) (**242**) proved to be a strong AChE inhibitor at 0.1 μg. Its isomer 2-oxokovalenic acid (**243**) showed a slight inhibition at 1 μg, while 3,4-epoxyclerodan-13*E*-en-15-oic acid (**244**) and 3,4-dihydroxyclerodan-13*Z*-en-15-oic acid (**245**) exhibited weak activity only at 10 μg [112].

Slavins A (**245**) and B (**246**), isolated from *Salvia santolinifolia* (Lamiaceae), displayed selective significant inhibitory potential against BChE with IC_{50} values of 12.5 and 65.5 μM, respectively [113]. Dihydrotanshinone I, isolated from *Salvia miltiorrhiza* (Lamiaceae) with an observed binding constant (KI) of 750 nM, provides a good lead compound for the development of high-affinity, uncharged inhibitors with specificity for the P-site [114]. Bioassay-guided fractionation of *Lycopodium clavatum* (Lycopodiaceae) resulted in the isolation of lyclavatol (**247**) that showed a weak, but dose-dependent inhibitory effect on AChE [115].

From the branches and roots of *Malpighia emarginata* (Malpighiaceae) norfriedelanes A and B were isolated. Norfriedelin A (with a α-oxo-β-lactone group) and norfriedelin B (with a keto-lactone group) showed AChE inhibitory effects with the IC_{50} values of 10.3 and 28.7 μM, respectively [116]. Bengalensinone and benganoic acid were isolated from the roots of *Ficus bengalensis* (Moraceae). The lupine triterpene and the apocarotenoid displayed inhibitory potential against cholinesterase in a concentration-dependent manner with IC_{50} values 194.5 and 154.5 μM for AChE and 224.9 and 120.0 μM for BChE, respectively [117].

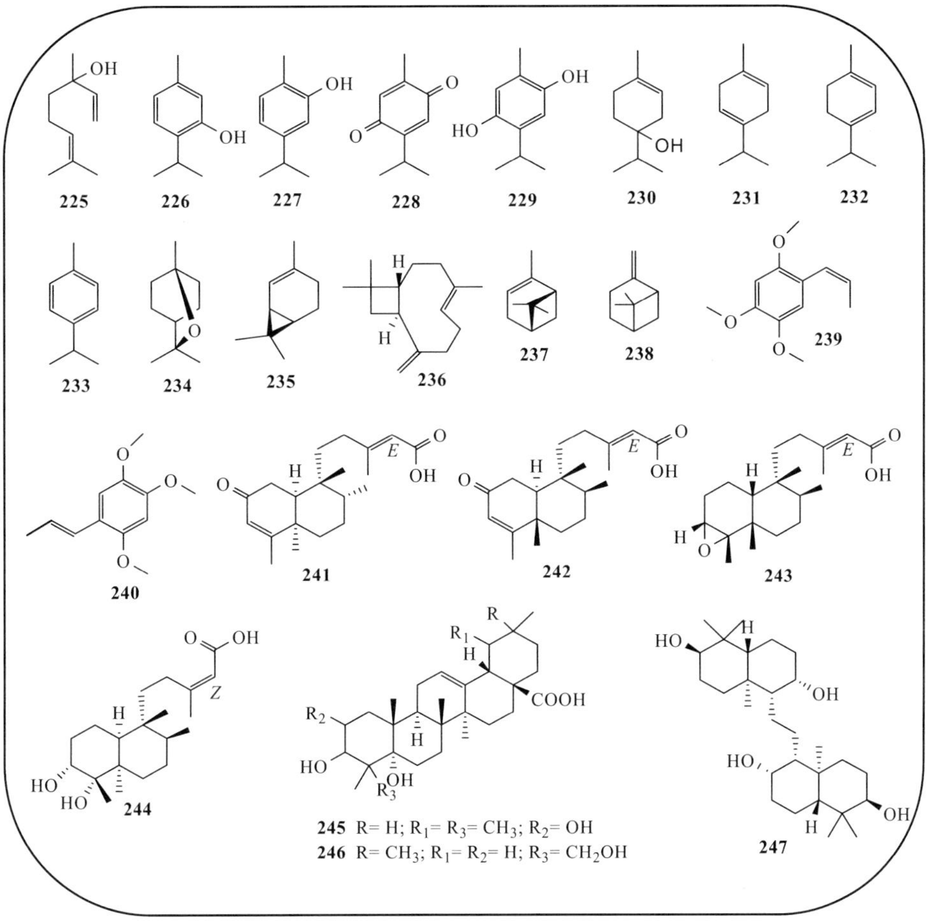

Haloxysterols A-D (**248-251**), 5α,8α-epidioxy-(24*S*)-ethylcholesta-6,9(11),22(*E*)-triene-3β-ol (**252**), (24*S*)-ethylcholesta-7,9(11),22(*E*)-triene-3β-ol (**253**), lawsaritol (**254**), 24-ethyl-cholest-7-ene-3,5,6-triol (**255**) and 24-ethylcholest-6-ene-3,5-diol (**256**), isolated from *Haloxylon recurvum* (Chenopodiaceae), inhibited cholinesterase with K_i values ranging from 0.85 to 25.5 against AChE and ranging from 1.0 to 19.0 μM against BChE [118]. Compound **249** displayed highest inhibitory potential against AChE probably for one extra double-bond in the ring A that can be involved in the π-π stacking interactions, and these interactions are energetically very favorable to make the AChE-**249** complex stable. The oxygen of carbonyl moiety of **250** (K_i 1.1 μM) can make a strong hydrogen bond with the amino acid residues on the active site, additional π-π stacking interactions of **250** with the active site amino acid residues of AChE and operational π-π stacking interactions can further strengthen the AChE-**250** complex. Compound **252** displayed least inhibitory potential among the series of sterols reported in this work which may be due to the oxygen bridging that can shield the π-π stacking interaction of the amino acid residues in the active center of AChE, the only driving force in the **252**-AChE complex seems to be the hydrogen bonding. The potency of enzyme inhibition was lower for **253** (K_i 20.6 μM) probably due to the reduction of the number of hydroxyl groups and olefinic double bonds. There was a slight increase in the activity with the availability of hydroxyl group and olefinic double bond **254** (K_i 16.1 μM). An increase in the number of hydroxyl groups further increased the activity **256** (K_i 12.6 μM). A sharp increase in the inhibitory potential occurred when both hydrogen bonding and π-π stacking interaction were operative **256** (K_i 4.0 μM). The potential of **248** (K_i 4.5 μM) to inhibit BChE might be due to the presence of hydroxyl group in ring A. The inhibitory potential was enhanced by the increased activity of double bond as in the case of **249** (K_i 2.5 μM). The potency of **250** (K_i 18.0 μM) decreased rapidly with the decrease of hydroxyl groups. Again, the potency increased sharply with the increase in the number of hydroxyl groups **251** (K_i 2.2 μM) as compared to **250** with the decrease of hydroxyl groups. There is a decrease in the potency of **252** (K_i 5.0 μM) where there is no hydroxyl group on the ring A or B. The potency gradually increased when there is an increase of hydroxyl groups in **253** (K_i 4.6 μM). Decrease in the number of hydroxyl moieties and olefinic bonds reduced the potency in the case of **254** (K_i 3.7 μM) and **256** (K_i 4.2 μM). A slight increase in the potency occurred with the increase of hydroxyl groups and olefinic bonds in **255** (K_i 2.5 μM). The withanolides **257-259** and **260-261**, isolated from *Ajuga bracteosa* (Lamiaceae) and *Withania somnifera* (Solanaceae), respectively, inhibited both cholinesterases with IC_{50} values ranging from 20.5 to 49.2 μM and 29.0 to 85.2 μM for AChE and BChE, respectively [119]. Lineweaver-Burk as well as Dixon plots and their secondary replots indicated that compounds **257**, **259** and **261** are the linear mixed-type inhibitors of AChE, while **258** and **259** are non-competitive inhibitors of AChE with K_i values ranging from 20.0 to 45.0 μM. All compounds were found to be non-competitive inhibitors of BChE with K_i values ranging from 27.7 to 90.6 μM.

From the fractionation of *Inula oculus-christi* and *I. aucheriana* (Asteraceae) three sesquiterpene lactones namely gaillardin, britannin and pulchellin C were isolated. These compounds tested against AChE demonstrated a inhibition percentage of 67.0, 25.2 and 10.9% at 300 μg/l, respectively [120].

SHIKIMATE DERIVED COMPOUNDS

Linarin (**262**), isolated from *Mentha arvensis* (Lamiaceae), exhibited twice as potent as galanthamine (**1**), while acacetin (**263**) inhibited AChE or BChE activities [121]. Related to linarin (**262**) are linariin (**264**), isolinariins A (**265**) and B (**266**), isolated from *Linaria reflexa* (Scrophulariaceae). These compounds were able to inhibit AChE with IC_{50} values of 0.27, 0.28 and 0.30 μM for **265**, **266** and **264**, respectively. Structurally, **264** has the same rhamnoglycoside moiety as **262**, but has an additional methoxyl group in position C6 in the aglycone. The presence of rhamnoglycoside in **262** has been shown to play a part in its interaction with the enzyme [122]. Leufolins A (**267**) and B (**268**), isolated from *Leucas urticifolia* (Labiateae), exhibited an interesting activity against BChE with IC_{50} values of 1.6 and 3.6 μM, respectively. On the contrary, both (**267**) and (**268**) exhibited a very weak activity against AChE (IC_{50} values of 74.5 and 72.3 μM, respectively) [123]. From *Zygophyllum eurypterum* (Zygophyllaceae), atricarpan A (**269**), B (**270**), C (**271**) and D (**272**) were isolated and tested for their ability to inhibit cholinesterase. Atricarpan A (**269**) exhibited the highest *Hs*BChE inhibitory activity (IC_{50} 12.5 μM) probably for the presence in ring A of hydroxyl and methoxy groups. Hydroxyl groups can form H-bonds with amino acid residues at the active site of the enzyme. Compound **270** exerted a less inhibitory activity with an IC_{50} value of 65.5 μM, that may due to the presence of ethylenedioxy substituent that can hinder the access of hydroxyl groups in the active site of *Hs*BChE. The activity of atricarpan C (**271**) (IC_{50} value of 19.5 μM) increased sharply due to the presence of methoxyl group as substituent. The ability of **272** (IC_{50} value of 20.5 μM) to inhibit *Ee*AChE could be explained through the possibility that oxygen atom of carboxy can form strong H-bonds with the active site of the enzyme. At the same time **272** could inhibit *Hs*BChE (IC_{50} value of 30.5 μM) [124]. A series of chalcones was synthesized and investigated for its biological activity against *Ee*AChE and *Hs*BChE. The hydroxyl group present in ring A has a vital role in the inhibition of AChE in fact compounds **273-277** and **278** exhibited IC_{50} values in the range from 28.2 to 213.5 μM. Compounds **279-283**, lacking of a hydroxyl group in ring A, are completely inactive. Compound **282** closely resembles compound **275** but only lacks a hydroxyl group in ring A. It is totally inactive whereas **275** showed weak activity. Although compounds **280**, **281** and **282** have a hydroxyl group in the molecule they are inactive, because they have a hydroxyl group in ring B and not in ring A. Chalcone **278** exhibited lesser activity than **276** probably due to the presence of a chloro substituent in ring A. Compound **284** was totally inactive, probably due to the presence of two chloro substituents, one in each ring. In addition, the methyl group may be another substituent that further deactivates the chalcone molecule against AChE, as observed in **278**. For AChE inhibitory activity it may be concluded that a hydroxyl group *ortho* to the side chain is very much responsible for AChE inhibitory activity. Substituents like methoxy, chloro or methyl group may increase or decrease the AChE activity of chalcones in the present set of compounds. Interestingly, when the present series of compounds was screened for its BChE inhibition activity, it was found that all those compounds, which exhibited good activities against AChE, were inactive against BChE. Compound **279** was found to be a potent and selective BChE inhibitor. It has a hydroxyl group on ring B *ortho* to the side chain, which indicates that the hydroxyl group at the same position may be responsible for the BChE inhibitory activity. It is also obvious from the activity of **280** and **281**, which exhibited lower activity than **279**, though both have hydroxyl group on ring B but *meta* or *para* to the side chain. Another substituent effect was also found in the series due to the presence of a methoxy group on ring B, which may also be responsible for BChE activity, but to a lesser extent as compared to a hydroxyl group *ortho* to the side chain. These effects can be observed in **282** and **283**. A very weak BChE inhibitory activity was shown by **285**, which is an unsubstituted chalcone [125].

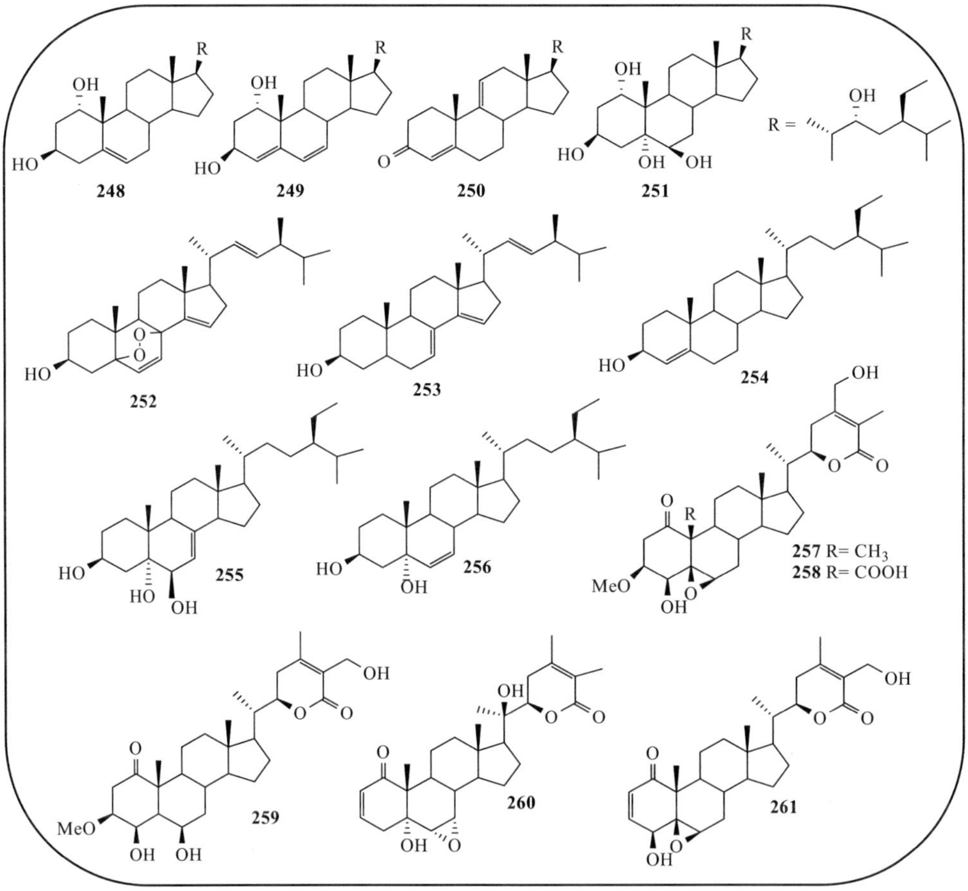

From *Seriphidium stenocephalum* (Asteraceae) stenocepflavone, cirsimaritin, 5,7,5'-trihydroxy-3',4',6-trimethoxyflavone, 5,6,7,5'-tetrahydroxy-4'-methoxyflavone, and axillaroside were isolated. All compounds exhibited a moderate cholinesterase inhibitory activity in a range from 49.27 and 59.11 μM for AChE and in the range from 46.31 to 49.71 μM for BChE, respectively [126].

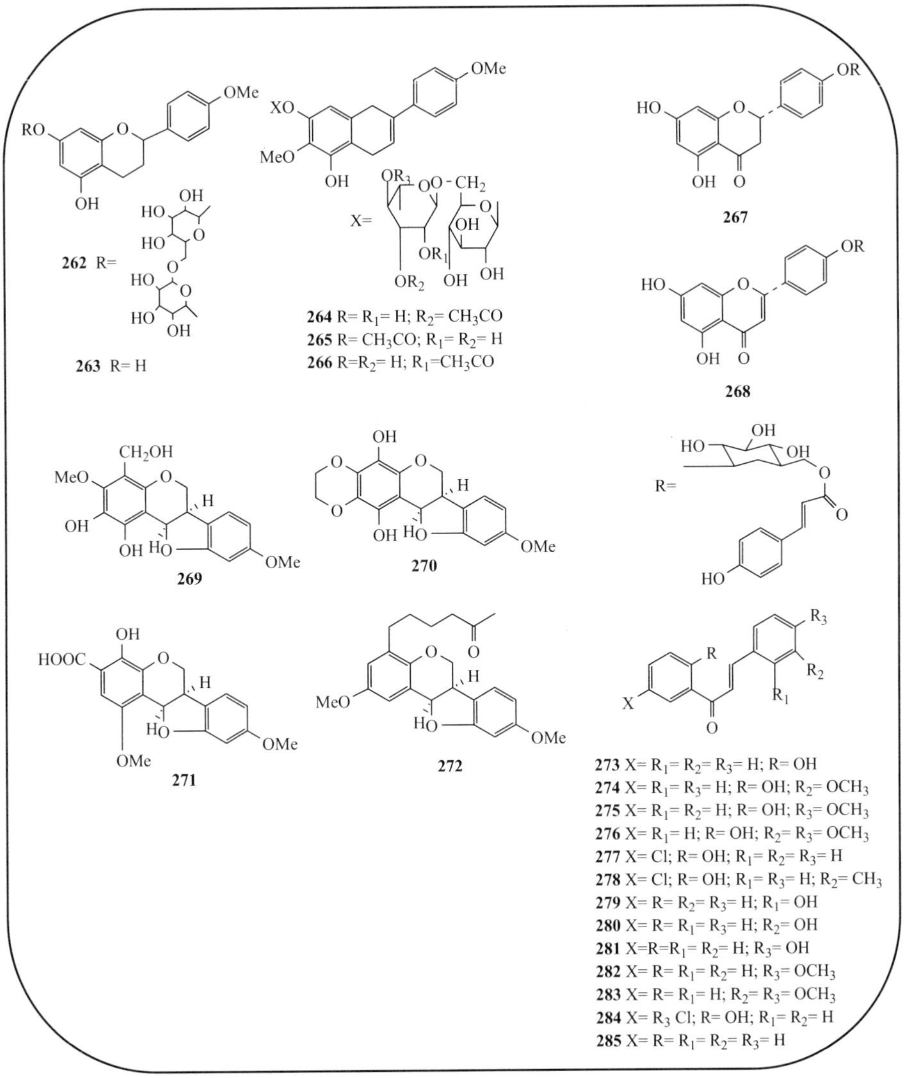

Bioassay-guided investigation of the stem bark of *Hopea chinensis* (Dipterocarpaceae) led to the isolation of hopeachinols C that exhibited significant AChE activity with IC_{50} value of 4.81 μM [127]. Recently, Tundis *et al.* [128] isolated from *Globularia meridionalis* (Globulariaceae) three flavonoids with cholinesterase inhibitory activity, namely pectolinarigenin 7-*O*-β-glucoside, apigenin 7-*O*-β-glucoside and luteolin. The most potent was luteolin with IC_{50} values of 25.2 and 37.2 μg/ml against AChE and BChE, respectively.

Saffron has been used in traditional medicine for the treatment of AD. The saffron extract showed moderate AChE inhibitory activity (up to 30%), while isolated compounds, namely crocetin, dimethylcrocetin, and safranal exhibited IC_{50} values of 96.33, 107.1, and 21.09 μM, respectively. Kinetic analysis showed mixed-type inhibition. Safranal interacts only with the binding site of the AChE, but crocetin and dimethylcrocetin bind simultaneously to the catalytic and peripheral anionic sites [129].

The bio-fractionation of *Calceolaria talcana* and *C. integrifolia* (Scrophulariaceae) led to isolation of several cholinesterase inhibitors. Both ursolic acid and 2-hydroxy-3(1,1-dimethylallyl-1,4-naphthoquinone) showed a selective BChE inhibition with IC_{50} values of 168.1 and 142.4 μg/ml, respectively, while verbascoside exhibited IC_{50} values of 189.9 and 105.9 μg/ml against AChE and BChE, respectively. All compounds act as competitive type inhibitors [130].

From *Ficus foveolata* (Moraceae) a series of stilbenes was isolated. Gnetol had the lowest IC_{50} value of 1.3 μM towards BChE and showed a reversible and competitive inhibition [131]. Another interesting stilbene compound namely *trans*-3,5-dimethoxystilbene was isolated from *Macaranga kurzii* (Euphorbiaceae). This compound showed an IC_{50} value of 9 μM against AChE followed by 5,7-dihydroxy-6-prenylflavanone (IC_{50} 18 μM), *trans*-3,5-dimethoxy-2-prenylstilbene (IC_{50} 19 μM), macakurzin C (IC_{50} 20 μM), glabranin (IC_{50} 23 μM), and furanokurzin (IC_{50} 42 μM). SAR evidenced that *trans*-configuration of the double bond may be an important factor for activity of *trans*-3,5-dimethoxystilbene, as cis-3,5-dimethoxystilbene was less active. SAR analysis also evidenced the crucial role of the double-bond of the pyran ring of macakurzin C [132].

The investigation on *Gelsemium elegans* (Gelsemiaceae) and *Aglaia odorata* (Meliaceae) lead the isolation of ethyl caffeate and paeonol. Ethyl caffeate was promising as a dual inhibitor of AChE and BChE, while paeonol exhibited BChE-selective inhibition [133]. Two new benzoates, asiatoate A and asiatoate B, have been isolated from *Buddleja asiatica* (Buddlejaceae) whole plant. Both compounds showed significant inhibitory effects on AChE and BChE in a dose-dependent manner with IC_{50} values 5.54 and 8.34 μM against AChE, and 30.94

and 35.94 μM against BChE, for asiatoate A and B, respectively [134]. From *Ferula lutea* roots (Apiaceae) two well known coumarins, psoralen and xanthotoxin exhibited IC_{50} values of 6.48 and 0.76 μM against AChE [135].

Huang *et al.* [23] reported a complete review of cholinesterase inhibitors from natural sources and among analyzed compounds the most potent AChE inhibitor was 2-hydroxy-6(10'(Z)-heptadecenyl)benzoic acid with an IC_{50} value of 0.573 μM.

Certain coumarin derivatives have been proposed for the treatment of AD [136]. Studies showed that benzylamine substructure was found in many potent AChE inhibitors [137]. On the basis of these considerations several coumarin and their derivatives were investigated toward cholinesterase. Ensaculin (**286**) inhibited the activity of AChE with an IC_{50} value of 0.36 μM [138]. Coumarins substituted at position C7 by a benzyloxy group **287-300**, and compound **301** were non competitive inhibitors of AChE [139]. These observations suggest that the compounds bind to the peripheral anionic site of AChE as proposed for 3-chloro-7-hydroxy-4-methylcoumarin (**302**) [140]. The two coumarins substituted with smaller groups at position C7 (*i.e.*, hydroxy **303** and methoxy **304**) were mixed AChE inhibitors. SAR analysis demonstrated that the substitution at position C3 of the coumarin nucleus modulated AChE inhibitory activity. Substitution with methyl groups at positions C3 and C4 led to more active compounds (**287** *vs* **291**). Substitution of the phenyl ring by a methyl group at *ortho* (**288**), *meta* (**289**) or *para* (**295**) positions slightly decreased AChE activity as compared to **287**, where *ortho* substitution being the least favorable. The introduction *meta* position of electron-donating substituents, *i.e.*, OH (**290**), OCH_3 (**291**), or NH_2 (**292**), produced a decrease in AChE activity. Except for **294**, which was the best AChE inhibitor, substitution the *meta* and/or *para* position by a halogen also produced a loss of activity toward AChE, as for another 7-substituents on the coumarin nucleus, a significant drop in AChE inhibition was observed for compounds bearing a small substituent, **303** and **304**. Increasing the length of the ether bridge with an additional methylene group also produced a drop of AChE activity (**287** *vs.* **301**). Recently, Shen *et al.* [141] reported the synthesis and cholinesterase inhibitory activity of different coumarin derivatives with different skeleton type. Analogs of **305** which possessed strong electron-donating groups, such as OCH_3 in C2′ and C4′, NH_2 in C2′, OH in C2′, exhibited anti-AChE activity. Compound **306** characterized by an OCH_3 group in C2′ exhibited the highest activity against AChE. Analogs of **305** with weaker electron-donating group CH_3 in C2′ and C4′ showed a decreasing inhibitory activity, which suggests that strong electron-donating group in these compounds could help them to access the enzyme-binding site. Compounds which possess two substituent groups in C2′ and C4′ exhibited a decrease in AChE inhibitory effect compared to those bearing either one group in C2′or C4′. The congeners with bulkier substituent **307** led to a huge decrease in activity than that bearing the smaller one **308**, indicating that the steric hindrance could limit the access of compounds to the enzyme-binding site. Compounds **309**, **310** and **311** showed significantly increased inhibitory activity compared with their unsaturated counterparts **305**, **306** and **312**. Comparing the selectivity of **309**, **310** and **311** for AChE/BChE, it is clear that the substitution of methoxy group in C2 on the aromatic ring is endowed with optimal electronic properties, that showed agreement with above SARs results. Compound **313** showed less activity than **306** and **308** as an AChE inhibitor, suggested that the C-type compound has a weaker interaction with AChE, due to its poorer binding affinity to the enzyme. Compounds **309**, **310** and **311** showed significantly increased activity compared with their unsaturated counterparts **305**, **306** and **312**. This strong activity could be probably explained by a better affinity of these compounds to the enzyme. Compound **313** showed a slightly (two to 3-fold) less activity than **305** as an AChE inhibitor, indicating that the base skeleton of this compound has weaker binding affinity to AChE compared with others.

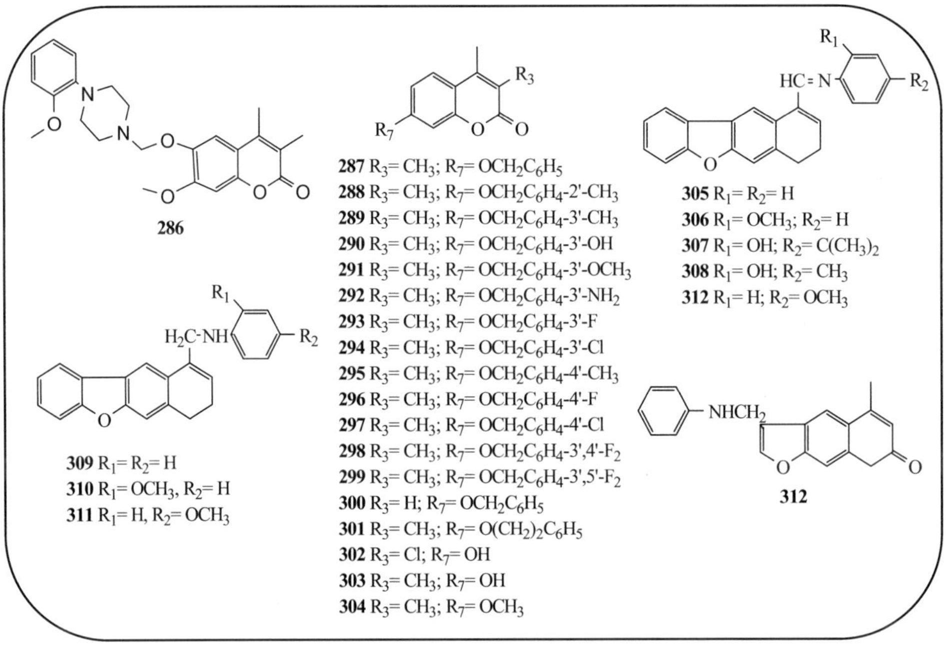

Changwong *et al.* [142] reported the cholinesterase inhibitory activities of three coumarins (mansorins A-C) and five naphthoquinones (mansonone C, E, G and

H) and evaluated the relationships between the AChE and BChE inhibitory effects and the core structures of these compounds. The order of the AChE inhibitory potency for the tested compounds is mansonone E > mansonone H > mansorin C > mansonone C > mansorin A ~ mansorin B ~ mansonone G > dehydrooxoperezinone. The BChE inhibitory potency of these compounds was also found to follow the same trend as that of AChE, except for the BChE inhibitory activity of mansonone H, which was lower than that of mansonone C but higher than that of mansorin A.

The enhanced activity observed in mansonone E compared with mansonone C could be correlated with the presence of a fused pyran ring. The same trend is also observed in a comparison of the inhibitory activities of mansorin C with those of mansorins A and B. Moreover, replacing the fused pyran ring of mansonone H with a fused furan ring (in dehydrooxoperezinone) led to a decrease in the observed ChE inhibitory activity against both ChEs.

Several lignans **314-318**, isolated from *Vitex negundo* (Verbenaceae), showed a weak AChE inhibitory activity with IC_{50} values ranging from 106.5 to 263.3 μM [143]. Vitrofolal E (**317**) displayed a moderate activity against BChE with an IC_{50} value of 35 μM. Compounds **314-316** showed a weak activity (IC_{50} values ranging from 85 to 300 μM) compared to galanthamine (**1**). Syringaresinol (**319**) isolated from the aerial part of *Leptadenia arborea* (Asclepiadaceae) demonstrated an inhibitory effect against AChE (IC_{50} value of 200 μg/ml) [144]. Interestingly, pinoresinol (**320**), that missed a methoxy group in C5' and C5", did not show any activity. The lignin gomisin A (**321**), a component of the fruits of *Schizandra chinesis* (Magnoliaceae) inhibited AChE activity with an IC_{50} value of 15.5 μM [145]. Gomisin C (**322**), gomisin G (**323**), gomisin D (**324**), schisandrol B (**325**) and gomisin A (**326**), isolated from the same species, entirely inhibited AChE in dose-dependent manners, with IC_{50} values of 6.71, 6.55, 7.84, 12.57 and 13.28 μM, respectively. Generally, the numbers and positions of the methylenedioxy groups on the aromatic rings, and the hydroxyl and benzoyl groups on the cyclooctadiene ring, influenced the biological activities.

Compound **324** substituted with α-hydroxyl group at C7, side chains at C6 and C14, and methylenedioxy groups at C12 and C13 was somewhat less active than **322** and **323**. Both **322** and **323** possessed aromatic methylenedioxy, α-hydroxyl (C7) and β-benzoyl (C6) groups. Interestingly, **325**, which possessed a β-hydroxyl group at C8 and methylenedioxy groups at C2 and C3, and **326**, which possessed a β-hydroxyl group at C7 and methylenedioxy groups at C12 and C13, exhibited significantly lower inhibitory activities than compounds **322-324**. Thus, in the presence of a hydroxy group, the β orientation might exhibit a lesser inhibitory effect than α orientation. These results indicate that aromatic methylenedioxy and cyclooctadiene hydroxyl groups are important for the AChE inhibitory activity; however, the methylenedioxy groups, whether at the C2 and C3 or C12 and C13, do not seem to be critical for the activity [146].

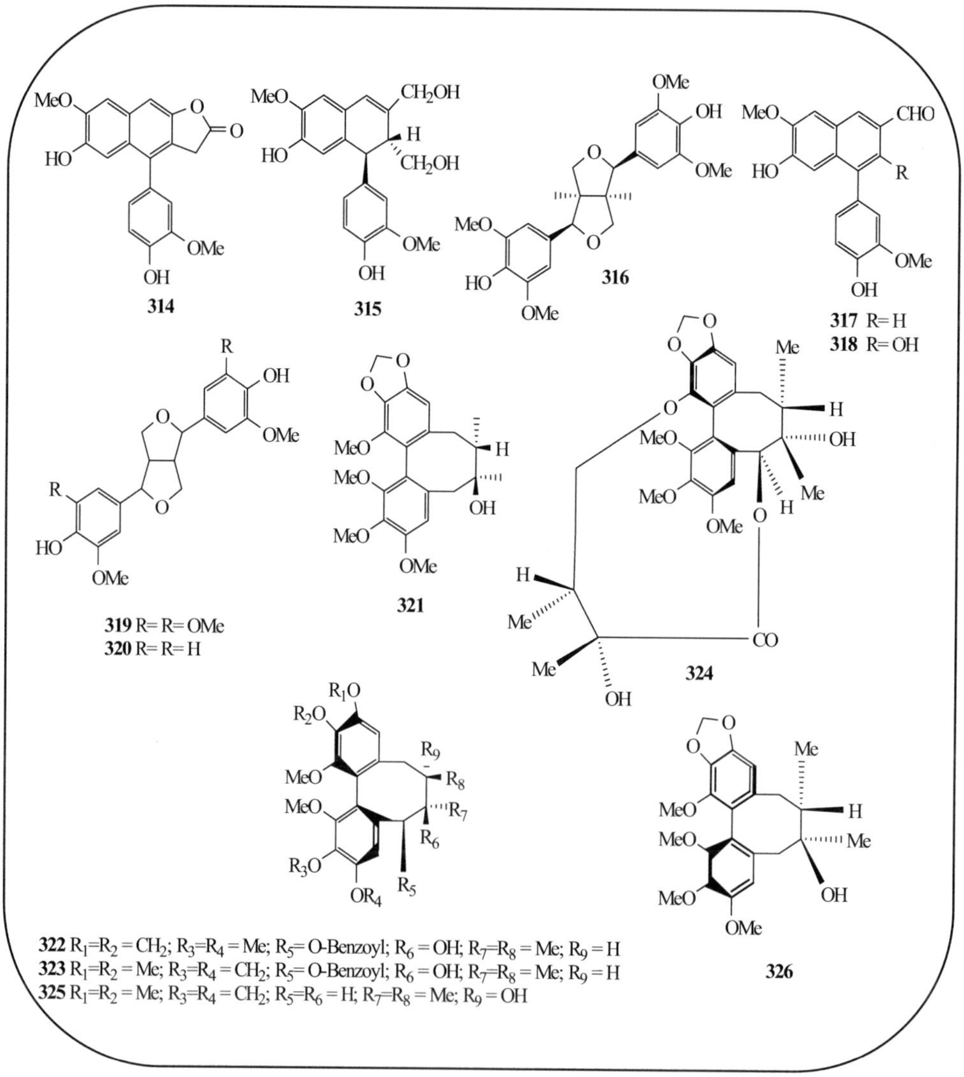

The *in vivo* effects of magnolol at different doses (15 mg/kg i.g Mag-L, 25 mg/kg i.g Mag-M and 35 mg/kg i.g Mag-H) were studied [147]. Interestingly, the AChE activity in the Mag-H treated group was restored, nearly to the normal level, indicating that magnolol may have the effect to reverse the dysregulation of central cholinergic nerves. The AChE activity in the Mag-M treated group was also significantly improved as compared to the Mag-L treated group. Two, terpene-sesquineolignans were isolated from *Illicium simonsii* (Illiciaceae) clovanedunnianol, and *p*-menthadunnianol, exhibited moderate AChE inhibitory activity with IC_{50} values of 4.58 and 6.55 μM, respectively [148].

Chuanxiongdiolide A and chuanxiongdiolide B, isolated from *Ligusticum chuanxiong* (Umbelliferae), exhibited a moderate anti-BChE activity (percentage of inhibition of 63.0 and 21.7%, respectively, at 50 μM) [149].

CONCLUDING REMARKS

AD is a devastating chronic disease that significantly increases healthcare costs and affects the quality of life of the afflicted patients and their caregivers. Population aging and other demographic changes may further increase the already staggering costs of this devastating disease. The cholinergic hypothesis of cognitive impairment and AD has been for decades a "polar star" for studies on dementia and neurodegenerative diseases [150]. Actually, cholinomimetics appeared as potentially useful drugs for symptomatic treatment of AD and, in particular, of the cognitive decline associated with it. The conceptual consistency and the therapeutic potential of the cholinergic hypothesis have continued to be intensively investigated during the recent years. For this reason the search of cholinesterase inhibitors has accelerated. The fact that AD is associated with a number of physical and systemic manifestations suggests that AD is a multifactorial disease that affects both the Central Nervous System and periphery. This syndrome is histopathologically characterized by Aβ plaques and neurofibrillary degeneration together with inflammation associated with an excessive generation of reactive oxygen species. For the above mentioned reason a new therapeutic approach is based on the use of drugs which act in different ways. Certainly, the evidence that cholinergic up-regulation also reduced the inflammatory process and β-amyloid peptide deposition through an influence on β-amyloid precursor protein processing *in vivo* confer to cholinesterase inhibitors a new interesting role [16, 151]. In this chapter we have focused attention on natural products and their derivatives as cholinesterase inhibitors. The large number of compounds from natural product sources, that are currently undergoing evaluation in clinical trials, is another positive indicator that natural product discovery provides good value for human medicine with particular attention to neurodegenerative diseases.

ACKNOWLEDGEMENTS

Declared None.

CONFLICT OF INTEREST

The authors confirm that this chapter contents have no conflict of interest.

DISCLOSURE

This is an updated report on the earlier review article by Loizzo MR, Tundis R, Menichini F, Menichini F. "Natural products and their derivatives as cholinesterase inhibitors in the treatment of neurodegenerative disorders: an update" Curr Med Chem 2008; 15(12): 1209-28.

ABBREVIATIONS

ACh	=	Acetylcholine
AChE	=	Acetylcholinesterase
AChEI	=	Acetylcholinesterase Inhibitors
AD	=	Alzheimer's Disease
BChE	=	Butyrylcholinesterase
CNS	=	Central Nervous System
DHED	=	Dehydroevodiamine Hydrochloride alkaloid
*Ee*AChE	=	*Electrophorus electricus* AChE
*Hs*BChE	=	Horse serum BChE
(-)-Hup A	=	(-)-Huperzine A
(-)-Hup B	=	(-)-Huperzine B
IC_{50}	=	50% Inhibitory Concentration
MAO-B	=	Monoamine Oxidase B

PAS	=	Peripheral Anionic Site
*Rb*AChE	=	Rat brain AChE
SAR	=	Structure Activity Relationships
t-Boc	=	tert-Butoxycarbonyl Group
*Tc*AChE	=	*Torpedo californica* AChE

REFERENCES

[1] Geldmacher DS. Alzheimer's Disease: Current Pharmacotherapy in the Context of Patient and Family Needs. J Am Geriatr Soc 2003; 51(5): S289-95.

[2] Jacobsen JS, Reinhart P, Pangalos MN. Current concepts in therapeutic strategie targeting cognitive decline and disease modification in Alzheimer's Disease. NeuroRx 2005; 2(4): 612-26.

[3] World Health Organization. Dementia: a public health priority. WHO Press, 2012; Switzerland, pp. 5-80.

[4] Hodges JR. Alzheimer's centennial legacy: origins, landmarks and the current status of knowledge concerning cognitive aspects. Brain 2006; 129(11): 2811- 22.

[5] Hasselmo ME, Giocomo LM. Cholinergic modulation of cortical function. J Mol Neurosci 2006; 30(1-2): 133-5.

[6] Perry EK, Perry RH, Blessed G, Tomlinson BE. Changes in brain cholinesterases in senile dementia of Alzheimer type. App Neurobiol 1978; 4(4): 273-7.

[7] Darvesh S, Hopkins DA, Geula C. Neurobiology of butyrylcholinesterase. Nat Rev Neurosci, 2003; 4(2): 131-8.

[8] Greig NH, Sambamurti K, Yu Q, *et al.* In: E. Giacobini Ed. Butyrylcholinesterase: Its Function and Inhibition. Martin Dunitz Ltd: London, 2003; pp. 69–90.

[9] Tasker A, Perry EK, Ballard CG. Butyrylcholinesterase: impact on symptoms and progression of cognitive impairment. Expert Rev Neurotherapeutics 2005; 5(1): 101-6.

[10] Ballard CG, Greig NH, Guillozet-Bongaarts AL, *et al.* Cholinesterases: roles in the brain during health and disease. Curr Alzheimer Res 2005; 2(3): 307-18.

[11] Giacobini E. Cholinesterases: new roles in brain function and in Alzheimer's disease. Neurochem Res 2003; 3(4): 515-22.

[12] Houghton PJ, Ren Y, Howes MJ. Acetylcholinesterase inhibitors from plants and fungi. Nat Prod Rep 2006; 23(2): 181-99.

[13] Munoz FJ, Aldunate R, Inestrosa NC. Peripheral binding site is involved in the neurotrophic activity of acetylcholinesterase. Neuroreport 1999; 10(17): 3621-5.

[14] Hardy JA, Higgins GA. Alzheimer's disease: the amyloid cascade hypothesis. Science 1992; 256(5054):184-5.

[15] He X, Feng S, Wang ZF, *et al.* Study on dual-site inhibitors of acetylcholinesterase: Highly potent derivatives of bis- and bifunctional huperzine B. Bioorg Med Chem Lett 2007; 15(3): 1394-408.

[16] Nizri E, Hamra-Amitay Y, Sicsic C, *et al.* Anti-inflammatory properties of cholinergic up-regulation: A new role for acetylcholinesterase inhibitors. Neuropharmacology 2006; 50(5): 540-7.

[17] Howes MJ, Perry NS, Houghton PJ. Plants with traditional uses and activities, relevant to the management of Alzheimer's disease and other cognitive disorders. Phytother Res 2003; 17(1): 1-18.

[18] Houghton PJ, Howes MJ. Natural products and derivatives affecting neurotransmission relevant to Alzheimer's and Parkinson's disease. Neurosignals 2005; 14(1-2): 6-22.

[19] Mukherjee PK, Kumar V, Mal M, Houghton PJ. Acetylcholinesterase inhibitors from plants. Phytomedicine, 2007; 14(4): 289-300.

[20] Francotte P, Graindorge E, Boverie S, de Tullio P, Pirotte B. New trends in the design of drugs against Alzheimer's disease. Curr Med Chem 2004; 11(3): 1757-78.

[21] Kingston DG, Newman DJ. Natural products as drug leads: an old process or the new hope for drug discovery? IDrugs 2005; 8(12): 990-2.

[22] Park SY. Potential therapeutic agents against Alzheimer's disease from natural sources. Arch Pharm Res 2010; 33(10): 1589-609.

[23] Mukherjee PK, Satheeshkumar N, Venkatesh P, Venkatesh M. Lead finding for acetyl cholinesterase inhibitors from natural origin: structure activity relationship and scope. Mini Rev Med Chem 2011; 11(3): 247-62.

[24] Uriarte-Pueyo I, Calvo MI. Flavonoids as acetylcholinesterase inhibitors. Curr Med Chem 2011; 18(34): 5289-302.

[25] Ebrahimi A, Schluesener H. Natural polyphenols against neurodegenerative disorders: potentials and pitfalls. Ageing Res Rev 2012; 11(2): 329-45.

[26] Fang L, Gou S, Fang X, Cheng L, Fleck C. Current progresses of novel natural products and their derivatives/ analogs as anti-Alzheimer candidates: an update. Mini Rev Med Chem 2013; 13(6): 870-87.

[27] Huang L, Su T, Li X. Natural products as sources of new lead compounds for the treatment of Alzheimer's disease. Curr Top Med Chem 2013; 13(15): 1864-78.

[28] Orhan IE. Nature: a substantial source of auspicious substances with acetylcholinesterase inhibitory action. Curr Neuropharmacol 2013; 11(4): 379-87.

[29] Ahmed F, Ghalib RM, Sasikala P, Ahmed KK. Cholinesterase inhibitors from botanicals. Pharmacogn Rev 2013; 7(14): 121-30.

[30] Konrath EL, Passos CS, Klein LC, Henriques AT. Alkaloids as a source of potential anticholinesterase inhibitors for the treatment of Alzheimer's disease. J Pharm Pharmacol 2013; 65(12): 1701-25.

[31] Russo P, Frustaci A, Del Bufalo A, Fini M, Cesario A. Multitarget drugs of plants origin acting on Alzheimer's disease. Curr Med Chem 2013; 20(13):1686-93.

[32] Russo P, Frustaci A, Del Bufalo A, Fini M, Cesario A. From traditional European medicine to discovery of new drug candidates for the treatment of dementia and Alzheimer's disease: acetylcholinesterase inhibitors. Curr Med Chem 2013; 20(8): 976-83.

[33] Pinho BR, Ferreres F, Valentão P, Andrade PB. Nature as a source of metabolites with cholinesterase-inhibitory activity: an approach to Alzheimer's disease treatment. J Pharm Pharmacol 2013; 65(12): 1681-700.

[34] Natarajan S, Shunmugiah KP, Kasi PD. Plants traditionally used in age-related brain disorders (dementia): an ethanopharmacological survey. Pharm Biol 2013; 51(4), 492-23.

[35] Murray AP, Faraoni MB, Castro MJ, Alza NP, Cavallaro V. Natural AChE inhibitors from plants and their contribution to Alzheimer's Disease Therapy. Curr Neuropharmacol 2013; 11(4): 388-13.

[36] Sweeney T, Kewitz H. Selective inhibition of human acetylcholinesterase by galanthamine *in vitro* and *in vivo*. Life Sci. 1990; 46(21): 1553-8.

[37] Fulton B, Benfield P. Galanthamine. Drugs Aging 1996; 9(1): 60-5.

[38] Maelicke A, Samochocki M, Jostock R, *et al.* Allosteric sensitization of nicotinic receptors by galantamine, a new treatment strategy for Alzheimer's disease. Biol Psychiatry 2001; 49(3): 279-88.

[39] Han SY, Sweeney JE, Bachman ES, *et al.* Chemical and pharmacological characterization of galanthamine. Eur. J. Med. Chem., 1992; 27(2), 673-87.

[40] Poschalko A, Welzig S, Treu M, Nerdinger S, Mereiter K, Jordis U. Synthesis of (+/-)-6H-benzofuro[3a,3,2,ef][3]benzazepine: an unnatural analog of (-)-galanthamine. Tetrahedron 2002; 58(8): 1513-8.

[41] Treu M, Jordis U, Mereiter K. 12H-[2]-Benzothiepino[6,5a,5-bc]benzofuran: Synthesis of a Sulfur-Analog of Galanthamine. Heterocycles 2001; 55: 1727-35.

[42] Bores GM, Kosley RW. Galanthamine derivatives for the treatment of Alzheimer's disease. Drugs Future, 1996, 21, 621-35.

[43] Thal, C.; Guillou, C.; Mary, A.; Renko, D.Z.; Potier, P.; Christen, Y. PCT Int. Appl. WO 97 03,987 (1996), GB Appl. 95/14821.9 (1995).

[44] Guillou C, Mary A, Renko DZ, Thal C. Potent acetylcholinesterase inhibitors: design, synthesis and structure–activity relationships of alkylene linked bis-galanthamine and galanthamine–galanthaminium salts. Bioorg Med Chem Lett 2000; 10(7): 637-9.

[45] Mary A, Renko DZ, Guillou C, Thal C. Potent acetylcholinesterase inhibitors : Design, synthesis, and structure-activity relationships of bis-interacting ligands in the galanthamine series. Bioorg Med Chem 1998; 6(10): 1835-50.

[46] Herlem D, Martin MT, Thal C, Guillou C. Synthesis and structure-activity relationships of open D-ring galanthamine analogues. Bioorg Med Chem Lett 2003; 13(14): 2389-93.

[47] Rhee IK, Appels N, Hofte B, *et al.* Isolation of the acetylcholinesterase inhibitor ungeremine from *Nerine bowdenii* by preparative HPLC Ccoupled on-line to a flow assay system. Biol Pharm Bull 2004; 27(11): 1804-9.

[48] Houghton PJ, Agbedahunsi JM, Adegbulugbe A. Choline esterase inhibitory properties of alkaloids from two Nigerian *Crinum* species. Phytochemistry 2004; 65(21): 2893-6.

[49] Lopez S, Bastida J, Viladomat F. Acetylcholinesterase inhibitory activity of some Amaryllidaceae alkaloids and *Narcissus* extracts. Life Sci 2002; 71(21): 2521-9.

[50] Lee SS, Venkatesham U, Rao CP, Lam SH, Lin JH. Preparation of secolycorines against acetylcholinesterase. Bioorg Med Chem 2007; 15(2): 1034-43.

[51] Wang YH, Wan QL, Gu CD, Luo HR, Long CL. Synthesis and biological evaluation of lycorine derivatives as dual inhibitors of human acetylcholinesterase and butyrylcholinesterase. Chem Cent J 2012; 6(1): 96-9.

[52] Elgorashi EE, Stafford GI, Van Staden J. Acetylcholinesterase enzyme inhibitory effects of amaryllidaceae alkaloids. Planta Med 2004; 70(3): 260-2.

[53] Adewusi EA, Fouche G, Steenkamp V. Cytotoxicity and acetylcholinesterase inhibitory activity of an isolated crinine alkaloid from *Boophane disticha* (Amaryllidaceae). J Ethnopharmacol 2012; 143(2): 572-8.

[54] Nair JJ, Aremu AO, van Staden J. Isolation of narciprimine from *Cyrtanthus contractus* (Amaryllidaceae) and evaluation of its acetylcholinesterase inhibitory activity. J Ethnopharmacol 2011; 137(3): 1102-6.

[55] Cavallaro V, Alza NP, Murray MG, Murray AP. Alkaloids from Habranthus tubispathus and H. jamesonii, two amaryllidaceae with acetyl- and butyrylcholinesterase inhibition activity. Nat Prod Commun 2014; 9(2): 159-62.

[56] Ingkaninan K, Phengpa P, Yuenyongsawad S, Khorana N. Acetylcholinesterase inhibitors from *Stephania venosa* tuber. J Pharm Pharmacol 2006; 58(5): 695-700.

[57] Cho KM, Yoo ID, Kim WG. 8-hydroxydihydrochelerythrine and 8-hydroxydihydrosanguinarine with a potent acetylcholinesterase inhibitory activity from *Chelidonium majus* L. Biol Pharm Bull 2006; 29(11): 2317-20.

[58] Bonesi M, Loizzo MR, Conforti F, *et al. Berberis aetnensis* and *B. libanotica*: a comparative study on the chemical composition, inhibitory effect on key enzymes linked to Alzheimer's disease and antioxidant activity. J Pharm Pharmacol 2013; 65(12): 1726-35.

[59] Bianchi DA, Hirschmann GS, Theoduloz C, Bracca ABJ, Kaufman T. Synthesis of tricyclic analogs of stephaoxocanidine and their evaluation as acetylcholinesterase inhibitors. Biorg Med Chem Lett 2005; 15(11): 2711-5.

[60] Novák Z, Chlebek J, Opletal L, *et al.* Corylucinine, a new alkaloid from *Corydalis cava* (Fumariaceae), and its cholinesterase activity. Nat Prod Commun 2012; 7(7): 859-60.

[61] Al-Jafari A, Kamal MA, Greig NH, Alhomida AS, Perry ER. Kinetics of human erythrocyte acetylcholinesterase inhibition by a novel derivative of physostigmine: phenserine. Biochem Biophys Res Commun 1998; 248(1): 180-5.

[62] Klein J. Phenserine. Expert Opin Investig Drugs, 2007; 16(7): 1087-97.

[63] Kamal MA, Greig NH, Alhomida AS, Al-Jafari AA. Kinetics of human acetylcholinesterase inhibition by the novel experimental Alzheimer therapeutic agent, tolserine. Biochem Pharmacol 2000; 60(4): 561-70.

[64] Yu Q, Holloway HW, Flippen-Anderson JL, Hoffman B, Brossi A, Greig NH. Methyl analogues of the experimental Alzheimer drug phenserine: synthesis and structure/activity relationships for acetyl- and butyrylcholinesterase inhibitory action. J Med Chem 2001; 44(24): 4062-71.

[65] Soreq H, Seidman S. Acetylcholinesterase-new roles for an old actor. Nat Rev Neurosci 2001; 2(9): 294-302.
[66] Kamal MA, Klein P, Yu QS, *et al.* Kinetics of human serum butyrylcholinesterase and its inhibition by a novel experimental Alzheimer therapeutic, bisnorcymserine. J Alzheimers Dis 2006; 10(1): 43-51.
[67] Ingkaninan K, Changwijit K, Suwanborirux K. Vobasinyl-iboga bisindole alkaloids, potent acetylcholinesterase inhibitors from *Tabernaemontana divaricata* root. J Pharm Pharmacol 2006; 58(6): 847-52.
[68] Chaiyana W, Schripsema J, Ingkaninan K, Okonogi S. 3'-R/S-hydroxyvoacamine, a potent acetylcholinesterase inhibitor from *Tabernaemontana divaricata*. Phytomedicine 2013; 20(6): 543-8.
[69] Passos CS, Simões-Pires CA, Nurisso A, *et al.* Indole alkaloids of *Psychotria* as multifunctional cholinesterases and monoamine oxidases inhibitors. Phytochemistry. 2013; 86: 8-20.
[70] Ingkaninan K, Choo CY, Hirasawa,Y, *et al.* Carinatumins A-C, new alkaloids from *Lycopodium carinatum* inhibiting acetylcholinesterase. Bioorg Med Chem 2007; 15(4): 1703-7.
[71] Halldorsdottir ES, Olafsdottir ES. Abstracts of the 54th Annual Congress on Medicinal Plant Research, 29 August-2 September 2006, Helsinki, Finland. Planta Med 2006; 72: 961-1083.
[72] Katakawa K, Nozoe, A, Kogure N, Kitajima M, Hosokawa M, Takayama HJ. Fawcettimine-related alkaloids from *Lycopodium serratum*. Nat Prod 2007; 70(6): 1024-8.
[73] Zhou GC, Zhu DY. Synthesis of 5-substituted analogues of huperzine A. Bioorg Med Chem Lett 2000; 10(18): 2055-7.
[74] Hogenauer K, Baumann K, Enz A, Mulzer J. Synthesis and acetylcholinesterase inhibition of 5-desamino huperzine A derivatives. Bioorg Med Chem Lett 2001; 11(19): 2627-30.
[75] Rajendran V, Prakash KR, Ved HS, Saxena A, Doctor BP, Kozikowski AP. Synthesis, chiral chromatographic separation, and biological activities of the enantiomers of 10,10-dimethylhuperzine A. Bioorg Med Chem Lett 2000; 10(21): 2467-9.
[76] Haviv H, Wong DM, Silman I, Sussman JL. Bivalent ligands derived from Huperzine A as acetylcholinesterase inhibitors. Curr Top Med Chem 2007; 7(4): 375-87.
[77] Du DM, Carlier PR. Development of bivalent acetylcholinesterase inhibitors as potential therapeutic drugs for Alzheimer's disease. Curr Pharm Des 2004; 10(25): 3141-56.
[78] Camps P, Gŏmez E, Muñoz-Torrero D, Font-Bardia M, Solans X. Synthesis of diastereomeric 13-amido-substituted huprines as potential high affinity acetylcholinesterase inhibitors. Tetrahedron 2003; 59: 4143-51.
[79] Zhu XD, Tang XC. Improvement of impaired memory in mice by huperzine A and huperzine B. Zhongguo Yao Li Xue Bao 1988; 9(6): 492-7.
[80] Yan XF, Lu WH, Lou WJ. Effects of huperzine A and B on skeletal muscle and electroencephalogram. Zhongguo Yao Li Xue Bao 1987; 8(2): 117-23.
[81] Rajendran V, Saxena A, Doctor BP, Kozikowski AP. Synthesis of more potent analogues of the acetylcholinesterase inhibitor, huperzine B. Bioorg Med Chem Lett 2002; 12(11): 1521-3.
[82] Feng S, Xia Y, Han D, *et al.* Synthesis and acetylcholinesterase inhibition of derivatives of huperzine B. Bioorg Med Chem Lett 2005; 15(3): 523-6.
[83] Feng S, Wang Z, He X, *et al.* Bis-huperzine B: highly potent and selective acetylcholinesterase inhibitors. J Med Chem 2005; 48(3): 655-7.
[84] Tan CH, Chen GF, Ma XQ, Jiang SH, Zhu DY. Huperzine R, a novel 15-carbon *Lycopodium* alkaloid from *Huperzia serrata*. J Nat Prod 2002; 65(7): 1021-2.
[85] Atta-ur-Rahman, Khalid A, Sultana N, *et al.* New natural cholinesterase inhibiting and calcium channel blocking quinoline alkaloids. J Enz Inhib Med Chem 2006; 21(6): 703-10.
[86] Tang H, Ning FX, Wei YB, *et al.* Derivatives of oxoisoaporphine alkaloids: a novel class of selective acetylcholinesterase inhibitors. Bioorg Med Chem Lett 2007; 17(13): 3765-8.
[87] Adsersen A, Kjølbye A, Dall O, Jäger AK. Acetylcholinesterase and butyrylcholinesterase inhibitory compounds from Corydalis cava Schweigg. & Kort. J Ethnopharmacol. 2007; 113(1):179-82.
[88] Markmee S, Ruchirawat S, Prachyawarakorn V, Ingkanian K, Khorana N. Isoquinoline derivatives as potential acetylcholinesterase inhibitors. Bioorg Med Chem Lett 2006; 16(8): 2170-2.
[89] Zhang X, Oh M, Kim S, *et al.* Epimediphine, a novel alkaloid from *Epimedium koreanum* inhibits acetylcholinesterase. Nat Prod Res 2013; 27(12): 1067-74.

[90] Choudhary MI, Shahnaz S, Parveen S, *et al.* New cholinesterase-inhibiting triterpenoid alkaloids from *Buxus hyrcana*. Chem Biodivers 2006; 3(9): 1039-52.
[91] Babar ZU, Ata A, Meshkatalsadat MH. New bioactive steroidal alkaloids from *Buxus hyrcana*. Steroids, 2006; 7(13-14): 1045-51.
[92] Atta-ur-Rahman, Feroz F, Naeem I. *et al.* New pregnane-type steroidal alkaloids from *Sarcocca saligna* and their cholinesterase inhibitory activity. Steroids 2004; 69(11-12): 735-41.
[93] Zaheer-Ul-Haq ZU, Wellenzohn B, Liedl KR, Rode BM. Molecular docking studies of natural cholinesterase-inhibiting steroidal alkaloids from *Sarcococca saligna*. J Med Chem 2003; 46(23): 5087-90.
[94] Atta-ur-Rahman, Zaheer-ul-Haq, Khalid A, Anjum S, Khan MR, Choudhary MI. Pregnane-type steroidal alkaloids of *Sarcococca saligna*: A new class of cholinesterases inhibitors. Helv Chim Acta 2002; 85: 678-88.
[95] Ferheen S, Ahmed E, Afza N, *et al.* Haloxylines A and B, antifungal and cholinesterase inhibiting piperidine alkaloids from *Haloxylon salicornicum*. Chem Pharm Bull 2005; 53(5): 570-2.
[96] Viegas CJr, Bolzani VS, Pimentel LS, *et al.* New selective acetylcholinesterase inhibitors designed from natural piperidine alkaloids. Bioorg Med Chem 2005; 13(13): 4184-90.
[97] Viegas CJr, da S Bolzani V, Fulraln M, *et al.* Further bioactive piperidine alkaloids from the flowers and green fruits of *Cassia spectabilis*. J Nat Prod 2004; 67(5): 908-10.
[98] Tang Y, Fu Y, Xiong J, *et al.* Casuarinines A-J, lycodine-type alkaloids from *Lycopodiastrum casuarinoides*. J Nat Prod 2013; 76(8): 1475-84.
[99] Park CH, Kim SH, Choi W, *et al.* Novel anticholinesterase and antiamnesic activities of dehydroevodiamine, a constituent of *Evodia rutaecarpa*. Planta Med 1996; 62(5), 405-9.
[100] Decker M. Novel inhibitors of acetyl- and butyrylcholinesterase derived from the alkaloids dehydroevodiamine and rutaecarpine. Eur J Med Chem 2005; 40(3): 305-13.
[101] Giacobini E. In: J.J. Buccafusco, Ed. Cognitive Enhancing Drugs. Birkhäuser Verlag: Basel, Boston, Berlin, 2004; pp. 11-36.
[102] Rollinger JM, Schuster D, Baier E, Ellmerer EP, Langer T, Stuppner H. Taspine: bioactivity-guided isolation and molecular ligand-target insight of a potent acetylcholinesterase inhibitor from *Magnolia x soulangiana*. J Nat Prod 2006; 69(9): 1341-6.
[103] Wang P, Liu AL, An Z, Li ZH, Du G, Qin HL. Novel alkaloids from the roots of *Stemona sessilifolia*. Chem Biodivers 2007; 4(3): 523-30.
[104] Lai DH, Yang ZD, Xue WW, Sheng J, Shi Y, Yao XJ. Isolation, characterization and acetylcholinesterase inhibitory activity of alkaloids from roots of *Stemona sessilifolia*. Fitoterapia 2013; 89: 257-64.
[105] Kulhánková A, Cahlíková L, Novák Z, Macáková K, Kuneš J, Opletal L. Alkaloids from *Zephyranthes robusta* BAKER and their acetylcholinesterase- and butyrylcholinesterase-inhibitory activity. Chem Biodivers 2013; 10(16): 1120-7.
[106] Mollataghi A, Coudiere E, Hadi AH, *et al.* Anti-acetylcholinesterase, anti-α-glucosidase, anti-leishmanial and anti-fungal activities of chemical constituents of *Beilschmiedia* species. Fitoterapia 2012; 83: 298-302.
[107] Jukic M, Politeo O, Maksimovic M, Milos M, Milos M. *In vitro* acetylcholinesterase inhibitory properties of thymol, carvacrol and their derivatives thymoquinone and thymohydroquinone. Phytoth Res 2007; 21(3): 259-61.
[108] Miyazawa M, Yamafuji C. Inhibition of acetylcholinesterase activity by tea tree oil and constitu- ent terpenoids. Flav Fragr J 2005; 20(2), 617-20.
[109] Savelev SU, Okello EJ, Perry EK. Butyryl- and acetyl-cholinesterase inhibitory activities in essential oils of *Salvia* species and their constituents. Phytoth Res 2004; 18(4): 315-24.
[110] Kundu A, Mitra A. Flavoring extracts of *Hemidesmus indicus* roots and *Vanilla planifolia* pods exhibit *in vitro* acetylcholinesterase inhibitory activities. Plant Foods Hum Nutr 2013; 68(3): 247-53.
[111] Menichini F, Tundis R, Loizzo MR, *et al.* Acetylcholinesterase and butyrylcholinesterase inhibition of ethanolic extract and monoterpenes from *Pimpinella anisoides* V Brig. (Apiaceae). Fitoterapia 2009; 80(5): 297-300.
[112] Cavin AL, Hay AE, Marston A, *et al.* Bioactive diterpenes from the fruits of *Detarium microcarpum*. J Nat Prod 2006; 69(5): 768-73.

[113] Mehmood S, Riaz N, Nawaz SA, Afza N, Malik A, Choudhary MI. New butyrylcholinesterase inhibitory triterpenes from *Salvia santolinifolia.* Arch Pharm Res 2006; 29(3): 195-8.
[114] Beri V, Wildman SA, Shiomi K, Al-Rashid ZF, Cheung J, Rosenberry TL. The natural product dihydrotanshinone I provides a prototype for uncharged inhibitors that bind specifically to the acetylcholinesterase peripheral site with nanomolar affinity. Biochemistry 2013; 52(42), 7486-99.
[115] Rollinger JM, Ewelt J, Seger C, Sturm S, Ellmerer EP, Stuppner H. New insights into the acetylcholinesterase inhibitory activity of *Lycopodium clavatum.* Planta Med 2005; 71(11): 1040-3.
[116] Liu JQ, Peng XR, Li XY, *et al.* Norfriedelins A-C with acetylcholinesterase inhibitory activity from acerola tree (*Malpighia emarginata*). Org Lett 2013; 15(7): 1580-3.
[117] Riaz N, Naveed MA, Saleem M, *et al.* Cholinesterase inhibitory constituents from *Ficus bengalensis.* J Asian Nat Prod Res 2012; 14(12): 1149-55.
[118] Ahmed E, Nawaz SA, Malik A, Choudhary MI. Isolation and cholinesterase-inhibition studies of sterols from *Haloxylon recurvum.* Bioorg Med Chem Lett 2006; 16(3): 573-80.
[119] Choudhary MI, Nawaz SA, Zaheer-ul-Haq, *et al.* Withanolides, a new class of natural cholinesterase inhibitors with calcium antagonistic properties. Biochem Biophys Res Comm 2005; 334(1): 276-87.
[120] Hajimehdipoor H, Mosaddegh M, Naghibi F, Haeri A, Hamzeloo-Moghadam M. Natural sesquiterpen lactones as acetylcholinesterase inhibitors. An Acad Bras Cienc 2014; http://dx.doi.org/10.1590/0001-3765201420130005
[121] Oinonen PP, Jokela JK, Hatakka AI, Vuorela PM. Linarin, a selective acetylcholinesterase inhibitor from *Mentha arvensis.* Fitoterapia 2006; 77(6): 429-34.
[122] Loizzo MR, Tundis R, Menichini F, *et al.* Acetyl-cholinesterase inhibition by extracts and isolated flavones from *Linaria reflexa* Desf. (Scrophulariaceae). Nat Prod Commun 2007; 2: 759-63.
[123] Atia-tun-Noor, Fatima I, Ahmad I, *et al.* Leufolins A and B, potent butyrylcholinesterase-inhibiting flavonoid glucosides from *Leucas urticifolia.* Molecules 2007; 12(7): 1447-54.
[124] Ahmad VU, Iqbal S, Nawaz SA, *et al.* Isolation of four new pterocarpans from *Zygophyllum eurypterum* (Syn. *Z. atriplicoides*) with enzyme-inhibition properties. Chem Biodivers 2006; 3(9): 996-03.
[125] Hasan A, Khan KM, Sher M, *et al.* Synthesis and inhibitory potential towards acetylcholinesterase, butyrylcholinesterase and lipoxygenase of some variably substituted chalcones. J Enz Inhib Med Chem 2005; 20(1): 41-7.
[126] Shafiq N, Riaz N, Ahmed S, *et al.* Bioactive phenolics from *Seriphidium stenocephalum.* J Asian Nat Prod Res 2013; 15(3): 286-93.
[127] Yan T, Wang T, Wei W, *et al.* Polyphenolic acetylcholinesterase inhibitors from *Hopea chinensis.* Planta Med 2012; 78(10): 1015-9.
[128] Tundis R, Bonesi M, Menichini F, *et al.* Antioxidant and anti-cholinesterase activity of *Globularia meridionalis* extracts and isolated constituents. Nat Prod Commun 2012; 7: 1015-20.
[129] Geromichalos GD, Lamari FN, Papandreou MA, *et al.* Saffron as a source of novel acetylcholinesterase inhibitors: molecular docking and *in vitro* enzymatic studies. J Agric Food Chem 2012; 60(24): 6131-8.
[130] Cespedes CL, Muñoz E, Salazar JR, *et al.* Inhibition of cholinesterase activity by extracts, fractions and compounds from *Calceolaria talcana* and *C. integrifolia* (Calceolariaceae: Scrophulariaceae). Food Chem Toxicol 2013; 62: 919-26.
[131] Sermboonpaisarn T, Sawasdee P. Potent and selective butyrylcholinesterase inhibitors from *Ficus foveolata.* Fitoterapia 2012; 83(4),780-4.
[132] Trinh Thi Thanh V, Doan Thi Mai H, Pham VC, *et al.* Acetylcholinesterase inhibitors from the leaves of *Macaranga kurzii.* J Nat Prod 2012; 75(12): 2012-5.
[133] Liu B, Yang L, Xu YK, Liao SG, Luo HR, Na Z. Two new triterpenoids from *Gelsemium elegans* and *Aglaia odorata.* Nat Prod Commun 2013; 8: 1373-6.
[134] Ali F, Khan HU, Afzal M, Samad A, Khan SU, Ali I. Two new cholinesterase inhibitors asiatoates A and B from *Buddleja asiatica.* J Asian Nat Prod Res 2013; 15(6): 631-7.
[135] Ben Salem S, Jabrane A, Harzallah-Skhiri F, Ben Jannet H. New bioactive dihydrofuranocoumarins from the roots of the Tunisian *Ferula lutea* (Poir.) Maire. Bioorg Med Chem Lett 2013; 23(14): 4248-52.
[136] Boyd B. Ongoing progress in the Alzheimer's disease arena. Drug News Perspect 2000; 13(7): 425-38.

[137] Sugimoto H, Yamanishi Y, Iimura Y, Kawakami Y. Donepezil hydrochloride (E2020) and other acetylcholinesterase inhibitors. Curr Med Chem 2000, 7(3): 303-39.
[138] Hilgert M, Nöldner M, Chatterjee SS, Klein J. KA-672 inhibits rat brain acetylcholinesterase *in vitro* but not *in vivo*. Neurosci Lett 1999; 263(1-2): 193-6.
[139] Brülmann C, Ooms F, Carrupt PA, *et al.* Coumarins derivatives as dual inhibitors of acetylcholinesterase and monoamine oxidase. J Med Chem 2001; 44(19): 3195-8.
[140] Simeon-Rudolf R, Kovarik Z, Radic Z, Reiner E. Reversible inhibition of acetylcholinesterase and butyrylcholinesterase by 4,4'-bipyridine and by a coumarin derivative. Chem Biol Interact 1999; 14: 119-20.
[141] Shen Q, Peng Q, Shao J, *et al.* Synthesis and biological evaluation of functionalized coumarins as acetylcholinesterase inhibitors. Eur J Med Chem 2005; 40(12): 1307-15.
[142] Changwong N, Sabphon C, Ingkaninan K, Sawasdee P. Acetyl- and butyryl-cholinesterase inhibitory activities of mansorins and mansonones. Phytother Res 2012; 26(3): 392-6.
[143] Azhar-Ul-Haq, Malik A, Anis I, *et al.* Enzyme inhibiting lignans from *Vitex negundo*. Chem Pharm Bull 2004; 52(11): 1269-72.
[144] El-Hassan A, El-Sayed M, Hamed AI, *et al.* Bioactive constituents of *Leptadenia arborea*. Fitoterapia 2003; 74(1-2): 184-7.
[145] Kim DH, Hung TM, Bae KH, *et al.* Gomisin A improves scopolamine-induced memory impairment in mice. Eur J Pharm 2006; 542(1-3): 129-35.
[146] Hung TM, Na M, Min BS, *et al.* Acetylcholinesterase inhibitory effect of lignans isolated from *Schizandra chinensis*. Arch Pharm Res 2007; 30(6): 685-90.
[147] Li YS, Hong YF, He J, *et al.* Effects of magnolol on impairment of learning and memory abilities induced by scopolamine in mice. Biol Pharm Bull 2013; 36(5): 764-71.
[148] Dong C, Liu L, Li X, Guan Z, Luo H, Wang Y. Sesquineolignans and terpene-sesquineolignans: anti-acetylcholinesterase constituents from *Illicium simonsii*. Planta Med 2013; 79(5): 338-47.
[149] Huang J, Lu XQ, Lu J, *et al.* Two new phthalides with BuChE inhibitory activity from *Ligusticum chuanxiong*. J Asian Nat Prod Res 2013; 15(12): 1237-42.
[150] Contestabile A. The history of the cholinergic hypothesis. Behav Brain Res 2011; 221(12): 334-40.
[151] Racchi M, Mazzucchelli M, Porrello E, Lanni C, Govoni S. Acetylcholinesterase inhibitors: novel activities of old molecules. Pharmacol Res 2004; 50(4): 441-51.

CHAPTER 7

*Pf*DHFR Enzyme Inhibitors: Rational Design Using Pharmacoinformatic Tools

Sheenu Abbat[1], Shweta Bhagat[2] and Prasad V. Bharatam[2,*]

[1]Department of Pharmacoinformatics and [2]Department of Medicinal Chemistry, National Institute of Pharmaceutical Education and Research (NIPER), S.A.S Nagar-160 062, Punjab, India

Abstract: Malaria is recognized as the most dangerous disease originating from parasites. *Plasmodium falciparum* dihydrofolate reductase (*Pf*DHFR) is one of the important targets for anti-malarial drug design. All the reported anti-folates show drug resistance. Mutations in the active-site of *Pf*DHFR enzyme, particularly at 16, 51, 59, 108, and 164 residues, are responsible for the observed resistance. The new strategies for anti-malarial drug discovery include design of leads which are effective against both wild type and mutant *Pf*DHFR. Recent efforts in this direction lead to the identification of P218 as an important lead compound. In addition a set of guanylthiourea derivatives were also reported. Rational lead design strategies are being adopted for the design of *Pf*DHFR inhibitors as anti-malarial agents. In this update, a review of the computer-aided drug design which was adopted over the past decade, in the design of *Pf*DHFR inhibitors, are elaborated. Efforts based on molecular docking, pharmacophore mapping, QSAR, homology modeling, and quantum chemical studies are discussed in detail. Special emphasis is given to the importance of drug action while discussing the pharmacoinformatic approaches. In future the rational methods of lead design and lead optimization are expected to increase because sufficient knowledge is already generated on the structural aspects of *Pf*DHFR and its possible inhibitors are already available.

Keywords: Antifolates, anti-malarial, molecular docking, molecular dynamics, *P. falciparum* DHFR inhibitors, pharmacoinformatic tools and X-ray crystallographic analysis, pharmacophore mapping, QSAR, quadruple mutant enzyme, quantum medicinal chemistry, structure based drug design.

INTRODUCTION

Malaria is a parasitic disease which is prevalent throughout the world's tropical and sub-tropical areas, mainly in Central and South America, Africa, Asia and the Indo-Pacific region. About 3.4 billion people are estimated to be at risk of malaria, of which 1.2 billion are reported to be at high risk. In the year 2012, 207

***Corresponding author Prasad V. Bharatam:** Department of Medicinal Chemistry, National Institute of Pharmaceutical Education and Research (NIPER), S.A.S Nagar-160 062, Punjab, India; Tel: +91-172-2292018; Fax: +91-172-2214692; E-mail: pvbharatam@niper.ac.in

Atta-ur-Rahman, Allen B. Reitz, M. Iqbal Choudhary and Jizhou Wang (Eds.)

million cases of malaria with an uncertainty range: 135-287 million and an estimated 6,27, 000 deaths with an uncertainty range: 4,73,000-7,89,000 were estimated. 90% of all malaria deaths are reported to occur in sub-Saharan Africa. Malaria widespread may be attributed to many factors such as (i) resistance of the mosquito to insecticides (such as DDT) (ii) favorable climatic conditions for mosquito reproduction as created by global warming (iii) non-availability of malaria vaccine and (iv) resistance being shown by the parasite towards the known anti-malarial drugs such as chloroquine, artemisinin, cycloguanil and pyrimethamine. Among all the plasmodium species that cause malaria, *P. falciparum* is the most hazardous parasite, being responsible for most of the morbidity and mortality due to malaria and associated health problems particularly in tropical countries [1-12]. Moreover, this parasite has developed resistance for almost every known drug used for the treatment of malaria. Antimalarial drug resistance can be defined as 'the parasite ability to survive and/or reproduce despite the administration and absorption of a drug given in doses equal or higher than those of usually recommended but within the tolerance limit of the subject. So the drug resistance by the parasite can be easily observed when a formerly effective drug dose is no longer effective to cure this disease [1, 9, 13]. Since the emergence of resistance by the parasite, various novel strategies have been adopted to control malaria. These include combination therapy [14-16], identification and validation of novel and specific targets in the parasitic cells [17-20], and designing new potential antimalarial drugs [21-23]. Recent trends in the designing novel anti-folates employ pharmacoinformatic tools, which are explained in the next section.

Pharmacoinformatics is an emerging computational research topic, which may be broadly defined as the application of information technology in drug discovery and development. It offers several Computer-aided tools for the drug discovery process [24, 25], and all the tools are being employed heavily for the purpose of anti-malarial agent design [26, 27]. Quantum chemical methods can be employed to obtain intricate electronic structure details associated with the drugs/leads. *Ab initio* MO/Density functional (DFT) methods can be employed to understand molecular electrostatic potentials, conformational *vs.* tautomeric stabilities, estimate ionization potentials, proton affinities, and polarizabilities of drugs/leads [28]. Quantitative Structure Activity Relationship (QSAR) methods can be employed to study the correlation between physico-chemical properties of a series of molecules and their therapeutic activities [29-32]. Pharmacophore mapping methods are employed to identify the minimum common pharmacophoric features required to produce the desired therapeutic action. This mapping may be ligand based (done purely on the basis of ligands), structure based (done on the basis of receptors) or by the combination of both strategies [33-35]. Molecular docking methods can be employed to explore drug-receptor, enzyme-substrate and drug-DNA interactions to understand the basic electronic/steric features required for therapeutic action and to design new leads with improved features that may enhance their therapeutic action [34, 36-38]. Molecular dynamics simulation study can be used to identify the interaction profile of a ligand-receptor over a period of time. The simulation details provide the stability of complex in the dynamics phase which to some extent mimics the natural body system [39, 40]. X-ray crystallography [41, 42]. homology modeling [43-46], and NMR spectroscopy [47, 48] analyses are being employed to determine 3D-structures of macromolecules, and their complex with natural substrates and inhibitors. Chemoinformatics methods mainly led by virtual screening are being employed to identify new scaffolds in lead identification [49, 50]. Many of the above techniques are commonly being employed in the design of antimalarial agents. In this review, we present some specific applications of these techniques to show their potential in designing novel and specific anti-malarial agents. The scope of this review includes pharamacoinformatics techniques used to study *P. falciparum* DHFR enzyme and its inhibitors. Modeling studies on DHFR enzyme of other species are included only when necessary. The purpose of this review article is to present (i) the medicinal chemistry and (ii) the drug action information emerging from such studies rather than to discuss the methodologies and their advantages.

Plasmodium falciparum DHFR Enzyme as a Target for Anti-Malarial Drug Design

The dihydrofolate reductase (DHFR; EC. 1.5.1.3) is an interesting and widely explored enzyme from the pharmacological point of view. This enzyme was discovered in 1950s and since then, many antifolate drugs were designed to inhibit this enzyme in various species, leading to the treatment of malaria as well as other protozoan diseases and cancer [2, 51]. Compounds such as trimethoprim (anti-bacterial), piritrexim and trimetrexate (anti-fungal), methotrexate (anti-cancer), cycloguanil (anti-malarial) and pyrimethamine (anti-malarial) are some of the known DHFR inhibiting agents [28, 52-54].

Different research teams have identified potential drug targets in *P. falciparum* parasite which may aid in the discovery of anti-malarial drugs. The identified targets include folate metabolism, hemoglobin metabolism, glycolysis, purine salvage, pyrimethamine metabolism and electron transport, oxidation stress and damage, phospholipid metabolism, artemisinin receptors, and other essential enzymes of the parasite [4, 17-19, 23, 55-59]. *Plasmodium falciparam* dihydrofolate reductase (*Pf*DHFR) is one of the most validated targets, in malaria chemotherapy and is being widely used for designing of most of the type-2 antifolates to treat this deadly disease. Its role is mainly associated with the biochemical conversion of dihydrofolate to tetrahydrofolate, which is further required for the production of DNA in the parasitic cells. Inhibition of DHFR by type 2 antifolates prevents the biosynthesis of thymidine, and consequently interrupts DNA biosynthesis. This impairment of DNA synthesis along with the accumulation of deoxyuridine triphosphate leads to cell death in the parasite [51, 60, 61].

Most important concern in any chemotherapy is that there should be good selectivity of the drug between the pathogen and the host organism. Though DHFR plays an essential role in the folate biosynthesis of all species, many differences are noticeable in metabolic pathways and their regulation among species which leads to the selectivity and specificity of antifolates for this enzyme. This fact has been largely exploited to develop species-specific DHFR inhibitors which are highly potent against the parasitic enzyme, *e.g. P. falciparum* DHFR enzyme over the corresponding enzyme in human. Since the details of these differences (i-iii) are beyond the scope of this review, readers can refer the original reports for more information. The focus of this review article will be on the drug action at the *Pf*DHFR, its resistance and the rational methods of designing new compounds, using pharmacoinformatic tools.

Plasmodium falciparum Resistance to Type-2 Antifolates

Antifolates are among the most widely used anti-malarial agents in the malarial chemotherapy. They are further divided in two groups: the first type is type-1 antifolates or non-classical antifolates. Some of the examples in this category include sulfonamides (sulfadoxine, sulfadiazine and sulfalene) and sulfones. They are known to be competitive inhibitors of *para*-amino-benzoic acid for the active-site of dihydropteroate synthase, and prevent the formation of dihydropteroate which is an immediate precursor of Dihydrofolate (DHF). Type-2 antifolates act as antagonists at the DHFR domain of *P. falciparum* DHFR-TS enzyme. These antifolates compete with the natural substrate, dihydrofolate (**1**), thus inhibiting the normal function of the cell to synthesize DNA. Cycloguanil (**2**), chlorcycloguanil (**3**), pyrimethamine (**4**), methotrexate (**5**), trimethoprim (**6**); WR99210 (**7**), biguanide derivatives, PS-15 (pro-drug of WR99210) (**8**), 1-[3-(3,4-dichlorophenoxy)propyloxy]-5-isopropylbiguanide (PS-16) (**9**), 1-[3-(4-chlorophenoxy)propyloxy]-5-isopropylbiguanide (PS-33) (**10**), proguanil (prodrug of cycloguanil) (**11**) P218 (**12**) and guanylthiourea derivatives (**13**) are some of the compounds which belong to type-2 antifolates (Fig. **1**). Two of the most important type-2 antifolates that act against DHFR enzyme are cycloguanil (**2**) and pyrimethamine (**3**) which are now reported to be of much less use in the treatment of malaria due to the development of parasitic resistance in many areas of the world [2, 51-53, 58, 60-62].

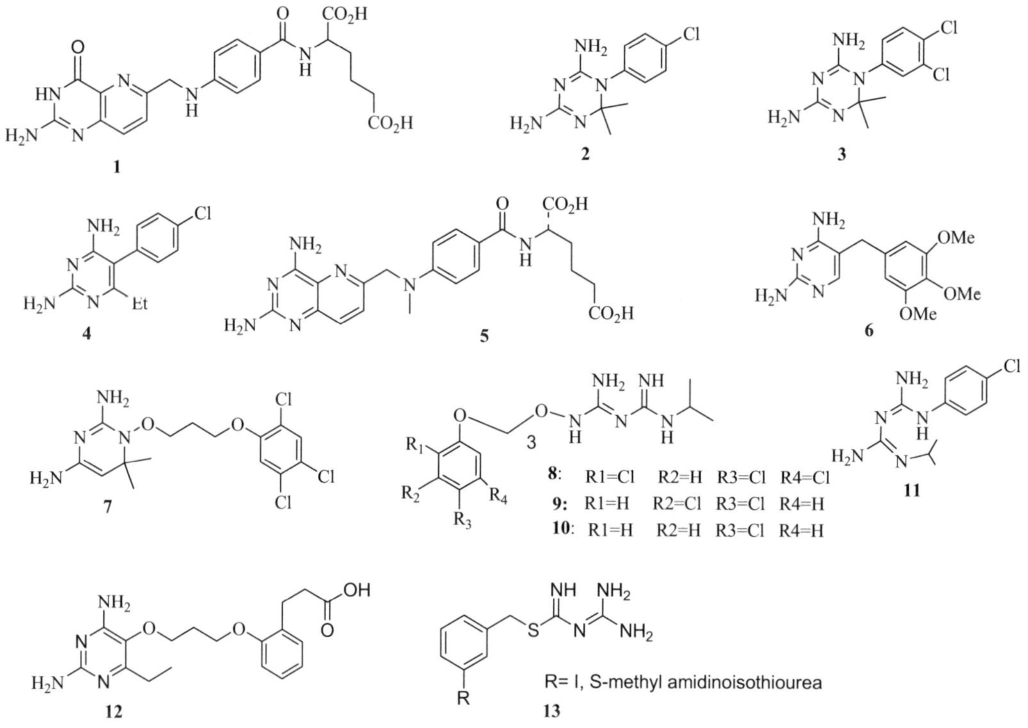

Figure 1: Chemical structures of some antifolates and prodrug biguanide derivatives.

The emergence of drug resistance in *Plasmodium falciparum* continues to create hindrance for malaria control and elimination efforts. Drug resistance can arise by different mechanisms. These mechanisms include altered drug uptake, overproduction of the target enzyme, altered target enzyme (or site of action), production of drug destroying enzyme, deletion of a prodrug-activating enzyme, overproduction of the substrate for the target enzyme, new pathway for formation of the product of the target enzyme, efflux pump, and genetic mutation in the target enzyme/protein. This in turn leads to reduction in therapeutic effect of a drug [4, 8, 63]. However, molecular level investigations of *P. falciparum* parasite resistance to type-2 antifolate drugs have identified amino acid mutations in the active-site of DHFR enzyme as the cause of resistance. The mutations occur most commonly at one or more amino acid residues in the active-site of the parasitic DHFR enzyme. A16V, N51I, C59R, S108T and S108N are the commonly observed mutations which are considered responsible for the development of resistance to type-2 antifolates in the active-site of an enzyme. Pyrimethamine is known to induce specific substitutions at amino acid residue S108N, N51I, and C59R, as well as double or triple mutants [64-67], whereas, Cycloguanil, the active metabolite of proguanil, is known to induce a different set of mutations in the gene encoding parasite DHFR which leads to mutations at positions S108T and A16V, as well as the double mutant [64, 68, 69]. A combination of mutations S108N and I164L is known to confer cross resistance to Pyr and Cyc, which is heightened when the C59R or N51I is also present. Overall, mutations causing Pyr resistance are all non-conservative substitution replacement of a Ser with an Asn, an Asn with an Ile, or a Cys with an Arg. In contrast, mutations causing cycloguanil resistance are conservative replacement of a Ser with a Thr or an Ala with a Val. These changes lead to a decrease in the therapeutic effects of these antifolates by either increasing the volume of the active-site by weakening binding of the drugs with the active-sites or by creating steric clashes between bulky amino acid residues of the mutant parasite at the point of mutation and the anti-malarial drugs. In some cases a combination of both effects could happen resulting in a dramatic loss of inhibitory activities of type-2 antifolates. Cumulative effect of mutations has been observed where multiple mutations increase the level of resistance by ten-to hundred-folds as compared to that of wild type *P. falciparum* parasite [4, 8, 58, 60, 61, 69-72]. Thus, the above mentioned point mutations have enabled the parasite resistant towards all known antifolates such as pyrimethamine, cycloguanil and other antifolate drugs. Based on the current and previous trends, it is expected that in the near future multiple mutant *P. falciparum* will be prevalent in most regions of our planet. So focus should be made on designing of novel antifolate drugs targeting multiple mutant *P. falciparum* parasite. In the following sections we present some developments and methods so far used in designing of DHFR inhibitors particularly *P. falciparum* DHFR enzyme inhibitors using computational methods.

Plasmodium falciparum DHFR Polymorphisms in Humans and Anopheles Mosquitoes

Polymorphism in molecular biology is described as point mutations in the organism's genotype. Mharakurwa *et al.* have explored the existence of *Pf*DHFR polymorphism in humans and Anopheles mosquitoes [73]. The study which was specifically carried in southern Zambia revealed that the parasites in human blood demonstrated high levels of pyrimethamine resistant mutants (mutations at positions

S108N, N51I, and C59R), whereas cycloguanil resistant mutants were detected in very low frequency in parasites from human blood samples. This observed *P. falciparum* polymorphisms in humans may be attributed to selection pressure from sulfadoxine/Pyrimethamine use for the past few decades. In contrast the authors observed significant reduction (as much as 100-fold) of *P. falciparum* pyrimethamine resistant alleles in *Anopheles arabiensis* (primary host of *P. falciaprum*), whereas, it exhibited high levels of cycloguanil resistant mutants (mutations at positions S108T and A16V). The underlying basis for *P. falciparum* polymorphisms in mosquitoes still remains to be unexplored. Previous kinetic studies reported that S108N is a loss-of-function mutation, whereas, S108T is a gain-of-function mutation [71, 74]. These studies to some extent explain why S108N mutation is not prevalent in mosquitoes as this mutation might lead to loss of catalytic activity of an enzyme. The specific activity of S108T has been reported to be approximately twice that of the wild type, which suggests that this mutation confers an added advantage for the catalytic function of an enzyme and thus prevails in the mosquito phase of the parasite. However, it does not explain why S108T prevalence is not elevated in human infections during absence of sulfadoxine/pyrimethamine selection pressure and also why this favorable mutation does not overtake the wild type form of an enzyme. The authors also carried out structural investigation of the dihydrofolate reductase by *in silico* modeling and they reported that an there exists an important 2.56 Å hydrogen bond linking the side chain of S108 to the pyrophosphate moiety of NADPH in the NADPH binding pocket within the wild type and cycloguanil mutant (S108T) *Pf*DHFR enzyme. The extra methyl group of threonine in S108T does not cause any steric collisions. In contrast, side chain of N108 in the pyrimethamine resistant mutant cannot form an effective hydrogen bond, as it lies 5.40 Å away from the pyrophosphate group of NADPH. From this study the authors conclude that due to different selective pressures in humans and mosquitoes, *P. falciparum* can bear host-specific drug resistant polymorphisms. Thus, it is suggested to sample both human and mosquito vector infections to precisely determine the epidemiological status of drug resistant alleles. Future studies are still required to elucidate the factors behind these observed allele patterns.

STRUCTURE OF *P. falciparum* DHFR ENZYME AND ITS MUTANTS

Rational drug design approach involves target identification, target structure determination, computer-aided inhibitor design and synthesis and testing of compounds for their therapeutic activity [24]. Thus, 3D-structure of a target is one of the key factors in this strategy which helps in designing potent and species-specific/selective drugs. In this section, we discuss the importance of X-ray crystallography in understanding the interactions of active-site amino acid residues with the antifolate drugs.

Crystal Structure of DHFR Enzyme of *Plasmodium falciparum*

Structures of many *Pf*DHFR enzymes and their complexes with ligands have been determined by X-ray crystallography, and these structures have been utilized in designing of *Pf*DHFR inhibitors, and also in predicting the binding modes of these inhibitors in the active-site of *Pf*DHFR enzyme [75-77]. Antifolate drugs such as pyrimethamine and cycloguanil act on *Pf*DHFR, but resistance to these antifolates noticed within a very short span of time. WR99210 (lead molecule) has been

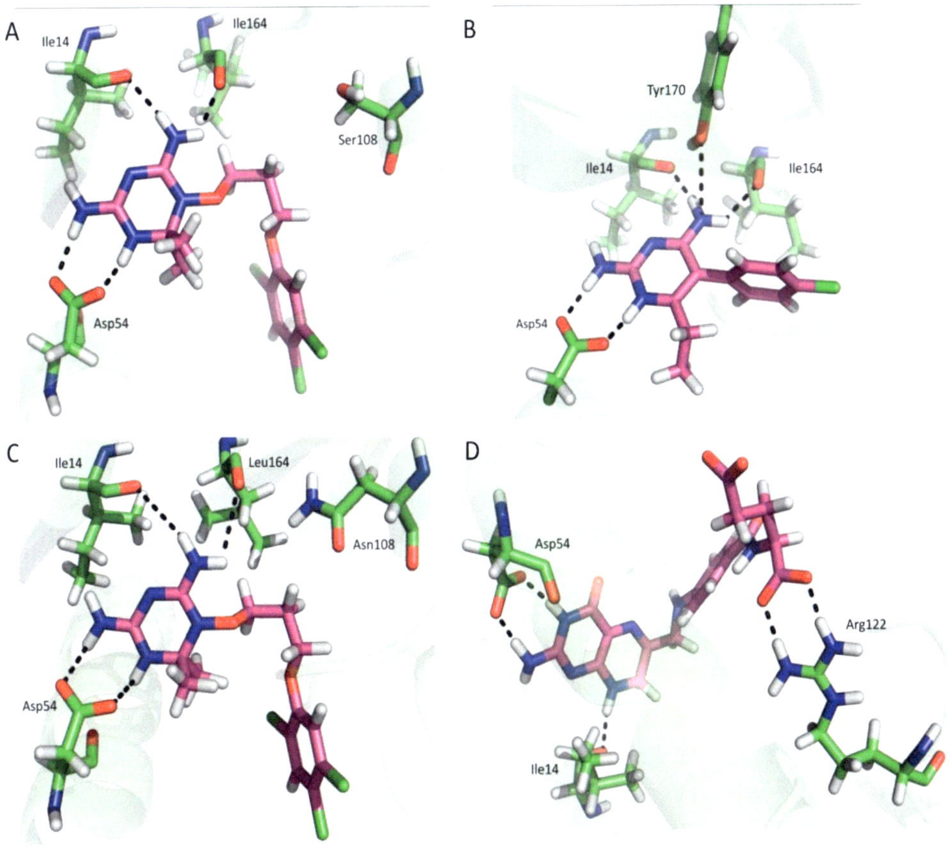

Figure 2: Stereoviews of the *P. falciparum* DHFR active-site residues of wild type bound to WR99210 (A), double mutant bound to pyrimethamine (B) quadruple mutant bound to WR99210 (C) and wild type bound to DHF (D). For the sake of clarity, only important amino acid residues are shown.

identified as *Pf*DHFR inhibitor that inhibits both wild type and mutant *Pf*DHFR, but due to low bioavailability and toxicity issues WR99210 cannot be used as an anti-malarial drug. Hence researchers have solved X-ray crystal structures of *Pf*DHFR bound to its inhibitors and natural substrate (dihydrofolate) to identify the binding mode of these ligands and their important molecular recognition interactions in wild type and mutant *Pf*DHFR and also to identify the key factors that render WR99210 equally effective in both wild type and mutant forms of *Pf*DHFR. A major breakthrough in X-ray crystallographic analysis of *P. falciparum* DHFR enzyme was reported by Yuvaniyama *et al.* in the year 2003 [78]. The authors carried out crystallographic analysis for three strains of *P. falciparum*: the wild type (PDB code: 1J3I), double mutant (C59R+S108N) (PDB code: 1J3K) and quadruple mutant (PDB code: 1J3J). The stereoviews of the active-sites of these enzymes are given in Fig. **2**(**A-C**).

The authors reported that the crystal structures of both the wild type and mutant *P. falciparum* DHFR-TS enzymes were similar with isomorphous orthorhombic form containing two *P. falciparum* DHFR-TS molecules per asymmetric unit related by pseudo two-fold symmetry. The two folate binding sites in TS domain are far apart by 32 Å each composed of residues from both units whereas the corresponding sites in the two DHFR domains are not in contact, and are 65Å apart from each other (Fig. **3**). The major force that facilitates active substrate transport from TS to DHFR domain in homo dimeric DHFR-TS enzyme was assumed to be the electrostatic interaction. Franca *et al.* confirmed the role of this interaction by calculating electrostatic contour maps of *P. falciparum* DHFR-TS model. The authors suggested that as TS processes its substrate producing Dihydrofolate (DHF), the negative electrostatic potential of the TS active-site repels DHF into the positive electrostatic potential zone and from there to the DHFR active-site [62]. WR99210 is an antifolate with a long and flexible side-chain which can bind to wild type and mutant *Pf*DHFR enzyme with no loss in its inhibiting activity. WR99210 in the active-site of mutant enzyme adopts an orientation such that it avoids steric clash with N108. On the other hand, pyrimethamine and cycloguanil have rigid structures, and thus cannot avoid steric clash with N108 which results in poor binding and reduced inhibiting activity of these drugs against mutant *P. falciparum* DHFR enzyme.

Later many crystal structures of *Pf*DHFR in its wild and mutant forms were solved and till date about fourteen crystal structures of *Plasmodium falciparum* DHFR are available in Protein Data Bank and their details are given in Table **1**.

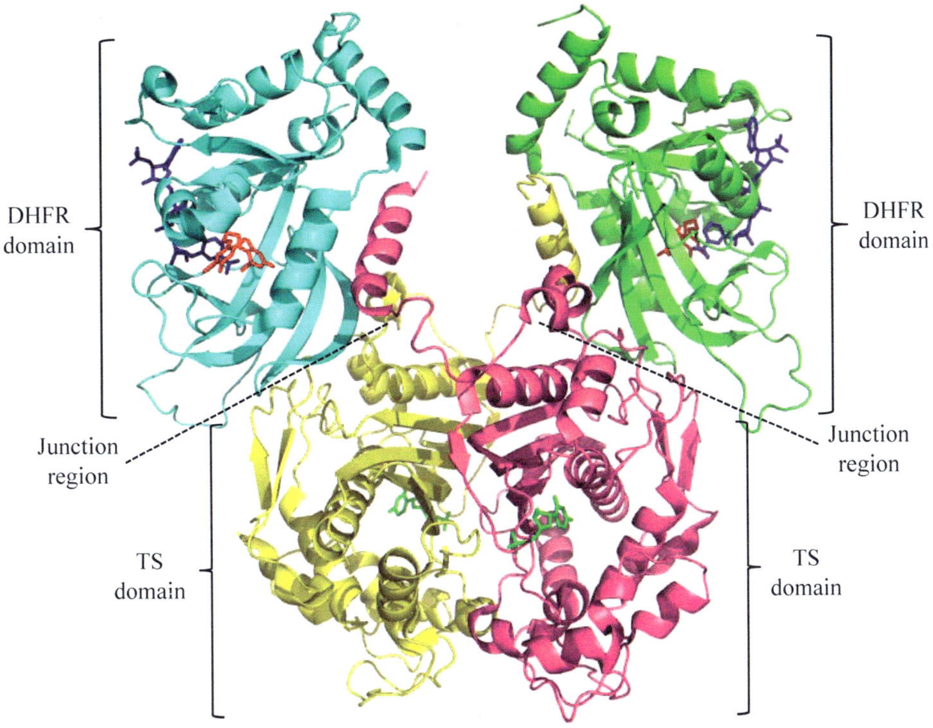

Figure 3: Illustration of homodimeric *P. falciparum* DHFR-TS enzymes.

Table 1: Information on the reported *Pf*DHFR crystal structures complexed with its ligands

No.	PDB	Mutations	Ligand	Resolution (Å)	Authors
1	1J3I		WR99210	2.33	Yuvaniyama *et al.* (2003) [78]
2	1J3J	C59R+S108N	Pyrimethamine	2.30	Yuvaniyama *et al.* (2003) [78]
3	1J3K	N51I+C59R+S108N+I164L	WR99210	2.10	Yuvaniyama *et al.* (2003) [78]
4	3DGA	-----	RJF01302	2.70	Dasgupta *et al.* (2009) [79]
5	3DG8	N51I+C59R+S108N+I164L	RJF670	2.58	Dasgupta *et al.* (2009) [79]
6	3JSU	N51I+C59R+S108N+I164L	QN254	2.70	Chitnumsub *et al.* (2010) [80]
7	3QGT	-----	Pyrimethamine	2.30	Vanichtanankul *et al.* (2011) [81]
8	3QG2	N51I+C59R+S108N+I164L	Pyrimethamine	2.30	Vanichtanankul *et al.* (2011) [81]
9	3UM5	A16V+S108T	Pyrimethamine	2.40	Vanichtanankul *et al.* (2012) [82]
10	3UM6	A16V+S108T	Cycloguanil	2.65	Vanichtanankul *et al.* (2012) [82]
11	3UM8	-----	Cycloguanil	2.60	Vanichtanankul *et al.* (2012) [82]
12	4DPH	N51I+C59R+S108N+I164L	P65	2.38	Yuthavong *et al.* (2012) [83]
13	4DP3	N51I+C59R+S108N+I164L	P218	2.40	Yuthavong *et al.* (2012) [83]
14	4DPD	-----	Dihydrofolate	2.50	Yuthavong *et al.* (2012) [83]

Recently, Yuthavong *et al.* solved the crystal structure of wild type *Pf*DHFR bound to its natural substrate dihydrofolate (DHF) [83]. This information on the binding mode of DHF (in addition to the information on binding modes of different inhibitors) may be exploited for the structure based drug design approach. As per the report of these authors the pteridine ring of DHF interacts with Ile14, Asp54 and Ile164 to form hydrogen bonds and the flexible linker chain binds in the hydrophobic cavity of *Pf*DHFR active-site. In addition to these interactions it was studied that at the opposite end of DHF pteridine ring is the carboxylic group which forms two hydrogen bonds with Arg122. Further authors report that due to the tight packing of aliphatic portion of the Arg122, its side chain mobility is fixed hence it may provide a rigid docking site for carboxylic group of DHF. The stereoview of the active-site of this enzyme is given in (Fig. **2D**).

Importance of Junctional Region of *Plasmodium falciparum* Dihydrofolate Reductase-Thymidylate Synthase

The X-ray crystal structures reveals that *Plasmodium falciparum* DHFR is a bifunctional enzyme having DHFR and TS domain as one unit (monomer) and it exists as dimer. The two domains DHFR and TS are joined together by linker sequence known as the junction region. The analysis carried out by Yuvaniyama *et al.* [78] provided some information on the roles of inter-domain interactions between DHFR and TS domains of the parasite which affect their activities. But still the role and importance of this junction region was not clear until Chaianantakul *et al.* [84] provided some light on the importance of this region in the enzyme. The authors investigated the interactions of *Pf*DHFR and *Pf*TS domains of the bifunctional enzyme and also the role of juctional region on the activity of this enzyme. A series of recombinant *Pf*TS with truncated junctional regions were developed and protein-protein interactions between *Pf*DHFR and the *Pf*TS domains were demonstrated by using a bacterial two-hybrid system. The authors reported that the deletion of juctional region affect the activities of both DHFR and TS domains of *Pf*DHFR and at least 44 amino acids (D283-N296) or longer junctional region was necessary for the *Pf*TS domain to be active for the synthesis of thymidylate for the cells. Based on their studies the authors also reported that the *Pf*DHFR domain is essential for *Pf*TS catalytic activity. This finding could be explored to design novel strategies to develop anti-malarial agents targeting the non-active-site (domain-domain interaction site) of *Pf*DHFR-*Pf*TS bifunctional enzyme.

Novel Target sites in *Plasmodium falciparum* DHFR

Plasmodium falciparum Dihydrofolate Reductase-Thymidylate Synthase (DHFR-TS) is a validated drug target and is been widely used for treatment of malaria. Much of the work has been carried out on the active-site of *Pf*DHFR (site of DHF binding) as a potential target site. Many drugs (Pyrimethamine and Cycloguanil), lead molecules (WR99210 and P218) and inhibitors have been designed to bind in this site and inhibit the normal functioning of this essential enzyme. *Pf*DHFR has developed resistance to all the known antifolate drugs in a very short span of time. Hence there is a need to design novel inhibitors of *Pf*DHFR and also novel target sites in the enzyme. Dasgupta *et al.* have reported that apart from the known active-site of *P. falciparum* DHFR-TS, it also contains unique distant or 'non-active' regions which might influence the catalytic activity of this enzyme [85]. The identified distant regions studied by the authors were N-terminal tail, and a crossover helix of the 'linker' region joining DHFR to TS, which also forms crucial electrostatic interactions with the DHFR active-site. Initially the authors studied the DHFR domain and TS domain interactions in this enzyme to find the influence of one domain on the activity of the other. The authors reported that indeed the domains affect the activity of each other and the overall rate-limiting step in this enzyme was found to be TS catalysis. It was noted that DHFR rate is 2-fold activated in the wild type enzyme when TS is in an 'activated' (ligand bound) conformation. On a comparative scale TS activation rate has reduced activation ~1.5-fold when DHFR is in its ligand-bound conformation. Conformational changes to the enzyme active-site initiated by ligand-binding have already been reported in literature and the authors suggest that this probably might be the reason for the studied communication between DHFR and TS active-sites [48, 86-89]. Further the authors studied the role of crossover helix (residues 284 to 295) in the linker region. This 15 amino acids long linker region (mainly composed of acidic residues) has 5 amino acids at its one face which are reported to interact with positively charged amino acids at the back side of enzyme active-site. To study the kinetic consequences of disrupting the electrostatic interactions of this helix the authors generated mutants and the results obtained suggested that the crossover helix does not play a role in modulating *P. falciparum* TS-DHFR catalytic activity. Finally the authors explored the importance of 6 amino acids long N-terminal tail which is oriented away from the DHFR, in a position remote from active-site. Some previous studies done on the monofunctional *P. falciparum* DHFR and TS enzymes also suggested that deletions of the N-terminal region decrease activity of DHFR and also affects the interaction between TS and DHFR domain [90, 91]. The authors studied the role of the N-terminal tail in modulating catalysis in the full-length, bifunctional TS-DHFR enzyme and the results confirmed that deleting the N-terminal tail in *P. falciparum* TS-DHFR decreases DHFR catalytic rate by a factor of 2. The study suggests that the N-terminal tail of *P. falciparum* TS-DHFR is a highly selective (no homologous counterpart in humans) and can be explored as novel target for potential antifolate development in malaria.

Dasgupta *et al.* in their study reported that the mutations in the linker region neither affect catalytic rate nor domain-domain communication [85]. Recently, Chaianantakul *et al.* investigated the role of junction/linker region on the activity of this enzyme [84]. They developed series of recombinant *Pf*TS with truncated junctional regions and studied protein-protein interactions between *Pf*DHFR and the *Pf*TS. In contrary to what is reported by Dasgupta *et al.* Chaianantakul *et al.* reported that the deletion of juctional region affect the activities of both DHFR and TS domains of *Pf*DHFR and at least 44 amino acids (D283-N296) or longer junctional region was necessary for the *Pf*TS domain to be active for the synthesis of thymidylate for the cells. Overall, understanding the importance of these distinct non active structural regions may aid in rational drug design approach for the development of more targeted and less toxic therapies for malaria.

QSAR STUDIES OF *P. falciparum* DHFR ENZYME INHIBITORS

The data from pharmacological experiments can be directly employed in obtaining Quantitative Structure-Activity Relationships (QSAR), and hence, QSAR is one of the powerful and most widely used tools in drug design and discovery process. It is used to predict biological activity of a molecule from a set of structure-based descriptors and statistical methods [29, 30, 92, 93]. There are different types of QSAR methods: 2D-QSAR (classical QSAR'), 3D-QSAR, 4D-QSARs and 5D-QSAR [8, 92-95]. 3D-QSAR is a statistical approach that correlates chemical 3D-structural parameters of structurally related compounds with their biological activities for a particular target. 3D-QSAR methods can be further sub-divided into ligand-based 3D-QSAR and receptor-based 3D-QSAR. The former is commonly used to predict properties of compounds from the models developed using a training set of series of molecules. This approach mainly depends on 3D-structures of ligands, and knowledge of 3D-structures of macromolecular target is not required. In the case of receptor-based 3D-QSAR analysis (also known as receptor-dependent 3D-QSAR analysis), the 3D-structure of the receptor is used along with corresponding ligand structure-activity data [29-31, 96-102].

Several QSAR models of DHFR enzyme inhibitors of various species have been developed that gave valuable information in design and development of effective and species-selective inhibitors. Some of the models include 4D-QSAR [96-99, 101, 102], (FEEF) RD-3D-QSAR [101], QSAR formulated with multiple regression (MLR) and genetic algorithm (GA) [103], QSAR with evolved neural networks [101, 103-106], QSAR using mathematical molecular descriptors [107], QSAR/optimization correlation weights of local graph invariants (OCWLGI) based on nearest neighboring codes (NNC) [108], Self-organizing maps (SOM) 4D-QSAR [109, 110], QSAR/CoMSIA [111-113], QSAR/CoMFA [113, 114], QSAR/All-Oriented Search (AOS)/CoMFA [115], and Most of the above models were developed based on DHFR enzymes of species (other than *P. falciparum* DHFR enzyme) such as *E. coli*, *L. casei*, rat liver, *P. carinini* and *T. gondii*. Understanding QSAR models has helped, and will also help in developing QSAR models for different series of compounds that inhibit both wild type and mutant *P. falciparum* DHFR enzyme [102]. Two QSAR models which were developed in the recent years based on *P. falciparum* DHFR enzyme are presented below.

Three-dimensional quantitative structure-activity relationship (3D-QSAR) models are used to study favorable and unfavorable pharmacophoric features of molecules which play an important role in the ligand-protein interaction. Two most widely used 3D-QSAR approaches are Comparative Molecular Field Analysis (CoMFA) [29] and Comparative Molecular Similarity Index Analysis (CoMSIA) [32]. CoMFA computes the steric and electrostatic fields based on 3D-structures of ligands. Different color-coded contour maps surrounding the ligands are generated which gives insights about the favorable and unfavorable ligand receptor interactions. CoMSIA model uses physicochemical properties such as steric, electrostatic hydrogen bond donor, hydrogen bond acceptor and hydrophobicity to calculate similarity indices around the molecules. Adane *et al.* carried out 3D-QSAR study on 42 cycloguanil derivatives which were reported as growth inhibitors of *Plasmodium falciparum* clone (T9/94 RC17) [113]. This clone harbors A16V + S108T mutant dihydrofolate reductase (DHFR) enzyme. The authors carried out CoMFA and CoMSIA to investigate the structural requirements for the activities of these compounds and also to derive predictive models that may be used for designing novel *Pf*DHFR enzyme inhibitors. Different combinations of steric, electrostatic, hydrophobic, H-bond donor and H-bond acceptor fields were used to develop CoMSIA model. pIC_{50} values were used as dependent variables for generation of the CoMFA and CoMSIA models. The CoMFA model developed by the authors gave statistically significant results with cross-validation coefficient (r^2_{cv}) and non cross-validate coefficient (r^2_{ncv})

values of 0.654 and 0.951, respectively. The best CoMSIA model also gave significantly good r^2_{cv} and r^2_{ncv} values of 0.669 and 0.907, respectively. The evaluation of predictive abilities of the CoMFA and CoMSIA was carried out using test compounds which gave r^2_{pred} values of 0.735 and 0.557, respectively.

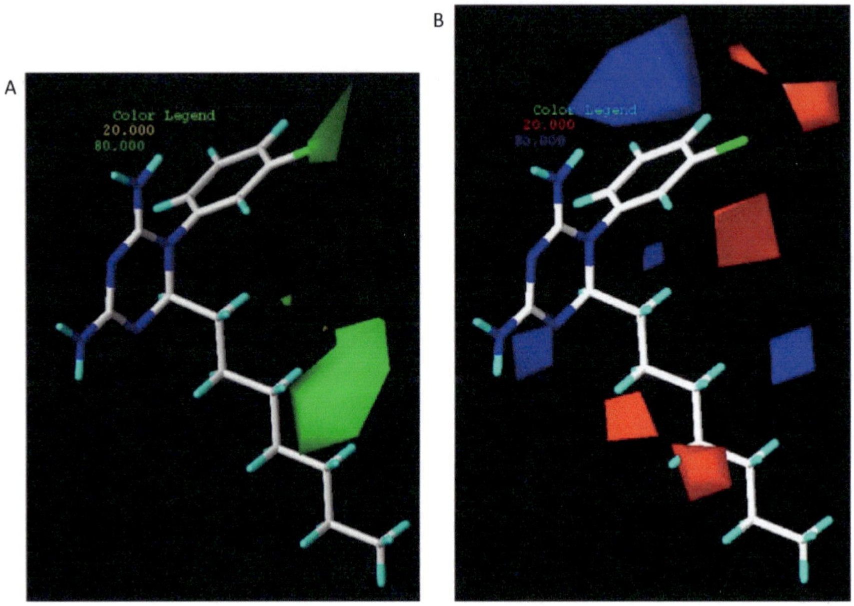

Figure 4: STDDEV*COEFF plots of the CoMFA steric **(A)** and electrostatic **(B)** contour maps. The most active molecule of the series is displayed in the background. Green region: sterically favored; yellow region: sterically disfavored; red region: negatively charged favored; blue region: positively charged favored. (Reproduced from Ref. [113] with permission from Elsevier).

In the CoMFA steric contour maps the green contours depicts the regions where sterically bulky groups are linked with increase in biological activity; yellow contours signify the regions where such bulky groups lower biological activities of ligands. In the best CoMFA model, green contours were observed near 3'-substituent of 5-aryl ring and at about three C-C bonds away from C6 atom of the 1,3,5-triazine ring of the most active compound (Fig. **4A**). This observation indicates that increasing steric groups at these positions result in increased inhibitory activities of cycloguanil derivatives. No yellow colored contour was observed in this model. In general the CoMFA electrostatic interaction contour maps are represented by red and blue regions. The regions where electron-rich (electronegative) groups enhance bioactivity are represented by red contours whereas the regions where increase in bioactivity is observed by increasing positive charge (or decreased negative charge) are represented by blue contours. Three small red contours near 3'-Cl substituent; two small red contours on C6-substituent were observed (Fig. **4B**). In the CoMFA model, one big blue contour (Fig. **4B**) near 5'-position of the aryl ring (of the most active compound), and a group of other small blue contour maps one near N1-and C6-substituents were observed. This indicates that introduction of electropositive groups at these positions would increase inhibitory activities of cycloguanil derivatives. Based on relative contributions of steric (76.7%) and electrostatic fields (23.3%), as calculated from CoMFA model the authors reported that steric interactions were the major contributors for the variation of biological activities of cycloguanil derivatives used in the study. Overall analysis of the CoMFA contour maps indicated that the most active compounds have bulky and electronegative substituents or at least one of them which is properly oriented at the sterically and electrostatically favorable regions. Combination of three physicochemical properties, *viz*. steric, electrostatic and hydrophobic fields was used to generate CoMSIA model. A group of small yellow contours were observed near C6-position of the 1,3,5-triazine ring as opposed to CoMFA model where no yellow contour was observed. Big green contours were observed on the C6-aliphatic substituent of most active compound similar to in CoMFA model (Fig **5A**). Only one big red contour map was observed in the CoMSIA model near C6-atom (Fig. **5B**) whereas a group of small red contours were observed in the CoMFA model near 3'-Cl substituent and C-6 substituent of the most active compound (Fig. **4b**). In the hydrophobic contour map of CoMSIA model, white contours indicate that regions where hydrophobic substituents decrease bioactivities of compounds whereas the yellow contours indicate regions where hydrophobic substituents increase bioactivities of compounds. In the generated hydrophobic contour map (Fig. **5C**) with the most active compound white contours were observed near the 1,3,5-triazine ring. This indicates that hydrophobicity is not required in these regions particularly near N1 of the triazine ring. It has been reported that N1 of triazine gets protonated at physiological pH in order to form a bidentate H-bond interaction with the amino acid residue Asp54 in the active site of *Pf*DHFR enzyme. Such H-bonding interactions are known to be fundamental for *Pf*DHFR inhibitory activities of antifolates such as cycloguanil [34, 116]. Hence, the authors report that hydrophobicity needs to be reduced near the 1,3,5-triazine ring to enhance inhibitory activities of cycloguanil analogues. A big yellow contour was observed near the 5-phenyl ring indicating that hydrophobicity is required in this region to increase activity. This contour is found near the 3'-Cl substituent of the template most active molecule suggesting significant contribution of 3'-Cl substituent to the hydrophobic interaction. The results obtained from this study could further be utilized for rational design of potent inhibitors against A16V + S108T mutant *Pf*DHFR enzyme.

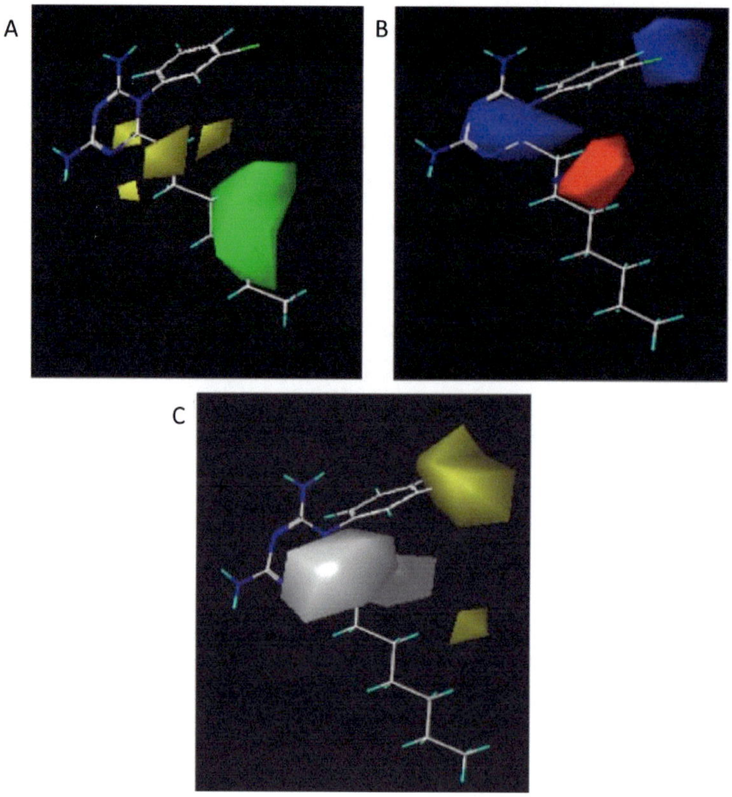

Figure 5: The CoMSIA steric **(A)** and electrostatic **(B)** contour maps and hydrophobic field **(C)**. Green region: sterically favored; yellow region: sterically disfavored; red region: negatively charged favored; blue region: positively charged favored; yellow region: favored; white: disfavored). The most active compound of the series is shown in the background. (Reproduced from Ref. [113] with permission from Elsevier).

Basak *et al.* used mathematical molecular descriptors to perform QSAR on 58 cycloguanil analogs reported as wild type and double mutant (A16V+S108N) *Pf*DHFR inhibitors [107]. The authors utilized two molecular descriptors topological indices (TIs) and atom pairs (APs), as independent variables in their study. TIs were derived from graph theoretical methods and APs were taken as molecular substructures. An AP represents any two atoms in the molecule which also includes their path-wise interatomic separation and also the information on the presence or absence of π-orbitals. Both these descriptors were solely based on compounds chemical structure and no input from any experimental method was used to develop the model. POLLY v2.3, Triplet and Molconn-Z were used to calculate a set of 369 TIs. Additionally, a set of 527 APs were also calculated and included in the set of independent variables. Results were obtained from three comparative regression methods-Ridge Regression (RR), Principal Component Regression (PCR), and Partial Least Square (PLS). The authors reported that in the wild type *Pf*DHFR, both TIs and TI plus AP combination give good predictive models for the binding affinities (K_i) of the cycloguanil derivatives. The addition of APs to the set of TIs improved the model as indicated on the basis of increase in q^2 from 0.674 to 0.790. For the mutant type *Pf*DHFR, also the authors found that the addition of APs to the TIs made some improvement in model quality, with a resulting q^2 of 0.874. In terms of the quality of models it was found that RR statistical method outperformed the other two calculated statistical models PLS and PCR. Based on the developed models, the authors reported that the descriptors influencing QSAR models for wild type and double mutant *Pf*DHFR were non-overlapping. Based on their analysis they also suggested that inhibitors are recognized distinctly in the active-site of wild type and double mutant *Pf*DHFR enzymes.

MOLECULAR DOCKING STUDIES OF *P. falciparum* DHFR INHIBITORS IN THE ACTIVE-SITE OF THE ENZYME

Molecular docking is one of the structure-based drug design approaches, which is widely and routinely applied to discover novel and potential enzyme inhibitors with specific and desired properties. This tool is mainly useful in predicting the binding affinity (binding constant or free energy of binding) of ligands or inhibitors to a particular target/enzyme prior to their synthesis [117-121]. Various types of molecular docking algorithms that could generate and identify complementary match between ligands and macromolecular targets are available [117, 122]. It involves a combination of methods for locating binding sites and evaluating the strength/weakness of the binding or energy of binding [117, 118, 123-134]. Molecular docking is also been employed in the virtual screening of compounds during drug discovery process in which a virtual library of ligands are screened to identify their binding affinities through docking approach using protein target of known 3D-structure [123, 135, 136]. Molecular docking analysis also provides insight into the ligand binding domain of macromolecules. In this section the discussion is limited to the application of docking approach in the discovery and development of *P. falciparum* DHFR inhibitors.

Reports revealed that docking has been widely used in molecular level understanding of resistance of *P. falciparum* to antifolates such as pyrimethamine and cycloguanil. This understanding led to the discovery and development of potential antimalarial drug candidates effective against both wild type and mutant *P. falciparum* DHFR enzyme [40, 137]. Molecular docking analysis has been carried out by researchers to gain better insight into the binding mode of *Pf*DHFR inhibitors in the active-site of an enzyme and also to design novel *Pf*DHFR inhibitors [34, 40, 79, 138-146]. Using Glide, GOLD and FlexX docking softwares, Adane *et al.* performed molecular docking studies on 28 compounds (2,4-diaminoquinazoline and 2,4-diaminopteridine analogs) with wild type and quadruple mutant *Pf*DHFR [140]. These compounds were reported to have anti-plasmodial activities in both wild type and quadruple mutant enzymes. The authors reported that most of these compounds docked into the active-site of both wild type and mutant *Pf*DHFR. Visual inspection of the binding modes illustrated that most of these compounds showed H-bond interactions with the key amino acid residues (Asp54, Ile14 and Leu/Ile164) and had better docking scores than the bound compound (WR99210). Long side chains of these compounds were reported to be settled in the hydrophobic portion of the active-site as similar to WR99210 whose trichloro aryloxy side chain occupies the same region. Thus, these molecules are proposed to avoid potential steric clashes with N108 (mutated from S108). Recently, the same group utilized molecular electrostatic potential analysis and molecular docking approach to design a new class of compounds based on S-benzylated guanylthiourea (mono-and bi-substituted) as potential *Pf*DHFR inhibitors [144]. The designed molecule had guanylthiourea (GTU) and hydrophobic aromatic tail joined together by linker region with 1, 2 or 3 carbon atoms. The Glide docking program was used to study the binding poses of the designed compounds. The authors reported that similar to cycloguanil, pyrimethamine, WR99210 and other biguanide derivatives, GTU moiety also formed the expected hydrogen bond interactions with Ile14, Asp54, Leu164 and hydrophobic interactions with Phe58 and Phe116 in the active-site of *Pf*DHFR enzymes. (Fig. **6**) The steric clash with Asn108 in the active-site of the mutant *Pf*DHFR enzyme was prevented due to flexible linker unit between the hydrophobic aromatic tail and GTU moiety. Out of all the studied compounds the bi-substituted GTU structures showed G score comparable to that of WR99210. Top scoring 20 compounds which also showed important molecular recognition interactions were further synthesized and *in vitro* studies were carried to access their *Pf*DHFR inhibition potency. Based on *in vitro* results, the authors reported that two compounds from this class were active with the IC_{50} value of 400 nM and 100 μM. Overall it was noticed that bi-substituted guanylthiourea compound showed good activity. The probable reason for this as suggested by the authors may be increase in hydrogen bond interactions in the active-site of *Pf*DHFR enzyme due to two GTU moieties. This additional feature of bi-substituted GTU derivatives can be further explored to improve the activity of this novel class of compounds and also to develop novel potential *Pf*DHFR inhibitors.

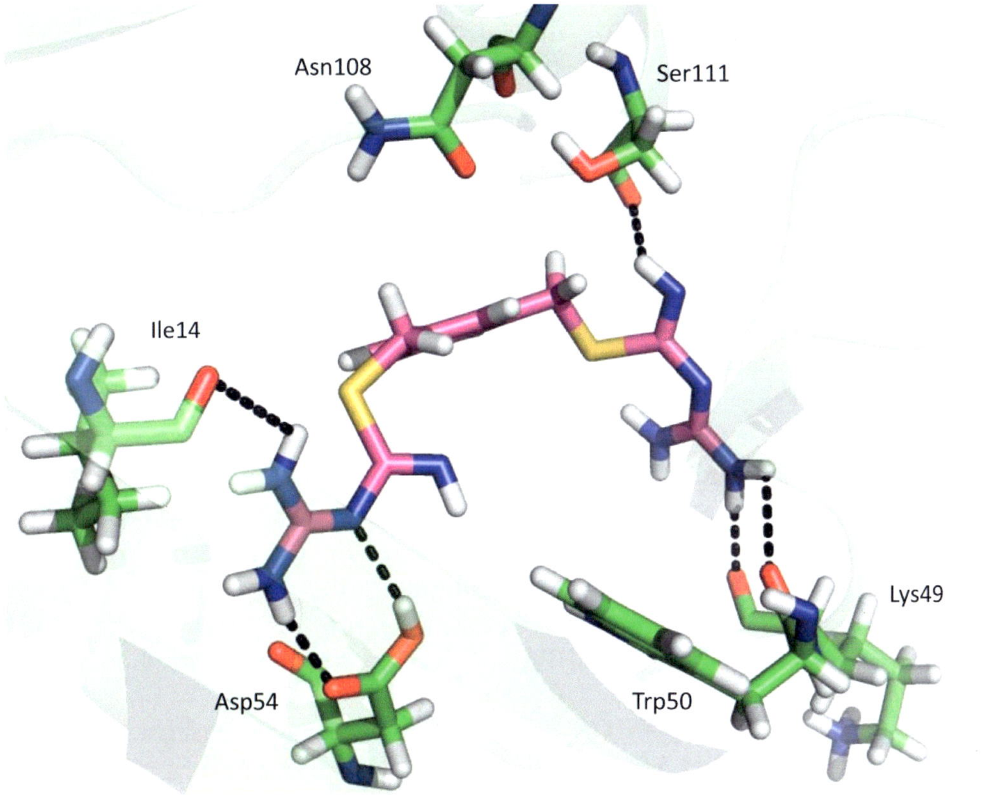

Figure 6: Docked pose of most active bi-substituted GTU derivative in the active-site of quadruple mutant *Pf*DHFR (PDB code: 1J3K).

Much of the molecular docking studies have been carried out to aid design of novel *Pf*DHFR active-site inhibitors which compete with the normal binding of natural substrate DHF. Researchers have also identified some other critical non active-site regions in *Pf*DHFR which can be explored for the design of novel and specific *Pf*DHFR inhibitors [84, 85]. To explore this area Dasgupta *et al.* performed *in silico* virtual screening to identify *Pf*DHFR inhibitors which can bind in the non active-site of an enzyme [79]. The non-active-site pocket selected by the authors was the 89 residues long linker region (residues 232-320) in the bifunctional enzyme. Virtual screening studies were performed on 16000 diverse, drug-like molecules from Maybridge database using Glide docking software. 14 compounds were identified as probable non-active-site *Pf*DHFR inhibitors from the virtual screeing analysis. *In vitro* enzyme screening, as well as malarial cell culture studies identified three biguanide containing lead compounds (RJF 00670, RJF 01302, and RJF 00719) with similar activity in μM range against wild type and antifolate-resistant *P. falciparum* parasites. Free energy of binding calculations, as calculated by MM-GBSA method, predicted that these lead compounds would actually preferentially interact with the DHFR active-site, instead of the linker region (as screened by molecular docking studies). To confirm the binding mode of these inhibitors, structural studies were carried out to obtain the structure of the inhibitors (RJF 00670 and RJF01302) bound to the bifunctional *P. falciparum* TS-DHFR enzyme and also to distinguish between the two possible sites for inhibitor binding. Co-crystallography studies confirmed that these lead inhibitors bind to the DHFR active-site for wild type and drug resistant TS-DHFR enzyme rather than the non-active pocket. The structural studies of the co-crystallized ligands (in both wild type and mutant enzyme) revealed that both RJF 00670 and RJF 01302 are not sterically hindered by S108N mutation. Both RJF 00670 and RJF 01302 showed hydrogen bonds with the carboxyl side chain of Asp54 and the backbone carbonyl of Ile164. Furthermore, the authors reported that RJF compounds also make polar interactions with Ile14, Tyr170, and Cys15 and van der Waals contact with residues Phe58, Ile112, Leu119, Met55, Cys15, and Ala16 at the DHFR active-site. Serendipity in this work, lead to the design of novel active-site *Pf*DHFR inhibitors but the design of novel inhibitors specific for the non active-site region of *Pf*DHFR still remains to be unexplored.

MOLECULAR DYNAMICS STUDIES OF *P. falciparum* DHFR INHIBITORS IN THE ACTIVE-SITE OF THE ENZYME

Molecular dynamics (MD) is a computer simulation of physical movements of atoms and molecules, which gives time dependent behavior of a system. It is one of the principle tools for studying theoretical behavior of biological systems. The atoms and molecules are allowed to interact for a period of time, giving a view of the motion of the atoms. Newton equations of motion are solved for the system to determine the trajectories of atoms and molecules. Molecular mechanics force fields are utilized to define forces between all the particles and the potential energy of the system. In biophysics and structural biology, the method is frequently applied for ligand docking, simulations of lipid bilayers, homology modeling and even *ab initio* prediction of protein structure by simulating folding of the polypeptide chain from random coil.

Molecular dynamics remained unexplored for the study of *Pf*DHFR-ligand interactions until recently when Mokmak *et al.* 2014 carried out molecular dynamics of pyrimethamine and WR99210 bound to wild type and mutant enzymes to understand the factors contributing to binding affinity differences of these inhibitors to the wild type and quadruple mutant enzyme [39]. The authors found that Ile14 and Asp54 were crucial for pyrimethamine and WR99210 binding to *Pf*DHFR as they form strong hydrogen bonds with the ligands. The quadruple mutations in *Pf*DHFR leads to reduced hydrogen bonding capacity of pyrimethamine (with Ile14 and Leu164), and also results in its altered orientation because of which it cannot efficiently interact with Asp54. N1 of pyrimethamine and WR99210 were modeled in their protonated states, as it has been reported that all biguanides act in their protonated states. The authors reported that WR99210 stably binds to both wild type and quadruple mutant enzyme and pyrimethamine binds stably to wild type *Pf*DHFR whereas strong changes were observed for the binding mode of pyrimethamine in the active-site of quadruple mutant *Pf*DHFR. In the case of WR99210 bound to both wild type and mutant enzyme, strong H-bonds were formed with Asp54 and weak H-bonds with the amino acid residues Ile164/Leu164 were observed. Similarly pyrimethamine also formed strong hydrogen bond with Asp54; additionally it also formed strong hydrogen bond with Ile164 in the case of wild type *Pf*DHFR. Pyrimethamine formed weak H-bonds with Ile14 and Ile164 of quadruple mutant active-site and hydrogen bonds with Asp54 were also not stable over a period of time. The binding of WR99210 to wild type and quadruple mutant enzyme as well as the binding of pyrimethamine to wild type was reported to be favorable, with ΔG_{bind} values of-56.5,-57.3 and-41.0 kcal/mol, respectively. It was further reported that the van der Waals and electrostatic contributions favor binding of WR99210 to wild type and mutant enzymes and binding of pyrimethamine to wild type enzyme. The binding of PYR to quadruple mutant was found to be very unfavorable with ΔG_{bind} of 87.6 kcal/mol. The authors reported that unfavorable desolvation of polar groups (155.5 kcal/mol) was the main factor opposing the pyrimethamine binding to quadruple mutant enzyme and the electrostatic interactions were also less favorable, as compared to other three complexes. Per residue energy decomposition analysis carried out by the authors revealed that in the case of WR99210 bound to wild type *Pf*DHFR, S108 had a slightly favorable binding energy which provided some degree of ligand stabilization. However, N108 had an unfavorable binding energy and its side chain slightly destabilized WR99210 binding in the case of quadruple mutant enzyme. The patterns of the per residue energy decomposition of pyrimethamine bound to wild type and quadruple mutant enzyme were found to be very different among themselves, especially at the residues positions14, 54, 55, 108 and 164. The per residue binding energies of these residues in wild type were reported to be favorable while those of quadruple mutant were unfavorable. The binding energy contribution of Asp54 of quadruple mutant was found to be > 40 kcal/mol which is highly unfavorable. The distances between the centers of masses of Asp54 and pyrimethamine as well as those between the centers of masses of Asn108 and pyrimethamine were observed throughout the simulation and it was found that there occurs dramatic change in the two distances during the simulation run. It was found that pyrimethamine cannot reside stably in the same location of the active-site of quadruple mutant in contrast to what is observed for wild type. These results provide fundamental information that could be useful for the design of new drugs that may overcome PYR resistance.

Pharmacophore Mapping and Discovery of *P. falciparum* DHFR Inhibitors

A pharmacophore is an important concept in rational drug design. This can be defined as the 3D arrangement of important chemical features that is required for a molecule to bind to a receptor in order to elicit biological activity [33, 119-121, 136, 147, 148]. The chemical features in 3D-pharmacophore models may include hydrophobic groups, charged/ionizable groups, hydrogen bond donors/acceptors, and any other groups properly assembled in 3D-orientation which can be recognized by the receptor, and interact favorably with it. Thus, their 3D-orientation is expected to be geometrically complementary to the receptor. Pharmacophore mapping is defined as the process of generation of 3D-pharmacophore models using computational methods. There are two commonly used ways to generate pharmacophore maps in rational drug design process: i) by analogy to a natural substrate or known ligand. This method is applied to a series of molecules which bind to the same receptor and when 3D-structure of the receptor is unknown or ii) by direct analysis of the 3D-structure of a target protein and its ligand binding domain. Having derived a pharmacophore model there are two generally used ways to identify molecules which share common features with the pharmacophore model, and might elicit a desired biological response. These are i) *de novo* design which seeks to link the disjoint parts of the pharmacophore together with fragments in order to generate hypothetical structures that are chemically reasonable yet typically wholly novel and ii) "3D-database searching" which involves use of large databases comprising 3D-structures of compounds. These databases are searched for those compounds that match to the pharmacophoric pattern of the model. 3D database searching is more advantageous than *de novo* design. This is because it allows the ready identification of existing molecules which are either easily available or have a known synthesis mechanism. Moreover, its speed made 3D-database searching to be a method of choice in drug design and discovery processes, and with proven success in practical applications [136]. Several reports also indicated that pharmacophore mapping or 3D-pharmacophore models to be used as a tool in designing anti-malarial inhibitors [34, 119, 149, 150].

Using Accelry's catalyst software, Adane *et al.* generated a shape and chemical feature based 3D-pharmacophore model by using bioactive conformation of WR99210 extracted from the X-ray crystal structure of quadruple mutant *Pf*DHFR [151]. The chemical features selected for the model generation were hydrogen bond donor, hydrophobic aromatic, hydrophobic aliphatic and positive ionizable groups in the bioactive conformation of WR99210. Using the same bioactive conformation, a shape based query was also generated and finally single 3D-pharmacophore model was developed by merging chemical feature based and shape based queries. (Fig. **7A**) The model was validated by the presence of important chemical features required for interaction with key amino acid residues (Asp54, Ile14, Leu164) in the active-site; by comparing its shape and features with that of the bioactive conformation of WR99210 extracted from quadruple mutant *Pf*DHFR enzyme (Fig. **7A**) and also by the ability of the model to pick known actives from in-house database of 134 compounds whose anti-malarial activities are reported in literature. The generated 3D-pharmacophore was then used as a query for virtually screening of Maybridge2004 database (59,652 compounds) and NCI2000 database (238,819 compounds) to identify potential hits. The obtained hits from each database were filtered based on Lipinski's rule-of-five and by restricting the number of rotatable bonds to < 7. Further screening was carried out by molecular docking analysis using Glide, GOLD and FlexX softwares. The results indicated that, scoring functions from FlexX and GOLD resulted in poor correlation ($R^2 < 0.500$) with the reported binding data of the compounds, whereas Gscore function of the Glide-XP program resulted in a comparatively better correlation ($R^2 = 0.790$). Thus the authors utilized Gscore scoring function of Glide-XP program to decide the final selection of hits from the virtual screening study. Glide-XP molecular docking studies identified 73 hits that showed similar binding mode and binding interactions to that of WR99210 in the active-sites of both wild type and quadruple mutant *Pf*DHFR. The binding pose of WR99210 in the active-site of quadruple mutant *Pf*DHFR is given in Fig. **7B**. The docking poses of these compounds indicated that there are no potential steric clashes with Asn108 of the mutant *Pf*DHFR. It has been reported that the hits have all the properties required by drug-like molecules. Hence these identified hits can be considered for further pharmacological studies to study their anti-malarial activities.

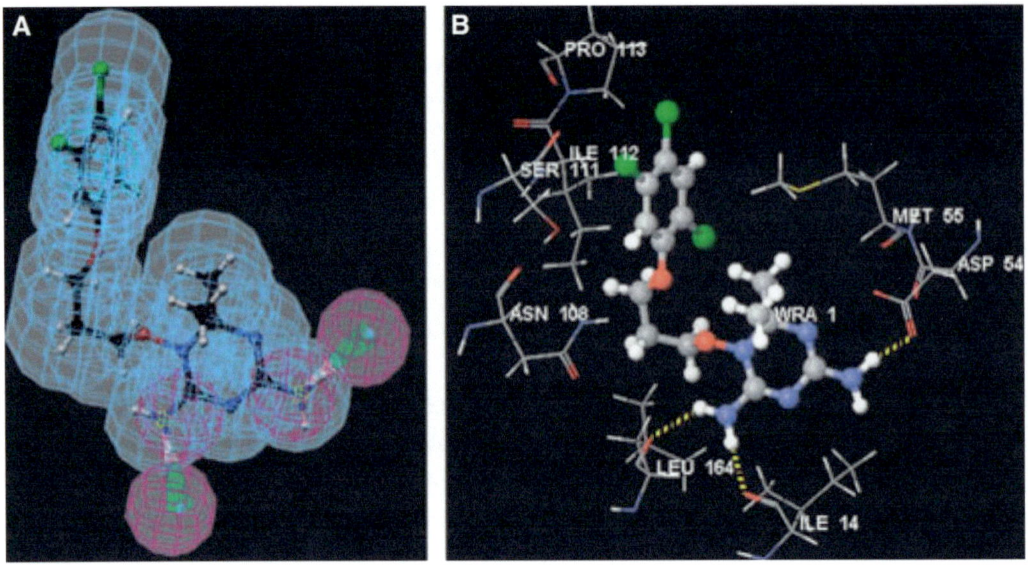

Figure 7: (A) The bioactive conformation of WR99210 mapped onto the 3D-pharmacophore model used in the study. **(B)** The docking pose of WR99210 in the active-site of quadruple mutant *Pf*DHFR. (For the sake of clarity, only important amino acid residues are given). (Reproduced from Ref. [151], from Wiley publishers).

Efforts have also been made to generate a common feature-based 3D-pharmacophore model to aid in identification of potential *Pf*DHFR inhibitors. Adane *et al.* developed a four-feature based 3D-pharmacophore model from a set of 24 compounds whose activities were reported against the V1/S strain of the *Plasmodium falciparum* dihydrofolate reductase (*Pf*DHFR) enzyme [152]. This enzyme is reported to have Asn51Ile + Cys59Arg + Ser108Asn + Ile164Leu mutations. The data set of compounds consisted of 2,4-diaminoquinazoline, 2,4-diamino 5,6,7,8-teterahydroquinazoline, 2,4-diamonopteridine derivatives. The HipHop module of the Catalyst program was used to generate the pharmacophore model. Best model among the generated hypotheses by HipHop was selected on the basis of rank and best-fit values. The best model (hypo1) selected by the authors consisted of two H-bond donors, one hydrophobic aromatic, and one hydrophobic aliphatic features. Hypo1 was then used as a query to virtually screen Maybridge2004 and NCI2000 databases. The hits obtained from the search were subsequently subjected to FlexX and Glide docking studies. A set of nine hits were identified as potential *Pf*DHFR inhibitors on the basis of binding scores and interactions in the active-site of quadruple mutant *Pf*DHFR. The identified hits showed all the necessary interactions required by antifolate *Pf*DHFR inhibitors to bind in the active-site of an enzyme. The hydrogen bond donor features interacted with Asp54 and Ile14 or/and Leu164, whereas the hydrophobic portions were reported to interact with amino acid residues Met55, Ser111, Pro113. Some of the hits also showed H-bond interaction with Asn108. Their binding modes also indicated that there is no unfavorable steric clash with Asn108 (mutated from Ser108). Hence this study revealed novel compounds which could be potential *Pf*DHFR inhibitors.

QUANTUM CHEMICAL STUDIES OF DHFR ENZYME-ANTIFOLATE INTERACTIONS

Computational analyses including quantum mechanical calculations are usually carried out to predict mechanism of interaction of drugs in the active-sites of target proteins [153]. Moreover, the results of these studies also serve as supportive evidences for some experimental observations such as NMR and crystallographic studies. For instance, protonation and hydride transfer are important reactions in the catalytic reduction of 7,8-dihydrofolate (DHF) to 5,6,7,8-tetrahydrofolate (THF) by DHFR enzymes [2, 154, 155]. Computational approaches have been used to predict plausible mechanism of this reduction reaction, and also to understand and predict drug resistance by pathogens [156]. The electronic structure information obtained from *ab initio* MO/DFT methods is quite useful in understanding the internal as well as surface properties of drugs or/and leads [157].

Quantum chemical calculations have been employed to predict the mechanism of catalytic reduction of DHFR enzyme. Most of the studies are done on catalytic reduction mechanism of DHFR enzymes of other species (other than *P. falciparum* DHFR). However, because of similarity in active-site of *P. falciparum* DHFR enzyme with the corresponding active-site of DHFR enzymes of other species such as *E. coli* and *L. casei*, understanding of these discussions would be helpful in predicting the interaction of newly designed antifolate drug candidates with *P. falciparum* DHFR enzyme. The reports published in the past decades pointed out that there are two steps in these enzyme-catalyzed reductions. The first step involves protonation of the substrate whereas the second step involves transfer of hydride ion from NADPH to DHF [154, 158]. Polshakov *et al.* [159] employed a quantum chemical semi-empirical method, PM3 SCF MO, in order to refine NMR-determined complex of antifolate drugs and *L. casei* DHFR enzyme particularly interaction of protonated nitrogen (N_1H) of the antifolate and carboxylate groups of the conserved acid residue, Asp26, of the enzyme in solution. The authors observed that increasing the number of water molecules in the calculations favors the ionization state interaction in the antifolate-DHFR enzyme complex models [159].

Polshakov *et al.* [160] also carried out QM calculations at Hartree-Fock (HF/3-21G) and B3LYP/6-311G levels in order to investigate the origin of the differences in the binding interactions of NADPH and $NADP^+$ in the active-site of *L. casei* DHFR enzyme. The QM calculations demonstrated the variation in H-bond strength of binding amino acid residues at the active-site of DHFR bound to NADPH and $NADP^+$. Cannon *et al.* [161] carried out Poison-Boltzman electrostatic and quantum chemical (HF/3-21G and HF/6-31G) vibrational frequency calculations in order to evaluate the mechanism of enzyme assisted-catalysis of DHFR. The authors proposed a pH-dependent hydride transfer mechanism which is consistent with experimental observations. Cummins *et al.* [162] in their attempt to describe details of catalytic mechanism of DHFR enzyme, carried out QM/MM and *ab initio* SCRF calculations to demonstrate the presence of polar low-barrier H-bonding (LBHB) which is presumed to be essential to activate C-7 of the pterine ring for transfer of hydride ion from NADPH during DHFR-catalyzed reduction of folates. The results obtained from these calculations that involve use of hybrid QM/MM model, including solvent effects and free energy perturbation (FEP/MDS), confirmed the existence of an LBHB in antifolate-DHFR enzyme complex [163]. Cummins and Greedy [155] carried out *ab initio* quantum chemical calculations (B3LYP, HF and MP2 levels) in order to investigate plausible mechanism of reduction of DHF to THF. Cummins and Gready [164] also reported a new method known as overlapping mechanically embedded (OME) method for hybrid semi-empirical-QM/MM calculations in order to overcome some of the limitations prevailed in semi-empirical QM (AM1 or PM3) level calculations in describing enzyme-catalyzed reduction of DHF to THF. Viloca *et al.* [165] employed a combined QM and MM potential in molecular dynamic (MD) computations in order to determine the effect of electric field of the enzyme on molecular polarization of cofactor, NADPH and substrate at various steps of chemical transformations. Pang *et al.* [166] from investigation of *Thermotoga maritime* DHFR and using QM/MM calculations at different temperatures, and ensemble-averaged variational transition state theory with multidimensional tunneling (EA-VTST/MT) found that QM tunneling has a significant contribution in influencing the rate of hydrogen transfer during the enzymatic catalyzed reaction. Quantum chemical methods have also been employed to understand the structure of *Pf*DHFR inhibitors. In this section, without going to the details of computational methods, we present some studies on electronic structure analysis of *Pf*DHFR inhibitors.

All the known antifolate drugs such as pyrimethamine, cycloguanil and proguanil (prodrug of cycloguanil) and the lead compound WR99210 possess biguanide moiety. Biguanides (carbamimidoylguanidines) are neutral species which are characterized by 4π electron conjugation, an intramolecular hydrogen bond, and resonance stabilization due to low 1,5-H shift barriers. It has been reported that that these biguanide derivatives adopt a tautomeric state in which an H atom is absent at the central nitrogen [157, 167]. Patel *et al.* carried out quantum chemical studies to explore the Divalent N(I) character in these species. Divalent N(I) systems are characterized by low oxidation state N(I), two lone pairs on the central nitrogen, and nitrogen involved in coordination bonds with electron donating groups. All biguanide drugs/leads are reported to act in their protonated state (for oral bioavailability). The authors revealed that the preferred site of protonation in biguanides is N6 rather than the central nitrogen N1 in these species. Reported *ab initio* MO and DFT calculations indicated that charge localization at nitrogen, accumulation of two lone pairs at the central nitrogen, increased flexibility due to easy C–N rotations, and N-inversion process are the advantages experienced by the biguanide derivatives during the protonation at N6 [168]. MO analysis of cyclic biguanides indicated that two lone pairs on central nitrogen are characterized as σ (HOMO-3) and π (HOMO) type of orbitals. NBO analysis also supported the presence of two lone pairs on the central nitrogen that occupy σ-and π-type orbitals with electron occupancies of 1.91 and 1.51, respectively. This electronic structure analysis indicated that the bridging nitrogen (N1) in protonated cycloguanil can be considered as a divalent N(I) system. Even though the protonated biguanides contain two lone pairs of electrons on N1, they are weakly nucleophilic as they carry positive charge. The authors suggest that this property of protonated biguanides may help in proton exchange in body fluids while remaining mildly active. Hence the authors report that biguanides (cycloguanil, pyrimethamine and WR99210) have divalent N(I) character.

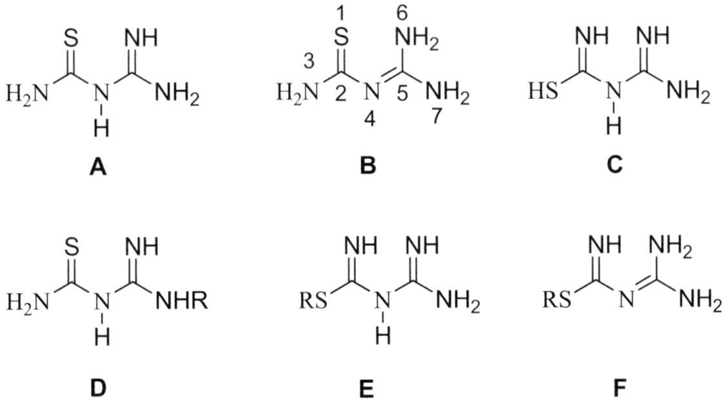

Figure 8: Possible structural representations of Guanylthiourea (**A-C**) and its alkylated derivatives (**D-F**).

Guanylthiourea derivatives have recently been reported as potential *Pf*DHFR inhibitors [144]. In an effort to understand the appropriate structural and electronic representations of guanylthiourea, Mehdi *et al.* carried out electronic structure analysis of guanylthiourea (GTU) and its isomers using quantum chemical methods [169]. Possible structural representations of GTU and its alkylated derivatives as reported by the authors and previous works on GTU are given in (Fig. **8**). Previous reports on potential DHFR inhibitor compound with two guanylthiourea moieties represented its structure as D (Fig. **8**) [170] which was later corrected by Summerfield *et al.* to E (Fig. **8**) [171]. In the present study the authors have identified F to be the preferred structure of such inhibitor. The detailed analysis carried out using G2MP2 method revealed that out of five possible thione isomers of GTU the most stable isomer was found to have no hydrogen atom on the central nitrogen (N4); was characterized by intramolecular H-bonding interaction (S... H-N, 2.205 Å) and conjugating π-electron system. The authors reported that the presence of hydrogen atom on N4 was found to destabilize structures due to lack of conjugative delocalization of π-electron density. Further, NBO analysis, at MP2(full)/6-31+G* level, employed to estimate second-order interactions in thione form of GTU indicated large delocalization energy values. This observation again indicated the significance of lone pair delocalization on the π-electron system in the stabilization. High rotational barriers of 8.54, 11.3, and 17.72 kcal/mol (G2MP2 level) across C5-N7, C5-N6, and C2-N4 bonds, respectively in GTU also indicated the presence of strong electron delocalization which provides the partial double-bond characters to the above considered bonds. Thiol based isomers are also possible stable structures on the PE surface. The most stable thiol isomer was also characterized by intramolecular interaction and conjugating π-electron system, but this isomer was found to be less stable by 11.3 kcal/mol compared to the most stable thione isomer. The authors reported that the interconversion from the most stable thione conformer to the most stable thiol conformer was found to take place *via* bimolecular process which involves protonation at sulfur atom of GTU followed by subsequent C-N bond rotation and deprotonation. The preferred protonation sites for both isomers, and their alkylated derivatives in gaseous and aqueous conditions were also investigated in this study. In thione isomers the terminal atoms S1 and N3 and in thiol isomers N1 and N4 were found to be preferred sites of protonation. Protonated biguanides have been reported as a novel $^{\oplus}N(\leftarrow L)_2$ species which showed new bonding environment for nitrogen with two lone pairs on nitrogen [168]. The central core of biguanide and guanylthiourea are similar, hence the authors investigated this character in the protonated GTU. S1 protonated GTU are expected to fall into $^{\oplus}N(\leftarrow L)_2$ configuration. Most of the organic reactions in GTU involve alkyl cation attack on S atom followed by subsequent C-N bond rotation and deprotonation. The intermediates in such reactions were characterized by two lone pairs of electrons on N4 which facilitates alkylation reactions. The formation of alkylated analogs of thiol isomers (alkylated guanylthiourea) is believed to take place *via* bimolecular process which involves alkyl cation attack at S atom followed by C-N bond rotation and deprotonation. The reactive intermediate $RS(NH_2)C\text{-}N\text{-}C(NH_2)_2^+$ belongs to the newly identified $^{\oplus}N(\leftarrow L)_2$ class of species and provides the necessary dynamism for easy conversion of thione to thiol.

Proguanil is an anti-malarial prodrug which is reported to undergo cytochrome P450 catalyzed bioactivation to a pharmacologically active triazine metabolite (cycloguanil), which ultimately leads to *Pf*DHFR inhibition [172]. CYP2C19 is reported to catalyze this cyclization process but the details of this pathway are still unanswered. Recently, quantum chemical studies have been carried out to elucidate the mechanism of this cyclization process. The reaction pathways for the metabolism of model compound (N4-isopropyl-N6-methylbiguanide for proguanil) were explored under non-enzymatic conditions as well as under enzymatic conditions. Under non-enzymatic conditions, overall endoergic value for the model reaction was found to be 13.89 kcal/mol. Alternate path involving the initial formation of hydroxyl intermediate involved a barrier of 5.67 kcal/mol for the cyclization of a triazatriene moiety. The potential energy surface of radical and hydroxyl pathways were also explored under enzymatic conditions using Cpd I as a model for cytochrome. It was found that the role of CYP was mainly in the generation of radical *via* H radical abstraction from the isopropyl group. Overall the reaction under enzymatic conditions was reported to be exothermic by ~55 kcal/mol, which involves a barrier of ~13 kcal/mol and also loss of a water molecule. Hence, this study provides useful information to understand the molecular mechanism for the metabolic conversion of isopropyl biguanide based antimalarial prodrugs to their cyclic active metabolites under cytochrome mediated condition [173].

STRUCTURE BASED DRUG DESIGN APPROACH

Drug design involves the design of small molecules (drugs) that are complementary in shape and charge to the biomolecular target with which they interact and therefore will bind to it. The drug design process that relies on the knowledge of the three-dimensional structure of the biomolecular target is known as structure-based drug design. Using the structure of the biological target, candidate drugs that are predicted to bind with high affinity and selectivity to the target may be designed. Current methods for structure-based drug design can be divided roughly into two categories. Till date many X-ray crystal structures of *Pf*DHFR complexed with its inhibitors and substrate are available. Yuthavong *et al.* utilized X-ray structures of *Pf*DHFR and human DHFR co-crystallized with substrates and inhibitors in the design of a selective, highly efficacious antimalarial agent [83]. The authors studied that the petridine ring of DHF interacts with Ile14, Asp54 and Ile164 and on the opposite end its carboxylic group interacts with Arg122. Based on this information they designed 14 novel *Pf*DHFR inhibitors which may bind in a similar fashion as that of DHF in the active-site *Pf*DHFR. Out of these designed compounds, P218 (**12**, Fig. **1**) was reported to have significant inhibitory potency for wild type and mutant *Pf*DHFR. P218 was designed such that its diaminopyrimidine ring interacts with Ile14, Asp45 and Ile164 and its terminal carboxyl group interacts with Arg122 and fits rigidly in the active-site of *Pf*DHFR. Due to flexible linker it can also avoid steric hindrance due to S108N mutation and also it was found to bind in a similar fashion as that of natural substrate. The authors also studied the specificity of this inhibitor for *Pf*DHFR by studying the inhibitory potency of P218 for human DHFR (*h*DHFR). The results suggested that P218 inhibition for *h*DHFR is significantly lower than *Pf*DHFR. The calculated binding constant values for wild type *Pf*DHFR (0.51 ± 0.032); quadruple mutant *Pf*DHFR (0.53 ± 0.13) and *h*DHFR (2,841 ± 319) reveal the binding affinity difference among species. The authors also solved crystal structures of P218 bound to quadruple mutant *Pf*DHFR and *h*DHFR. They found that due to substitution of Met55, Cys/Arg59, and Phe116 in *Pf*DHFR by Phe31, Gln35, and Asn64 at structurally equivalent positions in human DHFR there exist a difference in the active-site of these species. This change in active-site pocket alters P218 binding to *h*DHFR such that carboxylic group of P218 does not interact with *h*DHFR Arg70 (structural homolog of Arg122) which leads to poor binding of P218. Authors report that P218 not only has potent *in vitro* and *in vivo* antimalarial activity against wild type and pyrimethamine-resistant *Pf*DHFR, but it also possesses suitable pharmacological, metabolic, and safety profiles for further consideration towards clinical development.

PRACTICAL SOLUTION ARISING FROM PHARMACOINFORMATIC APPROACH

Many of the above modeling studies have been carried out to understand the mechanism of resistance *P. falciparum* parasite to type-2 antifolate drugs, and mechanism of interactions of antifolates with the amino acid residues of active-site of DHFR enzyme. The results of most of these studies are found to be consistent with experimental observations, and being used to design lead compounds with promising inhibitory activities against *P. falciparum* DHFR enzyme. Recently, Guanylthiourea (GTU) derivatives are reported to be novel class of compounds as *Pf*DHFR inhibitors. In this section, practical application of the results of molecular docking studies in the design of GTU derivatives will be discussed.

Antifolates are type-2 anti-malarial drugs which included pyrimethamine, proguanil and cycloguanil. The molecular modeling studies of these marketed drugs and reference ligand (WR99210) pointed out the main structural requirements of the ligands which are essential for the activity. These include (i) H-bond donor head group that can form H-bond interaction with Asp54, Ile14 and Leu164, (ii) hydrophobic aromatic tail which occupies the hydrophobic pocket of the active-site (Phe58, Met55, Phe116, Pro113, Ile112 and Ser111) to enhance inhibitory activity, and (iii) linker unit between the H-bond donor head groups and hydrophobic aromatic tail to provide flexibility, in order to avoid unfavorable steric clashes with Asn108 in the active-site of the mutant *Pf*DHFR enzyme. This hypothesis was proved by Dasgupta *et al.* who showed the activity of biguanide derivatives as potential inhibitors of *Pf*DHFR enzyme which bind in the active cavity of the enzyme showing similar interactions as that of other ligands. Subsequent studies on DHFR enzyme of *E. coli* showed that S-substituted guanylthiourea derivatives inhibit this enzyme.

This led to a further study on this type of molecules for their activity on *Pf*DHFR enzyme. Molecular electrostatic potential studies were thus carried out on the guanylthiourea moiety and the results were compared with that of pyrimidine moiety of pyrimethamine. These studies showed that under protonated condition the surface of pyrimethamine and GTU are similar in nature with hydrogen bond donor and acceptor regions at relative positions in the two structures. These results proved that the GTU moiety has the potential to bind in the active-site showing similar interactions that are shown by pyrimethamine.

Further docking studies were carried out on the several designed GTU derivatives to study their interactions in the active-site of *Pf*DHFR enzyme (Fig. **9**). To study the binding poses and docking scores of the designed GTU derivatives Glide docking program was used. The docking calculations were carried out using X-ray crystal structures of wild type (PDB code: 1J3I) and quadruple mutant type (PDB code: 1J3K) *Pf*DHFR enzymes with bound reference ligand WR99210. The crystal structures are in the dimeric form of the DHFR-TS complex and they are co-crystals with cofactors and ligands. During the protein preparation step, only chain A (DHFR) with cofactor NDP610 was retained; all water molecules and the rest of the chains were removed. A radius of 15 Å was selected for active-site cavity during receptor grid generation. The reproducibility of the docking calculation was evaluated by docking the bound reference ligand into the prepared active-sites. The results were observed keeping in view both the docking scores and desired interactions.

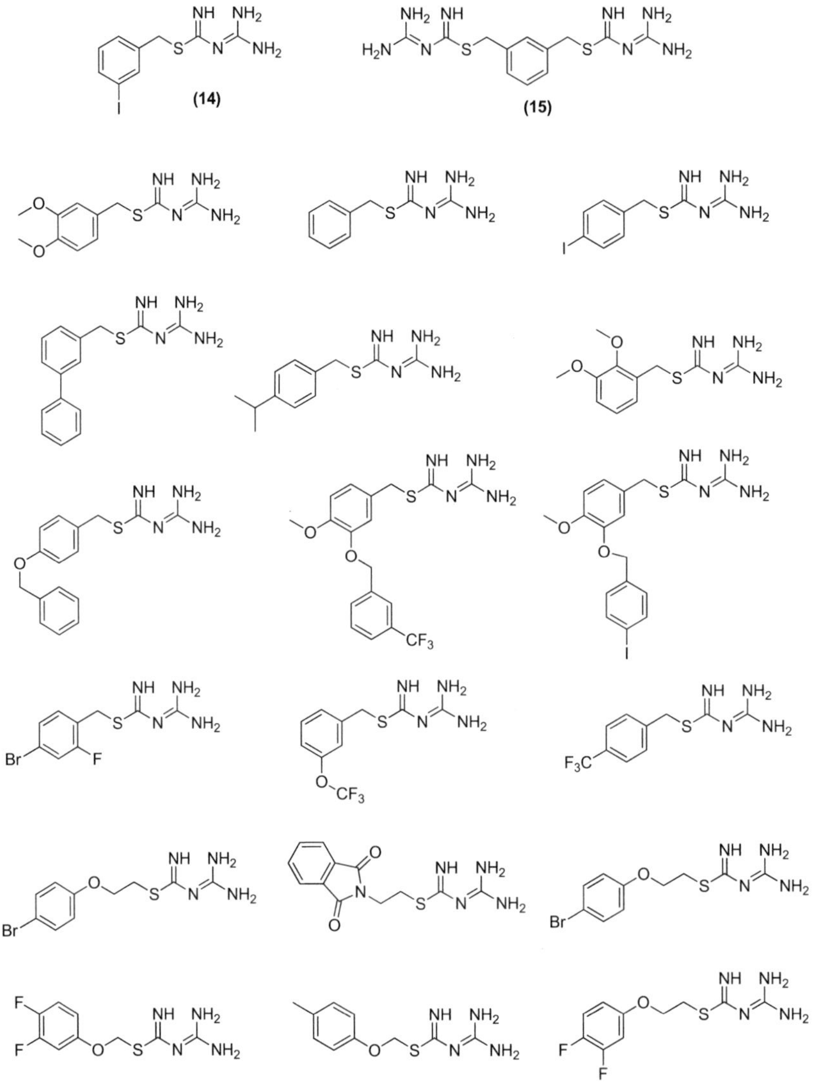

Figure 9: Newly identified derivatives of guanylthiourea from docking studies with promising antimalarial activity against both wild type and mutant *P. falciparum* DHFR enzyme.

Analysis of the docking result proved following points: First, the interactions shown by S-substituted GTU derivatives are similar to the interaction shown by other drugs and ligands already known to be potential inhibitors of the enzyme (*i.e.* pyrimethamine, WR99210 and biguanide derivatives). GTU moiety was found to form the expected hydrogen bond interactions with Asp54, Leu164, Ile14 and hydrophobic interactions with Phe58 and Phe116 in the active-site of *Pf*DHFR enzymes. Second, introducing a linker unit with 1, 2 or 3 carbon atoms between the hydrophobic aromatic tail and GTU moiety provided sufficient flexibility in order to prevent potential steric clashes with the amino acid residue Asn108 in the active-site of the mutant *Pf*DHFR enzyme. Third, introducing a two carbon chain acidic group at the *ortho*-position of the aromatic ring in the hydrophobic region of the active-site increases the docking score and improves the interactions by binding with basic amino acids like Arg122 and Arg59.

Molecular docking interactions suggest that the aromatic ring in the tail region show π-π interactions with Phe58 and in some cases with Phe116. Bi-substituted guanylthiourea structure (compd. **15**) showed Gscore (-8.754; PDBID: 1J3K) comparable to that of WR99210 (-8.708 in 1J3K, mutant type) and showed additional interactions with Ser111, Lys49 and Trp48. These additional interactions are possible only because of the six rotatable bonds between the two guanylthiourea moieties, facilitating stronger interactions.

In vitro studies on crude enzyme showed two compounds (compd. **14** and **15**) could inhibit *Pf*DHFR enzyme at μM range with IC_{50} values ~100 μM and 400 nm respectively. Thus, pharmacoinformatics and *in vitro* analysis suggested that derivatives with bulky group at the *meta*-position of the phenyl ring has efficacy against the *Pf*DHFR enzyme. These studies also indicated that bi-substituted GTU derivative in which the phenyl ring is present in the linker region can be developed further to improve the activity against *Pf*DHFR enzyme.

CONCLUSION AND FUTURE PERSPECTIVES

Inhibition of *Plasmodium falciparum* DHFR enzyme is one of the important strategies in antimalarial drug design. Antifolate drugs such as pyrimethamine and cycloguanil are known to inhibit this enzyme in its wild form. Due to amino acid mutations at specific positions, *Pf*DHFR has acquired resistance to these antifolate drugs. Structural information of this enzyme in its wild type and mutant forms (double and quadruple mutants) provides crucial details regarding antifolate drug resistance acquired by the enzyme. The 3D-structures of Enzyme-Substrate and Enzyme-Inhibitor complexes provide details regarding key molecular level interactions which are extremely useful in elucidating the mechanism of action. Mutations at positions 16, 51, 59, 108 and 164 accounts for the *P. falciparum* resistance to antifolate drugs, and the mutations are known to have cumulative effect *i.e.* resistance increases with number of mutations. Many *Pf*DHFR inhibitors have been designed based on the known structural information of *Pf*DHFR. For instance, P218 (a potent antimalarial agent, currently in preclinical trials) has been designed by carefully investigating the binding mode of dihydrofolate (DHF, natural substrate) in the active-site of *Pf*DHFR. This molecule is reported be specific for *Pf*DHFR; have sub-nanomolar inhibition for both wild type and mutant forms of enzyme and its binding mode in the active-site of *Pf*DHFR is similar to that of the natural substrate. Due to similar binding fashion of P218 and DHF, P218 may face less resistance as compared to other antifolates. Such drug design strategies based on structural information may be of immense use in future.

Moreover, several potent DHFR inhibitors have been designed based on pharmacophore mapping and virtual screening followed by molecular docking and molecular dynamics. Quantitative structure-activity relationship (QSAR) models have also been utilized to understand important structural/pharmacophoric features required for the enzyme inhibition, this information has further been used in designing (and screening) *P. falciparum* DHFR inhibitors. Molecular docking methods are being used to get insights about binding modes and binding affinities of *P. falciparum* DHFR inhibitors towards active-site. Several compounds have been identified as potent *P. falciparum* DHFR inhibitors based on their docking score (binding affinity values) and favorable interactions in the active-site of the enzyme. For instance, guanylthiourea derivatives have been identified as *Pf*DHFR inhibitors using this approach and further *in vitro* analysis indeed proved these compounds as potential *Pf*DHFR inhibitors. Molecular dynamics approach is being utilized to analyze the binding mode and affinity of ligand towards active-site of an enzyme in dynamic phase. For instance, molecular dynamics analysis of pyrimethamine and WR99210 bound to wild type and mutant *Pf*DHFR enzyme revealed important structural details which cause decreased affinity of pyrimethamine towards quadruple mutant enzyme as compared to its wild form. Pharmacophore mapping approach has also been widely utilized in designing several *P. falciparum* DHFR inhibitors. For instance, this method was used to retrieve compounds-from NCI and Maybridge databases showing promising inhibitory activities (as found by molecular docking analysis).

Quantum chemical calculations have been employed to get detailed electronic structural information on catalytic reduction mechanisms of dihydrofolate to tetrahydrofolate reduction and also to study interactions of antifolates with amino acid residues in the active-site. Based on this method it has been found that catalytic reduction process involves the activation of substrate (DHF) by protonation in the active-site of *Pf*DHFR followed by hydride ion transfer from cofactor (NADPH) to the activated substrate. Moreover, catalytic roles of water molecule in the active-site, electric field (explicit polarization) and quantum mechanical tunneling process have been well understood from these calculations. The antifolate drugs (pyrimethamine and cycloguanil) and lead compound (WR99210) are known to be biguanides and all of them act in their protonated state. Quantum chemical calculations have been employed to characterize N(I) divalent character in biguanides and the study also revealed that even though these compounds carry two lone pairs they are weakly nucleophilic as they carry positive charge. Quantum chemical studies are also being done to elucidate the reaction mechanism of proguanil (prodrug of cycloguanil) to cycloguanil conversion under non enzymatic and enzymatic conditions to understand the action of cytochromes in the metabolism of this important antimalarial drug. In future, the above mentioned pharmacoinformatic approaches towards drug discovery are expected to be even more effective in the design of potential and selective antimalarial agents due to the advent of genomic and proteomic tools. Modeling tools based on molecular dynamics and QM/MM methods need to be explored further to design new molecules with confidence gained from these powerful methods. These pharmacoinformatic tools have proven to be quite helpful in the effective design of new anti-malarial agents, but there still lies scope for extensive exploration of existing tools and also to design new tools that may help in drug discovery process.

ACKNOWLEDGEMENTS

The University Grants Commission is greatly acknowledged for the SRF fellowship to Sheenu Abbat. Shweta Bhagat thanks Department of Biotechnology for JRF fellowship. PVB thanks Department of Biotechnology (DBT) for financial assistance in setting up Bioinformatics Infrastructure Facility (BIF) at NIPER.

CONFLICT OF INTEREST

The authors confirm that this chapter contents have no conflict of interest.

DISCLOSURE

This is an updated report of the earlier review article by P. V. Bharatam "Modelling and Informatics in the Analysis of *P. falciparum* DHFR Enzyme Inhibitors (review)". Curr Med Chem 2008; 15(16): 1552-1569.

ABBREVIATIONS

3D-QSAR	=	Three-dimensional quantitative structure-activity relationship.
4D-QSAR	=	Four-dimensional quantitative structure-activity relationship.
5D-QSAR	=	Five-dimensional quantitative structure-activity relationship.
A16V	=	Mutation of alanine amino acid to valine amino acid residue at position 16 in the active-site of *P. falciparum* DHFR enzyme.
ACD	=	Available chemical directory.
Asp54 (D54)	=	Aspartic acid residue at position 54 in the active-site of *P. falciparum* DHFR enzyme
C59R	=	Mutation of cysteine amino acid residue to arginine amino acid residue at position 59 in the active-site of *P. falciparum* DHFR enzyme.
CADD	=	Computer-aided drug design.
DHF	=	Dihydrofolate.
DHFR	=	Dihydrofolate reductase.
FEEF3D-QSAR	=	Free energy force field three-dimensional quantitative structure-activity relationship.
FEP	=	Free energy perturbation.
GTU	=	Guanylthiourea
I164L	=	Mutation of isoleucine to leucine amino acid residue at position 164 in the active-site in the active-site of *P. falciparum* DHFR enzyme.
LBHB	=	Low barrier hydrogen bonding.
MM	=	Molecular mechanics.

N51I	=	Mutation of asparagine amino acid to isoleucine amino acid residue at position 51 in the active-site of *P. falciparum* DHFR enzyme.
NADPH	=	Nicotine adenine dinucleotide phosphate
MO/DFT	=	Molecular Orbital/Density functional theory.
OME	=	Overlapping mechanically embedded method.
QM/MM	=	Hybrid model of quantum mechanics and molecular mechanics.
QM	=	Quantum mechanics.
S108N	=	mutation of serine amino acid residue to asparagine amino acid residue at position 108 in the active-site of *P. falciparum* DHFR enzyme.
S108T	=	Mutation of serine amino acid to threonine amino acid residue at position 108 in the active-site of *P. falciparum* DHFR enzyme.
SCRF	=	Solvent continuum reaction field.
THF	=	Tetrahydrofolate.
TS	=	Thymidylate synthase.

REFERENCES

[1] Bloland PB. Drug resistance in malaria. Switzerland: World Health Organization Geneva; 2001.
[2] Delfino RT, Santos-Filho OA, Figueroa-Villar JD. Type 2 antifolates in the chemotherapy of falciparum malaria. J Braz Chem Soc 2002; 13(6): 727-41.
[3] Fidock DA, Rosenthal PJ, Croft SL, Brun R, Nwaka S. Antimalarial drug discovery: efficacy models for compound screening. Nat Rev Drug Discovery 2004; 3(6): 509-20.
[4] Gregson A, Plowe CV. Mechanisms of resistance of malaria parasites to antifolates. Pharmacol Rev 2005; 57(1): 117-45.
[5] Jain R. Recent advancements in antimalarial drug development. CRIPS 2002; 3: 2-8.
[6] Olliaro P. Drug resistance hampers our capacity to roll back malaria. Clin Infect Dis 2005; 41(Supplement 4): S247-S57.
[7] Phillips R. Current status of malaria and potential for control. Clin Microbiol Rev 2001; 14(1): 208-26.
[8] Sharma Y. Genetic alteration in drug resistance markers of Plasmodium falciparum. Indian J Med Res 2005; 121(1): 13-22.
[9] Talisuna AO, Bloland P, D'Alessandro U. History, dynamics, and public health importance of malaria parasite resistance. Clin Microbiol Rev 2004; 17(1): 235-54.

[10] Winstanley P, Ward S, Snow R, Breckenridge A. Therapy of falciparum malaria in sub-saharan Africa: from molecule to policy. Clin Microbiol Rev 2004; 17(3): 612-37.

[11] Organization WH. Malaria and HIV/AIDS interactions and implications: conclusions of a technical consultation convened by WHO, 23-25 June 2004. 2004.

[12] Organization WH. World Malaria Report 2012. Geneva: World Health Organization; 2012. 2013.

[13] Kalra BS, Chawla S, Gupta P, Valecha N. Screening of antimalarial drugs: An overview. Indian J Pharmacol 2006; 38(1): 5-12.

[14] Nzila A. Inhibitors of *de novo* folate enzymes in *Plasmodium falciparum*. Drug Discov Today 2006; 11(19): 939-44.

[15] Yeo A, Rieckmann K. The activity of triple combinations of antifolate biguanides, with and without folinic acid, against Plasmodium falciparum *in vitro*. Ann Trop Med Parasitol 1997; 91(3): 247-52.

[16] Yeo AE, Seymour KK, Rieckmann KH, Christopherson RI. Effects of dual combinations of antifolates with atovaquone or dapsone on nucleotide levels in *Plasmodium falciparum*. Biochem Pharmacol 1997; 53(7): 943-50.

[17] Egan T. Haemozoin formation as a target for the rational design of new antimalarials. Drug Design Reviews-Online 2004; 1(1): 93-110.

[18] Egan TJ. Haemozoin (malaria pigment): a unique crystalline drug target. Targets 2003; 2(3): 115-24.

[19] Macreadie I, Ginsburg H, Sirawaraporn W, Tilley L. Antimalarial drug development and new targets. Parasitol Today 2000; 16(10): 438-44.

[20] Olliaro PL, Yuthavong Y. An overview of chemotherapeutic targets for antimalarial drug discovery. Pharmacol Ther 1999; 81(2): 91-110.

[21] Gangjee A, Yang J, McGuire JJ, Kisliuk RL. Synthesis and evaluation of a classical 2, 4-diamino-5-substituted-furo [2, 3-d] pyrimidine and a 2-amino-4-oxo-6-substituted-pyrrolo [2, 3-d] pyrimidine as antifolates. Bioorg Med Chem 2006; 14(24): 8590-8.

[22] Ren H, Grady S, Gamenara D, Heinzen H, Moyna P, Croft SL, *et al.* Design, synthesis, and biological evaluation of a series of simple and novel potential antimalarial compounds. Bioorg Med Chem Lett 2001; 11(14): 1851-4.

[23] Rosenthal PJ. Antimalarial drug discovery: old and new approaches. J Exp Biol 2003; 206(21): 3735-44.

[24] Frühbeis H, Klein R, Wallmeier H. Computer-Assisted Molecular Design (CAMD)—An Overview. Angew Chem Int Ed Engl 1987; 26(5): 403-18.

[25] Jorgensen WL. The many roles of computation in drug discovery. Science 2004; 303(5665): 1813-8.

[26] Portela C, Afonso CM, Pinto MM, Ramos MJ. Computational studies of new potential antimalarial compounds-Stereoelectronic complementarity with the receptor. J Comput Aided Mol Des 2003; 17(9): 583-95.

[27] Yeh I, Hanekamp T, Tsoka S, Karp PD, Altman RB. Computational analysis of Plasmodium falciparum metabolism: organizing genomic information to facilitate drug discovery. Genome Res 2004; 14(5): 917-24.

[28] Cavalli A, Carloni P, Recanatini M. Target-related applications of first principles quantum chemical methods in drug design. Chem Rev 2006; 106(9): 3497-519.

[29] Cramer RD, Patterson DE, Bunce JD. Comparative molecular field analysis (CoMFA). 1. Effect of shape on binding of steroids to carrier proteins. J Am Chem Soc 1988; 110(18): 5959-67.

[30] Green SM, Marshall GR. 3D-QSAR: a current perspective. Trends Pharmacol Sci 1995; 16(9): 285-91.

[31] Winkler DA. The role of quantitative structure-activity relationships (QSAR) in biomolecular discovery. Brief Bioinform 2002; 3(1): 73-86.

[32] Klebe G, Abraham U, Mietzner T. Molecular similarity indices in a comparative analysis (CoMSIA) of drug molecules to correlate and predict their biological activity. J Med Chem 1994; 37(24): 4130-46.

[33] Marriott DP, Dougall IG, Meghani P, Liu Y-J, Flower DR. Lead generation using pharmacophore mapping and three-dimensional database searching: application to muscarinic M3 receptor antagonists. J Med Chem 1999; 42(17): 3210-6.

[34] Rastelli G, Pacchioni S, Sirawaraporn W, Sirawaraporn R, Parenti MD, Ferrari AM. Docking and Database Screening Reveal New Classes of Plasmodium falciparum Dihydrofolate Reductase Inhibitors. J Med Chem 2003; 46(14): 2834-45.

[35] Richmond NJ, Abrams CA, Wolohan PR, Abrahamian E, Willett P, Clark RD. GALAHAD: 1. Pharmacophore identification by hypermolecular alignment of ligands in 3D. J Comput Aided Mol Des 2006; 20(9): 567-87.
[36] Anderson AC, Wright DL. The design and docking of virtual compound libraries to structures of drug targets. Curr Comput Aided Drug Des 2005; 1(1): 103-27.
[37] Kitchen DB, Decornez H, Furr JR, Bajorath J. Docking and scoring in virtual screening for drug discovery: methods and applications. Nat Rev Drug Discovery 2004; 3(11): 935-49.
[38] Valler MJ, Green D. Diversity screening *versus* focussed screening in drug discovery. Drug Discov Today 2000; 5(7): 286-93.
[39] Mokmak W, Chunsrivirot S, Hannongbua S, Yuthavong Y, Tongsima S, Kamchonwongpaisan S. Molecular dynamics of interactions between rigid and flexible antifolates and dihydrofolate reductase from pyrimethamine-sensitive and-resistant Plasmodium falciparum. Chem Biol Drug Des 2014.
[40] Rastelli G, Sirawaraporn W, Sompornpisut P, Vilaivan T, Kamchonwongpaisan S, Quarrell R, *et al.* Interaction of pyrimethamine, cycloguanil, WR99210 and their analogues with *Plasmodium falciparum* dihydrofolate reductase: structural basis of antifolate resistance. Bioorg Med Chem 2000; 8(5): 1117-28.
[41] Kapoor M, MUKHI P, Surolia N, Suguna K, Surolia A. Kinetic and structural analysis of the increased affinity of enoyl-ACP (acyl-carrier protein) reductase for triclosan in the presence of NAD+. Biochem J 2004; 381: 725-33.
[42] Miao J, Hodgson KO, Sayre D. An approach to three-dimensional structures of biomolecules by using single-molecule diffraction images. Proc Natl Acad Sci 2001; 98(12): 6641-5.
[43] Alifrangis M, Christensen IT, Jørgensen FS, Rønn AM, Weng JE, Chen M, *et al.* Homology building as a means to define antigenic epitopes on dihydrofolate reductase (DHFR) from Plasmodium falciparum. Malar J 2004; 3(1): 16-22.
[44] Krieger E, Nabuurs SB, Vriend G. Homology modeling. Methods Biochem Anal 2003; 44: 509-24.
[45] Marelius J, Graffner-Nordberg M, Hansson T, Hallberg A, Åqvist J. Computation of affinity and selectivity: binding of 2, 4-diaminopteridine and 2, 4-diaminoquinazoline inhibitors to dihydrofolate reductases. J Comput Aided Mol Des 1998; 12(2): 119-31.
[46] Wieman H, Kristin T, Anderssen E, Drablos F. Homology-based modelling of targets for rational drug design. Mini Rev Med Chem 2004; 4(7): 793-804.
[47] Feeney J. NMR studies of ligand binding to dihydrofolate reductase. Angew Chem Int Ed 2000; 39(2): 290-312.
[48] Schnell JR, Dyson HJ, Wright PE. Structure, dynamics, and catalytic function of dihydrofolate reductase. Annu Rev Biophys Biomol Struct 2004; 33: 119-40.
[49] Bajorath J. Understanding chemoinformatics: a unifying approach. Drug Discov Today 2004; 9(1): 13-4.
[50] Xu J, Hagler A. Chemoinformatics and drug discovery. Molecules 2002; 7(8): 566-600.
[51] da Cunha EFF, Ramalho TC, Maia ER, de Alencastro RB. The search for new DHFR inhibitors: a review of patents, January 2001-February 2005. Expert Opin Ther Pat 2005; 15: 967-86.
[52] Kompis IM, Islam K, Then RL. DNA and RNA synthesis: antifolates. Chem Rev 2005; 105(2): 593-620.
[53] Schweitzer BI, Dicker AP, Bertino JR. Dihydrofolate reductase as a therapeutic target. FASEB J 1990; 4(8): 2441-52.
[54] Warhurst DC. Antimalarial drug discovery: development of inhibitors of dihydrofolate reductase active in drug resistance. Drug Discov Today 1998; 3(12): 538-46.
[55] Becker K, Kirk K. Of malaria, metabolism and membrane transport. Trends Parasitol 2004; 20(12): 590-6.
[56] Jiang L, Lee P-C, White J, Rathod PK. Potent and selective activity of a combination of thymidine and 1843U89, a folate-based thymidylate synthase inhibitor, against Plasmodium falciparum. Antimicrob Agents Chemother 2000; 44(4): 1047-50.
[57] Keough DT, Skinner-Adams T, Jones MK, Ng A-L, Brereton IM, Guddat LW, *et al.* Lead compounds for antimalarial chemotherapy: purine base analogs discriminate between human and P. falciparum 6-oxopurine phosphoribosyltransferases. J Med Chem 2006; 49(25): 7479-86.
[58] Olliaro P. Mode of action and mechanisms of resistance for antimalarial drugs. Pharmacol Ther 2001; 89(2): 207-19.

[59] Wang P, Brobey RK, Horii T, Sims PF, Hyde JE. Utilization of exogenous folate in the human malaria parasite Plasmodium falciparum and its critical role in antifolate drug synergy. Mol Microbiol 1999; 32(6): 1254-62.
[60] Anderson AC. Targeting DHFR in parasitic protozoa. Drug Discov Today 2005; 10(2): 121-8.
[61] Roth B, Rauckman BS, Ferone R, Baccanari DP, Champness JN, Hyde RM. 2, 4-Diamino-5-benzylpyrimidines as antibacterial agents. 7. Analysis of the effect of 3, 5-dialkyl substituent size and shape on binding to four different dihydrofolate reductase enzymes. J Med Chem 1987; 30(2): 348-56.
[62] França TC, Wilter A, Ramalho TC, Pascutti PG, Figueroa-Villar JD. Molecular dynamics of the interaction of Plasmodium falciparum and human serine hydroxymethyltransferase with 5-formyl-6-hydrofolic acid analogues: design of new potential antimalarials. J Braz Chem Soc 2006; 17(7): 1383-92.
[63] Silverman RB. The organic chemistry of drug design and drug action. USA: Elsevier Academic press; 2004.
[64] Plowe CV. The evolution of drug-resistant malaria. Trans R Soc Trop Med Hyg 2009; 103(Supplement 1): S11-S4.
[65] Cowman AF, Morry MJ, Biggs BA, Cross G, Foote SJ. Amino acid changes linked to pyrimethamine resistance in the dihydrofolate reductase-thymidylate synthase gene of Plasmodium falciparum. Proc Natl Acad Sci 1988; 85(23): 9109-13.
[66] Peterson DS, Walliker D, Wellems TE. Evidence that a point mutation in dihydrofolate reductase-thymidylate synthase confers resistance to pyrimethamine in falciparum malaria. Proc Natl Acad Sci 1988; 85(23): 9114-8.
[67] Kublin JG, Dzinjalamala FK, Kamwendo DD, Malkin EM, Cortese JF, Martino LM, *et al.* Molecular markers for failure of sulfadoxine-pyrimethamine and chlorproguanil-dapsone treatment of Plasmodium falciparum malaria. J Infect Dis 2002; 185(3): 380-8.
[68] Foote SJ, Galatis D, Cowman AF. Amino acids in the dihydrofolate reductase-thymidylate synthase gene of Plasmodium falciparum involved in cycloguanil resistance differ from those involved in pyrimethamine resistance. Proc Natl Acad Sci 1990; 87(8): 3014-7.
[69] Peterson DS, Milhous WK, Wellems TE. Molecular basis of differential resistance to cycloguanil and pyrimethamine in Plasmodium falciparum malaria. Proc Natl Acad Sci 1990; 87(8): 3018-22.
[70] Reynolds MG, Roos DS. A biochemical and genetic model for parasite resistance to antifolates Toxoplasma gondii provides insights into pyrimethamine and cycloguanil resistance in Plasmodium falciparum. J Biol Chem 1998; 273(6): 3461-9.
[71] Sirawaraporn W, Prapunwattana P, Sirawaraporn R, Yuthavong Y, Santi D. The dihydrofolate reductase domain of Plasmodium falciparum thymidylate synthase-dihydrofolate reductase. Gene synthesis, expression, and anti-folate-resistant mutants. J Biol Chem 1993; 268(29): 21637-44.
[72] Sirawaraporn W, Sathitkul T, Sirawaraporn R, Yuthavong Y, Santi DV. Antifolate-resistant mutants of Plasmodium falciparum dihydrofolate reductase. Proc Natl Acad Sci 1997; 94(4): 1124-9.
[73] Mharakurwa S, Kumwenda T, Mkulama MA, Musapa M, Chishimba S, Shiff CJ, *et al.* Malaria antifolate resistance with contrasting Plasmodium falciparum dihydrofolate reductase (DHFR) polymorphisms in humans and Anopheles mosquitoes. Proc Natl Acad Sci 2011; 108(46): 18796-801.
[74] Sano G-i, Morimatsu K, Horii T. Purification and characterization of dihydrofolate reductase of *Plasmodium falciparum* expressed by a synthetic gene in *Escherichia coli*. Mol Biochem Parasitol 1994; 63: 265-73.
[75] Champness J, Stammers D, Beddell C. Crystallographic investigation of the cooperative interaction between trimethoprim, reduced cofactor and dihydrofolate reductase. FEBS Lett 1986; 199(1): 61-7.
[76] Hunt WE, Schwalbe CH, Bird K, Mallinson PD. Crystallographic and molecular-orbital studies on the geometry of antifolate drugs. Biochem J 1980; 187: 533-6.
[77] Mehlin C. Structure-based drug discovery for Plasmodium falciparum. Comb Chem High Throughput Screen 2005; 8(1): 5-14.
[78] Yuvaniyama J, Chitnumsub P, Kamchonwongpaisan S, Vanichtanankul J, Sirawaraporn W, Taylor P, *et al.* Insights into antifolate resistance from malarial DHFR-TS structures. Nat Struct Mol Biol 2003; 10(5): 357-65.
[79] Dasgupta T, Chitnumsub P, Kamchonwongpaisan S, Maneeruttanarungroj C, Nichols SE, Lyons TM, *et al.* Exploiting structural analysis, *in silico* screening, and serendipity to identify novel inhibitors of drug-resistant falciparum malaria. ACS Chem Biol 2009; 4(1): 29-40.

[80] Nzila A, Rottmann M, Chitnumsub P, Kiara SM, Kamchonwongpaisan S, Maneeruttanarungroj C, *et al.* Preclinical evaluation of the antifolate QN254, 5-chloro-N′ 6′-(2, 5-dimethoxy-benzyl)-quinazoline-2, 4, 6-triamine, as an antimalarial drug candidate. Antimicrob Agents Chemother 2010; 54(6): 2603-10.
[81] Vanichtanankul J, Taweechai S, Yuvaniyama J, Vilaivan T, Chitnumsub P, Kamchonwongpaisan S, *et al.* Trypanosomal dihydrofolate reductase reveals natural antifolate resistance. ACS Chem Biol 2011; 6(9): 905-11.
[82] Vanichtanankul J, Taweechai S, Uttamapinant C, Chitnumsub P, Vilaivan T, Yuthavong Y, *et al.* Combined spatial limitation around residues 16 and 108 of Plasmodium falciparum dihydrofolate reductase explains resistance to cycloguanil. Antimicrob Agents Chemother 2012; 56(7): 3928-35.
[83] Yuthavong Y, Tarnchompoo B, Vilaivan T, Chitnumsub P, Kamchonwongpaisan S, Charman SA, *et al.* Malarial dihydrofolate reductase as a paradigm for drug development against a resistance-compromised target. Proc Natl Acad Sci 2012; 109(42): 16823-8.
[84] Chaianantakul N, Sirawaraporn R, Sirawaraporn W. Insights into the role of the junctional region of Plasmodium falciparum dihydrofolate reductase-thymidylate synthase. Malar J 2013; 12(1): 91-103.
[85] Dasgupta T, Anderson KS. Probing the role of parasite-specific, distant structural regions on communication and catalysis in the bifunctional thymidylate synthase-dihydrofolate reductase from Plasmodium falciparum. Biochemistry 2008; 47(5): 1336-45.
[86] Rod TH, Radkiewicz JL, Brooks CL. Correlated motion and the effect of distal mutations in dihydrofolate reductase. Proc Natl Acad Sci 2003; 100(12): 6980-5.
[87] Stroud RM, Finer-Moore JS. Stereochemistry of a multistep/bipartite methyl transfer reaction: thymidylate synthase. FASEB J 1993; 7(8): 671-7.
[88] Matthews D, Villafranca J, Janson C, Smith W, Welsh K, Freer S. Stereochemical mechanism of action for thymidylate synthase based on the X-ray structure of the covalent inhibitory ternary complex with 5-fluoro-2′-deoxyuridylate and 5, 10-methylenetetrahydrofolate. J Mol Biol 1990; 214(4): 937-48.
[89] McElheny D, Schnell JR, Lansing JC, Dyson HJ, Wright PE. Defining the role of active-site loop fluctuations in dihydrofolate reductase catalysis. Proc Natl Acad Sci U S A 2005; 102(14): 5032-7.
[90] Wattanarangsan J, Chusacultanachai S, Yuvaniyama J, Kamchonwongpaisan S, Yuthavong Y. Effect of N-terminal truncation of *Plasmodium falciparum* dihydrofolate reductase on dihydrofolate reductase and thymidylate synthase activity. Mol Biochem Parasitol 2003; 126(1): 97-102.
[91] Shallom S, Zhang K, Jiang L, Rathod PK. Essential protein-protein interactions between Plasmodium falciparum thymidylate synthase and dihydrofolate reductase domains. J Biol Chem 1999; 274(53): 37781-6.
[92] Mattioni BE, Jurs PC. Prediction of dihydrofolate reductase inhibition and selectivity using computational neural networks and linear discriminant analysis. J Mol Graph Model 2003; 21(5): 391-419.
[93] Salama I, Schlotter K, Utz W, Hübner H, Gmeiner P, Boeckler F. CoMFA and CoMSIA investigations of dopamine D3 receptor ligands leading to the prediction, synthesis, and evaluation of rigidized FAUC 365 analogues. Bioorg Med Chem 2006; 14(17): 5898-912.
[94] Lill MA. Multi-dimensional QSAR in drug discovery. Drug Discov Today 2007; 12(23): 1013-7.
[95] Wade RC, Henrich S, Wang T. Using 3D protein structures to derive 3D-QSARs. Drug Discovery Today: Technologies 2004; 1(3): 241-6.
[96] Brobey RK, Sano G-i, Itoh F, Aso K, Kimura M, Mitamura T, *et al.* Recombinant *Plasmodium falciparum* dihydrofolate reductase-based *in vitro* screen for antifolate antimalarials. Mol Biochem Parasitol 1996; 81(2): 225-37.
[97] Head RD, Smythe ML, Oprea TI, Waller CL, Green SM, Marshall GR. VALIDATE: A new method for the receptor-based prediction of binding affinities of novel ligands. J Am Chem Soc 1996; 118(16): 3959-69.
[98] Hopfinger A, Wang S, Tokarski JS, Jin B, Albuquerque M, Madhav PJ, *et al.* Construction of 3D-QSAR models using the 4D-QSAR analysis formalism. J Am Chem Soc 1997; 119(43): 10509-24.
[99] Huang H, Pan X, Tan N, Zeng G, Ji C. 3D-QSAR study of sulfonamide inhibitors of human carbonic anhydrase II. Eur J Med Chem 2007; 42(3): 365-72.
[100] Marshall GR, Cramer III RD. Three-dimensional structure-activity relationships. Trends Pharmacol Sci 1988; 9(8): 285-9.

[101] Santos-Filho OA, Mishra RK, Hopfinger AJ. Free energy force field (FEFF) 3D-QSAR analysis of a set of Plasmodium falciparum dihydrofolate reductase inhibitors. J Comput Aided Mol Des 2001; 15(9): 787-810.
[102] Toyoda T, Brobey RK, Sano G-i, Horii T, Tomioka N, Itai A. Lead Discovery of Inhibitors of the Dihydrofolate Reductase Domain of *Plasmodium falciparum* Dihydrofolate Reductase-Thymidylate Synthase. Biochem Biophys Res Commun 1997; 235(3): 515-9.
[103] Ivanciuc O, Ivanciuc T, Cabrol-Bass D. QSAR for dihydrofolate reductase inhibitors with molecular graph structural descriptors. J Mol Struct Theochem 2002; 582(1): 39-51.
[104] Hirst JD, King RD, Sternberg MJ. Quantitative structure-activity relationships by neural networks and inductive logic programming. I. The inhibition of dihydrofolate reductase by pyrimidines. J Comput Aided Mol Des 1994; 8(4): 405-20.
[105] Landavazo DG, Fogel GB, Fogel DB. Quantitative structure-activity relationships by evolved neural networks for the inhibition of dihydrofolate reductase by pyrimidines. Biosystems 2002; 65(1): 37-47.
[106] Hecht D, Cheung M, Fogel GB, editors. Docking scores and QSAR using evolved neural networks for the pan-inhibition of wild-type and mutant PfDHFR by cycloguanil derivatives. Evolutionary Computation, 2009 CEC'09 IEEE Congress on; 2009. IEEE.
[107] Basak S, Mills D. Quantitative structure-activity relationships for cycloguanil analogs as PfDHFR inhibitors using mathematical molecular descriptors. SAR QSAR Environ Res 2010; 21(3-4): 215-29.
[108] Toropov AA, Nesterov IV, Nabiev OM. QSAR modeling of dihydrofolate reductase inhibitory activity by correlation weighting of nearest neighboring codes. J Mol Struct Theochem 2003; 622(3): 269-73.
[109] Guha R, Serra JR, Jurs PC. Generation of QSAR sets with a self-organizing map. J Mol Graph Model 2004; 23(1): 1-14.
[110] Polanski J, Bak A, Gieleciak R, Magdziarz T. Self-organizing neural networks for Modeling robust 3D and 4D QSAR: Application to dihydrofolate reductase inhibitors. Molecules 2004; 9(12): 1148-59.
[111] Cheng F, Shen J, Luo X, Zhu W, Gu J, Ji R, *et al.* Molecular docking and 3-D-QSAR studies on the possible antimalarial mechanism of artemisinin analogues. Bioorg Med Chem 2002; 10(9): 2883-91.
[112] Sutherland JJ, Weaver DF. Three-dimensional quantitative structure-activity and structure-selectivity relationships of dihydrofolate reductase inhibitors. J Comput Aided Mol Des 2004; 18(5): 309-31.
[113] Adane L, Bharatam PV. 3D-QSAR analysis of cycloguanil derivatives as inhibitors of A16V+ S108T mutant *Plasmodium falciparum* dihydrofolate reductase enzyme. J Mol Graph Model 2009; 28(4): 357-67.
[114] Gangjee A, Lin X, Queener SF. Design, Synthesis, and Biological Evaluation of 2, 4-Diamino-5-methyl-6-substituted-pyrrolo [2, 3-d] pyrimidines as Dihydrofolate Reductase Inhibitors 1. J Med Chem 2004; 47(14): 3689-92.
[115] Gangjee A, Lin X. CoMFA and CoMSIA analyses of Pneumocystis carinii dihydrofolate reductase, Toxoplasma gondii dihydrofolate reductase, and rat liver dihydrofolate reductase. J Med Chem 2005; 48(5): 1448-69.
[116] Parenti MD, Pacchioni S, Ferrari AM, Rastelli G. Three-dimensional quantitative structure-activity relationship analysis of a set of Plasmodium falciparum dihydrofolate reductase inhibitors using a pharmacophore generation approach. J Med Chem 2004; 47(17): 4258-67.
[117] Bliznyuk AA, Gready JE. Identification and energetic ranking of possible docking sites for pterin on dihydrofolate reductase. J Comput Aided Mol Des 1998; 12(4): 325-33.
[118] Claußen H, Buning C, Rarey M, Lengauer T. FlexE: efficient molecular docking considering protein structure variations. J Mol Biol 2001; 308(2): 377-95.
[119] Cottrell SJ, Gillet VJ, Taylor R. Incorporating partial matches within multiobjective pharmacophore identification. J Comput Aided Mol Des 2006; 20(12): 735-49.
[120] Eksterowicz JE, Evensen E, Lemmen C, Patrick Brady G, Kevin Lanctot J, Bradley EK, *et al.* Coupling structure-based design with combinatorial chemistry: application of active site derived pharmacophores with informative library design. J Mol Graph Model 2002; 20(6): 469-77.
[121] Liu J, Li Z, Yan H, Wang L, Chen J. The design and synthesis of ALS inhibitors from pharmacophore models. Bioorg Med Chem Lett 1999; 9(14): 1927-32.
[122] Mohan V, Gibbs AC, Cummings MD, Jaeger EP, DesJarlais RL. Docking: successes and challenges. Curr Pharm Des 2005; 11(3): 323-33.

[123] Alvarez JC. High-throughput docking as a source of novel drug leads. Curr Opin Chem Biol 2004; 8(4): 365-70.

[124] Bliznyuk AA, Gready JE. A New Approach to Estimation of the Electrostatic Component of the Solvation Energy in Molecular Mechanics Calculations. J Phys Chem 1995; 99(39): 14506-13.

[125] Broughton HB. A method for including protein flexibility in protein-ligand docking: improving tools for database mining and virtual screening. J Mol Graph Model 2000; 18(3): 247-57.

[126] Ewing TJ, Makino S, Skillman AG, Kuntz ID. DOCK 4.0: search strategies for automated molecular docking of flexible molecule databases. J Comput Aided Mol Des 2001; 15(5): 411-28.

[127] Gastreich M, Lilienthal M, Briem H, Claussen H. Ultrafast *de novo* docking combining pharmacophores and combinatorics. J Comput Aided Mol Des 2006; 20(12): 717-34.

[128] Gerber PR, Mark AE, van Gunsteren WF. An approximate but efficient method to calculate free energy trends by computer simulation: Application to dihydrofolate reductase-inhibitor complexes. J Comput Aided Mol Des 1993; 7(3): 305-23.

[129] Halperin I, Ma B, Wolfson H, Nussinov R. Principles of docking: an overview of search algorithms and a guide to scoring functions. Proteins: Struct, Funct, Bioinf 2002; 47(4): 409-43.

[130] Honig B, Sharp K, Yang AS. Macroscopic models of aqueous solutions: biological and chemical applications. J Phys Chem 1993; 97(6): 1101-9.

[131] Kollman P. Free energy calculations: applications to chemical and biochemical phenomena. Chem Rev 1993; 93(7): 2395-417.

[132] Merlitz H, Burghardt B, Wenzel W. Application of the stochastic tunneling method to high throughput database screening. Chem Phys Lett 2003; 370(1): 68-73.

[133] Wang R, Lai L, Wang S. Further development and validation of empirical scoring functions for structure-based binding affinity prediction. J Comput Aided Mol Des 2002; 16(1): 11-26.

[134] Wang R, Liu L, Lai L, Tang Y. SCORE: A new empirical method for estimating the binding affinity of a protein-ligand complex. Mol Mod Annual 1998; 4(12): 379-94.

[135] Lengauer T, Lemmen C, Rarey M, Zimmermann M. Novel technologies for virtual screening. Drug Discov Today 2004; 9(1): 27-34.

[136] Seifert MH, Wolf K, Vitt D. Virtual high-throughput *in silico* screening. Biosilico 2003; 1(4): 143-9.

[137] Lemcke T, Christensen IT, Jørgensen FS. Towards an understanding of drug resistance in malaria: Three-dimensional structure of *Plasmodium falciparum* dihydrofolate reductase by homology building. Bioorg Med Chem 1999; 7(6): 1003-11.

[138] Fogel GB, Cheung M, Pittman E, Hecht D. *In silico* screening against wild-type and mutant *Plasmodium falciparum* dihydrofolate reductase. J Mol Graph Model 2008; 26(7): 1145-52.

[139] Fogel GB, Cheung M, Pittman E, Hecht D. Modeling the inhibition of quadruple mutant Plasmodium falciparum dihydrofolate reductase by pyrimethamine derivatives. J Comput Aided Mol Des 2008; 22(1): 29-38.

[140] Adane L, Bharatam P. Binding modes of 2, 4-diaminoquinazoline and 2, 4-diaminopteridine analogs to P. falciparum dihydrofolate reductase enzyme: Molecular docking studies. Indian J Pharm Sci 2010; 72(3): 324-33.

[141] Adinarayana KPS, Devi RK. Protein-Ligand interaction studies on 2, 4, 6-trisubstituted triazine derivatives as anti-malarial DHFR agents using AutoDock. Bioinformation 2011; 6(2): 74-7.

[142] Bhat H, Ghosh S, Prakash A, Gogoi K, Singh U. *In vitro* antimalarial activity and molecular docking analysis of 4-aminoquinoline-clubbed 1, 3, 5-triazine derivatives. Lett Appl Microbiol 2012; 54(5): 483-6.

[143] Maitarad P, Kamchonwongpaisan S, Vanichtanankul J, Vilaivan T, Yuthavong Y, Hannongbua S. Interactions between cycloguanil derivatives and wild type and resistance-associated mutant Plasmodium falciparum dihydrofolate reductases. J Comput Aided Mol Des 2009; 23(4): 241-52.

[144] Adane L, Bhagat S, Arfeen M, Bhatia S, Sirawaraporn R, Sirawaraporn W, *et al.* Design and synthesis of guanylthiourea derivatives as potential inhibitors of *Plasmodium falciparum* dihydrofolate reductase enzyme. Bioorg Med Chem Lett 2014; 24(2): 613-7.

[145] Adane L, Bharatam PV. Computer-aided molecular design of 1H-imidazole-2, 4-diamine derivatives as potential inhibitors of Plasmodium falciparum DHFR enzyme. J Mol Model 2011; 17(4): 657-67.

[146] Gahtori P, Ghosh SK, Parida P, Prakash A, Gogoi K, Bhat HR, *et al.* Antimalarial evaluation and docking studies of hybrid phenylthiazolyl-1, 3, 5-triazine derivatives: A novel and potential antifolate lead for *Pf*-DHFR-TS inhibition. Exp Parasitol 2012; 130(3): 292-9.

[147] Güner OF. Pharmacophore perception, development, and use in drug design. La Jolla, CA: Internat'l University Line; 2000.
[148] Mannhold R, Kubinyi H, Folkers G, Langer T, Hoffmann RD. Pharmacophores and pharmacophore searches. Germany: John Wiley & Sons; 2006.
[149] Gemma S, Campiani G, Butini S, Kukreja G, Joshi BP, Persico M, *et al.* Design and synthesis of potent antimalarial agents based on clotrimazole scaffold: exploring an innovative pharmacophore. J Med Chem 2007; 50(4): 595-8.
[150] Riel MA, Kyle DE, Bhattacharjee AK, Milhous WK. Efficacy of proton pump inhibitor drugs against Plasmodium falciparum *in vitro* and their probable pharmacophores. Antimicrob Agents Chemother 2002; 46(8): 2627-32.
[151] Adane L, Patel D, Bharatam PV. Shape-and Chemical Feature-Based 3D-Pharmacophore Model Generation and Virtual Screening: Identification of Potential Leads for P. falciparum DHFR Enzyme Inhibition. Chem Biol Drug Des 2010; 75(1): 115-26.
[152] Adane L, Bharatam PV, Sharma V. A common feature-based 3D-pharmacophore model generation and virtual screening: identification of potential Pf DHFR inhibitors. J Enzyme Inhib Med Chem 2010; 25(5): 635-45.
[153] Mannhold R, Kubinyi H, Timmerman H, Folkers G, Höltje H, Vacca J. Methods and principles in medicinal chemistry. Weinheim: John Wiley and Sons, Inc; 2000.
[154] Andres J, Moliner V, Safont V, Domingo L, Picher M, Krechl J. On transition structures for hydride transfer step: a theoretical study of the reaction catalyzed by dihydrofolate reductase enzyme. Bioorg Chem 1996; 24(1): 10-8.
[155] Cummins PL, Gready JE. Energetically most likely substrate and active-site protonation sites and pathways in the catalytic mechanism of dihydrofolate reductase. J Am Chem Soc 2001; 123(15): 3418-28.
[156] Cao ZW, Han LY, Zheng CJ, Ji ZL, Chen X, Lin HH, *et al.* Computer prediction of drug resistance mutations in proteins. Drug Discov Today 2005; 10(7): 521-9.
[157] Bharatam PV, Patel DS, Iqbal P. Pharmacophoric features of biguanide derivatives: an electronic and structural analysis. J Med Chem 2005; 48(24): 7615-22.
[158] Bajorath J, Kraut J, Li Z, Kitson DH, Hagler AT. Theoretical studies on the dihydrofolate reductase mechanism: electronic polarization of bound substrates. Proc Natl Acad Sci 1991; 88(15): 6423-6.
[159] Polshakov VI, Birdsall B, Gradwell MJ, Feeney J. The use of PM3 SCF MO quantum mechanical calculations to refine NMR-determined structures of complexes of antifolate drugs with dihydrofolate reductase in solution. J Mol Struct Theochem 1995; 357(3): 207-16.
[160] Polshakov VI, Biekofsky RR, Birdsall B, Feeney J. Towards understanding the origins of the different specificities of binding the reduced (NADPH) and oxidised ($NADP^+$) forms of nicotinamide adenine dinucleotide phosphate coenzyme to dihydrofolate reductase. J Mol Struct 2002; 602: 257-67.
[161] Cannon WR, Garrison BJ, Benkovic SJ. Electrostatic characterization of enzyme complexes: Evaluation of the mechanism of catalysis of dihydrofolate reductase. J Am Chem Soc 1997; 119(10): 2386-95.
[162] Cummins PL, Gready JE. QM/MM and SCRF studies of the ionization state of 8-methylpterin substrate bound to dihydrofolate reductase: existence of a low-barrier hydrogen bond. J Mol Graph Model 2000; 18(1): 42-9.
[163] Cleland W, Kreevoy MM. Low-barrier hydrogen bonds and enzymic catalysis. Science 1994; 264(5167): 1887-90.
[164] Cummins PL, Gready JE. Computational methods for the study of enzymic reaction mechanisms. II. An overlapping mechanically embedded method for hybrid semi-empirical-QM/MM calculations. J Mol Struct Theochem 2003; 632(1): 247-57.
[165] Garcia-Viloca M, Truhlar DG, Gao J. Importance of substrate and cofactor polarization in the active site of dihydrofolate reductase. J Mol Biol 2003; 327(2): 549-60.
[166] Pang J, Pu J, Gao J, Truhlar DG, Allemann RK. Hydride transfer reaction catalyzed by hyperthermophilic dihydrofolate reductase is dominated by quantum mechanical tunneling and is promoted by both inter-and intramonomeric correlated motions. J Am Chem Soc 2006; 128(24): 8015-23.
[167] Sundriyal S, Khanna S, Saha R, Bharatam PV. Metformin and glitazones: does similarity in biomolecular mechanism originate from tautomerism in these drugs? J Phys Org Chem 2008; 21(1): 30-3.

[168] Patel DS, Bharatam PV. Novel ⊕N (← L) 2 species with two lone pairs on nitrogen: systems isoelectronic to carbodicarbenes. Chem Commun 2009; (9): 1064-6.
[169] Mehdi A, Adane L, Patel DS, Bharatam PV. Electronic structure and reactivity of guanylthiourea: a quantum chemical study. J Comput Chem 2010; 31(6): 1259-67.
[170] Zolli-Juran M, Cechetto JD, Hartlen R, Daigle DM, Brown ED. High throughput screening identifies novel inhibitors of *Escherichia coli* dihydrofolate reductase that are competitive with dihydrofolate. Bioorg Med Chem Lett 2003; 13(15): 2493-6.
[171] Summerfield RL, Daigle DM, Mayer S, Mallik D, Hughes DW, Jackson SG, *et al.* A 2.13 Å structure of E. coli dihydrofolate reductase bound to a novel competitive inhibitor reveals a new binding surface involving the M20 loop region. J Med Chem 2006; 49(24): 6977-86.
[172] Diaz DS, Kozar MP, Smith KS, Asher CO, Sousa JC, Schiehser GA, *et al.* Role of specific cytochrome P450 isoforms in the conversion of phenoxypropoxybiguanide analogs in human liver microsomes to potent antimalarial dihydrotriazines. Drug Metab Dispos 2008; 36(2): 380-5.
[173] Arfeen M, Patel DS, Abbat S, Taxak N, Bharatam PV. Importance of cytochromes in cyclization reactions: Quantum chemical study on a model reaction of proguanil to cycloguanil. J Comput Chem 2014; 35(28): 2047-55.

CHAPTER 8

Bacterial Protein Toxins: Current and Potential Clinical Use

Alessia Fabbri*, Francesca Rosadi, Giulia Ballan, Antonella Del Brocco, Sara Travaglione, Stefano Loizzo and Carla Fiorentini

Department of Therapeutic Research and Medicines Evaluation, Istituto Superiore di Sanità, Viale Regina Elena, 299.00161 Roma Italy

Abstract: Natural toxins are the product of a long-term evolution, and act on essential mechanisms in the most crucial and vital processes of living organisms. They can attack components of the protein synthesis machinery, actin polymerization, signal transduction pathways, intracellular trafficking of vesicles as well as immune and inflammatory responses. For this reason, toxins have increasingly been used as valuable tools for analysis of cellular physiology, and in the recent years, some of them are used medicinally for the treatment of human diseases.

This review is devoted to protein toxins of bacterial origin, specifically those toxins that are currently used in therapy or those under study for their potential clinical applications. Bacterial protein toxins are all characterized by a specific mechanism of action that involves the central molecular pathways in the eukaryotic cell. Knowledge of their properties could be used for medical purposes.

Keywords: Adenylate cyclase, anthrax toxin, bacterial protein toxins, botulinum toxin, cancer, cholera toxin, clinical application, cytotoxic necrotizing factor 1 (CNF1), drug delivery agent, fusion protein, heat-labile enterotoxin, human disease, immunotoxins, molecular mechanism, neurological disorders, pertussis toxin, tetanus toxin, therapeutic agents, vaccine adjuvant, viral infection, zonula occludens toxin (ZOT).

BACTERIAL PROTEIN TOXINS USED IN THERAPY

Bacterial protein toxins are the principal virulence factors of pathogenic bacteria and their production represents one of the main bacterial strategies to interact with mammalian cells. Bacterial toxins manipulate, by diverse and sophisticated mechanisms, the host cell functions in a way that can favor the survival and

***Corresponding author Alessia Fabbri:** Department of Therapeutic Research and Medicines Evaluation, Istituto Superiore di Sanità, Viale Regina Elena, 299. 00161 Roma Italy; Tel. +39-06-49902939; Fax: +39-06- 49903691; E-mail: alessia.fabbri@iss.it

Atta-ur-Rahman, Allen B. Reitz, M. Iqbal Choudhary and Jizhou Wang (Eds.)

spreading of the microbes. Defining the molecular activity of a toxin can also highlight pathways potentially useful to convert a toxin effect from detrimental to favorable. In fact, an increasing number of bacterial toxins are employed as new therapeutics or are regarded as possible innovative drugs.

In this chapter, we report a list of bacterial protein toxins used in therapy, briefly summarizing their main characteristics. The use of selected bacterial toxins as components of engineered molecules, referred to as immunotoxins, is also briefly presented (Table **1**).

Table 1: Bacterial protein toxins with a potential use in therapy

TOXIN	Therapeutic use:
Botulinum neurotoxin (BoNt) from *C. botulinum*	Dystonia Muscle tone disorders Autonomic nervous system disorders Cosmetic use Cancer therapy
Tetanus neurotoxin (TeNT) from *C. tetani*	Carrier for therapeutic molecules to the CNS Ischemia-induced neuronal damage repair
Lethal toxin (LF) from *B. anthracis*	Potential treatment of cancer
Pertussis toxin (PTX) from *B. pertussis*	Control of viral infection in HIV-1 and LCMV Neurons protection Prevention of gastric inflammation and associated carcinogenesis Control of blood pressure and vasoconstriction
Cytotoxic necrotizing factor 1 (CNF1) from *E. coli*	Effect on neurodegenerative diseases Effect on neurodevelopmental diseases Improvement of learning and memory use as vaccine adjuvant Antineoplastic effect
IMMUNOTOXINS	Cancer therapy

Botulinum Toxin from *Clostridium Botulinum*

C. botulinum are gram-positive, anaerobic and spore-forming bacteria widely distributed in soil and water. Strains of *C. botulinum* produce seven different neurotoxins designated BoNT/A to G. Although differing in their biosynthesis, size, structure, and specific site of action, all botulinum neurotoxins bind and enter inside peripheral cholinergic terminals and cause a sustained block of acetylcholine release, with ensuing flaccid paralysis [1-3].

Botulinum neurotoxins are synthesized as a single polypeptide protoxin chain with a molecular mass of about 150 kDa [4]. After bacterial death and lysis, the protoxin undergoes a process of activation by bacterial proteases to yield the 50 kDa light chain and the 100 kDa heavy chain, held together by a disulfide bond. One exception is BoNT/E, which is activated by splitting into light and heavy chains only by gut proteases [3, 5, 6]. The heavy chain of botulinum neurotoxins is responsible for binding to specific surface receptors, whereas the light chain is a zinc-dependent metalloprotease that enzymatically cleaves specific Soluble N-ethylmaleimide-sensitive factor-Attachment protein Receptor (SNARE) proteins, namely vesicle-associated membrane protein (VAMP), synaptosome-associated protein of 25 kDa (SNAP-25) and Syntaxin, before they assemble into the complex responsible for vesicle docking (Fig. **1**) [1, 2]. To exert its action, each active neurotoxin undergoes a 3-step process consisting of i) binding to a surface receptor; ii) internalization within an endosome with subsequent translocation of the light-chain portion of the toxin into the cytosol; iii) proteolysis of target proteins and the subsequent effect on neurotransmission [5, 7, 8]. However, the 7 mentioned toxins (BoNT/ A to G) act in a slightly different way on the targets [for more details see reference 9].

The enzymatic cleavage of SNARE proteins ultimately results in the observed clinical effects of muscle paralysis. The clinical effects include dysfunction of the autonomic nervous system and, in particular, limb paralysis, dysarthria, facial muscle weakness, dyspnea, constipation, urinary retention, and opthalmoplegia. Most patients survive botulism, but complete recovery is slow and may require mechanical ventilation [10].

Therapeutical Applications

Over the last 50 years, botulinum neurotoxin has been transformed from a cause of life-threatening disease to a medical therapy. In 1817 Justinus Kerner, who provided the earliest account of food borne botulism, recognized the potential for the therapeutic use for botulinum toxin [11]. After having been regarded for several decades as a prospective bioterrorism weapon [12], the botulinum toxin was first used by Scott [13] as a muscle-weakening agent for the correction of experimental

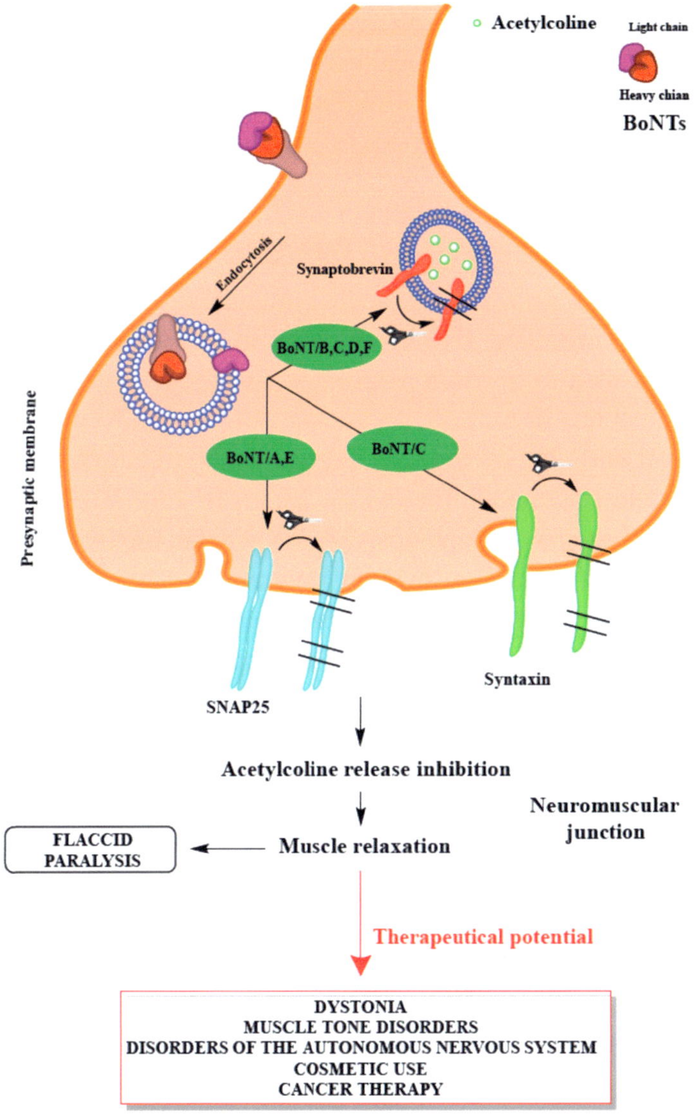

Figure 1: *C. botulinum* neurotoxins. Seven serotypes (A to G) of botulinic neurotoxins (BoNTs) exist, and all bind and enter inside peripheral cholinergic terminals causing a sustained block of acetylcholine release and ensuing muscle relaxation and flaccid paralysis. This occurs because the toxins cleave and inactive SNARE proteins required for the fusion of acetylcoline vesicles with the presynaptic membrane. In particular, BoNT B, G, D, F cleave Synaptobrevin, a VAMP protein localized on the vesicle membrane, BoNT C acts on Syntaxin and SNAP-25, two proteins of the pre-synaptic membrane, while BoNT A and E act on SNAP-25. The ability of BoNTs to induce muscle relaxation makes them potential therapeutic tools for treatment of different neurological disorders. In the huge spectrum of activities of BoNTs, surely the most popular is the ability to relax facial wrinkles and then counteract the aging signs. At least BoNTs are also largely used as a potent weapon in cancer therapy.

strabismus in monkeys. Since then, different commercial preparations were made available and the number of publications dedicated to the clinical uses of botulinum toxins has significantly increased [14]. Botulinum neurotoxin is employed in a range of neurological human pathologies characterized by hyperfunction of all types of peripheral cholinergic terminals but it has also acquired a large popularity as a cosmetic agent to reduce facial wrinkles [15].

Use in neurological disorders. Botox (onabotulinumtoxinA, Allergan) first gained Food and Drug Administration (FDA) approval in 1989 for its marked benefits produced in patients with dystonia, a neurological disorder dominated by repetitive and patterned contractions of muscles producing abnormal movements and postures [16]. FDA approval was also earned by Dysport (abobotulinumtoxinA, Medicis) in 2009 and by Xeomin (incobotulinumtoxinA, Merz) in July 2010 for treatment of adults with cervical dystonia or blepharospasm. In general, treatment with botulinum toxin has been reported to cause a marked improvement in a large percentage of patients with blepharospasm [17], oromandibular dystonia, manifested by involuntary jaw closure and bruxism [18, 19] hemifacial spasm [17], cervical dystonia, also known as torticollis [20], focal hand and cranial dystonia [21] and laryngeal dystonia manifested by spasmodic dysphonia [22]. Botulinum neurotoxins have been shown to be a safe and effective therapeutic option for most forms of dystonia, and are now considered as the best symptomatic treatment in these disorders with stable effects over time [23].

Botulinum toxin injections have been also found to be useful in patients with Parkinson's disease and other neurodegenerative disorders, stroke related hemiplegia [24, 25], disorders of muscle tone like spasticity associated with cerebral palsy [26, 27], brain trauma, multiple sclerosis [28], but also painful muscle spasms and other painful disorders including headaches [29-31]. Recently, botulinum neurotoxin has also been regarded as an effective adjunct to therapy and surgery in managing muscle imbalance, co-contractions, and contractures in Neonatal Brachial Plexus Palsy [32].

Furthermore, because interfering with transmission not only at the neuromuscular junction but also in the cholinergic autonomic parasympathetic and postganglionic sympathetic nervous system, botulinum toxin is also applied for the treatment of various disorders of the autonomic nervous system. In this context, an intradermal injection of botulinum toxin has provided a highly effective treatment for focal hyperhidrosis [33] (defined as excessive sweating of the palms, feet or axillae), or sialorrhoea, the last one associated with amyotrophic lateral sclerosis and other neurodegenerative disorders [34- 37].

The array of disorders treated with this toxin also includes various otolaryngology, gastroenterology and genitourinary disorders [38-42]. Recently, botulinum toxins have been used in cancer therapy to open tumor vessels, allowing a more effective destruction of cancer cells by radiotherapy and chemotherapy [43], and in preventing gastric stasis in esophageal cancer's patients [44]. Furthermore, in experimental epilepsy models, botulinum toxin delivery to the hippocampus showed anti-epileptogenic properties [45].

Cosmetic applications. There is growing interest in the use of botulinum toxin in cosmetics, in particular for the treatment of wrinkles. The excellent results in the treatment of furrows and frown lines on the forehead have been confirmed by several open and controlled studies [46]. In this context, botulinum toxin A was initially used for treatment of upper facial rhytides, but is now considered a key component of facial rejuvenation procedures as an adjunct to either non-surgical or surgical interventions, including in the lower face and neck. Side effects are mostly mild and always self-limited. More serious complications are associated with higher doses, improper injection techniques, and occur in patients with underlying (mostly cardiovascular or neurological) co-morbidities [for a review see 47]. However, inappropriate placement or dose of the toxin may result in loss of facial expression and disfiguring complications. Botulinum toxin A has also been used for the treatment of keloids by intralesional injection in a prospective, uncontrolled study [48] and to reduce the development of a widened scar [49].

As botulinum neurotoxin therapy is often continued over many years, some patients may develop detectable antibodies that may or may not affect their biological activity. Although botulinum neurotoxins are considered "lower risk" biologics, since antibodies that may develop are not likely to cross-react with endogenous proteins, it is possible that patients may lose their therapeutic response. Various factors impact the immunogenicity of botulinum neurotoxins, including product-related factors, such as the manufacturing process, the antigenic protein load, and the presence of accessory proteins, as well as treatment-related factors, such as the overall toxin dose, booster injections, and prior vaccination or exposure. Detection of antibodies by laboratory tests does not necessarily predict the clinical success or failure of treatment. Overall, botulinum neurotoxin type A products exhibit low clinically detectable levels of antibodies when compared with other approved biologic products [50].

Tetanus Toxin from *Clostridium Tetani*

Clostridium tetani, a gram-positive anaerobic bacterium whose spores are commonly found in soil and animal waste, produces the tetanus neurotoxin (TeNT, also known as TeTx or tetanospasmin), which is synthesized as a single polypeptide and post-translationally modified to produce light (L) and heavy (H) chains linked by disulfide bonds. The L chain exerts a neurotoxic zinc metalloprotease activity and its target is represented by the integral membrane protein VAMP/synaptobrevin, a synaptic membrane protein required for neurotransmitter release. The H chain consists of two non-toxic fragments, the first one involved in translocation of the L chain and the second one (called TTC) required for neuronal cell binding and retrograde transport. TeNT shows a strong affinity and specificity to neuronal terminals. It is primarily internalized by the motor neurons at the neuromuscular junction where it enters the axonal retrograde transport pathway and is transported to the neuronal soma in the central nervous system [51]. Once reached the neuronal synapses, TeNT L chain proteolytically cleaves VAMP/synaptobrevin proteins in the synapses of inhibitory interneurons, thus preventing the release of inhibitory neurotransmitters into the synaptic cleft (Fig. **2**). The result is a sustained motor neuron excitation causing tetanic spasms [52].

Therapeutical Applications

Carrier for therapeutic molecules to the central nervous system (CNS). The possible therapeutic activity of tetanus toxin resides in its TTC non-toxic fragment involved in retrograde transport. TTC was first proposed by Bizzini and coworkers in 1977 [53] as a carrier for therapeutic molecules to the CNS and, since then, a number of proteins have been efficiently internalized by TTC-mediated endocytosis. Various chemically or genetically TTC-conjugated molecules have been produced and used primarily as tracers for neuroanatomy and cellular physiology. Subsequently, molecules constructed for therapeutic approaches [54] provided promising results in those conditions where a single gene product is negatively affected and enzyme replacement therapy could possibly compensate insufficient biological activity. This is the case of Tay Sachs disease (TSD), in which an inactivating mutation in the gene for lysosomal GM2

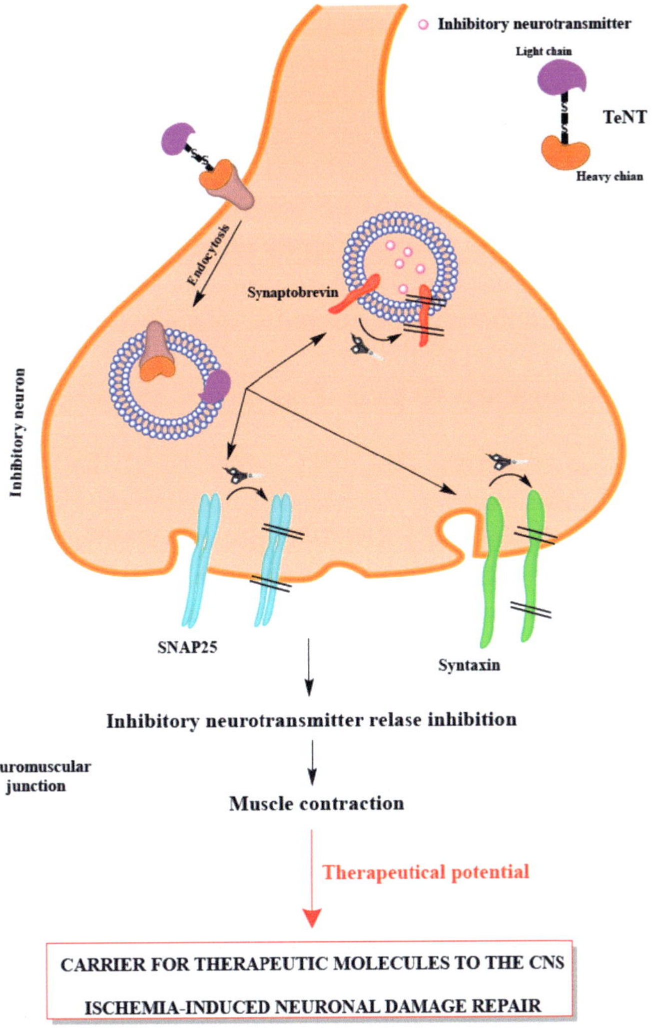

Figure 2: *C. tetani* neurotoxin. Tetanus neurotoxin (TeNT) is composed by two chains linked by disulfide bonds: the light chain (L), responsible for the proteolitic activity on SNARE proteins, and the heavy chain (H), consisting of two fragments involved in the translocation of the L chain and in the retrograde transport. TeNT is internalized primarily by the motor neurons at the neuromuscular junction (NMJ) where it enters the axonal retrograde transport pathway and is subsequently transported to the neuronal soma in the CNS. There, it becomes rapidly fixed to gangliosides at the presynaptic inhibitory motor nerve endings, and is taken up into the axon by endocytosis. The toxin, by cleaving SNARE proteins required for fusion between vesicle's membrane and pre-synaptic membrane, blocks the release of inhibitory neurotransmitters, glicine and gamma-aminobutyric acid, which are required to check the nervous impulse. The result is a sustained motor neuron excitation causing titanic spasms. The therapeutic potential of tetanus toxin resides in the H chain TTC-fragment, responsible for the retrograde axonal transport. Indeed TTC is used as a carrier for the transport of therapeutic molecules to the CNS.

ganglioside-degrading enzyme beta-hexosaminidase (HEXA) causes lipid accumulation in the lysosomes of the nervous tissue and progressive brain and spinal cord deterioration. Human placental HEXA conjugated with TTC is efficiently internalized by rat brain cells *in vitro* and could lessen the accumulation of lysosomal lipids [55]. Other recombinant proteins have been constructed with the same purpose of vehiculating altered metabolic enzyme activities [56-59], although their activities were impaired by either improper location or steric interference of the fused domains [59, 60]. TTC has also been used as carrier for a number of neurotrophins or antiapoptotic proteins obtaining encouraging data in the field of neurodegenerative disorders [61-65]. In some cases, however, investigations on their effect *in vivo* are still pending [66- 68].

Increasing evidence suggests that TTC may itself harbor neuroprotective properties, fueling those pathways implicated in neuronal apoptosis and survival [69-71]. The evidence suggesting that TTC possesses antiapoptotic properties *in vitro* is corroborated by studies conducted in a rat model of Parkinsonism, where TTC has been shown to ameliorate motor deficits and to restore dopamine levels, phenomena normally affected during apoptosis [72-74].

Very recently, intramuscular injection of nacked DNA-TTC has been proved to have beneficial effects on the development and expression of global cerebral ischemia/reperfusion-induced oxidative stress and motor hyperactivity [75]. In fact, several oxidative stress and motor behavioral parameters were assayed after reperfusion, and neuroprotective efficiency of TTC was detected in the forebrain cortex, striatum, hippocampus, and cerebellum. In addition, TTC significantly decreased ischemia-induced motor hyperactivity.

Altogether, these results suggest that the TTC fragment, with its capacity of being transported in a retrograde way through motoneurons and its nontoxic nature, can be considered as a promising therapeutic strategy for neuroprotection.

Anthrax Toxin from *Bacillus Anthracis*

Bacillus anthracis, the etiological agent of anthrax, is a gram-positive bacterium mainly present in the environment in the form of highly resistant spores [76]. The disease takes one of several forms: cutaneous, gastrointestinal or inhalational, depending on the route of entry of infectious spores. The cutaneous is the most common form, manageable with antibiotics, whereas the gastrointestinal anthrax is a much more serious systemic disease. However, the most deadly anthrax is the inhalational form, which is extremely rare and occurs when spores are taken up by alveolar macrophages and transported to lymph nodes, where germination occurs. The bacteria then produce anthrax toxins, which inhibit the innate and adaptive immune response. This enables the bacterium to escape the lymph-node defense barrier to reach the blood system causing bacteremia and toxemia, which rapidly kills the host [76, 77]. *B. anthracis* produces toxins that are crucial for the establishment of infection and pathogenesis and the anthrax toxins consist of three distinct proteins: the protective antigen (PA), the edema factor (EF) and the lethal factor (LF). Binary combination of these secreted proteins comprise the two anthrax toxins: PA combined with LF is called letal toxin (LeTx) and PA combined with EF is known as edema toxin (EdTx). LF and EF act enzymatically on intracellular substrates: LF is a zinc-dependent metallo-protease that cleaves and inactivates most isoforms of mitogen-activated protein kinase (MAPK) kinases (MEKS) whereas EF is a calcium- and calmodulin-dependent adenylate cyclase that causes a dramatic increase in cytoplasmic cAMP, leading to an imbalance of water homeostasis [78]. PA escort LF and EF to the interior of the host cell by binding to cell surface receptors [79] and then being cleaved by membrane endoproteases of the furin family. Two types of cell surface receptor have been identified: ANTXR1 (also known as tumor endothelial marker-8 (TEM8)), and ANTXR2 (also known as capillary morphogenesis protein 2 (CMG2)). These receptors are type 1 transmembrane proteins with a single membrane-spanning domain. ANTXR1 and ANTXR2 each contain an extracellular domain that is highly related to von Willebrand factor type A domains and integrin inserted domains [80]. Cleaved PA molecules assemble into heptamers, which then associate with EF and LF. PA heptamers enter lipid rafts and are quickly endocytosed into acidic compartments [81] where the low pH triggers a conformational change of the complex, allowing translocation of LF and EF into the cytosol where they exert their enzymatic activity (Fig. **3**). By acting on their cytosolic targets, the two toxins functionally impair cells of both the innate and adaptive immune systems. In particular, they i) block neutrophil chemotaxis and ROS production; ii) induce cell death in macrophages; iii) impair dendritic cells; iv) have effects on T and B cells, decreasing production of growth-promoting cytokines and blocking proliferation [for a review see 78]. In this way, both toxins contribute to the establishment of infection by promoting bacterial dissemination and causing disease.

A better understanding of the toxins will help in the design of much-needed anti-toxin drugs and the development of new toxin-based medical applications.

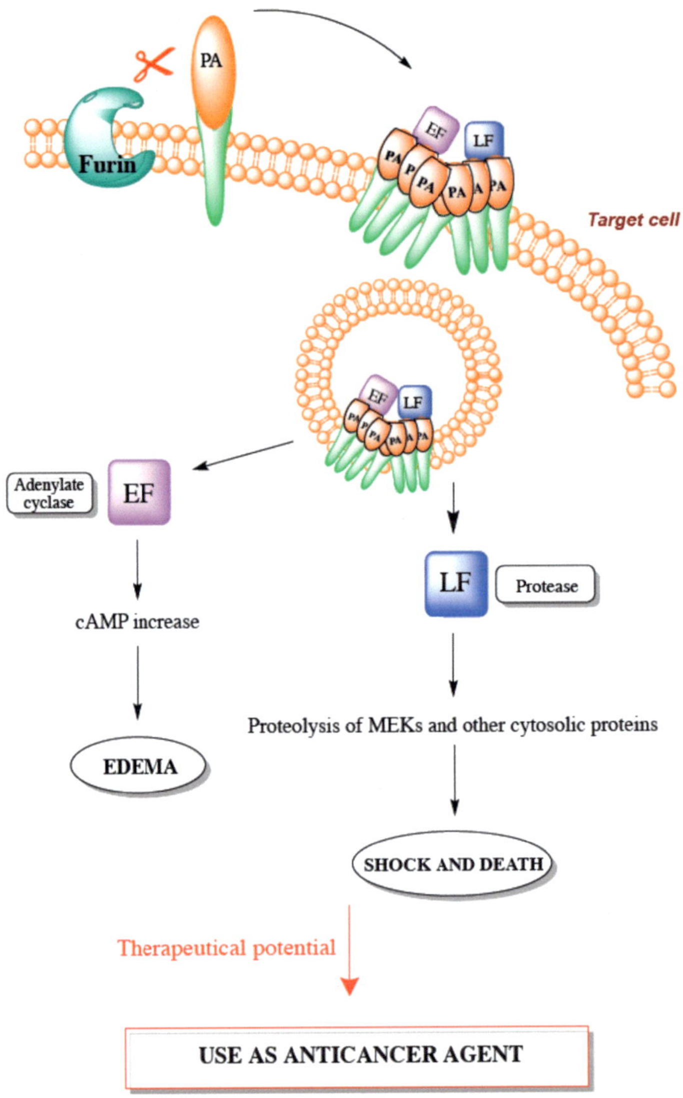

Figure 3: *B. anthracis* lethal factor. The anthrax toxin consists of three distinct proteins: the protective antigen (PA), the edema factor (EF) and the lethal factor (LF). After binding to a host cell surface receptor, PA is cleaved by membrane endoproteases of the furin family. Cleaved PA molecules assemble into heptamers, which then associate with EF or LF. PA heptamers are endocytosed into acidic compartments, where a conformational change of the complex allows translocation of LF and EF into the cytosol. LF and EF act enzimatically on intracellular substrates: LF is a zinc-dependent metallo-protease that cleaves and inactives most isoforms of mitogen-activated protein kinase (MAPK) kinases (MEKS), whereas EF is a calcium- and calmodulin-dependent adenylate cyclase that causes a dramatic increase in cytoplasmic cAMP, leading to an imbalance of water homeostasis. Anthrax LF might be used to counteract increased levels of MEKs activity in tumors, thus representing a novel potential chemotherapeutic agent.

Therapeutical Applications

Treatment of cancer. As previously stated, anthrax LF selectively inactivates MEKs. Inappropriate activation of these kinases has been implicated in tumorigenesis, tumor metastasis and tumor angiogenesis [for a review see 82]. This raised the possibility that anthrax LF might be used to counteract increased levels of MEKs activity in tumors, LF representing a novel and potentially useful chemoterapeutic agent. Analysis of different human tumor cell lines derived from melanomas, breast tumors and colon tumors showed that their growth was inhibited by LeTx treatment. In addition, when (V12) H-ras transformed NIH 3T3 cells were treated with LeTx *in vitro*, their transformed phenotype was reverted. These results show that LeTx is capable of inhibiting tumor-like properties of transformed cells *in vitro*. Initial observations indicate that LeTx effectively blocks tumor growth *in vivo*. Preclinical studies with anthrax LeTx show the sensitivity of malignant melanoma cell lines in tissue culture and anti-tumor efficacy in melanoma xenograft models [83]. Additional studies to define the maximal tolerated doses and dose-limiting toxicity of anthrax lethal toxin in rodent and primate models should pave the way for Phase I studies testing the efficacy of the anthrax lethal toxin in patients with metastatic melanoma.

The use of anthrax toxin will also be described in the section devoted to 'Immunotoxins'.

Pertussis Toxin from *Bordetella Pertussis*

B. pertussis, the aetiological agent of whooping cough, is a gram-negative aerobic bacterium that produces numerous toxins including the pertussis toxin (PTX). PTX is composed of different subunits organized in a typical A-B type structure. The B oligomer (PTX-B) is responsible for binding PTX to target cells, whereas the A moiety exhibits ADP-ribosyltransferase activity. After binding to the cell receptors, the toxin most likely follows the retrograde transport to the endoplasmic reticulum. Probably from there, the A moiety translocates into the cytosol where it expresses the ADP-ribosyltransferase activity using NAD as donor and trimeric G-proteins as acceptors [84]. ADP-ribosylated G-proteins loose their ability to transduce signals and this constitutes the molecular basis of most PTX actions. Recent evidence demonstrates that a p38 MAPK activity is possibly linked to the ADP-ribosylation of G-proteins, leading to alteration in lung permeability [85]. These studies provide mechanistic information on the pathogenesis of PTX-induced lung toxicity.

However, it is now clear that PTX-B alone directs cellular responses [86]. In particular, it is a rapid and robust stimulus for p42/p44 MAP kinase (ERK) activity in endothelium [85]. Other signaling pathways are stimulated by PTX-B and they involve PKC, Ca^{+2} and tyrosine kinases as well [87-89]. These phenomena lead to multiple effects including DNA synthesis in B cells, proliferation of T lymphocytes, potentiation of Th1-dependent immune responses *via* up-regulation of IFN and IL-2, and inhibition of IL-1-induced IL-2 expression and prostaglandin production (Fig. **4**) [86, 90].

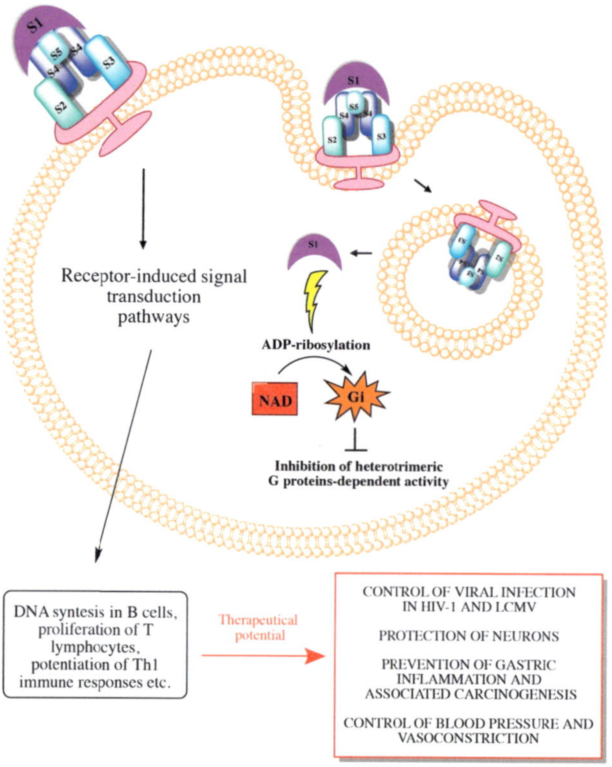

Figure 4: ***B. pertussis*** **toxin.** Pertussis toxin (PTX) is composed of a B-oligomer, consisting of four subunits (S2-S5), responsible for binding PTX to target cells, and a moiety that exhibits ADP-ribosyltransferase activity (S1). After binding to the cell receptors, the toxin is endocytosed and the S1 moiety translocates into the cytosol where it expresses the ADP-ribosyltransferase activity on eterotrimeric G proteins. ADP-ribosylated G-proteins loose their ability to transduce signals. Furthermore, the binding of the B-oligomer to the cell receptor is able to trigger different signal transduction pathways that, in turn, can lead to DNA synthesis in B cells, proliferation of T lymphocytes, and other responses. Such receptor-induced activity of PTX has been exploited to develop potential therapeutic tools to control viral infection of HIV-1 and LCMV1 viruses; to induce neurons' protection; to prevent gastric inflammation and associated carcinogenesis; and also to control blood pressure and vasoconstriction.

Therapeutical Applications

Protection of neurons from inflammatory insults. A possible application of PTX has been reported for multiple sclerosis (MS). PTX is commonly used as an adjuvant to induce experimental autoimmune encephalitis (EAE), an animal model of MS, and its effect in EAE is attributed to the ability of increasing permeability of the blood brain barrier that allows proinflammatory cells to enter the CNS [91, 92]. Surprisingly, Yin and coworkers (2010) [93] demonstrated that a single large dose of PTX intracerebroventricularly (icv) administered attenuates demyelination, axonal loss and neurological deficits in EAE. The same protective effect was confirmed by other experimental works [94, 95] that associated the protection from CNS autoimmune disease with immunological alterations, principally consisting in increased secretion of regulatory cytokines, expanded T-reg population and enhanced neuronal vascular endothelial growth factor. A very recent study [96] demonstrated that PTX reduces clinical deficits of EAE by 91.3%, mainly by attenuating demyelination by 75.5%. Furthermore, PTX influences the general inflammatory stutus of the CNS by reducing lymphocyte infiltration, deactivating microglia activation and changing T cell profile (increasing T helper and T regulatory cells).

Control of viral infection. The whole PTX as well as a non-toxic derivative of this molecule, has been reported to down-regulate the replication of the virus SIVmac251 in infected macaques by inducing a transient lymphocytosis [97]. However, the vast majority of reports highlighting a therapeutic potential of PTX against HIV-1, deals with PTX-B, the binding portion capable of transducing a variety of cellular signals. In particular, PTX-B can block the infectious process of HIV-1 *in vitro* at multiple levels [90, 98, 99]. It affects the entry of R5 but not X4 strains into primary T lymphocytes by cross-deactivation of one of the major HIV-1 co-receptor, CCR5, thus affecting infection by the viruses that use this receptor [90]. PTX-B also blocks the post-entry step of HIV-1 replication, using a mechanism that affects the transcription and stability of HIV-1 mRNA [98]. In addition, PTX-B potently inhibits cytokine-induced virus expression in chronically infected promonocytic U1 cells, thus suggesting the potential value of such a toxin as an agent that can prevent reactivation of latent HIV provirus [99, 100]. Finally, experiments with a surrogate model of HIV-1 infection, that is, SCID mice reconstituted with human peripheral blood leukocytes (hu-PBL-SCID) infected with a CCR5-dependent (R5) HIV-1 strain, demonstrated that PTX-B potently inhibits virus infection and replication [101]. No matter which molecular pathway is exploited by PTX-B, its described activities make it an attractive candidate for anti-HIV therapy as it can inhibit virus production both after re-infection and after reactivation of the latent virus.

The anti-viral potential of PTX is not restricted to HIV-1. In fact, PTX also delays lymphocytic choriomeningitis virus (LCMV) dissemination after intragastric inoculation in mice. PTX has no direct effect on LCMV replication, but it modulates integrin expression in part by blocking chemokine signals, thus inhibiting the G protein-coupled chemotactic response of macrophages infected during the initial exposure to the virus [102].

Few other studies have been performed where a putative therapeutic activity of PTX can be extrapolated. In particular, they concern the beneficial effects on the *H. pylori* –associated gastric inflammation [103], the up-regulation of angiotensin type 1 receptors (AT1R) thought to be involved in cardiac remodeling [104], the decrease of high blood pressure and the reduction of sympathetic vasoconstriction in spontaneously hypertensive rats [105].

Cytotoxic Necrotizing Factor 1 (CNF1) from *Escherichia coli*

Escherichia coli, a gram-negative aerobic bacterium that normally inhabits our intestine, becomes pathogenic following the acquisition of virulence factors. Some *E. coli* pathogenic strains mainly involved in extra-intestinal infections [106, 107] and, occasionally, detected in feces of children with diarrhea [108-110], produce a protein toxin named Cytotoxic Necrotizing Factor 1 (CNF1). The toxin is taken up into mammalian cells by receptor-mediated endocytosis and is delivered from late endosomes into the cytosol [111, 112] where it exerts its enzymatic activity. This consists in the deamidation of a pivotal glutamine residue of the GTP-binding proteins Rho, Rac, and Cdc42 (glutamine 63 of Rho or glutamine 61 of Rac and Cdc42) that gives rise to a glutamic acid [113-115]. The glutamine residue modified by CNF1 lies in the switch 2 domain of Rho proteins, which is involved in GTP hydrolysis; thus, the modification exerted by CNF1 blocks the Rho GTPases in their GTP-bound activated state. High levels of activated Rho GTPases are recognized by cells that ubiquitinate and degrade them into more physiological levels *in vitro* and *in vivo* [116, 117]. Rho, Rac and Cdc42 are involved in a huge number of crucial cellular processes, including actin cytoskeleton organization, gene transcription, cell proliferation and survival [118]. Most of the so far explored consequences of CNF1-induced Rho activation are either actin-dependent phenomena, such as contractility, cell spreading [119] and assembly of focal adhesion plaques [120] or stimulation of new activities in cells, comprising the induction of macropinocytosis [121, 122], the activation of the transcription factor NF-κB [123], the counteraction of apoptosis

[124, 125]. Very recently, we have reported the Rho-dependent ability of CNF1 to trigger the mitochondrial ATP production and to induce the formation of elongated and interconnected mitochondria [126]. This occurs *via* PKA-stimulated phosphorilation of Drp1, a protein member of the dynamin family of large GTPases that controls the final part of mitochondrial fission (Fig. **5**).

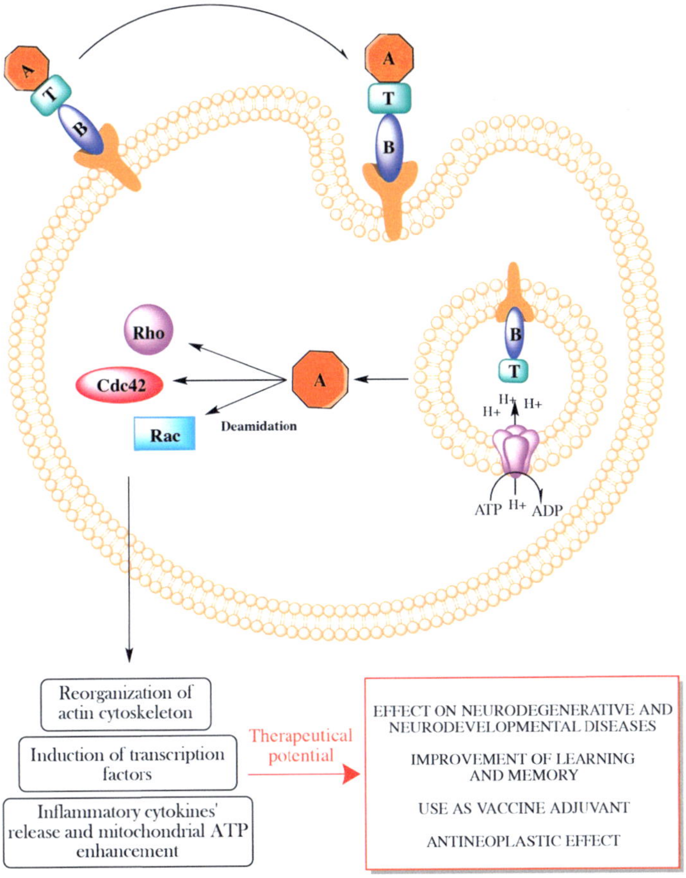

Figure 5: *E. coli* cytoxic necrotizing factor 1. The cytoxic necrotizing factor 1 (CNF1) is a monomeric A-B toxin composed of a receptor-binding (B) domain and an enzymatically active (A) one, separated by a hydrophobic domain (T) that allows the translocation of the A moiety across the endosome membrane. CNF1 acts by deamidating a specific aminoacidic residue of the small GTP-binding proteins Rho, Rac, and Cdc42. This modification blocks the Rho GTPases in their activated state, thus inducing different cell responses, among which the reorganization of actin cytoskeleton, the induction of transcription factors, the inflammatory cytokines' release and a mitochondrial ATP enhancement. The ability of CNF1 to modulate the actin network and the inflammatory response has been exploited *in vivo* to improve learning and memory and to design new vaccine adjuvants. Of importance, CNF1 shows a beneficial effect on neurodegenerative and neurodevelopmental diseases and has also possesses an antineoplastic property.

Therapeutical Applications

Effects on nervous system diseases. The ability of CNF1 to counteract the formalin-induced inflammatory pain in mice has been firstly reported in 2009 [127]. It is renowned that inflammatory pain is somehow regulated by Rho-dependent mechanisms [128, 129] and that Rho GTPases activate the Akt/NF-κB (nuclear factor kappa B)-dependent pathway that is involved in diverse opioid functions [130, 131]. Opioids are typically associated with pain and inflammation and, consequently, changes in opioid peptide levels as well as in the expression of their receptors at the cellular surface have been observed combined with these phenomena [132, 133]. Hence, it is reasonable that manipulation of the Rho GTPases would represent a strategy for treatments aimed at controlling the pain response. In keeping with these data, we have shown that the Rac GTPase-activating CNF1 induces analgesia by up-regulating μ-opioid receptors [127]. Preliminary data show that this response to CNF1 also occurs in a model of neuropathic pain [134].

Furthermore, we have demonstrated, for the first time, that the constitutive activation of Rho GTPases in mice, induced by a single intracerebroventricular (icv) injection of CNF1, results in rearrangements of actin in neural cells, enhanced synaptic transmission and improved learning and memory in young healthy mice [135]. These effects are persistent for months and are not observed in animals injected with an enzymatically inactive version of the toxin (CNF1 C866S), in which the enzymatic activity is abolished by substituting serine to cysteine at position 866. It is known that learning and memory depend on the morphology and dynamics of dendritic spines, their development and structural plasticity being controlled by Rho GTPases and actin dynamics. Interestingly, we have demonstrated that CNF1 can trigger structural remodeling and functional plasticity in the adult rat visual cortex [136], suggesting that the plasticizing effects of Rho GTPase activation could be beneficial in the treatment of those pathologies where spines are involved. Recent data support the hypothesis that enhanced synaptic plasticity and improved learning and memory described in CNF1-injected animals are mediated by astrocytes [137].

Overall, these results suggest that learning abilities can be improved through pharmacological manipulation of neural connectivity. In fact, using the toxin on animal models of CNS pathologies with cognitive impairment, we have recently demonstrated that a single icv inoculation of CNF1 can markedly improve the behavioral phenotype of a mouse model of Rett syndrome (RTT), a rare neurodevelopmental disorder and a genetic cause of mental retardation. In addition, CNF1 dramatically reverses the evident signs of atrophy in astrocytes of RTT mice, restoring wt-like levels of this cell population. Furthermore, CNF1 improves mitochondria functionality in CNF1-treated mice, indicating that modulation of brain Rho GTPases may constitute a totally innovative therapeutic approach for RTT [138]. Further exploring the CNF1 effect on learning, we have tested the toxin on an animal model of Alzheimer's Disease (AD) and atherosclerosis: the ApoE4 mouse. Transgenic mice homozygous for human ApoE4 gene show decreased levels of ATP, increased inflammatory cytokines level and accumulation of β-amyloid (Aβ) in the brain, all factors deemed responsible for triggering the cognitive decline. A single icv administration of the toxin is able to improve spatial and emotional memory, to increase the levels of Aβ and of the pro-inflammatory cytokine IL-1β, and to modify the cell energy, in terms of ATP content [117].

Treatment of cancer. Although CNF1 is apparently endowed with long-lasting cognition enhancing properties, its therapeutic potentiality is not limited to cognitive conditions. In fact, we have very recently shown an antineoplastic activity of CNF1, which is able to block proliferation and lead to death GL261 cells, a glioma cell model. *In vivo*, this CNF1 capacity was evaluated by an icv administration of the toxin five days after the injection of GL261 glioma cells into the adult mouse visual cortex. Even at high doses (up to 80 nM) CNF1 enhances survival of the experimental animals without causing obvious signs of toxicity. Indeed, 57% of the treated animals survived up to 60 days following GL261 glioma cell transplant [139].

Most CNF1 effects on brain of pathologic mouse models are probably mediated by astrocytes [137], which play a prominent role in pain modulation [140], in the regulation of synaptic plasticity and memory formation [141, 142], and are intimately involved in neuronal signaling processes and their energetic metabolism [143]. In our opinion, the ability of CNF1 to increase the mitochondrial energy production, observed both in *in vivo* and *in vitro*, could represent a key factor underlying the therapeutic property of the toxin in CNS disorders.

IMMUNOTOXINS

Immunotoxins (ITs) are hybrid molecules created by coupling an antibody or antigen with part or all of a toxin. The hybrid molecule combines the specificity of the antibody or antigen with the toxicity of the toxin. An IT can kill a cell through the binding of specific cell surface receptors, internalization and delivery of the toxin moiety to the cell cytoplasm. The first generation of ITs has been created by chemically conjugating an antibody to a protein toxin. Currently, ITs are constructed by using the recombinant DNA techniques and the principle of protein engineering, and the recombinant chimeric proteins are produced in *Escherichia coli* [144]. The cell binding ligand may be a growth factor, a cytokine, an antibody or a fragment of an antibody, and the toxin is usually the catalytic portion of either exotoxin A from *Pseudomonas aeruginosa* (PE) or diphteria toxin (DT) produced by *Corynebacterium diphteriae*. Although being different in their sequence and structure, both PE and DT ADP-ribosylate the Elongation Factor 2 (EF-2), thus blocking proteins synthesis and inducing cell death (Fig. **6**) [145-147]. Also, Shiga toxin that inactivates ribosomes by N-glycosidase activity, anthrax toxin that inactivates MAP kinases and cholera toxin, a toxin inhibiting by ADP ribosylation the Gs-α subunit of the heterotrimeric G protein, can be used in the drawing of the ITs. Furthermore, both fungal toxins, like α-sarcin and restrictocin, as well as toxins derived from plants, like ricin, abrin, modeccin, saporin (SAP), pokeweed antiviral protein (PAP) gelonin, bryodin 1, bouganin and tricosanthin can be conjugated with a specific target to construct an IT [148].

Therapeutical Applications

Treatment of cancer. The goal of the IT therapy is to target, to cell surface molecules, a cytotoxic agent whose subsequent internalization will cause cell death. Developing this type of therapy has aroused a great interest in the past two decades, especially in the field of cancer cure. The IT treatment greatly differs from that of chemotherapy, in its action and toxicity profile, and it has the potential to improve the systemic treatment of incurable tumors with existing modalities. They rely on tumor cells that often express surface receptors or other molecules, which distinguish them from their normal surrounding tissues. Thus, ITs are constructed by using ligands that recognize those surface receptors over expressed on tumor cells.

Different tumors are successfully treated with the use of ITs and an increasing amount of ligands are currently exploited to construct these efficient ITs. In particular, hematologic malignancies, like hairy cell leukemia (HCL), Hodgkin's lymphoma, T-cell lymphoma and chronic lymphocytic leukemia (CLL), are treated by means of ITs. In fact, immunotoxins are optimal for treating hematological malignancies, since malignant cells are often intravascular and accessible to intravenously administered drugs. Furthermore, patients often lack sufficient immunity to make antibodies against the toxin. On the other hand,

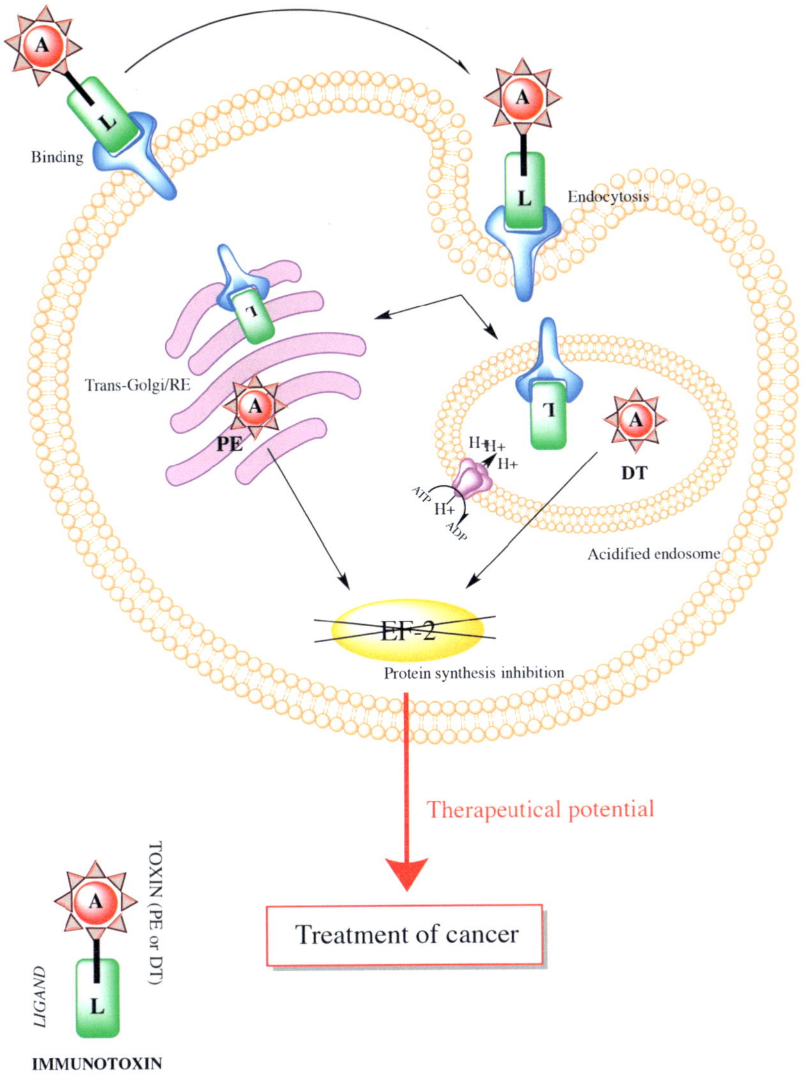

Figure 6: Mechanism of action of immunotoxins. The immunotoxins are chimeric proteins composed of a ligand subunit (L), which recognizes surface receptor over-expressed on target cells, fused to the catalytic portion of a toxin (A). The ligand subunit may be a growth factor, a cytokine, a monoclonal antibody or a fragment of an antibody, while the toxin moiety is commonly the enzymatic domain of the exotoxin A from *P. aeruginosa* (PE) or diphtheria toxin (DT) produced by *C. diphtheria*. Both PE and DT, ADP-ribosylate the Elongation Factor-2 (EF-2), thus blocking protein synthesis and inducing cell death. After the ligand subunit has bound to the surface of the target cell, the immunotoxin is internalized by endocytosis and the enzymatic fragment of the toxin translocates into the cytosol. Once into the cytosol, the toxin molecule blocks protein synthesis, thus killing the target cell. Immunotoxins that recognize antigens over-expressed on cancer cells can be used as anticancer therapeutics.

targeting solid tumors with ITs has been much more difficult because not only are the cellular junctions tighter and the tumor cells more packed, but also patients are less immuno-suppressed and thus likely to make neutralizing antibody to the toxin. However, currently a number of clinical trials, including those targeting the less accessible solid tumors such as breast, colon, stomach, pancreas, cervix and lung cancer, are in progress [a detailed description is out of the scope of this paper, for more details see references 149-151].

Regarding the anthrax toxin, the unique requirement of PA to be activated on the surface of the target cell has provided an opportunity to re-engineer this protein to make its activation dependent on either Matrix metalloproteinases (MMPs) [152] or urokinase Plasminogen Activator (uPA) [153]. Both molecules are over-expressed in a variety of tumor tissues and tumor cell lines, their expression correlating to tumor malignancy and metastasis. Mutated anthrax toxin protective antigen (PA) proteins in which the furin protease cleavage site is replaced by sequences selectively cleaved by MMPs have been constructed [152] or activated by uPA [153]. These PA proteins were rapidly and selectively activated on the surface of tumor cells over-expressing such molecules, thus causing the internalization of the recombinant anthrax toxin that selectively kills tumor cells (Fig. 7). Interestingly, the anti-tumor efficacy of the urokinase activation-dependent anthrax toxin has also been reported in *in vivo* human tumor model [154, 155]. The engineered toxins exhibit much lower toxicity than wild-type LT to mice but higher toxicity to human tumors in tumor xenograft models. This is due to a greater bioavailability of the engineered LT, a consequence of the limited uptake and clearance of the modified toxin by normal cells. Interestingly, the MMP-activated LT shows a potent anti-tumor activity to human melanomas but also to other tumor types, including lung and colon carcinomas. Its antitumor activity is also largely due to the indirect targeting of tumor vasculature and angiogenic processes [156]. Another approach developed with anthrax toxin was based on its highly efficient system for delivering cytotoxic enzymes to the cytosol of tumor cells. In fact, the protease-activated PA assembles into an oligomeric-protein-conducting channel that delivers the anthrax catalytic effectors to endosomes and then translocates them to the cytosol. The native anthrax effector proteins LF and EF can be replaced with a fusion protein containing the N-terminal 254 amino acids of anthrax toxin LF and the *Pseudomonas aeruginosa* exotoxin A (PE) catalytic domain (PEIII) [157] or the enzymatic subunit of the Cytolethal Distending Toxin (CdtB) produced by several Gram-negative bacteria [158]. The LF domain was used to deliver PEIII or CdtB to the cytosol where they act by causing cell death [158, 159].

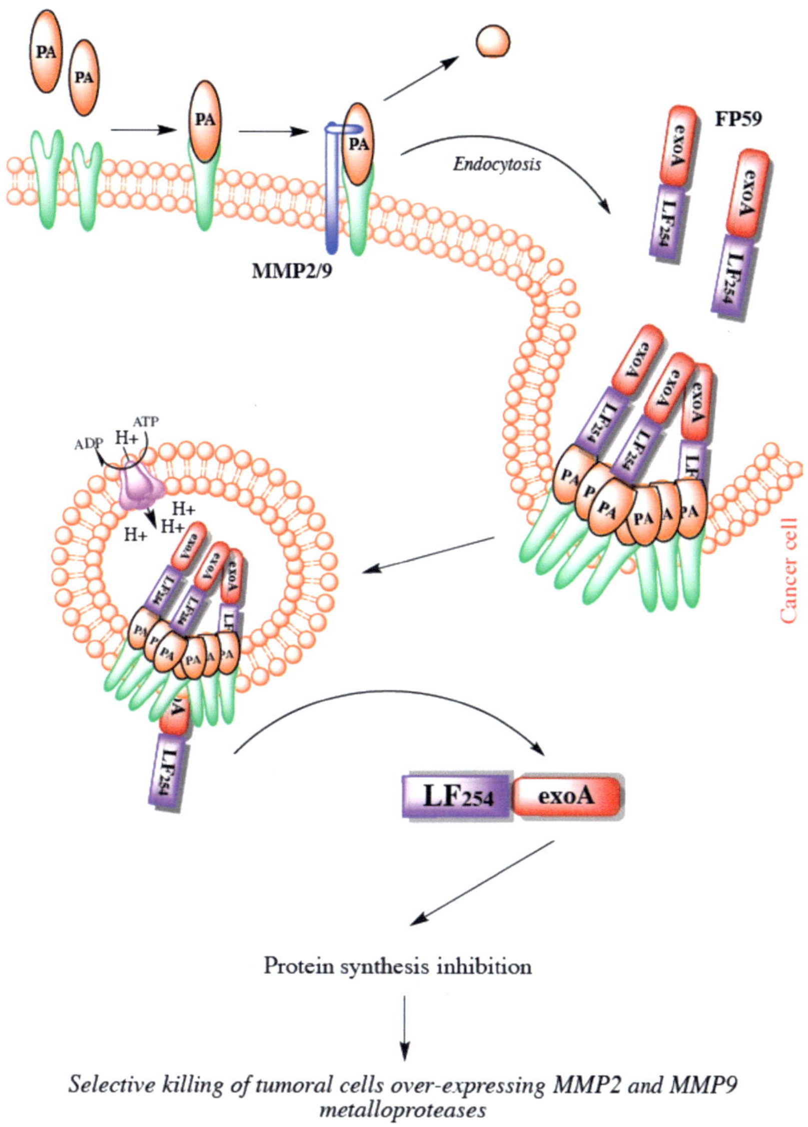

Figure 7: Engineered anthrax toxin. Mutated anthrax toxin protective antigen (PA) consists of PA proteins in which the furin protease cleavage site is replaced by sequences selectively cleaved by metalloproteases (MMP) MMP- 2 and MMP-9. These MMP-targeted PA proteins are activated rapidly and selectively on the surface of MMP-over-expressing tumor cells. The activated PA proteins permit internalization of a recombinant cytotoxin FP59, consisting of anthrax toxin lethal factor residues 1-254 fused to the ADP-rybosilation domain of PE that induces selective killing of MMP-over-expressing human tumor cell lines.

The major challenges faced in ITs' research are the immunogenicity, toxicity, specificity and stability. To minimize these problems, Pennell and Erickson [160] have suggested ways to design an ideal IT suggesting strategies to enlarge the IT therapeutic window. They emphasized on the reduction of size and humanization to reduce immunogenicity and toxicity while making divalent or multivalent target moieties for improving affinity and specificity. The fully humanized ITs carry an endogenous protein of human origin, like a pro-apoptotic protein or RNase as for the cytotoxic moiety. In this context, the pro-apoptotic proteins Bik, Bax and Bak have been successfully employed to target IL2R for CTCL, renal cancer and melanoma. Another important concern is the poor penetration in solid tumors, which is now defeated by generation of small antibody fragments domains and protein scaffolds, many of which have been tested and are in early clinical trials. For an exhaustive review on the potential of anticancer ITs see [160-162].

BACTERIAL PROTEIN TOXINS USED AS ADJUVANTS OR DRUG DELIVERY AGENTS

Bacterial toxins are among the strongest antigens and some of them have been described to act as adjuvant for vaccines or drug delivery agent (Table **2**). Adjuvants are vaccine additives that enhance the elicited levels of antibodies and T lymphocyte responses [163] in ways that presumably mimic natural infection, whereas drug delivery agents are compounds able to promote absorption of therapeutic agents, thus increasing their effectiveness [164].

Table 2: Bacterial protein toxins used as potentiation or delivery agents

TOXIN	Therapeutic use
Cholera toxin (CT) from *V. cholerae*	Mucosal adjuvant
Zonula occludens toxin (ZOT) from *V. cholerae*	Mucosal adjuvant Drug delivery agent
CTA1-DD fusion protein from CT and *Staphylococcus aureus* protein A	Mucosal adjuvant Enhancer of humoral immune responses
Cytotoxic necrotizing factor 1 (CNF1) from *E. coli*	Mucosal adjuvant
Adenylate Cyclase (AC) from *B. pertussis*	Mucosal adjuvant Immunotherapy adjuvant
Pertussis toxin (PTX) from *B. pertussis*	Mucosal adjuvant

Cholera Toxin (CT) from *Vibrio Cholerae* and Heat-Labile Enterotoxin (LT) from *E. coli*

Bacterially-derived enterotoxins, such as LT and the closely related CT are among the most powerful mucosal adjuvants identified so far [165]. They both promote secretory diarrhea during microbial infection. The induction of intestinal fluid secretion occurs after a series of events involving both changes of toxin structure and activation of intracellular signaling pathways. Each enterotoxin is composed of an A and B subunit of which the B subunit forms a homo-pentamer structure that binds to surface receptors of glycosphingolipid nature, virtually present in all cells, including enterocytes, dendritic cells, macrophages, B and T lymphocytes [166]. The toxin activity depends on the proteolytic cleavage and disulfide bond reduction that separate the A-subunit into two domains, the enzymatically active A1 subunit and a smaller A2 peptide. Transport of A1 into the cytoplasm results in ADP-ribosylation of Gsα, followed by irreversible activation of adenylate cyclase and increases in intracellular levels of cAMP. In intestinal epithelial cells, this induction of cAMP causes a disregulation of cAMP-sensitive ion transport mechanisms, inhibiting intracellular salt absorption, increasing electrolyte transport into the gut lumen, and creating an osmotic gradient favoring intestinal water secretion [167].

Vaccine adjuvants. Beside being potent immunogens, both CT and LT can act as adjuvants for the enhancement of mucosal and serum antibody responses to a mucosal co-administered antigen, resulting in long-term memory to antigens [168, 169]. Their use in humans is, however, limited by the fact that CT and LT are extremely toxic, and very low doses are sufficient to induce diarrhea in human volunteers. An immune response occurs against these toxins when used as adjuvant. However, high levels of toxin specific IgA at the inductive site, reduce but do not completely inhibit the mucosal adjuvanticity [170] so that CT and LT remain the most potent mucosal adjuvants known to date. CT and LT have immune-modulatory effects on different cell types that may facilitate adjuvanticity [171-173], although their interaction with dendritic cells is thought to be of major importance for mediating the adjuvant effects [174, 175].

Two different approaches have been followed in order to develop non-toxic CT and LT derivatives with strong mucosal adjuvanticity. The first has been based on the use of the non-toxic B subunits alone. The results obtained demonstrated that B-subunits are very poor mucosal adjuvants [165, 176]. The second approach has been based on the generation of genetically detoxified mutants of CT and LT where mutations in the enzymatically active A subunit were introduced. Several mutants were generated and all behave as strong mucosal adjuvants [177-180] for all antigens tested when given intranasally, orally or vaginally, inducing very strong systemic and mucosal antigen-specific antibody responses [181]. Interestingly, one of these mutants, LTK63, is being used in Phase I clinical studies where influenza vaccine is given intranasally with LTK63 as an adjuvant. Results are promising since LTK63 enhances mucosal IgA responses to the influenza virus [182, 183]. The fine molecular mechanisms of mucosal adjuvanticity of CT and LT mutants are still not fully understood, but it is reasonable to hypothesize that the enzymatic activity of the A subunit is not necessary in the adjuvanticity of mutants. Moreover, the holotoxin structure exerts an important role in adjuvanticity since mutants completely lacking the A subunit are weak adjuvants [181].

More recently, Norton and coworkers have characterized the structural, enzymatic, enterotoxic, and adjuvant properties of a novel mutant of LT, designated LT(R192G/L211A), or dmLT, obtained with the introduction of aminoacid substitutions in order to generate an adjuvant-active but nontoxic molecule. This mutant showed a reduced enzymatic activity and no enterotoxicity in the patent mouse assay. *In vitro* and *in vivo* data suggest that the reduced toxicity of this molecule is a consequence of increased sensitivity to proteolysis and rapid intracellular degradation in mammalian cells. In conclusion, dmLT is a safe and powerful detoxified enterotoxin with the potential to function as a mucosal adjuvant for co-administered antigens and to elicit anti-LT antibodies without undesirable side effects [184].

As regards CT, most studies have proposed that it promotes a strong T helper type 2 (Th2)-dominant response to bystander antigens, based on the production of IL-4, IL-5, and IL-10 but little IFN-γ. Other studies have reported mixed Th1/Th2 response with the production of both IFN- γ and IL-4 following oral and intranasal immunization with CT, and that CT also promotes the generation of regulatory T cells against bystander antigens. Furthermore, Lee and coworkers have reported the induction of Th17-dominated responses in mice by intranasal delivery of CT [185]. This may occur through activation of dendritic cells and lung epithelial cells *via* the GM1-ganglioside binding to produce and activate Th17-driving cytokines, such as IL-6 and TGF-β, which result in the skewed differentiation of Th17 cells during antigenic stimulation [185].

Zonula Occludens Toxin (ZOT) from *Vibrio Cholerae*

Pathogenic strains of *V. cholerae* produce several exotoxins, including the Zonula Occludens toxin (ZOT) [186]. ZOT is of bacteriophage origin and is originally composed by a single polypeptide chain of 45 kDa localized in the *V. cholerae* outer membrane. Following cleavage, a carboxyl terminal fragment of 12 kDa is excreted that is probably responsible for the biological effect of the toxin [187]. The 12 kDa fragment of ZOT binds to a specific receptor, whose surface expression differs on various cells, and activates a complex intracellular cascade of events that involve a dose and time-dependent protein kinase Cα - related polymerization of actin filaments, strategically localized to regulate the paracellular pathway [188]. The result is the modification of the cytoskeletal organization and the opening of the intercellular tight junctions (TJ), thus leading to a transient increase of the permeability of the paracellular route [186]. TJs are intercellular highly dynamic structures regulated in response to exogenous and endogenous stimuli that control paracellular permeability and epithelial polarity. Barrier function is a new target for drug development, drug delivery, and adjuvant development. The discovery of ZOT protein from *V. cholerae* has allowed to explore TJs as a drug development target since it is important to create TJ opening molecules that are able to open TJ in a reversible and non-toxic manner.

Drug delivery agent. The main route of absorption of hydrophilic drugs (proteins, petptides and others) is the paracellular route that is governed by TJ. *In vitro* experiments in rabbit ileum demonstrate that ZOT reversibly increased intestinal permeability of insulin by 72% and immunoglobulin by 52% in a time-dependent manner [189]. ZOT was also tested in an *in vivo* primate model of diabetes mellitus where insulin was intragastrically administered in combination with ZOT. Measurements of blood insulin levels reveal that insulin bioavailability increases from 5.4% in controls to 18% when ZOT was used as enhancer of drug delivery. Recently, the ability to enhance drug delivery has been demonstrated to reside in a small part of the ZOT molecule. In fact, a novel synthetic hexapeptide derived from ZOT (AT1002) enhances the absorption of both cyclosporine A [190] and heparin [191]. Interestingly, ZOT as well as its 12 kDa active fragment have been introduced as potential absorption enhancers for delivery of the therapeutic agents also to the central nervous system [192, 193].

The ability of ZOT to reversibly alter intestinal epithelial tight junctions was exploited to deliver soluble antigens through the nasal mucosa for the induction of antigen-specific systemic and mucosal immune responses. Intranasal immunization of mice with ovalbumin (Ova) and recombinant ZOT, induces anti-Ova serum immunoglobulin G (IgG) titers that are approximately 40-fold higher than those induced by immunization with antigen alone [194]. ZOT also interferes with tolerance induced by both nasal and oral administration of soluble protein antigens. Rossi and coworkers [195], using a model of systemic tolerance in which a reduced production of interferon-gamma (IFN-gamma) and interleukin-2 (IL-2) was induced upon administration of gliadin alone, demonstrated that this immune suppression is reverted in mice receiving gliadin together with ZOT. It is known that down-regulation of such cytokines is associated to a suppression of the T-cell proliferation, and ZOT co-administration with gliadin completely reverts this phenomenon. Recombinant fragments of ZOT are being tested for their mucosal adjuvant activity that can provide promising tools for the development of mucosal vaccine [196].

To avoid the technical hurdles of developing a complex protein as a drug to open TJs, Gopalakrishnan group have worked on the ZOT hexapeptide AT-1002 providing a mechanistic explanation for AT-1002-induced tight junction disassembly, and demonstrating that AT-1002 can be used for delivery of other agents *in vivo*. This peptide, in fact, beside retaining the TJ opening properties of the parent molecule, is able to induce several effects including redistribution of ZO-1 away from cell junctions, activation of *src* and MAP kinase pathways, increase of ZO-1 tyrosine phosphorylation, rearrangement of actin filaments, reduction in transepithelial electrical resistance (TEER) and increase in lucifer yellow permeability across Caco-2 cell monolayers [197]. Additional experiments showed that co-administration of AT-1002 with salmon calcitonin intratracheally increases the systemic exposure of salmon calcitonin by up to 5.2- fold, thus suggesting AT-1002 as a prototype of a new class of TJ modulators that represent a promising advancement in mucosal drug delivery [197].

(CTA1)–DD Fusion Protein from CT and *Staphylococcus Aureus* Protein A

Mast cell activation is one of the most dramatic immune-mediated responses that the body can encounter and, recently, attention has been given to the role of mast cells in regulating innate and adaptive immune responses. It has been reported that mast cell activation can be used as an adjuvant to promote Ag-specific humoral immune responses upon vaccination. Fang and coworkers have used a clinically relevant mucosal adjuvant, the cholera toxin A1 subunit (CTA1)–DD, which is a fusion protein composed of CTA1, the ADP-ribosylating part of cholera toxin, and DD, two Ig-binding domains derived from *Staphylococcus aureus* protein A [198].

Vaccine Adjuvant. Since mast cells reside at sites of pathogen entry, such as the skin and mucosal surfaces, including the gastrointestinal and respiratory tracts, they are particularly relevant from a vaccine development point of view. Fang and coworkers (2010) [198] demonstrated that CTA1–DD could target mast cells causing the TNF-α- release. *In vitro* data showed that binding of CTA1–DD to IgG is necessary to activate mouse mast cells, and *in vivo* observations revealed that serum IgG could form immune complexes with CTA1–DD (*e.g.*, at mucosal sites, such as the nasal mucosa). Hence, CTA1–DD mouse ear skin intradermal administration triggers increased vascular permeability, which is generally considered a mast cell-mediated effect. Altogether, these data suggest that mast cell activation may be part of the mechanisms that render CTA1–DD effective mucosal adjuvant functions and that CTA1–DD could use mast cell activation and TNF-α release to enhance humoral immune responses following intranasal immunizations.

These observations are encouraging and suggest that mast cells could be exploited to further increase humoral immune responses following mucosal immunizations. Importantly, mice immunized with CTA1–DD do not exhibit augmented levels of total IgE or IgG Abs in their serum, clearly in favor of its safe and nontoxic use as a clinical vaccine adjuvant [198].

CNF1 from *E. coli*

We have already described the *E. coli* toxin CNF1 in the "Bacterial protein toxins used in therapy" section. Now we report its use as a vaccine adjuvant, a characteristic that was first described a few years ago [199].

Previous studies have demonstrated that CNF1 induces secretion of pro-inflammatory and immunomodulatory cytokines *via* the activation of Rho proteins [117, 118]. The subsequent investigation of the toxin immunomodulatory effects in a mouse model of oral administration showed that, similarly to cholera toxin, CNF1 elicits adjuvanticity anti-OVA responses, both systemic and mucosal [196]. Animals that were orally immunized with OVA, a prototype soluble antigen, co-administered with CNF1 displayed serum IgG anti-OVA antibody responses comparable to those elicited by cholera toxin. Interestingly, immunization with CNF1 C866S failed to enhance serum anti-OVA responses in animals co-fed OVA, demonstrating that the enzymatic activity of CNF1 is necessary for adjuvanticity. Afterwards, Munro and coworkers also reported the ability of CNF1 to act as an adjuvant in immunization against tetanus toxoid through nasal route [200], pointing at CNF1 as an attractive mucosal adjuvant candidate for nasal vaccines. In the same vein, the *Bordetella dermonecrotic* toxin (DNT), a closely related Rho activating toxin, also showed adjuvant properties. These findings strongly suggest that the host response to CNF1, in addition to promoting innate immune effectors and regulators, may also stimulate the adaptative immune system. Manipulation of Rho proteins thus provides a possible new approach for the development of effective mucosal immunoadjuvants.

Adenylate Cyclase (AC) Toxin and PTX from *B. Pertussis*

In addition to PTX, *B. pertussis* produce a second important protein toxin of 1706 aminoacids known as Adenylate Cyclase (AC) toxin [201]. It expresses haemolytic and calmodulin-dependent adenylate cyclase activities, both important at the early stages of infection. The catalytic and calmodulin binding domain is located in 400 amino-terminal residues whereas hemolysis resides in the carboxy-terminal domain. This domain also mediates binding and internalization of the toxin into target cells. The translocation of the catalytic domain appears to directly occur at the plasma membrane and does not require endocytosis and vesicular trafficking.

Adjuvant for vaccines and in immunotherapy. AC toxin was shown to be highly antigenic either in mice or in humans [202, 203]. Some studies report that AC acts as mucosal adjuvant for other antigens when coadministered by the intra-nasal route in mice. For this purpose, two forms of AC have been used, the enzymatically active form and the toxin lacking the enzymatic adenylate cyclase activity. In all cases the adjuvanticity effect was greater with the enzymatically inactive form [202, 204].

Also a genetically modified form of PTX (already described in the "Bacterial protein toxins used in therapy" section), PT- 9K/129G, devoid of the A-protomer-specific enzymatic activity but retaining a functional PTX-B subunit, has successfully been used as a mucosal adjuvant for a tetanus vaccine in mice [205]. The PTX ability to inhibit G-protein-coupled chemokine receptors by the ADP-ribosylation of Gi proteins, affects the innate immune response, including increased antibody production and Th1/Th2 cytokine production [206]. This toxin property permits its use as adjuvant for vaccines and in immunotherapy. Very recent studies, in fact, indicate PTX as a good adjuvant for vaccines against *Streptococcus pneumoniae*. These vaccines, based on the Pneumococcal Surface protein A (PspA), if combined with the whole cell pertussis vaccine (wP), induce high levels of antibodies and protection against different pneumococcal strains when administered to mice *via* the nasal route. The formulation composed of PspA from clade 5 (PspA5) and wP (PspA5-wP) also induces secretion of IL-17 by splenocytes and infiltration of leukocytes in the lungs after the challenge. Moreover, combination of PspA5 with wild type or mutant *B. pertussis* strains or with purified components, showed that the PTX-containing formulations induce the highest levels of antibodies and protection [207].

Another aspect of the adjuvant potential of PTX concerns the immunotherapy. In fact, PTX has been used as adjuvant immunotherapy in a C6 glioma model. Glioblastoma multiforme is the most frequent primary brain tumor. It has poor prognosis, and it remains refractory to current treatment, since temozolomide (TMZ) activity appears to be limited by the occurrence of chemoresistance. The cytotoxic effect of PTX in combination with TMZ for glioma treatment, both *in vitro* and *in vivo* RG2 glioma model, suggests that the concomitant use of PTX plus TMZ could represent an advantage to ameliorate the glioma treatment [208]. Therefore, the two protein toxins CNF1 and PTX apparently offer a novel effective approach to attack glioma cells. Also, PTX, by inhibiting G(αi)PCRs, protects hepatocytes, but not hepatocellular carcinoma cells, against bile acid- and cytokine-induced apoptosis and possesses a therapeutic potential as a primary hepatoprotective drug, as well as an adjuvant in anti-cancer therapy [209].

CONCLUSION

Bacterial toxins, capable of interfering with many essential functions and pathways of eukaryotic cells, have evolved along with the carrier bacterium, probably because conferring some benefit for the bacterial spread and survival. However, even if harmful to the susceptible host during an infection, the activities of several bacterial toxins have been exploited as tools for cell biology studies and, also for medical purposes. In fact, the scientific research has permitted the conversion of some of these harmful bacterial molecules into powerful and beneficial drugs, "magic bullets" that are increasingly used for the treatment of human diseases. In this Chapter, we have briefly drawn the many potentialities of toxins, some of which are now approved for the treatment of some types of cancer or some forms of neurological disorders. Although the current industry model for drug discovery shows a certain resistance to accept bacterial factors as putative new therapeutics, the resource is so inexhaustible and promising that will certainly provide clues in the near future for the development of new medicines.

ACKNOWLEDGEMENTS

The Authors are indebted to Rossella Di Nallo, for her invaluable help in editing the manuscript.

CONFLICT OF INTEREST

The authors confirm that this chapter contents have no conflict of interest.

DISCLOSURE

This is an updated report of the earlier review article by A. Fabbri, S. Travaglione, L. Falzano, C. Fiorentini. "Bacterial protein toxins: current and potential clinical use" Curr Med Chem 2008; 15(11): 1116-1125.

REFERENCES

[1] Aoki KR. Botulinum toxin: a successful therapeutic protein. Curr MedChem 2004; 11: 3085-92.
[2] Wenzel RG. Pharmacology of botulinum neurotoxin serotype A. Am J Health Syst Pharm 2004; 61: S5-10.
[3] Scott AB. Development of botulinum toxin therapy. Dermatol Clin 2004; 22: 131-3.
[4] Simpson LL. Identification of the major steps in botulinum toxin action. Annu Rev Pharmacol Toxicol 2004; 44: 167-93.
[5] Thakker MM, Rubin PA. Pharmacology and clinical applications of botulinum toxins A and B. Int Ophthalmol Clin 2004; 44: 147-63.
[6] Aoki KR, Guyer B. Botulinum toxin type A and other botulinum toxin serotypes: a comparative review of biochemical and pharmacological actions. Eur J Neurol 2001; 8 (S5): 21-9.
[7] Setler PE. Therapeutic use of botulinum toxins: background and history. Clin J Pain 2002; 18: S119-24.
[8] Gonnering RS. Pharmacology of botulinum toxin. Int Ophthalmol Clin 1993; 33: 203-26.
[9] Rossetto O, Megighian A, Scorzeto M, *et al.* Botulinum neurotoxins. Toxicon 2013; 67: 31-6.
[10] Schiavo G, Matteoli M, Montecucco C. Neurotoxins affecting neuroexocytosis. Physiol Rev 2000; 80: 717-66.
[11] Erbguth FJ, Naumann M. Historical aspects of botulinum toxin: Justinus Kerner (1786-1862) and the "sausage poison". Neurol 1999; 53: 1850-3.
[12] Villar RG, Elliott SP, Davenport KM. Botulism: the many faces of botulinum toxin and its potential for bioterrorism. Infect Dis Clin North Am 2006; 20: 313-27.
[13] Scott AB. In Botulinum Neurotoxin and Tetanus Toxin; Simpson L.L., Ed. Academic Pres: San Diego 1989; pp. 399-412.
[14] Truong DD, Jost WH. Botulinum toxin: clinical use. Parkin Rel Disorders 2006; 12: 331-55.
[15] Hackett R, Kam PC. Botulinum toxin: pharmacology and clinical developments: a literature review. Med Chem 2007; 3: 333-45.
[16] Papapetropoulos S, Singer C. Botulinum toxin in movement disorders. Semin Neurol 2007; 27: 183-94.
[17] Kenney C, Jankovic J. Botulinum toxin in the treatment of blepharospasm and hemifacial spasm. J Neural Transm 2007; 18: 233-9.
[18] Clark GT, Ram S. Four oral motor disorders: bruxism, dystonia, dyskinesia and drug-induced dystonic extrapyramidal reactions. Dent Clin North Am 2007; 51: 225-43.
[19] Long H, Liao Z, Wang Y,*et al.*Efficacy of botulinum toxins on bruxism: an evidence-based review. Int Dent J 2012; 62: 1-5.
[20] Ferreira JJ, Costa J, Coelho M, *et al.*The management of cervical dystonia. Expert Opin Pharmacother 2007; 8: 129-40.
[21] Karp BI. Botulinum toxin physiology in focal hand and cranial dystonia. Toxins (Basel) 2012; 4: 1404-14.
[22] Watts CR, Truong D, Nye C. Evidence for the effectiveness of botulinum toxin for spasmodic dysphonia from high-quality research designs. J Neural Transm 2008; 115: 625-30.
[23] Colosimo C, Tiple D, Berardelli A. Efficacy and safety of long-term botulinum toxin treatment in craniocervical dystonia: a systematic review. Neurotox Res 2012; 22: 265-73.
[24] Cordivari C, Misra VP, Catania S, *et al.*Treatment of dystonic clenched fist with botulinum toxin. Mov Disord 2001; 16: 907-13.

[25] Rahimi B, Bee C, Debicki D, *et al.*Effectiveness of BoNT A in Parkinson's disease upper limb tremor management. Can J Neurol Sci 2013; 40: 663-9.

[26] Hoare B. Rationale for Using Botulinum Toxin A as an Adjunct to Upper Limb Rehabilitation in Children With Cerebral Palsy. J Child Neurol 2014; May 11 [Epub ahead of print].

[27] Mesterman R, Gorter JW, Harvey A, *et al.* Botulinum toxin type A in children and adolescents with severe cerebral palsy: a retrospective chart review. J Child Neurol 2014; 29: 210-3.

[28] Baba Y, Osborne MD, Wszolek ZK, *et al.* Treatment of spasticity with botulinum toxin. Ortop Traumatol Rehabil 2004; 6: 665-72.

[29] Freitag FG. Botulinum toxin type A in chronic migraine. Expert Rev Neurother 2007; 7: 463-70.

[30] Ramsey RR, Ryan JL, Hershey AD, *et al.* Treatment adherence in patients with headache: a systematic review. Headache 2014; 54: 795-816.

[31] Shamliyan TA, Kane RL, Taylor FR. Migraine in Adults: Preventive Pharmacologic Treatments [Internet].: Agency for Healthcare Research and Quality. Rockville (MD) 13-EHC068-EF. AHRQ Comparative Effectiveness Reviews 2013.

[32] Michaud LJ, Louden EJ, Lippert WC, *et al.* Use of Botulinum Toxin Type A in the Management of Neonatal Brachial Plexus Palsy. PM R 2014 pii: S1934-1482(14)00200-7. doi: 10.1016/j.pmrj.2014.05.002.

[33] Glaser DA, Hebert AA, Pariser DM, Solish N. Primary focal hyperhidrosis: scope of the problem. Cutis 2007; 79 (S5): 5-17.

[34] Verma A, Steele J. Botulinum toxin improves sialorrhea and quality of living in bulbar amyotrophic lateral sclerosis. Muscle Nerve 2006; 34: 235-7.

[35] Svetel M, Vasic M, Dragasevic N, Pekmezovic, T.; Petrovic, I.; Kostic, V. Botulinum toxin in the treatment of sialorrhea. Vojnosanit Pregl 2009; 66: 9–12.

[36] Wortmann SB, De Brouwer APM, Wevers RA, *et al.* In GeneReviews® [Internet]; University of Washington Seattle 1993-2014.

[37] Lisan Q, Raynal M, Pons Y, Kossowski M. Catheterization of post-traumatic parotid duct sialocele., Eur. Ann. Otorhinolaryngol. Head Neck Dis 2014; pii, S1879-7296(14)00035-0.

[38] Antunes AA, Srougi M, Coelho RF, de Campos Freire G. Botulinum toxin for the treatment of lower urinary tract symptoms due to benign prostatic hyperplasia. Nat Clin Pract Urol 2007; 4: 155-60.

[39] Morrisroe SN, Chancellor MB. Botulinum toxin use in the lower urinary tract. ScientificWorldJournal 2007; 7: 808-17.

[40] Radwan MM, Ramdan K, Abu-Azab I, Abu-Zidan FM. Botulinum toxin treatment for anal fissure. Afr Health Sci 2007; 7: 14-7.

[41] Brashear A. Botulinum toxin type A: Exploring new indications. Drugs Today (Barc) 2010; 46: 671-82.

[42] Dellis A, Papatsoris AG. Expert. Intravesical treatment of bladder pain syndrome/interstitial cystitis: from the conventional regimens to the novel botulinum toxin injections. Opin Investig Drugs 2014; 23: 751-7.

[43] Ansiaux R, Gallez B. Use of botulinum toxins in cancer therapy. Expert Opin Investig Drugs 2007; 16: 209-18.

[44] Bagheri R, Fattahi SH, Haghi SZ, *et al.* Botulinum toxin for prevention of delayed gastric emptying after esophagectomy. Asian Cardiovasc Thorac Ann 2013; 21: 689-92.

[45] Costantin L, Bozzi Y, Richichi C, *et al.* Antiepileptic effects of botulinum neurotoxin E J Neurosci 2005; 25: 1943-51.

[46] De Boulle KL. Botulinum neurotoxin type A in facial aesthetics. Expert Opin Pharmacother 2007; 8: 1059-72.

[47] Walker TJ, Dayan SH. Comparison and Overview of Currently Available Neurotoxins. J. Clin Aesthet Dermatol 2014; 7: 31-39.

[48] Zhibo X, Miaobo Z. Intralesional botulinum toxin type A injection as a new treatment measure for keloids. Plast Reconstr Surg 2009; 124: 275–7.

[49] Xiao Z, Zhang F, Cui Z. Treatment of hypertrophic scars with intralesional botulinum toxin type A injections: a preliminary report. Aesthetic Plast Surg 2009; 33: 409–12.

[50] Naumann M, Boo LM, Ackerman AH, Gallagher CJ. Immunogenicity of botulinum toxins. Gallagher J Neural Transm 2013; 120: 275–90.

[51] Lalli G, Gschmeissner S, Schiavo G. Myosin Va and microtubule-based motors are required for fast axonal retrograde transport of tetanus toxin in motor neurons. J Cell Sci 2003; 116: 4639–50.

[52] Grumelli C, Verderio C, Pozzi D, Rossetto O, Montecucco C, Matteoli M. Internalization and mechanism of action of clostridial toxins in neurons. Neurotoxicol 2005; 26: 761–7.

[53] Bizzini B, Stoeckel K, Schwab M. An antigenic polypeptide fragment isolated from tetanus toxin: chemical characterization, binding to gangliosides and retrograde axonal transport in various neuron systems. J Neurochem 1977; 28: 529–42.

[54] Toivonen JM, Oliván S, Osta R. Tetanus toxin C-fragment: the courier and the cure? Toxins (Basel) 2010; 2: 2622-44.

[55] Dobrenis K, Joseph A, Rattazzi MC. Neuronal lysosomal enzyme replacement using fragment C of tetanus toxin. Proc Natl Acad Sci USA,1992;89: 2297–301.

[56] Figueiredo DM, Hallewell RA, Chen LL, *et al.* Delivery of recombinant tetanus-superoxide dismutase proteins to central nervous system neurons by retrograde axonal transport. Exp Neurol 1997;145: 546–54.

[57] Francis JW, Bastia E, Matthews CC, *et al.* Tetanus toxin fragment C as a vector to enhance delivery of proteins to the CNS. Brain Res 2004;1011: 7–13.

[58] Benn SC, Ay I, Bastia E, *et al.* Tetanus toxin fragment C fusion facilitates protein delivery to CNS neurons from cerebrospinal fluid in mice. J Neurochem 2005;95: 1118–31.

[59] Francis JW, Figueiredo D, vanderSpek JC, *et al.* A survival motor neuron: tetanus toxin fragment C fusion protein for the targeted delivery of SMN protein to neurons. Brain Res 2004;995: 84–96.

[60] Matthews CC, Figueiredo DM, Wollack JB, *et al.* Protective effect of supplemental superoxide dismutase on survival of neuronal cells during starvation. Requirement for cytosolicdistribution. J Mol Neurosci 2000;14: 155–66.

[61] Payne AM, Zheng Z, Messi ML, Milligan CE, Gonzalez E, Delbono O. Motor neurone targeting of IGF-1 prevents specific force decline in ageing mouse muscle. JPhysiol 2006;570: 283-94.

[62] Payne AM, Messi ML, Zheng Z, Delbono O. Motor neuron targeting of IGF-1 attenuates age-related external Ca^{2+}-dependent skeletal muscle contraction in senescent mice. Exp Gerontol 2007;42: 309-19.

[63] Larsen KE, Benn SC, Ay I, *et al.* A glial cell line-derived neurotrophic factor (GDNF):tetanus toxin fragment C protein conjugate improves delivery of GDNF to spinal cord motor neurons in mice. Brain Res 2006; 1120: 1-12.

[64] Li J. Chian RJ, Ay I, *et al.* Insect GDNF:TTC fusion proteinimproves delivery of GDNF to mouse CNS. Biochem Biophys Res Commun 2009;390: 947–51.

[65] Ciriza J, Moreno-Igoa M, Calvo AC, *et al.* A genetic fusion GDNF-C fragment of tetanustoxinprolongssurvivalinasymptomatic mouse ALS model. Restor Neurol Neurosci 2008;26: 459–65.

[66] Bordet T, Castelnau-Ptakhine L, Fauchereau F, Friocourt G, Kahn A, Haase G. Neuronal targeting of cardiotrophin-1 by coupling with tetanus toxin C fragment. Mol Cell Neurosci 2001;17: 842–854.

[67] Rind HB, Butowt R, von Bartheld, CS. Synaptic targeting of retrogradely transported trophic factors in motoneurons: Comparison of glial cell line-derived neurotrophic factor, brain- derived neurotrophic factor, and cardiotrophin-1 with tetanus toxin. J Neurosci 2005;25: 539–49.

[68] Carlton E, Teng Q, Federici T, Yang J, Riley J, Boulis NM. Fusion of the tetanus toxin C fragment binding domain and Bcl-xL for protection of peripheral nerve neurons. Neurosurgery 2008; 63: 1175-82, discussion 1182-4.

[69] Gil C, Chaib-Oukadour I, Blasi J, Aguilera J. HC fragment (C-terminal portion of the heavy chain) of tetanus toxin activates protein kinase C isoforms and phosphoproteins involved in signal transduction. Biochem J 2001; 356: 97–103.

[70] Gil C, Chaib-Oukadour I, Pelliccioni P, Aguilera J. Activation of signal transduction pathways involving trkA, PLCgamma-1, PKC isoforms and ERK-1/2 by tetanus toxin. FEBS Lett 2000; 481: 177–82.

[71] Chaib-Oukadour I, Gil C, Aguilera J. The C-terminal domain of the heavy chain of tetanus toxin rescues cerebellar granule neurones from apoptotic death: involvement of phosphatidylinositol 3-kinase and mitogen-activated protein kinase pathways. J Neurochem 2004; 90: 1227–36.

[72] Mendieta L, Venegas B, Moreno N, *et al.* The carboxyl-terminal domain of the heavy chain of tetanus toxin prevents dopaminergic degeneration and improves motor behavior in rats with striatal MPP(+)-lesions.Neurosci. Res 2009; 65: 98-106.

[73] Mendieta L, Bautista E, Sánchez A, *et al.* The C-terminal domain of the heavy chain of tetanus toxin given by intramuscular injection causes neuroprotection and improves the motor behavior in rats treated with 6-hydroxydopamine.Neurosci Res. 2012; 74: 156-67.
[74] Sánchez-González A, Mendieta L, Palafox V, *et al.* The restorative effect of intramuscular injection of tetanus toxin C-fragment in hemiparkinsonian rats. Neurosci Res 2014; pii, S0168-0102(14)00073-X.
[75] Radenovic L, Selakovic V, Olivan S, *et al.* Neuroprotective efficiency of tetanus toxin C fragment in model of global cerebral ischemia in Mongolian gerbils. Brain Res Bull 2014; 101: 37-44.
[76] Mock M, Fouet A. Anthrax. Annu. Rev. Microbiol 2001; 55: 647-71.
[77] Collier RJ, Young JA. Anthrax toxin. Annu Rev Cell Dev Biol 2003; 19: 45-70.
[78] Turk, BE. Manipulation of host signalling pathways by anthrax toxins. Biochem J 2007; 402: 405-17.
[79] Bradley KA, Young JA. Anthrax toxin receptor proteins. Biochem Pharmacol 2003; 65: 309-14.
[80] van der Goot G, Young JA. Receptors of anthrax toxin and cell entry. Mol Aspects Med 2009; 30: 406-12.
[81] Abrami L, Liu S, Cosson P, Leppla SH, van der Goot FG. Anthrax toxin triggers endocytosis of its receptor *via* a lipid raft-mediated clathrin-dependent process. J Cell Biol 2003; 160: 321-8.
[82] Bodart JF, Chopra A, Liang X, Duesbery N. Anthrax, MEK and cancer. Cell Cycle 2002; 1: 10-5.
[83] Koo HM, VanBrocklin M, McWilliams MJ, Leppla SH, Duesbery NS, Vande Woude GF. Apoptosis and melanogenesis in human melanoma cells induced by anthrax lethal factor inactivation of mitogen-activated protein kinase kinase. Proc Natl Acad Sci U S A 2002; 99: 3052-7.
[84] Locht C, Antoine R. A proposed mechanism of ADP-ribosylation catalyzed by the pertussis toxin S1 subunit. Biochimie 1995; 77: 333-40.
[85] Garcia JG, Wang P, Schaphorst KL, *et al.* Critical involvement of p38 MAP kinase in pertussis toxin-induced cytoskeletal reorganization and lung permeability. FASEB J 2002; 16: 1064-76.
[86] Wong WS, Rosoff PM. Pharmacology of pertussis toxin B-oligomer. Can J Physiol Pharmacol 1996; 74: 559-64.
[87] Thom RE, Casnellie JE. Pertussis toxin activates protein kinase C and a tyrosine protein kinase in the human T cell line Jurkat. FEBS Lett 1989; 244: 181-4.
[88] Wong WS, Luk JM. Signaling mechanisms of pertussis toxin-induced myelomonocytic cell adhesion: role of tyrosine phosphorylation. Biochem Biophys Res Commun 1997; 236: 479-82.
[89] Chen F, Vu ND, Wagner PD. Pertussis toxin modification of PC12 cells inhibits a protein phosphatase 2A-like phosphatase. J Neurochem 1998; 71: 248-57.
[90] Alfano M, Schmidtmayerova H, Amella CA, Pushkarsky T, Bukrinsky M. The B-oligomer of pertussis toxin deactivates CC chemokine receptor 5 and blocks entry of M-tropic HIV-1 strains. J Exp Med 1999; 190: 597-605.
[91] Bruckener KE, el Baya A, Galla HJ, Schmidt MA. Permeabilization in a cerebral endothelial barrier model by pertussis toxin involves the PKC effector pathway and is abolished by elevated levels of cAMP. J Cell Sci 2003; 116: 1837-46.
[92] Wolf RL, Andreoni J. Soluble inhibitory factor (SIF) in normal human serum. Cell Immunol 1982; 67: 299-311.
[93] Yin JX, Tu JL, Lin HJ, *et al.* Centrally administered pertussis toxin inhibits microglia migration to the spinal cord and prevents dissemination of disease in an EAE mouse model. PLoS One 2010; 5: e12400.
[94] Weber MS, Benkhoucha M, Lehmann-Horn K, *et al.* Repetitive pertussis toxin promotes development of regulatory T cells and prevents central nervous system autoimmune disease. PLoS One 2010; 5: e16009.
[95] Tang Z, Yin JX, Han P, *et al.* Pertussis toxin attenuates experimental autoimmune encephalomyelitis by upregulating neuronal vascular endothelial growth factor. Neuroreport 2013; 24: 469-75.
[96] Yin JX, Tang Z, Gan Y *et al.* Pertussis toxin modulates microglia and T cell profile to protect experimental autoimmune encephalomyelitis. Neuropharmacol 2014; 81: 1-5.
[97] Pauza CD, Hinds PW 2nd,Yin C, McKechnie TS, Hinds SB, Salvato M.S. The lymphocytosis-promoting agent pertussis toxin affects virus burden and lymphocyte distribution in the SIV-infected rhesus macaque. AIDS Res Hum Retrovir 1997; 13: 87-95.
[98] Alfano M, Pushkarsky T, Poli G, Bukrinsky M. The B-oligomer of pertussis toxin inhibits human immunodeficiency virus type 1 replication at multiple stages. J Virol 2000; 74: 8767-70.

[99] Alfano M, Vallanti G, Biswas P, *et al.* The binding subunit of pertussis toxin inhibits HIV replication in human macrophages and virus expression in chronically infected promonocytic U1 cells. J Immunol 2001; 166: 1863-70.

[100] Rizzi C, Crippa MP, Jeeninga RE, *et al.* Pertussis toxin B-oligomer suppresses IL-6 induced HIV-1 and chemokine expression in chronically infected U1 cells *via* inhibition of activator protein 1. J Immunol 2006; 176: 999-1006.

[101] Lapenta C, Spada M, Santini SM, *et al.* Pertussis toxin B-oligomer inhibits HIV infection and replication in hu-PBL-SCID mice. Int Immunol 2005; 17: 469-75.

[102] Yin C, Djavani M, Schenkel AR, Schmidt DS, Pauza CD, Salvato MS. Dissemination of lymphocytic choriomeningitis virus from the gastric mucosa requires G protein-coupled signaling. J Virol 1998; 72: 8613-9.

[103] Seo JH, Seo JY, Chung HY, Kim H. Effect of pertussis toxin and herbimycin A on proteinase-activated receptor 2-mediated cyclooxygenase 2 expression in Helicobacter pylori-infected gastric epithelial AGS cells. Yonsei Med J 2011; 52: 522-6.

[104] Nishida M, Suda R, Nagamatsu Y, *et al.* Pertussis toxin up-regulates angiotensin type 1 receptors through Toll-like receptor 4-mediated Rac activation. J Biol Chem 2010; 285: 15268-77.

[105] Pintérová M, Karen P, Kunes J, Zicha J. Role of nifedipine-sensitive sympathetic vasoconstriction in maintenance of high blood pressure in spontaneously hypertensive rats: effect of Gi-protein inactivation by pertussis toxin. J Hypertens 2010; 28: 969-78.

[106] De Rycke J, Milon A, Oswald E. Necrotoxic Escherichia coli (NTEC): two emerging categories of human and animal pathogens. Vet Res 1999; 30: 221-33.

[107] Landraud L, Gauthier M, Fosse T, Boquet P. Frequency of Escherichia coli strains producing the cytotoxic necrotizing factor (CNF1) in nosocomial urinary tract infections. Lett Appl Microbiol 2000; 30: 213-6.

[108] Elliott SJ, Srinivas S, Albert MJ, *et al.* Characterization of the roles of hemolysin and other toxins in enteropathy caused by alpha-hemolytic Escherichia coli linked to human diarrhea. Infect Immun 1998; 66: 2040-51.

[109] Okeke IN, Fayinka ST, Lamikanra A. Antibiotic resistance in Escherichia coli from Nigerian students, 1986 1998. Emerg Infect Dis 2000; 6: 393-6.

[110] Paciorek J. Virulence properties of Escherichia coli faecal strains isolated in Poland from healthy children and strains belonging to serogroups O18, O26, O44, O86, O126 and O127 isolated from children with diarrhoea. J Med Microbiol 2002; 51: 548-56.

[111] Contamin S, Galmiche A, Doye A, Flatau G, Benmerah A, Boquet P. The p21 Rho-activating toxin cytotoxic necrotizing factor 1 is endocytosed by a clathrin-independent mechanism and enters the cytosol by an acidic-dependent membrane translocation step. Mol Biol Cell 2000; 11: 1775-87.

[112] Kim KJ, Chung JW, Kim KS. 67-kDa laminin receptor promotes internalization of cytotoxic necrotizing factor 1-expressing Escherichia coli K1 into human brain microvascular endothelial cells. J Biol Chem 2005; 280: 1360-8.

[113] Flatau G, Lemichez E, Gauthier M, *et al.*Toxin-induced activation of the G protein p21 Rho by deamidation of glutamine. Nature 1997; 387: 729-33.

[114] Schmidt G, Sehr P, Wilm M, Selzer J, Mann M, Aktories K. Gln 63 of Rho is deamidated by Escherichia coli cytotoxic necrotizing factor-1. Nature 1997; 387: 725-9.

[115] Lerm M, Selzer J, Hoffmeyer A, Rapp UR, Aktories K, Schmidt G. Deamidation of Cdc42 and Rac by Escherichia coli cytotoxic necrotizing factor 1: activation of c-Jun N-terminal kinase in HeLa cells. Infect Immun 1999; 67: 496-503.

[116] Doye A, Mettouchi A, Bossis G, *et al.* CNF1 exploits the ubiquitin-proteasome machinery to restrict Rho GTPase activation for bacterial host cell invasion. Cell 2002; 111: 553-64.

[117] Loizzo S, Rimondini R, Travaglione S, *et al.* CNF1 increases brain energy level, counteracts neuroinflammatory markers and rescues cognitive deficits in a murine model of Alzheimer's disease. PLoS One 2013; 8: e65898.

[118] Hall A. Rho GTPases and the control of cell behaviour. Biochem. Soc Trans 2005; 33: 891-95.

[119] Fiorentini C, Arancia G, Caprioli A, Falbo V, Ruggeri FM, Donelli G. Cytoskeletal changes induced in HEp-2 cells by the cytotoxic necrotizing factor of Escherichia coli. Toxicon 1988; 11: 1047-56.

[120] Lacerda HM, Pullinger GD, Lax AJ, Rozengurt E. Cytotoxic necrotizing factor 1 from Escherichia coli and dermonecrotic toxin from Bordetella bronchiseptica induce p21(rho)-dependent tyrosine

phosphorylation of focal adhesion kinase and paxillin in Swiss 3T3 cells. J Biol Chem 1997; 14: 9587-96.

[121] Falzano L, Fiorentini C, Donelli G, *et al.* Induction of phagocytic behaviour in human epithelial cells by Escherichia coli cytotoxic necrotizing factor type 1. Mol Microbiol 1993; 6: 1247-54.

[122] Fiorentini C, Falzano L, Fabbri A, *et al.* Activation of rho GTPases by cytotoxic necrotizing factor 1 induces macropinocytosis and scavenging activity in epithelial cells. Mol. Biol Cell 2001; 12: 2061-73.

[123] Giamboi Miraglia A, Travaglione S, Meschini S, *et al.* Cytotoxic necrotizing factor 1 prevents apoptosis *via* the Akt/IkappaB kinase pathway: role of nuclear factor-kappaB and Bcl-2. Mol Biol Cell 2007; 18: 2735-44.

[124] Fiorentini C, Matarrese P, Straface E, *et al.* Rho-dependent cell spreading activated by E.coli cytotoxic necrotizing factor 1 hinders apoptosis in epithelial cells. Cell Death Differ 1998; 11: 921-29.

[125] Fiorentini C, Matarrese P, Straface E, *et al.* Toxin-induced activation of Rho GTP-binding protein increases Bcl-2 expression and influences mitochondrial homeostasis. Exp Cell Res 1998; 1: 341-50.

[126] Travaglione S, Loizzo S, Rizza T, *et al.* Enhancement of mitochondrial ATP production by the *E. coli* cytotoxic necrotizing factor 1. FEBS J 2014; in press.

[127] Pavone F, Luvisetto S, Marinelli S, *et al.* The Rac GTPase-activating bacterial protein toxin CNF1 induces analgesia up-regulating mu-opioid receptors. Pain 2009; 145: 219-29.

[128] Dina OA, McCarter GC, de Coupade C, Levine JD. Role of the sensory neuron cytoskeleton in second messenger signaling for inflammatory pain. Neuron 2003; 39: 613–24.

[129] Tatsumi S, Mabuchi T, Katano T, Matsumura S, Abe T, Hidaka H. Involvement of Rho-kinase in inflammatory and neuropathic pain through phosphorylation of myristoylated alanine-rich C-kinase substrate (MARCKS). Neuroscience 2005; 131: 491–8.

[130] Chen YL, Law PY, Loh HH. Nuclear factor jB signaling in opioid functions and receptor gene expression. J Neuroimmune Pharmacol 2006; 1: 270–9.

[131] Kraus J, Borner C, Giannini E, Hollt V. The role of nuclear factor kappaB in tumor necrosis factor-regulated transcription of the human mu-opioid receptor gene. Mol Pharmacol 2003; 64: 876–84.

[132] Cahill CM, Morinville A, Hoffert C, O'Donnell D, Beaudet A. Up-regulation and trafficking of delta opioid receptor in a model of chronic inflammation: implications for pain control. Pain 2003; 101: 199–208.

[133] Przewlocki R, Przewlocka B. Opioids in chronic pain. Eur J Pharmacol 2001; 429:79–91.

[134] Gozzo P, Marinelli S, Luvisetto S, *et al.* The bacterial protein toxin CNF1 induces analgesic effecs and upregulates Mu opioid receptors in a model of neuropathic pain. Eur J Pain Suppl 2011; 5 (S1), 25.

[135] Diana G, Valentini G, Travaglione S,*et al.*Enhancement of learning and memory after activation of cerebral Rho GTPases. Proc Natl Acad Sci US A 2007; 104: 636-41.

[136] Cerri C, Fabbri A, Vannini E, *et al.*Activation of Rho GTPases triggers structural remodeling and functional plasticity in the adult rat visual cortex. J Neurosci. 2011; 31: 15163-72.

[137] Malchiodi-Albedi F, Paradisi S, Di Nottia M, *et al.* CNF1 improves astrocytic ability to support neuronal growth and differentiation *in vitro*. PLoS One 2012; 7: e34115.

[138] De FilippisB,Fabbri A,Simone D, *et al.*Modulation of RhoGTPases improves the behavioral phenotype and reverses astrocytic deficits in a mouse model of Rett syndrome. Neuropsychopharmacology 2012; 37(5): 1152-63.

[139] Vannini E, Panighini A, Cerri C, *et al.*The bacterial protein toxin, cytotoxic necrotizing factor 1 (CNF1) provides long-term survival in a murine glioma model. BMC Cancer 2014; in press.

[140] Ren K., Dubner R. Interactions between the immune and nervous systems in pain. Nat Med 2010; 16: 1267-76.

[141] Ota Y, Zanetti AT, Hallock RM. The Role of Astrocytes in the Regulation of Synaptic Plasticity and Memory Formation. Neural Plast 2013; 2013: 185463.

[142] Gibbs ME, Hutchinson D, Hertz L. Astrocytic involvement in learning and memory consolidation. Neurosci Biobehav Rev 2008; 32: 927-44.

[143] Kreft M, Bak LK, Waagepetersen HS, Schousboe A. Aspects of astrocyte energy metabolism, amino acid neurotransmitter homoeostasis and metabolic compartmentation. ASN Neuro 2012; 4: e00086.

[144] Pastan I, Hassan R, Fitzgerald DJ, Kreitman RJ. Immunotoxin therapy of cancer. Nat Rev Cancer 2006; 6: 559-65.
[145] Wilson BA, Blanke SR, Reich KA, Collier RJ. Active-site mutations of diphtheria toxin. Tryptophan 50 is a major determinant of NAD affinity. J Biol Chem 1994; 269: 23296-301.
[146] Carroll SF, Collier RJ. Active site of Pseudomonas aeruginosa exotoxin A. Glutamic acid 553 is photolabeled by NAD and shows functional homology with glutamic acid 148 of diphtheria toxin. J Biol Chem 1987; 262: 8707-11.
[147] Collier RJ. Understanding the mode of action of diphtheria toxin: a perspective on progress during the 20th century. Toxicon 2001; 39: 1793-803.
[148] Choudhary S, Mathew M, Verma RS. Therapeutic potential of anticancer ITs. Drug Discov Today 2011; 16: 495-503.
[149] Kreitman RJ. Immunotoxins for targeted cancer therapy. AAPS J 2006; 8: E532-51.
[150] Li, Z.; Yu, T.; Zhao, P.; Ma, J. Immunotoxins and cancer therapy. Cell Mol Immunol 2005; 2: 106-12.
[151] Yang ZR, Wang HF, Zhao J, *et al.* Recent developments in the use of adenoviruses and immunotoxins in cancer gene therapy. Cancer Gene Ther 2007; 14: 599-615.
[152] Liu S, Netzel-Arnett S, Birkedal-Hansen H,Leppla SH. Tumor cell-selective cytotoxicity of matrix metalloproteinase-activated anthrax toxin. Cancer Res 2000; 60: 6061-7.
[153] Liu S, Aaronson H, Mitola DJ, Leppla SH, Bugge TH. Potent antitumor activity of a urokinase-activated engineered anthrax toxin. Proc Natl Acad Sci U S A 2003; 100: 657-62.
[154] Su Y, Ortiz J, Liu S, *et al.*Systematic urokinase-activated anthrax toxin therapy produces regressions of subcutaneous human non-small cell lung tumor in athymic nude mice. Cancer Res 2007; 67: 3329-36.
[155] Rønø B, Rømer J, Liu S, Bugge TH, Leppla SH, Kristjansen PE. Antitumor efficacy of a urokinase activation-dependent anthrax toxin. Mol. Cancer Ther 2006; 5: 89-96.
[156] Liu S, Wang H, Currie BM, *et al.*Matrix metalloproteinase-activated anthrax lethal toxin demonstrates high potency in targeting tumor vasculature. J Biol Chem 2008; 283: 529–40.
[157] Park S, Leppla SH. Optimized production and purification of Bacillus anthracis lethal factor. Protein Expr Purif 2000; 18: 293–302.
[158] Bachran C, Hasikova R, Leysath CE,*et al.*Cytolethal distending toxin B as a cell-killing component of tumor-targeted anthrax toxin fusion proteins. Cell Death Dis 2014; 5: e1003.
[159] Abi-Habib RJ, Singh R, Liu S, Bugge TH,Leppla SH, Frankel AE. A urokinase-activated recombinant anthrax toxin is selectively cytotoxic to many human tumor cell types. Mol Cancer Ther 2006; 5: 2556–62.
[160] Pennell CA, Erickson HA. Designing immunotoxinsITs for cancer therapy. Immunol Res 2002; 25: 177-91.
[161] Mathew M, Verma RS. Humanized immunotoxins: a new generation of immunotoxins for targeted cancer therapy. Cancer Sci 2009; 100: 1359-65.
[162] Holliger P, Hudson PJ. Engineered antibody fragments and the rise of single domains. Nat Biotechnol 2005; 23: 1126-36.
[163] Chiarella P, Massi E, De Robertis M, Signori E, Fazio VM. Adjuvants in vaccines and for immunisation: current trends. Expert Opin Biol Ther 2007; 7: 1551-62.
[164] Salama NN, Eddington ND, Fasano A. Tight junction modulation and its relationship to drug delivery. Adv Drug Deliv Rev 2006; 58: 15-28.
[165] Rappuoli R, Pizza M, Douce G, Dougan G. Structure and mucosal adjuvanticity of cholera and Escherichia coli heat-labile enterotoxins. Immunol Today 1999; 20: 493-500.
[166] Holmgren J, Lonnroth I, Svennrholm L. Tissue receptor for cholera exotoxin: postulated structure from studies with GM1 ganglioside and related glycolipids. Infect Immun 1973; 8: 208-14.
[167] Muanprasat C, Chatsudthipong V. Cholera: pathophysiology and emerging therapeutic targets. Future Med Chem 2013; 5: 781-98.
[168] Elson CO, Ealding W. Generalized systemic and mucosal immunity in mice after mucosal stimulation with cholera toxin. J Immunol 1984; 132: 2736-41.
[169] Spangler BD. Structure and function of cholera toxin and the related Escherichia coli heat-labile enterotoxin. Microbiol Rev 1992; 56: 622-47.

[170] Tamura S, Yajima A, Hatori E, *et al.*Effects of frequent intranasal administration of adjuvant-combined influenza vaccine on the protection against virus infection. Vaccine 1997; 15: 1784-90.
[171] Anastassiou ED, Yamada H, Francis ML, Mond JJ, Tsokos GC. Effects of cholera toxin on human B cells. Cholera toxin induces B cell surface DR expression while it inhibits anti-mu antibody-induced cell proliferation. J Immunol 1990; 145: 2375-80.
[172] Hornquist E, Lycke N. Cholera toxin adjuvant greatly promotes antigen priming of T cells. Eur J Immunol 1993; 23: 2136-43.
[173] Foss DL, Zilliox MJ, Murtaugh MP. Differential regulation of macrophage interleukin-1 (IL-1), IL-12, and CD80-CD86 by two bacterial toxins. Infect Immun 1999; 67: 5275-81.
[174] Rescigno M, Urbano M, Valzasina B, *et al.* Dendritic cells express tight junction proteins and penetrate gut epithelial monolayers to sample bacteria. Nat Immunol 2001; 2: 361-7.
[175] Gagliardi MC, Sallusto F, Marinaro M, Vendetti S, Riccomi A, De Magistris MT. Effects of the adjuvant cholera toxin on dendritic cells: stimulatory and inhibitory signals that result in the amplification of immune responses. Int J Med Microbiol 2002; 291: 571-5.
[176] Lycke N, Tsuji T, Holmgren J. The adjuvant effect of Vibrio cholerae and Escherichia coli heat-labile enterotoxins is linked to their ADP-ribosyltransferase activity. Eur J Immunol 1992; 22: 2277-81.
[177] Yokomizo Y, Watanabe F, Imada Y, Inumaru S, Yanaka T, Tsuji T. Mucosal immunoadjuvant activity of the low toxic recombinant Escherichia coli heat-labile enterotoxin produced by Bacillus brevis for the bacterial subunit or component vaccine in pigs and cattle. Vet Immunol Immunopathol 2002; 87: 291-300.
[178] Barackman JD, Ott G, Pine S, O'Hagan DT. Oral administration of influenza vaccine in combination with the adjuvants LT-K63 and LT-R72 induces potent immune responses comparable to or stronger than traditional intramuscular immunization. Clin Diagn Lab Immunol 2001; 8: 652-7.
[179] Chong C, Friberg M, Clements JD. LT(R192G), a non-toxic mutant of the heat-labile enterotoxin of Escherichia coli, elicits enhanced humoral and cellular immune responses associated with protection against lethal oral challenge with Salmonella spp. Vaccine 1998; 16: 732-40.
[180] Cox E, Verdonck F, Vanrompay D, Goddeeris B. Adjuvants modulating mucosal immune responses or directing systemic responses towards the mucosa. Vet Res 2006; 37: 511-39.
[181] Del Giudice G, Rappuoli R. Genetically derived toxoids for use as vaccines and adjuvants. Vaccine 1999; 17: S44-52.
[182] Stephenson I, Zambon MC, Rudin A, *et al.*Phase I evaluation of intranasal trivalent inactivated influenza vaccine with nontoxigenic Escherichia coli enterotoxin and novel biovector as mucosal adjuvants, using adult volunteers. J Virol 2006; 80: 4962-70.
[183] Pizza M, Giuliani MM, Fontana MR, Douce G, Dougan G, Rappuoli R. LTK63 and LTR72, two mucosal adjuvants ready for clinical trials. Int J Med Microbiol 2000; 290: 455-61.
[184] Norton EB, Lawson LB, Freytag LC, Clements JD.Characterization of a mutantEscherichia coli heat-labile toxin, LT(R192G/L211A), as a safe and effective oral adjuvant. Clin Vaccine Immunol2011; 18: 546-51.
[185] Lee JB, Jang JE, Song MK, Chang J. Intranasal delivery of cholera toxin induces th17-dominated T-cell response to bystander antigens. PLoS One 2009; 4: e5190.
[186] Fasano A, Baudry B, Pumplin DW,*et al.* Vibrio cholerae produces a second enterotoxin, which affects intestinal tight junctions. Proc Natl Acad Sci USA 1991; 88: 5242-6.
[187] Di Pierro M, Lu R, Uzzau S, *et al.*Zonula occludens toxin structure-function analysis. Identification of the fragment biologically active on tight junctions and of the zonulin receptor binding domain. J Biol Chem 2001; 276: 19160-5.
[188] Fasano A, Fiorentini C, Donelli G,*et al.*Zonula occludens toxin modulates tight junctions through protein kinase C-dependent actin reorganization, *in vitro*. J Clin Invest 1995; 96: 710-20.
[189] Fasano A, Uzzau S. Modulation of intestinal tight junctions by Zonula occludens toxin permits enteral administration of insulin and other macromolecules in an animal model. J Clin Invest 1997; 99: 1158-64.
[190] Song KH, Fasano A, Eddington ND. Effect of the six-mer synthetic peptide (AT1002) fragment of zonula occludens toxin on the intestinal absorption of cyclosporin A. Int J Pharm 2007; 351: 8-14.
[191] Motlekar NA, Fasano A, Wachtel MS, Youan BB. Zonula occludens toxin synthetic peptide derivative AT1002 enhances *in vitro* and *in vivo* intestinal absorption of low molecular weight heparin. J Drug Target 2006; 14: 321-9.

[192] Karyekar CS, Fasano A, Raje S, Lu R, Dowling TC, Eddington ND. Zonula occludens toxin increases the permeability of molecular weight markers and chemotherapeutic agents across the bovine brain microvessel endothelial cells. J Pharm Sci 2003; 92: 414-23.
[193] Menon D, Karyekar CS, Fasano A, Lu R, Eddington ND. Enhancement of brain distribution of anticancer agents using DeltaG, the 12 kDa active fragment of ZOT. Int J Pharmaceutics 2005; 306: 122-31.
[194] Marinaro M, Di Tommaso A, Uzzau S, Fasano A, De Magistris MT. Zonula occludens toxin is a powerful mucosal adjuvant for intranasally delivered antigens. Infect Immun 1999; 67: 1287-91.
[195] Rossi M, Maurano F, Luongo D,*et al.*Zonula occludens toxin (Zot) interferes with the induction of nasal tolerance to gliadin. Immunol Lett 2002; 81: 217-21.
[196] De Magistris MT. Zonula occludens toxin as a new promising adjuvant for mucosal vaccines. Vaccine 2006; 24 (Suppl 2): 60-1.
[197] Gopalakrishnan S, Pandey N, Tamiz AP,*et al.*Mechanism of action of ZOT-derived peptide AT-1002, a tight junction regulator and absorption enhancer. Int J Pharm 2009; 365: 121-30.
[198] Fang Y, Larsson L, Mattsson J, Lycke N, Xiang Z. Mast cells contribute to the mucosal adjuvant effect of CTA1-DD after IgG-complex formation. J Immunol 2010; 185: 2935-41.
[199] Munro P, Flatau G, Anjuere F, Hofman V, Czerkinsky C, Lemichez E. The Rho GTPase activators CNF1 and DNT bacterial toxins have mucosal adjuvant properties. Vaccine 2005; 23: 2551-6.
[200] Munro P, Flatau G, Lemichez E. Intranasal immunization with tetanus toxoid and CNF1 as a new mucosal adjuvant protects BALB/c mice against lethal challenge. Vaccine 2007; 25: 8702-6.
[201] Ladant D, Ullmann A. Bordatella pertussis adenylate cyclase: a toxin with multiple talents. Trends Microbiol 1999; 7: 172-6.
[202] Macdonald-Fyall J, Xing D, Corbel M, Baillie S, Parton R, Coote J. Adjuvanticity of native and detoxified adenylate cyclase toxin of Bordetella pertussis towards co-administered antigens. Vaccine 2004; 22: 4270-81.
[203] Arciniega JL, Hewlett EL, Johnson FD, *et al.* Human serologic response to envelope-associated proteins and adenylate cyclase toxin of Bordetella pertussis. J Infect Dis 1991; 163: 135-42.
[204] Orr B, Douce G, Baillie S, Parton R, Coote J. Adjuvant effects of adenylate cyclase toxin of Bordetella pertussis after intranasal immunisation of mice. Vaccine 2007; 25: 64-71.
[205] Roberts M, Bacon A, Rappuoli R, *et al.* A mutant pertussis toxin molecule that lacks ADP-ribosyltransferase activity, PT-9K/129G, is an effective mucosal adjuvant for intranasally delivered proteins. Infect Immun 1995; 63: 2100-8.
[206] Murphey C,Chang S,Zhang X,Arulanandam B,Forsthuber TG. Induction of polyclonal CD8+ T cell activation and effector function by Pertussis toxin.Cell Immunol. 2011; 267: 50-5.
[207] Salcedo-Rivillas C,Debrie AS,Miyaji EN,*et al.*Pertussis toxin improves the immune responses to a combined pneumococcal antigen and lead to enhanced protection against Streptococcus pneumoniae. Clin Vaccine Immunol. 2014; May 7 [Epub ahead of print].
[208] Magaña-Maldonado R, Manoutcharian K, Hernández-Pedro NY, *et al.* Concomitant treatment with pertussis toxin plus temozolomide increases the survival of rats bearing intracerebral RG2 glioma. J Cancer Res Clin Oncol 2014; 140: 291-301.
[209] Karimian G, Buist-Homan M, Faber KN, Moshage H. Pertussis toxin, an inhibitor of G(αi) PCR, inhibits bile acid- and cytokine-induced apoptosis in primary rat hepatocytes. PLoS One 2012; 7: e43156.

CHAPTER 9

Vaccines against Cysticercosis: A Revisited Review Focusing on *Taenia solium*

Edda Sciutto[1,*], Gladis Fragoso[1], Marisela Hernández[1], Gabriela Rosas[2], Carlos Larralde[1] and Sergio Rosales-Mendoza[3]

[1]Instituto de Investigaciones Biomédicas, UNAM; [2]Facultad de Medicina, Universidad Autónoma del Estado de Morelos, Cuernavaca, Morelos, México and [3]Laboratorio de Biofarmacéuticos Recombinantes, Facultad de Ciencias Químicas, Universidad Autónoma de San Luis Potosí, México

Abstract: *Taenia solium* cysticercosis is a major parasitic disease that severely affects human health and the economy of undeveloped countries. Since pigs are obligate intermediate hosts, it is plausible to curb human transmission by reducing pig cysticercosis through effective vaccination. This article reviews the current knowledge on the development of vaccines against porcine cysticercosis. It highlights the development of plant-based, needle-free vaccine versions as a feasible and affordable approach to reduce the logistic costs of injectable vaccines, while preventing the digestive hydrolysis of vaccine antigens.

Keywords: Anti-cestodiasis vaccine, cestode prevention, cysticercosis, infectious diseases, multiepitope vaccines, neurocysticercosis, oral vaccination, poverty, recombinant vaccines, rural communities, S3Pvac, subcutaneous vaccination, synthetic vaccines, *Taenia solium,* transgenic plants, transmission control, transplantomic cells, TSOL18, vaccine delivery, vaccine field trial, vaccines.

INTRODUCTION

Vaccination is one of the most promising biotechnological approaches to control infectious diseases. Recent progress in molecular biology and immunology has produced safer and more effective vaccines against diverse viral and bacterial pathogens [1-4]. However, vaccination has been less successful in preventing human infections caused by protozoan parasites, such as malaria, leishmaniasis, and trypanosomiasis, in spite of the huge efforts and resources invested to develop

***Corresponding author Edda Sciutto:** Instituto de Investigaciones Biomédicas, A.P. 70228; 04510 México, D.F., México; Tel: (5255) 5622-3153; Fax: (5255) 5622-3369; E-mail: edda@servidor.unam.mx

Atta-ur-Rahman, Allen B. Reitz, M. Iqbal Choudhary and Jizhou Wang (Eds.)

effective vaccines [5]. In contrast, the situation seems exceptionally different for cestode parasites, in particular for *Taenia solium* cysticercosis, against which several effective vaccines are available [6-10].

THE CASE OF *Taenia solium*

Human and pig infection by *T. solium* seems to be susceptible to immunological intervention. It is noteworthy how common is to find spontaneously destroyed cysticerci, surrounded by several immune cell types in activated states and expressing molecules with immune activity [7, 11-16]. Additionally, some epidemiological clues point to immune involvement in cysticercosis. In rural communities at endemic areas, cranial axial computerized tomography (CTscan) often finds single nodular brain lesions, compatible with a cysticercus located in the central nervous system (CNS). These putative cysticerci are usually calcified, their destruction not being associated to any symptoms or therapeutic interventions [17-19]. This capability to destroy the parasite does not depend solely on the time of evolution, since most brain cysticerci in children are found calcified as well [19, 20]. On the other hand, when experimentally challenged pigs are well-nourished and kept in clean and stress-free facilities, parasites begin to disappear months after the challenge. Within one year, most cysts in skeletal muscles are calcified [21, 22]. In contrast, CNS cysticerci remain vesicular, apparently undamaged one year after challenge [23]. Possibly, brain-located cysticerci in pigs require more time to be destroyed than those located in skeletal muscle. Whether or not this occurs in human infections as well, it would be hard to sustain such a thesis, since human and porcine cysticercosis exhibit significant differences in absolute duration (many years in humans and usually less than one year in pigs) and in parasite loads (rarely more than ten parasites in humans *versus* tens to several hundred in the pigs), as well as differences in the immune response at different times post-infection [24].

Nevertheless, other immunological findings in human neurocysticercosis (NC) strongly suggest a role of the immune response in the evolution of *T. solium* infection. Notwithstanding the fact that the parasite itself may have an internal biological clock timing its death, the involvement of the host's immune system is supported by the differences observed in the immune profiles between those individuals who destroyed the parasite and those who were unable to damage it [25]. In humans, a specific systemic Th2-type response associated to calcified NC is well defined, while symptomatic NC is associated to a depressed systemic specific response [26-28]. Additional evidence suggests the existence of acquired immunity and a temporary protection induced by primo-infection [22], a fact that supports the potential benefit of vaccination against *T. solium* cysticercosis. These observations do not conclusively prove that immunity is responsible of cysticerci destruction, but they add up to a strong inference, which satisfactorily explains the success of different immunogens to inhibit the establishment of *T. solium* cysticerci and/or to promote their destruction.

DEVELOPMENT OF VACCINES AGAINST PORCINE CYSTICERCOSIS

A number of parasite antigens, originally derived from related cestodes (*T. solium, T. crassiceps, T. saginata, T. ovis*) at different developmental stages (*i.e.*, oncospheres, cysticerci) (Table **1** [7, 29-37]), or of recombinant or synthetic origin, have been tested as vaccines (Table **2** [38-45]) and have proved to reduce the infection rate and parasite load in vaccinated pigs challenged with *T. solium* eggs under experimental conditions (Tables **1** and **2**). Different vaccine candidates were also tested against *T. solium* pig cysticercosis using DNA vaccination (Table **3**) [46-50].

Table 1: Experiences in vaccination against *T. solium* pig cysticercosis using native antigens

Source of Antigens + Adjuvant	**Protection (% Reduction)**		**Trial Experimental/Field**	**References**
	Cysticercotic Pigs	**# Cysticerci**		
T. solium cysticerci extract		74%	F	[7]
T. solium excretory-secretory antigens + FCA		95%	E	[29]
T. solium cysticerci extract	100%		F	[30]
T. solium scolex extract + FIA		71%	E	[31]
T. solium scolex extract + *Corynebacterium parvum*		75%	E	[31]
T. crassiceps extract + FCA		50%	E	[32]
T. solium cysticerci extract	82%		F	[33]
T. crassiceps 56+66+74kDa proteins + FCA		97% 86%	E	[34]
T. saginata asiatica (Indonesian) frozen oncospheres + FCA	100%	98% – 99%	E	[35]
T. saginata asiatica (Taiwan) frozen oncospheres + FCA	66%	99%	E	[35]
T. solium frozen oncospheres + FCA	100%	99% – 100%	E	[35]
T. solium cysticerci crude extract + adjuvant ISA 206		89%	E	[36]
T. solium cysticerci crude extract + adjuvant ISA 206		95.9%	E	[37]

In Mexico, the first report of effective vaccination against porcine cysticercosis was based on a total extract of cysticercal antigens from *T. solium* cysticerci drawn out from naturally infected pigs [7]. Further studies were conducted to identify, isolate, and produce the antigens responsible for the observed protection, to eliminate irrelevant and potentially parasite-promoting factors [51]. Among the 12 different antigen fractions obtained, a 56-kDa and a 74-kDa fraction induced high protection levels against murine cysticercosis [51]. Five different recombinant antigens were identified in a *T. crassiceps* cDNA library using specific polyclonal rabbit antibodies against the 56- and the 74-kDa antigen fractions. These antigens were pooled with anti-*T. solium* sera from naturally infected pigs [34]. Three of them, designated as KETc1, -7, and -12, reduced the expected parasite load in vaccinated mice by 53% - 84% [52]. Structural analysis of the KETc7 sequence [53, 54] led to the identification of an 18-amino acid immunogenic peptide. The latter, named GK1, induced high protection levels against experimental murine cysticercosis both in its linear form and coupled to the core matrix of multiple antigen peptides, or glutaraldehyde-linked to bovine serum albumin. Noticeably, the highest protection levels in mice (89% - 94%) were induced by synthetic GK1 in its uncoupled form [54]. Two protective peptides, 12- and 8-amino acid long, were also identified in the KETc1 and KETc12 recombinant antigens. These peptides were given the same name as the antigens from which they were derived [34]. KETc1 and KETc12 significantly protected mice against murine cysticercosis (KETc1, 66.7% - 100%, and KETc12, 52.7% - 88.1%) [52].

The three peptides mentioned above, that constitute the S3Pvac vaccine, were administered together to increase the protection level and override an eventual parasite genetic variation. As shown in (Table **2**), S3Pvac proved to be protective against *T. solium* porcine cysticercosis in natural conditions of transmission [38, 42], and also when expressed as a DNA vaccine in a modified hepatitis B virus core [49].

Table 2: Experiences in vaccination against *T. solium* pig cysticercosis using subunitary vaccines

Source of antigens + adjuvant	Protection (% reduction)		Trial Experimental/Field	Reference
	Cysticercotic pigs	# Cysticerci		
S3Pvac-synthetic +saponin		98%	F	[38]
TSO18-GST + Quil A		100%	E	[39]
TSOL45-1A-GST + Quil A		0%	E	[39]
TSO18-GST + TSOL45-1A-GST + Quil A		95%	E	[39]
TSO18-GST + Quil A		99%	E	[39]
TSOL45-1A-GST + Quil A		97%	E	[39]

S3Pvac-phage (KETc1+KETc7+KETc12+GK1)		97% 95%	E	[40]
S3Pvac-phage (KETc1+KETc7+KETc12+GK1)		89% 42%	E	[40]
TSO18-GST + Quil A		99.9%	E	[41]
TSOL45-1A-GST + Quil A		97%	E	[41]
S3Pvac-synthetic +saponin	70% – 80%		F	[42]
TSOL18 + ISA 206		94% / 100%*	E	[36]
S3Pvac-phage inactivated (KETc1+KETc7+KETc12+GK1) 0.05% of formaldehyde	54.2%	87.1%	F	[43]
TSOL45W-4B		97% / 98.4%*	E	[37]
TSOL18+ Quil A+ oxfendazole	100%	100%	F	[44]
TSOL18-GST +TSOL16-GST+ QuilA	63%	99.8%	F	[45]

*Repeated trial; GST: Glutathione S-transferase; MBP: Maltose binding protein.

Vaccine candidates against *T. solium* include some immunogens originally aimed to prevent diseases caused by other cestodes closely related to *T. solium*, such as *T. ovis* and *T. saginata* [10, 55]. The high protective capacity of three *T. ovis* recombinant antigens (45WB/X-GST, 16K-GST, and 18K-GST) was demonstrated in experimental porcine cysticercosis [39, 55, 56]. The HP6 antigen from *T. saginata* [57, 58], which induced high protection levels against cysticercosis in cows, was also found in *T. solium* cysticerci [10]. Considering the high protective capacity of the 18- and 45-kDa recombinant antigens from *T. ovis* [55], the corresponding *T. solium* antigens were isolated as TSOL18 and TSOL45-1A and proved to be more efficient to protect against experimental infection in pigs [39, 41]. Additional homologous recombinant antigens have confirmed the protective capacity of antigen B against experimental *T. solium* porcine cysticercosis [50], as previously reported against murine *T. crassiceps* [59]. More recently, another antigen named TSOL 16, which as the two above mentioned antigens includes a fibronectin type III (FnIII) domain, has been reported to protect pigs under experimental and natural conditions of infection [45, 60]. A new recombinant protein, named cC1, also induces high protection levels against pig cysticercosis when administered as a DNA vaccine [48, 50].

The distribution of these candidate vaccine antigens among regional or global populations of taenia specimens is not known. Considering the low genetic differences reported among cysticerci recovered from pigs in different regions of Mexico and other countries [61, 62], the high rate of common antigens among taenids will be of great interest for vaccine design, since it implies that a single vaccine could prevent a wide spectrum of cestode diseases. Indeed, immunological data support the existence of numerous antigens common to different *Taenia* species, and even among other cestodes [9, 63, 64], as seen with the antigens used for immunodiagnosis [64-66]. In this respect, the now available tapeworm genome sequences [67] provide a rich resource to search for vaccine antigens conserved among these related parasites [68]. The ongoing Mexican and Chinese *T. solium* comparative genome analysis (manuscript in preparation) will also contribute to the identification of other highly conserved protective antigens.

THE IMPORTANCE OF TESTING A VACCINE'S EFFICIENCY AGAINST THE REAL THREAT POSED BY THE NATURAL ENVIRONMENTAL CONDITIONS OF TRANSMISSION

For an antigen to be promoted from a "vaccine candidate" status to "vaccine", it is required to demonstrate its effectiveness in preventing the disease in a large and varied population of the real hosts under risk. Hosts must be challenged in the uncontrollably multiple combinations that only nature can provide, and they must live in their natural environment and dwell in their peculiar lifestyles. While convenient, it does not suffice that the antigen is effective in highly controlled experimental conditions. These usually employ a small number of pigs, all of the same breed, age, and gender, healthy and well-nourished; the animals are experimentally challenged only once with eggs originating from the same taenia, and let to establish for a short time after challenge. The realistic conditions on the field differ enormously from the experimental ones, and imply great differences in the chances of infection, in the time left for the cysticercus to develop, and in the presence of an adequate immune reaction by the infected pig. Genetically, pigs on the field are very heterogeneous; in addition, they are malnourished, chronically stressed, and exposed to other diseases and to mistaken official programs of mandatory vaccination against other infections. They are also exposed to ingest variable amounts of taenia eggs in multiple occasions, probably shed from different taenias, and let to live for about a year before being inspected for cysticercosis. Clearly, these circumstances are impossible to be experimentally reproduced and are likely to be related to the effectiveness of the immune response in a changing and threatening environment.

FIELD TRIAL DESIGN

Then, a realistic design of the field trial to evaluate the effectiveness of a vaccine is essential. The issues to be considered are numerous: 1) the area or village selected should have previously shown to have a porcine cysticercosis prevalence level of about 10%, as measured with an accurate and cost-effective method; 2) the age of pigs to be included in the trial (*e.g.*, a 2 month-old piglet has minimal risk of infection previous to vaccination, but it is old enough to be immunocompetent) [69]; 3) the genetic background of the host population to be vaccinated is suspected to yield differences in the effectiveness of vaccination [70]; 4) biological circumstances that could impact on the efficiency of vaccination are present (*e.g.*, sex, castration, and pregnancy are associated to an increase in susceptibility to infection) [71, 72]; 5) the regime of pig upbringing (there are remarkable differences in the risk to infection among confined, half-confined, and wandering pigs [71, 72]; 6) different occurrences and sizes of challenge [73].

All aforementioned variables should be duly balanced when making the pig groups included in the field trial as vaccinated and non-vaccinated pigs. Prevalence levels and parasite loads in sentinel pigs are critical, as they sense a possible active transmission in the selected area at the time of the trial and may serve as an additional contrast for the vaccinated group. Unfortunately, while tongue inspection is quite sensitive and cost-effective to estimate pig cysticercosis, it misses many truly cysticercotic pigs with low parasite loads [74-77]. Thus, whenever possible, tongue inspection should be performed several times during the trial (to spot any pig that got infected but apparently solved the infection) and at the end of the trial, usually after pigs are 12 month-old or at least 4 months post-vaccination. As many of the included pigs as possible should also be thoroughly examined post-mortem, to strengthen the credibility of tongue inspection by a correlative analysis.

Table 3: Experimental assays in DNA vaccination against *T. solium* pig cysticercosis

Source of antigens + adjuvant	**Protection (% reduction)**		**Reference**
	Cysticercotic pigs	**# Cysticerci**	
T. solium B + PV93		85% – 98%	[46]
cC1 + pcDNA3		73%	[47]
pcDNA3-cC1 + (GST-cC1+ FIA)		Challenged (weeks after vaccination)	[48]
pVAX-S-deltaC-3n: hepatitis B core antigen particle + S3Pvac + IL2 signal peptide		83% 6 wk, 85% 12 wk, 77% 20 wk, 72% – 79%	[49]
pcDNA3-B *T. solium:* antigen B + pcDNA3.1	80%	92.6%	[50]

A necropsy exam, where the entire muscle mass of each pig is examined and dissected centimeter by centimeter with microscopic corroboration, yields the most precise estimates of group prevalence, as well as parasite load and parasite integrity in each pig. However, these studies cannot be carried out in large numbers (say hundreds) of pigs. Additionally, it is not feasible to improvise a morgue in a rural spot when pig owners decide to sacrifice them. Tongue inspection in live pigs, looking for sub-epithelial cysticerci, while not a simple intervention, is more feasible to effect and program in large numbers of pigs than complete necropsies. The main drawback of tongue inspection is that it underestimates the prevalence and little or nothing reveals of parasite load [75-77]. On the other hand, antibody-based serum tests could be of use in experimental controlled conditions, but in real life situations they provide evidence of contact with the parasite, not necessarily of infection. As for antigen detection, many naturally infected pigs carry only a few parasites [21, 75] and it is uncertain that the currently available technology could detect those circulating antigens [78] or they could be overpowered by circulating antibodies [78]. Diagnosing porcine cysticercosis by detecting antibodies against *T. solium* is further complicated by the many cross-reactions between related cestodes that free-ranging pigs in endemic countries encounter in their hazardous lifestyle (*e.g.*, *Echinococcus spp.*) [64]. In addition, pigs in endemic areas are highly exposed to the parasite, and they often exhibit high antibody levels, even in the absence of any parasite as verified by necropsy [76].

VACCINE EFFICACY IN NATURAL CONDITIONS OF TRANSMISSION

S3Pvac-synthetic was the first candidate to the porcine anti-cysticercosis vaccine whose efficacy was critically assessed in real-life conditions, *i.e.*, in rustically-reared pigs from highly endemic rural locations in Mexico. S3Pvac-synthetic is constituted by three synthetically produced peptides, and it is owned by the National Autonomous University of Mexico. The three protective peptides in S3Pvac, as discussed earlier, were named KETc12, KETc1, and GK1. They are 8-, 13-, and 18-amino acid long, respectively. They were first identified by their protective capacity against experimental murine *T. crassiceps* cysticercosis [34]. These peptide sequences are part of native antigens present in various anatomical structures and developmental stages of *T. solium*. S3Pvac was tested on the field in communities of the State of Puebla [38]. Vaccination reduced by 50% the number of infected pigs and by 98% the amount of detected cysticerci (and therefore the amount of cysticerci able to transform into taenias). Similar results were obtained in a different community of the State of Morelos [42]. Thus, the effectiveness of S3Pvac to prevent naturally acquired porcine cysticercosis has been demonstrated.

S3Pvac could also have therapeutic properties in early infections [23]. Immunizing experimentally infected pigs with S3Pvac macroscopically decreased the percentage of vesicular cysticerci from 94.4% to 38.9%, and increased the percent of calcified parasites in muscle from 5.6% to 61.1% (Table **4**) [23]. The possible therapeutic activity S3Pvac would add interest to its use in controlling taeniasis/cysticercosis. In additional therapeutic experiments, we found S3Pvac ineffective in damaging cysticerci in naturally infected pigs with high parasite loads for an unknown number of months [79].

S3Pvac-synthetic has shown to be effective in preventing porcine cysticercosis to some extent; however, its high production costs and the logistic difficulties of parenteral application to rural pigs render it inadequate for the target population. Thus, we decided to express the peptides recombinantly in filamentous phages (S3Pvac-phage). S3Pvac-phage was also tested in rural pigs under natural conditions of transmission [43].

The three peptides were expressed in M13 filamentous phage proved effective against naturally acquired *T. solium* porcine cysticercosis. This new version included KETc7, a 100-amino acid long protective antigen that includes the GK1 peptide. The efficacy of S3Pvac-phage with respect to placebo was evaluated in a randomized trial. A total of 1,047 backyard pigs from 16 villages at the State of Morelos, in Central Mexico, were included in the study. Three to five months after vaccination, 530 pigs were examined by tongue inspection. Later, when the animals were sacrificed for consumption (pigs ranged 5-27 months of age), 331 pigs (197 vaccinated/134 controls) were inspected. All pigs in which one or more cysticerci were found were considered as cysticercotic. S3Pvac-phage vaccination reduced the frequency of tongue cysticercosis by 62%, muscle-cysticercosis by 54%, and the number of cysticerci by 87% [43].

Comparisons of the nucleotide and amino acid sequences coding for two of the S3Pvac components, KETc1 and GK1, revealed significant homologies in *T. crassiceps WFU, T. solium, T. saginata, Echinococcus granulosus* and *E. multilocularis*. Altogether, these results indicate that both peptides may be regarded as candidates to be included in the formulation of a multivalent, multistage vaccine against these cestodiases [68]. Additionally, S3Pvac-phage showed some effectiveness against hydatidosis, a parasitic disease that frequently affects backyard pigs in Mexico and other non-developed countries where poor sanitary conditions prevailed [80]. Pigs are essential in the transmission of cysticercosis and hydatidosis to humans. Considering this, the effect of S3Pvac-phage vaccine on hydatidosis was analyzed in the field trial described above. A total of 391 necropsies in 187 vaccinated and 204 non-vaccinated pigs were included: 60 pigs more than those included in the first report of this study [43]. Masseters, tongue, diaphragm, heart, and liver were dissected and sliced, and the number of cysticerci found in each pig was recorded. Owners were instructed not to eat the meat if pigs were found infected. Liver and the peritoneal cavity were visually inspected in search of hydatid cysts or other visceral helminths. Vaccination reduced the prevalence of cysticercosis in pigs by 61.7%, and the prevalence of *E. granulosus* hydatidosis by 56.1% [81].

Table 4: Therapeutic effect of the S3Pvac on the number and the viability of cysticerci recovered from pigs infected with *T. solium*

Status of the infected pigs	Macroscopically damaged cysticerci (%)	†Evagination	††Tapeworm transformation
Saponine	5.6% (59/1039)	70% (188/269)	58% (44/75)
S3Pvac †Saponine	61.1% (344/563)	38% (19/50)	19% (9/50)

5 piglets per group were orally infected with 20,000 *T. solium* eggs and 1 month later treated with saponine (control) or S3Pvac, three times each at a 30-day interval. Four months after the last immunization, the animals were euthanized and the total number of cysticerci found in half of the carcass plus heart, diaphragm, and brain and their macroscopical aspect were recorded. †*In vitro* evagination capacity from a total of 269 (controls) and 50 (S3Pvac) cysticerci sampled. ††*In vivo* transformation of cysticerci to tapeworms in immunodepressed hamsters orally infected with 5 cysticerci each in a total of 20 (control pigs) and 10 (S3Pvac) hamsters [23].

The TSOL18 vaccine was tested on the field in combination with cysticidal treatment in Cameroon; 240 backyard pigs were distributed in pairs to 114 individual households. Pigs were immunized thrice with TSOL18 and also treated with oxfendazole along with the second immunization. When pigs were 12 month-old, the effect of these interventions was measured. Twenty control pigs presented viable *T. solium* cysticerci (19.6%), whilst no cysticerci were found in those vaccinated [44].

In view of these amazing results, an additional oncospheral antigen (TSOL16) was added to TSOL18, and the efficacy of conjoint application was evaluated on the field in 274 animals distributed in pairs (one vaccinated and one control) in 86 households of the Department of Piura, Peru. Twelve months later, the effect of the vaccination was measured. Vaccination reduced from 16.8% (18/107) to 6.2% (7/113) the percent of cysticercotic pigs, and from 34,081 to 83 the number of cyst recovered [45].

ALTERNATIVES FOR PRODUCING A MASSIVE APPLICATION VACCINE

Several pathogens, including *T. solium,* enter the body through mucosal surfaces. Thus, it is likely that an improved mucosal immunity elicited by oral immunization would offer a first barrier to prevent infection. In addition, since oral immunization is also capable of eliciting systemic immunity, those oncospheres that eluded this first barrier could be damaged once established in the tissues. Needle-free oral vaccines are by far the most attractive immunization method, since they avoid costly logistics problems and are suitable for mass vaccination programs. For this purpose, the use of a delivery system that protects the vaccine against digestive hydrolysis is desirable [88]. Oral vaccines formulated from genetically engineered plant biomass provide a mean for easy and convenient immunization, since plant cells can be grown at low costs, have a high biosynthetic capacity, provide a delivery system that may enhance immunogenicity due to the presence of endogenous secondary metabolites, offer a particulate delivery system, and can be engineered by a set of versatile molecular tools [89].

Plant-based antigens were extensively promoted in the 1990s as an innovative source of low-cost vaccines. A number of edible crops such as potato, tomato, lettuce, bananas, corn, and rice were used to express antigens from various pathogens [90-92]. Although fresh plant materials were used in the initial efforts to test the concept, a more realistic approach to this vaccine comprise relatively simple processing, such as freeze-drying, to ensure the stability of plant biomass and a proper vaccine dosage. As an additional concept refinement, the use of bioreactors for biomass production was proposed to alleviate biosafety concerns, and expression strategies have been improved to increase productivity. The potential of plant-based vaccines is reflected by the fact that the pharmaceutical industry is adopting this technology at a faster pace than in previous years [93]. Given the advantages of plant-based vaccines, our research group has performed a number of efforts to assess the expression of *T. solium* antigens with distinct molecular approaches in various plant host species, and we are evaluating the potential of these as anti-cysticercosis vaccines.

S3Pvac-PAPAYA

The synthetic peptides (KETc1, KETc12, KETc7) were successfully expressed in transgenic papaya clones, and they were found to be immunogenic in mice [82]. High protection levels were induced by soluble extracts from calli of the different identified clones. Later on, a rabbit experimental model of *T. pisiformis* cysticercosis was used to compare the effectiveness of oral S3Pvac-papaya vaccine and the parenteral versions, S3Pvac-synthetic and S3Pvac-phage. Oral

S3Pvac-papaya significantly reduced the expected number of hepatic lesions and peritoneal cysticerci to a similar extent as the injectable vaccines. The effectiveness of this oral vaccine against *T. pisiformis* cysticercosis could make it useful against porcine *T. solium* cysticercosis [86].

MULTICOMPONENT VACCINES TROUGH TRANSPLASTOMIC AND PICORNAVIRAL 2A SEQUENCE-BASED EXPRESSION APPROACHES

One approach that is expected to significantly improve cysticercosis vaccines is the production of multicomponent vaccines in a robust platform that could facilitate formulation and delivery. Currently available molecular tools offer the possibility to express multiple open reading frames from a single DNA insert; therefore, our group is aiming to produce multicomponent plant-based cysticercosis vaccines.

Transplastomic technologies refer to genetic engineering approaches at the chloroplast genome level. They offer substantial advantages, such as a generally recognized absence of gene silencing, high productivity due to high transgene copy numbers, enhanced biosafety given by the strict maternal inheritance of chloroplasts in most plant species, and the possibility of expressing multiple open reading frames trough operon-like arrangements [94, 95]. Our group has applied transplastomic technologies to develop a vaccine candidate consisting of tobacco cells expressing simultaneously the S3Pvac antigens by means of a polycistronic arrangement encoded by a single DNA insertion at the chloroplast genome. Some of the protective peptides have been detected in transplastomic tobacco leaf extracts. These multicistronic tobacco candidate clones are intended to serve in the evaluation of a new multicomponent vaccine formulated with a single transformed clone, which would render a straightforward strategy to produce highly effective vaccines (Table **5**).

Table 5: Systems used to express protective antigens against swine cysticercosis

Vaccine	Expression system	Reference
KETc1, KETc12, and KETc7	Embryogenic callus of papaya	[82]
TSOL18	*Pichia pastoris*	[36]
Phagemids FK7, FGK1, FKETc12, and FKETc1	*Escherichia coli* TG1	[43]
Asiatic *T. solium* 45W-4B	*E. coli* BL21	[37]
TSOL16, TSOL45-1A, and TSOL45-1B antigens fused to GST and MBP	*E. coli* JM109	[60]
TSOL16-GST and TSOL18-GST	*E. coli* JM109	[45]
TSOL18	Recombinant attenuated *Salmonella typhimurium* live (strain x 4558)	[83]

GST: Glutathione S-transferase; MBP: Maltose binding protein.

The picornaviral 2A endopeptidase sequence, that mediates a translational skip mechanism, allows the production of different polypeptides from a single open reading frame, where target antigens are linked by a 2A-sequence [96, 97]. Since approaches based on the 2A-sequence to achieve multiple antigen expression have not been explored, our group is working on it to simultaneously express several *T. solium* antigens trough a single transformation event at the nuclear level. There are evidences that independent antigens were successfully expressed, and their immunogenicity has been found promising in mice; the immunoprotective capacity of this vaccine candidate will be assessed in the near future. This approach has a good potential to produce improved versions of plant-based cysticercosis vaccines (manuscript submitted for publication).

IMMUNITY AND VACCINATION

While several vaccine candidates have been reported as successful and their anti-parasite immunity has been assessed, much of their action mechanisms remain obscure or unexplored. Immunization with many of these proteins induces specific, complement-fixing antibodies against the total, purified, recombinant, or synthetic antigens that are believed to be associated with the vaccine-induced protection against *T. solium* infection. It has been argued that vaccine-elicited antibodies may destroy *T. solium* oncospheres by fixing complement [30, 56], but the vulnerability of these early larval phases of the parasite has not been explored yet. By evaluating the antibody subclasses and isotypes induced by vaccination, one may explore the possibility that oncosphere destruction could be also mediated by other effector functions of the antibodies, *e.g.*, ADCC (Antibody-dependent Cellular Cytotoxicity) by NK cells *via* Fcγ receptor. Fcγ receptors could also mediate the parasite destruction by eosinophil and basophil degranulation; these can liberate a mix of pro-inflammatory substances that can affect the onscophere viability (*i.e.*, peroxidase, histamine, proteoglycans, serotonin, serine proteases, *etc.*). Fcγ receptors can also mediate macrophage and neutrophil activation; once activated, both cells of the innate immune response can produce free radicals that can alter the parasite homeostasis. Finally, circulating antibodies elicited by systemic immunization could neutralize surface molecules that mediate the parasite invasion to other tissues, including the entrance to the central nervous system. In addition, as reported by our group, vaccination of mice and pigs with a parasite epitope (GK1) induced the synthesis of antibodies that are capable of interfering with the transformation of cysticerci into tapeworms in hamster intestine [98, 99].

On the other hand, vaccine antigens can also induce a protective cellular immune response. The cellular immunity induced by the TSOL18 antigen vaccine remains to be studied. In contrast, experimental evidences support the possibility that an exacerbated inflammatory response may damage the larval stage of the parasite and/or interfere with the cysticerci establishment [99]. However, their relevance in the establishment and/or destruction of the parasite has not been explored yet.

Other finding meriting comments is the protection induced by the S3Pvac vaccine peptides GK1 and KETc1, employed separately, with no adjuvant (64.7% and 38.7% of protection, respectively) [84, 85]. This result supports their intrinsic capacity to active cells of the innate immune response to evolve their specific anti-parasite immunity. Indeed, GK1 was able to active dendritic cells, and this property underlies the adjuvant capacity of GK1 [100]. Thus, the inflammatory environment induced by these peptides, along with the effector functions of the induced antibodies as described above, could be mediating the parasite damage, preventing oncospheres from crossing the intestinal epithelium to reach other body tissues: CNS, skeletal muscle, eyes, heart, lung, *etc.* In addition, those oncospheres that escaped this immune response would be vulnerable later, at the cysticercus stage, to both the adaptive immune response and the persistent inflammatory environment.

POSSIBILITIES OF VACCINATION AGAINST CYSTICERCOSIS IN HUMANS

There is evidence hinting the feasibility of vaccinating humans against NC, particularly against severe forms. Among these hints stands out the very low incidence of symptomatic NC in endemic communities, even where 9% - 10% of the population has clinically silent NC [17-19] and over 90% of subjects show immunological traits compatible with previous contacts with the parasite antigens [25]. These observations suggest that the contact with the parasite may generate a certain degree or some type of immunity, capable of protecting the individual against severe forms of NC.

TOWARDS A MULTIEPITOPE VACCINE AGAINST CESTODIASIS

T. solium, T. saginata, and *Echinococcus spp.* are parasites of utmost medical and veterinary interest [101-104]. The extensive similarities in the natural history, pathology, and antigenic composition among cestodes [64, 105, 106] often allow the use of protective antigens developed for one species to be applied against a different one [9, 32, 55, 87]. These inter-species shared antigens offer a common target for vaccine design. The observed cross-immunity induced by the S3Pvacphage anti-cysticercosis vaccine, which also prevents swine hydatidosis under natural conditions of transmission, supports this possibility [81].

PREVENTION OF INTESTINAL TAENIASIS AND CYSTICERCOSIS BY VACCINATION

Several antigens are present throughout different cestode developmental stages (*i.e.*, eggs, cysticerci, tapeworms) [52, 54, 107, 108]. Some of these antigens are being considered to develop multipurpose vaccine to prevent both cysticercosis and the tapeworm intestinal infection [109]. A taeniasis experimental model in golden hamster was used to test possible vaccine candidates. Since antibodies against the S3Pvac components strongly recognized different structures of the *T. solium* tapeworm, this vaccine was tested under different conditions with experimental taeniasis. After promising initial results, where high protection levels were obtained [109], further experiments showed that the level of protection critically depend on the age of the host hamsters and the size of the cysticerci used for challenge [110]. These findings, along with the low prevalence of *T. solium* taeniasis in humans, make remote the development of an anti-taeniasis vaccine.

CAN CYSTICERCOSIS BE ERADICABLE?

In spite of the prevailing optimism that technology will be able to resolve many a disease [111], it should be considered that cysticercosis transmission is perpetuated by a complex network of biological, historical, cultural, and economical factors; it is clear, then, that its control may require an approach as complex as the factors that allow parasite survival in modern times [72].

In several instances, *T. solium* cysticercosis has been regarded as easy to control, and it has been declared an eradicable disease [112-116]. Such optimism is based on the capital advances in diagnosis [95], on the effectiveness of cysticidal treatment [117, 118] and on the high protective capacity of the various reported vaccine candidates (Tables **1** and **2**). Unfortunately, success on this is unwarranted [119]. The eradication of *T. solium* will likely remain as a major challenge for many years in underdeveloped countries, where the parasite life-cycle is firmly rooted in complex social, cultural, political, and economic contexts [72]. Some success has been attained by control programs focused in rural villages and aimed to locally improve health education and provide treatment to tapeworm carriers [111, 117, 120-126].

Unfortunately, experience shows that these actions are insufficient to achieve wide and long-term control of the disease due to the complex social and biological networks underlying cestode transmission, which rapidly re-establish the infection after such transient interventions in minute geographical sites come to an end [42, 121]. Moreover, prevention programs often disregard the growing and massive migration from rural communities to the cities. Rural inhabitants bring parasites with them into urban areas at endemic and non-endemic countries [127, 128] and then back; these carriers escape control, spread the disease to naïve sites and reinstall the parasite transmission when returning to their communities [129, 130]. To further complicate the picture, heath authorities of endemic countries are too busy and overdrawn by many other diseases affecting the population to invest in preventing ancient and embarrassing "poverty diseases". So, notwithstanding the potential usefulness of cost-effective immunization technology, *T. solium* will not be under control until the affected civil societies at endemic countries strengthen their will to live healthy and change their infection-prone behaviors and environmental conditions. "Pigs are not the culprits; humans are" (Aline S. de Aluja).

The first successful extensive control program was recently reported in Mexico. Lasting from 2009 to 2013, it combined educational measures and swine vaccination with the S3Pvac-phage injectable vaccine. The program was systematically applied in three regions of the state of Guerrero, based on education. The impact of the program was gauged by porcine cysticercosis prevalence [131]. Cysticercosis prevalence was estimated by tongue inspection and ultrasound diagnosis. Three years after the application of the program, a significant reduction in the prevalence of porcine cysticercosis was observed, from 7% to 0.8% and from 3.6% to 0.4%, as estimated by tongue examination or ultrasound, respectively ($P < 0.01$). This establishes the effectiveness of the program in preventing infection. A sustained presence of antibodies, compatible with a contact to *T. solium* or other related helminthes, underlines the importance of continuing interventions to achieve eradication.

The multinational efforts to eradicate the disease are noteworthy. The recent establishment of network organizations in Africa, Asia, and Europe funded by the European Community, and in Latin-America by CYTED, is a first positive step in this direction. However, since economic, sociological, cultural, and political factors are of crucial importance in the continued persistence of *T. solium*, the governments of the affected countries should recognize the eradication of this parasite as a public health priority. Also, they should admit that a serious and effective control of the disease will critically depend on their commitment and support.

ACKNOWLEDGEMENTS

This work was supported by CONACyT (CB-152793 and CB-102109), Fundación Miguel Alemán, SAGARPA, SENASICA, and Programa de investigación para el desarrollo y la optimización de vacunas, adyuvantes y métodos diagnósticos del Instituto de Investigaciones Biomédicas, UNAM.

We thank Juan Francisco Rodriguez for copyediting the manuscript in its English version.

CONFLICTS OF INTEREST

The authors confirm that this chapter contents have no conflict of interest.

DISCLOSURE

This is an updated report of the earlier review article by Sciutto E, Fragoso G, de Aluja AS, Hernández M, Rosas G, Larralde C. "Vaccines against cysticercosis" Curr Top Med Chem 2008; 8(5): 415-23.

REFERENCES

[1] Joseph J, Saubi N, Pezzat E, Gatell JM. Progress towards an HIV vaccine based on recombinant bacillus Calmette-Guerin: failures and challenges. Expert Rev Vaccines 2006; 5(6): 827-38.

[2] Okada M. Novel vaccines against *M. tuberculosis*. Kekkaku 2006; 81(12): 745-51.

[3] Pathak CG, Kuhn RJ. Development of novel antivirals against flaviviruses. Novartis Found Symp 2006; 277: 41-52.

[4] Tettelin H, Medini D, Donati C, Masignani V. Towards a universal group B *Streptococcus* vaccine using multistrain genome analysis. Expert Rev Vaccines 2006; 5(5): 687-94.

[5] Tongren JE, Zavala F, Roos DS, Riley EM. Malaria vaccines: if at first you don't succeed. Trends Parasitol 2004; 20(12): 604-10.

[6] Kwa BH, Liew FY. Immunity in taeniasis-cysticercosis. Vaccination against *Taenia taeniaeformis* in rats using purified antigen. J Exp Med 1977; 146(1): 118-31.

[7] Molinari JL, Meza R, Suarez B, Palacios S, Tato P, Retana A. *Taenia solium*: immunity in hogs to the cysticercus. Exp Parasitol 1983; 55(3): 340-57.

[8] Johnson KS, Harrison GB, Lightowlers MW, O'Hoy KL, Cougle, WG, Dempster RP, *et al.* Vaccination against ovine cysticercosis using a defined recombinant antigen. Nature 1989; 338(6216): 585-7.

[9] Sciutto E, Fragoso G, Trueba L, Lemus D, Montoya RM, Diaz ML, *et al.* Cysticercosis vaccine: cross protecting immunity with *T. solium* antigens against experimental murine *T. crassiceps* cysticercosis. Parasite Immunol 1990; 12(6): 687-96.

[10] Harrison LJ, Garate T, Bryce DM, Gonzalez LM, Foster-Cuevas M, Wamae LW, *et al.* Ag-ELISA and PCR for monitoring the vaccination of cattle against *T. saginata* cysticercosis using an

oncospheral adhesion protein (HP6) with surface and secreted localization. Trop Anim Health Prod 2005; 37(2): 103-20.
[11] Aluja A, Vargas G. The histopathology of porcine cysticercosis. Vet Parasitol 1988; 28(1-2): 65-77.
[12] Restrepo BI, Llaguno P, Sandoval MA, Enciso JA, Teale JM. Analysis of immune lesions in neurocysticercosis patients: central nervous system response to helminth appears Th1-like instead of Th2. J Neuroimmunol 1998; 89(1-2): 64-72.
[13] Restrepo BI, Alvarez JI, Castano JA, Arias LF, Restrepo M, Trujillo J, *et al.* Brain granulomas in neurocysticercosis patients are associated with a Th1 and Th2 profile. Infect Immun 2001; 69(7): 4554-60.
[14] Alvarez JI, Colegial CH, Castano CA, Trujillo J, Teale JM, Restrepo BI. The human nervous tissue in proximity to granulomatous lesions induced by *T. solium* metacestodes displays an active response. J Neuroimmunol; 127(1-2): 139-44.
[15] Londono DP, Alvarez JI, Trujillo J, Jaramill MM, Restrepo BI. The inflammatory cell infiltrates in porcine cysticercosis: immunohistochemical analysis during various stages of infection. Vet Parasitol 2002; 109(3-4): 249-59.
[16] Sanzon F, Osorio AM, Morales JP, Isaza R, Cardona E, Moncayo LC, *et al.* Serological screening for cysticercosis in mentally altered individuals. Trop Med Int Health 2002; 7(6): 532-8.
[17] Sanchez AL, Lindback J, Schantz PM, Sone M, Sakai H, Medina, MT, *et al.* A population-based, case-control study of *Taenia solium* taeniasis and cysticercosis. Ann Trop Med Parasitol 1999; 93(3): 247-58.
[18] Fleury A, Gomez T, Alvarez I, Meza D, Huerta M, Chavarria, A, *et al.* High prevalence of calcified silent neuocysticercosis in a rural village of Mexico. Neuroepidemiology 2003; 22(2): 139-45.
[19] Fleury A, Morales J, Bobes RJ, Dumas M, Yanez O, Pina J, *et al.* An epidemiological study of familial neurocysticercosis in an endemic Mexican community. Trans R Soc Trop Med Hyg 2006; 100(6): 551-8.
[20] Saenz B, Ruiz-Garcia M, Jiménez E, Hernández-Aguilar J, Suastegui R, Larralde C, *et al.* Neurocysticercosis: clinical, radiologic, and inflammatory differences between children and adults. Pediatr Infect Dis J 2006; 25(9): 801-3.
[21] Aluja AS, Villalobos AN, Plancarte A, Rodarte LF, Hernandez M, Sciutto E. Experimental *Taenia solium* cysticercosis in pigs: characteristics of the infection and antibody response. Vet Parasitol 1996; 61(1-2): 49-59.
[22] Aluja AS, Villalobos AN, Plancarte A, Rodarte LF, Hernandez M, Zamora C, *et al. Taenia solium* cysticercosis: immunity in pigs induced by primary infection. Vet Parasitol 1999; 81(2): 129-35.
[23] Aluja AS, Villalobos N, Nava G, Toledo A, Martínez JJ, Plancarte A, *et al.* Therapeutic capacity of the synthetic peptide-based vaccine against *Taenia solium* cysticercosis in pigs. Vaccine 2005; 23(31): 4062-9.
[24] Yang H, Parkhouse RM. Phenotypic classification of porcine lymphocyte subpopulations in blood and lymphoid tissues. Immunology 1996; 89(1): 76-83.
[25] Chavarria A, Roger B, Fragoso G, Tapia G, Fleury A, Dumas M, *et al.* TH2 profile in asymptomatic *Taenia solium* human neurocysticercosis. Microbes Infect 2003; 5(12): 1109-15.
[26] Bueno EC, Vaz AJ, Machado LR, Livramento JA, Avila SL, Ferreira AW. Antigen-specific suppression of cultured lymphocytes from patients with neurocysticercosis. Clin Exp Immunol 2001; 126(2): 304-10.
[27] Chavarria A, Fleury A, Bobes RJ, Morales J, Fragoso G, Sciutto E. A depressed peripheral cellular immune response is related to symptomatic neurocisticercosis. Microbes Infect 2006; 8(4):1082-9.
[28] Sciutto E, Cárdenas G, Adalid-Peralta L, Fragoso G, Larralde C, Fleury A. Human neuocisticercosis: immunological features involved in the host's susceptibility to become infected and to develop disease. Microbes Infect 2013; 15(6-7): 524-30.
[29] Pathak KM, Gaur SN. Immunization of pigs with culture antigens of *Taenia solium.* Vet Parasitol 1990; 34(4): 353-6.
[30] Molinari JL, Soto R, Tato P, Rodriguez D, Retana A, Sepulveda J, *et al.* Immunization against porcine cysticercosis in an endemic area in Mexico: a field and laboratory study. Am J Trop Med Hyg 1993; 49(4): 502-12.
[31] Nascimento E, Costa JO, Guimaraes MP, Tavares CA. Effective immune protection of pigs against cysticercosis. Vet Immunol Immunopathol 1995;45(1-2): 127-37.

[32] Sciutto E, Aluja A, Fragoso G, Rodarte LF, Hernández M, Villalobos MN, *et al.* Immunization of pigs against *Taenia solium* cysticercosis: factors related to effective protection. Vet Parasitol 1995; 60(1-2): 53-6.
[33] Molinari JL, Rodriguez D, Tato P, Soto R, Arechavaleta F, Solano S. Field trial for reducing porcine *Taenia solium* cysticercosis in Mexico by systematic vaccination of pigs. Vet Parasitol 1997; 69(1-2): 55-63.
[34] Manoutcharian K, Rosas G, Hernandez M, Fragoso G, Aluja A, Villalobos N, *et al.* Cysticercosis: identification and cloning of protective recombinant antigens. J Parasitol 1996; 82(2): 250-4.
[35] Fan PC, Chung WC, Lin CY, Wu CC. Vaccination trials against *Taenia solium* eggs in pigs injected with frozen oncospheres of *T. solium* or *T. saginata asiatica.* J Microbiol Immunol Infect 2003; 36(2): 96-100.
[36] Cai X, Yuan G, Zheng Y, Luo X, Zhang S, Ding J, *et al.* Effective production and purification of the glycosylated TSOL18 antigen, which is protective against pig cysticercosis. Infect Immun 2008; 76(2): 767-70.
[37] Luo X, Zheng Y, Hou J, Zhang S, Cai X. Protection against Asiatic *Taenia solium* induced by recombinant 45W-4B protein. Clin Vaccine Immunol 2009; 16(2): 230-2.
[38] Huerta M, de Aluja AS, Fragoso G, Toledo A, Villalobos N, Hernandez M, *et al.* Synthetic peptide vaccine against *Taenia solium* pig cysticercosis: successful vaccination in a controlled field trial in rural Mexico. Vaccine 2001; 20(1-2): 262-6.
[39] Flisser A, Gauci CG, Zoli A, Martinez-Ocana J, Garza-Rodriguez A, Dominguez-Alpizar JL, *et al.* Induction of protection against porcine cysticercosis by vaccination with recombinant oncosphere antigens. Infect Immun 2004; 72(9): 5292-7.
[40] Manoutcharian K, Diaz-Orea A, Gevorkian G, Fragoso G, Acero G, Gonzalez, E, *et al.* Recombinant bacteriophage-based multiepitope vaccine against *Taenia solium* pig cysticercosis. Vet Immunol Immunopathol 2004; 99(1-2): 11-24.
[41] Gonzalez AE, Gauci CG, Barber D, Gilman RH, Tsang VC, Garcia HH, *et al.* Vaccination of pigs to control human neurocysticercosis. Am J Trop Med Hyg 2005; 72(6): 837-9.
[42] Sciutto E, Morales J, Martinez JJ, Toledo A, Villalobos MN, Cruz-Revilla C, *et al.* Further evaluation of the synthetic peptide vaccine S3Pvac against *Taenia solium* cysticercosis in pigs in an endemic town of Mexico. Parasitology 2007; 134(Pt 1): 129-33.
[43] Morales J, Martínez JJ, Manoutcharian K, Hernández M, Fleury A, Gevorkian G, *et al.* Inexpensive anti-cysticercosis vaccine: S3Pvac expressed in heat inactivated M13 filamentous phage proves effective against naturally acquired *Taenia solium* porcine cysticercosis. Vaccine 2008; 26(23): 2899-905.
[44] Assana E, Kyngdon CT, Gauci CG, Geerts S, Dorny P, De Deken R, *et al.* Elimination of *Taenia solium* transmission to pigs in a field trial of the TSOL18 vaccine in Cameroon. Int J Parasitol 2010; 40(5): 515-9.
[45] Jayashi CM, Kyngdon CT, Gauci CG, Gonzalez AE, Lightowlers MW. Successful immunization of naturally reared pigs against porcine cysticercosis with a recombinant oncosphere antigen vaccine. Vet Parasitol 2012; 188(3-4): 261-7.
[46] Cai X, Chai Z, Jing Z, Wang P, Luo X, Chen J, *et al.* Studies on the development of DNA vaccine against *Cysticercus cellulosae* infection and its efficacy. Southeast Asian J Trop Med Public Health 2001; 32(Suppl 2): 105-10.
[47] Wang QM, Sun SH, Hu ZL, Wu D, Wang ZC. Immune response and protection elicited by DNA immunization against *Taenia* cysticercosis. Vaccine 2003; 21(15): 1672-80.
[48] Guo YJ, Sun SH, Zhang Y, Chen ZH, Wang KY, Huang L, *et al.* Protection of pigs against *Taenia solium* cysticercosis using recombinant antigen or in combination with DNA vaccine. Vaccine 2004; 22(29-30): 3841-7.
[49] Wu L, Diao Z, Dengm X, Gao J, Zhou Z, Liu Y, *et al.* DNA vaccine against *Taenia solium* cysticercosis expressed as a modified hepatitis B virus core particle containing three epitopes shared by *Taenia crassiceps* and *Taenia solium.* J Nanosci Nanotechnol 2005; 5(8): 1204-10.
[50] Guo A, Jin Z, Zheng Y, Hai G, Yuan G, Li H. Induction of protection against porcine cysticercosis in growing pigs by DNA vaccination. Vaccine 2007; 25(1): 170-5.
[51] Valdez F, Hernandez M, Govezensky T, Fragoso G, Sciutto E. Immunization against *Taenia crassiceps* cysticercosis: identification of the most promising antigens in the induction of protective immunity. J Parasitol 1994; 80(6): 931-6.

[52] Toledo A, Fragoso G, Rosas G, Hernández M, Gevorkian G, Lopez-Casillas F, *et al.* Two epitopes shared by *Taenia crassiceps* and *Taenia solium* confer protection against murine *T. crassiceps* cysticercosis along with a prominent T1 response. Infect Immun 2001; 69(3): 1766-73.

[53] Wolf H, Modrow S, Motz M, Jameson BA, Hermann G, Förtsch B. An integrated family of amino acid sequence analysis programs. Comput Appl Biosci 1988; 4(1): 187-91.

[54] Toledo A, Larralde C, Fragoso G, Gevorkian G, Manoutcharian K, Hernández M. Towards a *Taenia solium* cysticercosis vaccine: an epitope shared by *Taenia crassiceps* and *Taenia solium* protects mice against experimental cysticercosis. Infect Immun 1999; 67(5): 2522-30.

[55] Plancarte A, Flisser A, Gauci CG, Lightowlers MW. Vaccination against *Taenia solium* cysticercosis in pigs using native and recombinant oncosphere antigens. Int J Parasitol 1999; 29(4): 643-7.

[56] Kyngdon CT, Gauci CG, Rolfe RA, Velasquez-Guzman JC, Farfan-Salazar MJ, Verastegui-Pimentel MR, *et al. In vitro* oncosphere killing assays to determine immunity to the larvae of *Taenia pisiformis*, *Taenia ovis*, *Taenia saginata*, and *Taenia solium*. J Parasitol 2006; 92(2): 273-81.

[57] Lightowlers MW, Rolfe R, Gauci CG. *Taenia saginata*: vaccination against cysticercosis in cattle with recombinant oncosphere antigens. Exp Parasitol 1996; 84(3): 330-8.

[58] Bonay P, Gonzalez LM, Benitez L, Foster M, Harrison LJ, Parkhouse RM, *et al.* Genomic and functional characterization of a secreted antigen of *Taenia saginata* oncospheres. Mol Biochem Parasitol 2002; 121(2): 269-73.

[59] Solis CF, Ostoa-Saloma P, Lugo-Martinez VH, Johnston SA, Laclette JP. Genetic vaccination against murine cysticercosis by using a plasmid vector carrying *Taenia solium* paramyosin. Infect Immun 2005; 73(3): 1895-7.

[60] Gauci CG, Jayash CM, Gonzalez AE, Lackenby J, Lightowlers MW. Protection of pigs against *Taenia solium* cysticercosis by immunization with novel recombinant antigens. Vaccine 2012; 30(26): 3824-8.

[61] Maravilla P, Souza V, Valera A, Romero-Valdovinos M, Lopez-Vidal Y, Dominguez-Alpizar JL, *et al.* Detection of genetic variation in *Taenia solium*. J Parasitol 2003; 89(6): 1250-4.

[62] Vega R, Pinero D, Ramanankandrasana B, Dumas M, Bouteille B, Fleury A, *et al.* Population genetic structure of *Taenia solium* from Madagascar and Mexico: implications for clinical profile diversity and immunological technology. Int J Parasitol 2003; 33(13): 1479-85.

[63] Cruz-Rivera M, Reyes-Torres A, Reynoso-Ducoing O, Flisser A, Ambrosio JR. Comparison of biochemical and immunochemical properties of myosin II in taeniid parasites. Cell Biol Int 2006; 30, 598-602.

[64] Larralde C, Montoya RM, Sciutto E, Diaz ML, Govezensky T, Coltorti E. Deciphering western blots of tapeworm antigens (*Taenia solium, Echinococcus granulosus* and *Taenia crassiceps*) reacting with sera from neurocysticercosis and hydatid disease patients. Am J Trop Med Hyg 1989; 40(3): 282-90.

[65] Arruda GC, da Silva AD, Quagliato EM, Maretti MA, Rossi CL. Evaluation of *Taenia solium* and *Taenia crassiceps* cysticercal antigens for the serodiagnosis of neurocysticercosis. Trop Med Int Health 2005; 10(10): 1005-12.

[66] Espindola NM, Iha AH, Vaz AJ, Fernandez I, Takayanagui OM, Machado LR, *et al.* Cysticercosis immunodiagnosis using 18- and 14-kilodalton proteins from *Taenia crassiceps* cysticercus antigens obtained by Immunoaffinity chromatography. J Clin Microbiol 2005; 43(7): 3178-84.

[67] Tsai IJ, Zarowiecki M, Holroyd N, Garciarrubio A, Sanchez-Flores A, Brooks KL, *et al.* The genomes of four tapeworm species reveal adaptations to parasitism. Nature 2013; 496(7443): 57-63.

[68] Rassy D, Bobes RJ, Rosas G, Anaya VH, Brehm K, Hernández B, *et al.* Characterization of S3Pvac anti-cysticercosis vaccine components: implications for the development of an anti-cestodiasis vaccine. PLoS One 2010; 5(6): e11287.

[69] Huerta M, Sciutto E, Garcia G, Villalobos N, Hernández M, Fragoso G, *et al.* Vaccination against *Taenia solium* cysticercosis in underfed rustic pigs of Mexico: roles of age, genetic background and antibody response. Vet Parasitol 2000; 90(3): 209-19.

[70] Sciutto E, Martinez JJ, Huerta M, Avila R, Fragoso G, Villalobos N, *et al.* Familial clustering of *Taenia solium* cysticercosis in the rural pigs of Mexico: hints of genetic determinants in innate and acquired resistance to infection. Vet Parasitol 2003; 116(3): 223-9.

[71] Morales J, Velasco T, Tovar V, Fragoso G, Fleury A, Beltran C, *et al.* Castration and pregnancy of rural pigs significantly increase the prevalence of naturally acquired *Taenia solium* cysticercosis. Vet Parasitol 2002; 108(1): 41-8.

[72] Morales J, Martinez JJ, Garcia-Castella J, Peña N, Maza V, Villalobos N, *et al*. *Taenia solium*: the complex interactions, of biological, social, geographical and commercial factors, involved in the transmission dynamics of pig cysticercosis in highly endemic areas. Ann Trop Med Parasitol 2006; 100(2): 123-35.
[73] Santamaría E, Plancarte A, de Aluja AS. The experimental infection of pigs with different numbers of *Taenia solium* eggs: immune response and efficiency of establishment. J Parasitol 2002; 88(1): 69-73.
[74] Viljoen NF. Cysticercosis in swine and bovines with special reference to South African conditions. Onderstepoort J Vet Res 1937; 9: 337-570.
[75] Gonzalez AE, Cama V, Gilman RH, Tsang VC, Pilcher JB, Chavera A, *et al*. Prevalence and comparison of serologic assays, necropsy, and tongue examination for the diagnosis of porcine cysticercosis in Peru. Am J Trop Med Hyg 1990; 43(2): 194-9.
[76] Sciutto E, Martinez JJ, Villalobos NM, Hernandez M, Jose MV, Beltran C, *et al*. Limitations of current diagnostic procedures for the diagnosis of *Taenia solium* cysticercosis in rural pigs. Vet Parasitol 1998; 79(4): 299-313.
[77] Quiroz RH. Parasitología y enfermedades parasitarias de animales domésticos. Eds. Limusa. ed. México 2002; pp. 335-65.
[78] Sciutto E, Hernandez M, Garcia G, de Aluja AS, Villalobos AN, Rodarte LF, *et al*. Diagnosis of porcine cysticercosis: a comparative study of serological tests for detection of circulating antibody and viable parasites. Vet Parasitol 1998; 78(3): 185-94.
[79] De Aluja AS, Herrera GS, Hernández M, Plancarte A, Fragoso G, Sciutto E. Limits of the therapeutic properties of synthetic S3Pvac anti-cysticercosis vaccine. Vet Parasitol 2011; 177(1-2): 90-6.
[80] Moro P, Schantz PM. Echinococcosis: a review. Int J Infect Dis 2009; 13(2): 125-33.
[81] Morales J, de Aluja AS, Martínez JJ, Hernández M, Rosas G, Villalobos N, *et al*. Recombinant S3Pvac-phage anticysticercosis vaccine: Simultaneous protection against cysticercosis and hydatid disease in rural pigs. Vet Parasitol 2011; 176(1): 53-8.
[82] Hernandez M, Cabrera-Ponce JL, Fragoso G, Lopez-Casillas F, Guevara-Garcia A, Rosas G, *et al*. A new highly effective anticysticercosis vaccine expressed in transgenic papaya. Vaccine 2007; 25(21): 4252-60.
[83] Ding J, Zheng Y, Wang Y, Dou Y, Chen X, Zhu X, *et al*. Immune responses to a recombinant attenuated *Salmonella typhimurium* strain expressing a *Taenia solium* oncosphere antigen TSOL18. Comp Immunol Microbiol Infect Dis 2013; 36(1): 17-23.
[84] Rosas G, Fragoso G, Ainciant N, Esquivel-Guadarrama F, Santana A, Bobes RJ, *et al*. *Brucella spp*. lumazine synthase: a novel adjuvant and antigen delivery system to effectively induce oral immunity. Microbes Infect 2006; 8(5): 1277-86.
[85] Fragoso G, Esquivel-Guadarrama F, Santana MA, Bobes RJ, Hernández B, Cervantes J, *et al*. Heterologous prime-boost oral immunization with GK1 peptide from *Taenia crassiceps* cysticerci induces protective immunity. Clin Vaccine Immunol 2011; 18(7): 1067-76.
[86] Betancourt MA, de Aluja AS, Sciutto E, Hernández M, Bobes R.J, Rosas G, *et al*. Effective protection induced by three different versions of the porcine S3Pvac anticysticercosis vaccine against rabbit experimental *Taenia pisiformis* cysticercosis. Vaccine 2012; 30(17): 2760-7.
[87] Rosas G, Fragoso G, Garate T, Hernandez B, Ferrero P, Foster-Cuevas M, *et al*. Protective immunity against *Taenia crassiceps* murine cysticercosis induced by DNA vaccination with a *T. saginata* tegument antigen. Microbes Infect 2002; 4(14): 1417-26.
[88] Cripps AW, Kyd JM, Foxwell AR. Vaccines and mucosal immunization. Vaccine 2001; 19(17-19): 2513-5.
[89] Rosales-Mendoza S, Salazar-González JA. Immunological aspects of using plant cells as delivery vehicles for oral vaccines. Expert Rev Vaccines 2014; 13(6): 737-49.
[90] Walmsley AM, Arntzen CJ. Plants for delivery of edible vaccines. Curr Opin Biotechnol 2000; 11(2): 126-9.
[91] Walmsley AM, Arntzen CJ. Plant cell factories and mucosal vaccines. Curr Opin Biotechnol 2003; 14(2): 145-50.
[92] Rybicki EP. Plant-made vaccines for humans and animals. Plant Biotechnol J 2010; 8(5): 620-37.
[93] Yusibov V, Streatfield SJ, Kushnir N. Clinical development of plant-produced recombinant pharmaceuticals: vaccines, antibodies and beyond. Hum Vaccin 2011; 7(3): 313-21.

[94] Bock R. Strategies for metabolic pathway engineering with multiple transgenes. Plant Mol Biol 2013; 83(1-2): 21-31.
[95] Bock R. Genetic engineering of the chloroplast: novel tools and new applications. Curr Opin Biotechnol 2014; 26: 7-13.
[96] Halpin C, Cooke SE, Barakate A, Amrani AE, Ryan MD. Self-processing 2A-polyproteins—a system for co-ordinate expression of multiple protein in transgenic plants. Plant J 1999; 17(4): 453-9.
[97] Ha SH, Liang YS, Jung H, Ahn MJ, Suh SC, Kweon SJ, *et al.* Application of two bicistronic systems involving 2A and IRES sequences to the biosynthesis of carotenoids in rice endosperm. Plant Biotechnol J 2010; 8(8): 928-38.
[98] Garcia G, Sciutto E, Fragoso G, Cruz-Revilla C, Toledo A, Villalobos N, *et al.* Inhibitory role of antibodies in the development of *Taenia solium* and *Taenia crassiceps* toward reproductive and pathogenic stages. J Parasitol 2001; 87(3): 582-6.
[99] Diaz MA, Villalobos N, de Aluja A, Rosas G, Gomez-Conde E, Hernandez P, *et al.* Th1 and Th2 indices of the immune response in pigs vaccinated against *Taenia solium* cysticercosis suggest various host immune strategies against the parasite. Vet Immunol Immunopathol 2003; 93(3-4): 81-90.
[100] Segura-Velázquez R, Fragoso G, Sciutto E, Sarukhan, A. Towards identification of the mechanisms of action of parasite-derived peptide GK1 on the immunogenicity of an influenza vaccine. Clin Vaccine Immunol 2009; 16(9): 1338-43.
[101] Sciutto E, Fragoso G, Fleury A, Laclette JP, Sotelo J, Aluja A, *et al. Taenia solium* disease in humans and pigs: an ancient parasitosis disease rooted in developing countries and emerging as a major health problem of global dimensions. Microbes Infect 2000; 2(15): 1875-90.
[102] Ito A, Wandra T, Yamasaki H, Nakao M, Sako Y, Nakaya K, *et al.* Cysticercosis/taeniasis in Asia and the Pacific. Vector Borne Zoonotic Dis 2004; 4(2): 95-107.
[103] Ito A, Nakao M, Wandra T, Suroso T, Okamoto M, Yamasaki H, *et al.* Taeniasis and cysticercosis in Asia and the Pacific: present state of knowledge and perspectives. Southeast Asian J Trop Med Public Health 2005; 36(Suppl 4): 123-30.
[104] Ito A. Brief historical remarks as an introduction for the international symposium on taeniasis/cysticercosis and echinococcosis. Parasitol Int 2006; (Supp l55): S3-5.
[105] Gemmell MA. Immunological responses of the mammalian host against tapeworm infections. XI. Antigen sharing among *Taenia pisiformis*, *T. hydatigena*, and *T. ovis*. Exp Parasitol 1969; 26(1): 67-72.
[106] Harrison LJ, Parkhouse RM. *T. saginata* and *Taenia solium*: reciprocal models. Acta Leiden 1989; 57(2): 143-52.
[107] Onyango-Abuje JA, Waiyaki PG, Gathuma JM. Specific and shared antigenic components of *T. saginata* oncospheres. Res Vet Sci 1989; 47(2): 236-40.
[108] Garcia-Allan C, Martinez N, Flisser A, Aluja A, Allan JC, Craig PS. Immunocharacterization of *Taenia solium* oncosphere and metacestode antigens. J. Helminthol 1996; 70(4): 271-80.
[109] Cruz-Revilla C, Toledo G, Rosas A, Huerta M, Flores-Perez I, Peña N, *et al.* Effective protection against experimental *Taenia solium* tapeworm infection in hamsters by primo-infection and by vaccination with recombinant or synthetic heterologous antigens. J Parasitol 2006; 92(4): 864-7.
[110] León-Cabrera S, Cruz-Rivera M, Mendlovic F, Avila-Ramírez G, Carrero JC, Laclette JP, *et al.* Standardization of an experimental model of human taeniosis for oral vaccination. Methods 2009; 49(4): 346-50.
[111] Gonzalez AE, Gavidia C, Falcon N, Bernal T, Verastegui M, Garcia HH, *et al.* Protection of pigs with cysticercosis from further infections after treatment with oxfendazole. Am J Trop Med Hyg 2001; 65(1): 15-8.
[112] Eddi C, Nari A, Amanfu W. *Taenia solium* cysticercosis/taeniosis: potential linkage with FAO activities; FAO support possibilities. Acta Trop 2003; 87(1): 145-8.
[113] Engels D, Urbani C, Beloto A, Meslin F, Savioli L. The control of human (neuro) cysticercosis: which way forward? Acta Trop 2003; 87(1): 177-82.
[114] Lightowlers MW. Vaccination for the prevention of cysticercosis. Dev Biol (Basel) 2004; 119: 361-8.
[115] Garcia HH, Del Brutto OH. Neurocysticercosis: updated concepts about an old disease. Lancet Neurol 2005; 4(10): 653-61.

[116] Willingham AL.3rd, Engels D. Control of *Taenia solium* cysticercosis/taeniosis. Adv Parasitol 2006; 61: 509-66.

[117] Pawlowski ZS. Control of *Taenia solium* taeniasis and cysticercosis by focus-oriented chemotherapy of taeniasis. Southeast Asian J Trop Med Public Health 1991; (22 Suppl): 284-6.

[118] Pawlowski ZS. Role of chemotherapy of taeniasis in prevention of neurocysticercosis. Parasitol Int 2006; (55 Suppl): S105-9.

[119] Pawlowski ZS, Allan J, Sarti E. Control of *Taenia solium* taeniasis/cysticercosis: from research towards implementation. Int J Parasitol 2005; 35(11-12): 1221-32.

[120] Sciutto E, Rosas G, Hernandez M, Morales J, Cruz-Revilla C, Toledo A, *et al.* Improvement of the synthetic tri-peptide vaccine (S3Pvac) against porcine *Taenia solium* cysticercosis in search of a more effective, inexpensive and manageable vaccine. Vaccine 2007; 25(8): 1368-78.

[121] Keilbach NM, de Aluja AS, Sarti-Gutierrez E. A programme to control taeniasis-cysticercosis (*T. solium*): experiences in a Mexican village. Acta Leiden 1989; 57(2): 181-9.

[122] Diaz-Camacho SP, Candil Ruiz A, Suate Peraza V, Zazueta Ramos ML Felix Medina M, Lozano R, *et al.* Epidemiologic study and control of *Taenia solium* infections with praziquantel in a rural village of Mexico. Am J Trop Med Hyg 1991; 45(4): 522-31.

[123] Allan JC, Velásquez-Tohom M, Fletes C, Torres-Alvarez R, Lopez-Virula G, Tyrritia P, *et al.* Mass chemotherapy for intestinal *Taenia solium* infection: effect on prevalence in humans and pigs. Trans R Soc Trop Med Hyg 1997; 91(5): 595-8.

[124] Sarti E, Flisser A, Schantz PM, Gleizer M, Loya M, Plancarte A, *et al.* Development and evaluation of a health education intervention against *Taenia solium* in a rural community in Mexico. Am J Trop Med Hyg 1997; 56(2): 127-32.

[125] Sarti E, Schantz PM, Avila G, Ambrosio J, Medina-Santillan R, Flisser A. Mass treatment against human taeniasis for the control of cysticercosis: a population-based intervention study. Trans R Soc Trop Med Hyg 2000; 94(1): 85-9.

[126] Boa M, Mukaratirwa S, Willingham AL, Johansen MV. Regional action plan for combating *Taenia solium* cysticercosis/taeniosis in Eastern and Southern Africa. Acta Trop 2003; 87(1): 183-6.

[127] Wallin MT, Kurtzke JF. Neurocysticercosis in the United States: review of an important emerging infection. Neurology 2004; 63(9): 1559-64.

[128] De Giorgio CM, Sorvillo F, Escueta SP. Neurocysticercosis in the United States: review of an important emerging infection. Neurology 2005; 64(8): 1486.

[129] Larralde C, Sciutto E. In: Larralde C, de Aluja A, Eds. Cisticercosis, guía para profesionales de la salud. ed. Mexico: Secretaría de Salud, Fundación Mexicana para la Salud, Instituto Nacional de Salud Pública, Fondo de Cultura Económica 2006; pp. 182-237.

[130] Cistimex. [Homepage on the Internet]. México: Red de científicos y profesionales encaminados al control de la *Taenia solium* en México [Cited on October 15th, 2006]. Available from: http://www-lab.biomedicas.unam.mx/cistimex/.

[131] Aluja AS, Suárez-Marín R, Sciutto-Conde E, Morales-Soto J, Martínez-Maya JJ, Martínez-Villalobos AN. Evaluación del impacto de un programa de control de la teniasis-cisticercosis (*Taenia solium*). Salud Pública Mex 2014; 56(3): 259-65.

CHAPTER 10

Therapeutic Approaches to Liver Fibrosis

Enrico Maria Zardi*

Department of Clinical Medicine, University "Campus Bio-Medico", Rome, Italy

Abstract: The advancements in the molecular pathogenesis and a deeper comprehension of cellular mechanisms of liver fibrosis promise to give new therapeutic approaches to the management of this disease in the future.

Apoptosis, cytokines, oxidants, and inflammosomes play a complex role in favouring or delaying the onset of liver fibrosis; sustained hepatic stellate cell activation and transformation, without forgetting the role of hepatocytes and other complex molecular and cellular mechanisms, are the direction that guides to fibrosis.

Therefore, in this review, after considering the causes of the development of fibrosis as well as the interrelation between molecular and cellular profibrotic mechanisms, we will focus on the part played in counteracting both of these actions by some anti-oxidants and anti-fibrotic molecules such as cytokines, prostacyclin and others.

Further validation in animal studies and confirmation in clinical trials are requested. Some data that are already available from *in vitro* and animal studies demonstrating the effectiveness of novel approaches to inhibiting or treating liver fibrosis can only offer moderate hope.

Keywords: Adiponectin, antifibrotic drugs, apoptosis, curcumine, cytokines, extracellular matrix, gene therapy, growth factors, hepatic stellate cells, inhibitors of metalloproteinases, interferon, liver fibrosis, metalloproteinases, N-acetyl-L-cysteine, peroxisome proliferator-activated receptors, prostacyclin, reactive oxygen species, relaxin, thiazolidinediones, transcription factors.

INTRODUCTION

Liver fibrosis is a severe consequence of the progression of chronic liver diseases and is also cause of hepatic vascular complications such as portal hypertension. In the past, many efforts have been dedicated to finding therapeutic cures to alleviate the symptoms of liver fibrosis. Recently, profound progress in our understanding

***Corresponding author Enrico Maria Zardi:** Università "Campus Bio-Medico" *Via* Àlvaro del Portillo 200 – 00128 Roma, Italy; E-mail: e.zardi@unicampus.it

Atta-ur-Rahman, Allen B. Reitz, M. Iqbal Choudhary and Jizhou Wang (Eds.)

of the underlying molecular and cellular mechanisms involved in the development of hepatic fibrosis suggest new perspectives for therapeutic strategies. At present, activation and transformation of sustained hepatic stellate cells (HSCs), seems to be the leading pathophysiological cause of liver fibrosis. HSCs, activated by several substances such as reactive oxygen species (ROS) and cytokines [tumor necrosis factor-α (TNF-α)], produce profibrotic mediators [transforming growth factor-β (TGF-β) and connective tissue growth factor (CTGF)] that cause expression of receptors, such as those for platelet-derived growth factor (PDGF) [1-5], that induce their phenotypic modification into myofibroblasts; in turn, the latter produce extracellular matrix (ECM) components and other profibrotic cytokines that also have autocrine effects [1]. However, not only are increases in ECM production implicated in the development of liver fibrosis, but also changes in the pattern of ECM degradation [6]: myofibroblasts change their expression of matrix metalloproteinases [MMPs (a family of extracellular zinc-dependent proteinases)], inducing the production of those that act against type I and IV collagen; on the other hand, they also enhance the expression of tissue inhibitors of metalloproteinases (TIMPs) [6] that inhibit matrix degradation.

Fibrosis stimulates neoangiogenesis *via* induction of vascular endothelial growth factor (VEGF) from HSCs; in turn VEGF may promote liver fibrosis activating HSCs *via* TGF-β activation by plasma kallikrein [7].

Other molecular and cellular mechanisms are also involved in liver fibrosis: for example, after their activation; HSCs upregulate and sensitizate endothelin-1 (ET_A) receptors, which elicit profibrotic and constrictive activity [7, 8]. Furthermore, macrophages also seem to promote liver fibrosis through the release of TGF-β and PDGF [6]. Hence, an exciting strategy would be to recognize the main causes underlying the development of liver fibrosis and to introduce new therapeutic options or combination therapy to oppose or to decrease it. Many compounds exist that have antifibrotic activity thanks to their actions during different steps of the fibrogenic cascade [modulation of HSCs activation and apoptosis, inhibition of endothelin-1 (ET-1), ECM and collagen synthesis, promoter of matrix degradation] [9, 10]. Other therapeutic opportunities might also derive from future research advancements in the fields of gene therapy, tissue engineering and stem cells. However, more research must still be done to find an effective method to use these techniques. Therefore, in this review, all these molecular aspects will be thoroughly analysed to gain a better understanding of the process leading to liver fibrosis and also to individuate potential therapeutic options.

LIVER FIBROSIS: MOLECULAR ASPECTS IN QUIESCENT HSCs

Functional Anatomy and Physiology of HSCs

HSCs, also known as fat storing cells or perisinusoidal lipocytes, represent 3–8% of all liver cells and are present at a ratio of about 3 6–6 cells/100 hepatocytes. They are strictly located close to hepatocyte laminates in the perisinusoidal space of Disse by means of star-like dendritic cytoplasmic processes extending along and around the liver endothelial cells [11]. The latter compose a discontinuous endothelial edge that presents multiple fenestrations without a diaphragm favouring the rapid transport of molecules to the space of Disse. Furthermore, HSCs and endothelial cells are separated by a basal lamina-like substance that is not a true basement membrane [12]. The membrane potential of quiescent phase HSCs is specifically determined by two distinct K+ channels and an intercellular mechanism involving gap-junctional communication, whereas no voltage-gated Ca2+ current is present [13]. HSCs have two types of ionic K+ channels: the voltage-dependent outward K+ channel (that plays an important role in the regulation of membrane potential and is activated by depolarization) and the voltage-dependent inward rectifier K+ channel (that regulates resting membrane potential, mediates k+ induced dilation and is activated by hyperpolarization) [13] (Fig. **1**).

Quiescent Phenotype Expression of HSCs

The embryonic origin of HSCs is unclear, being potentially of mesenchymal endodermal or neuroectodermal origin, although the mesenchymal hypothesis is the most accepted [14]. During the so-called ''quiescent'' phenotype, HSCs express adipocyte specific genes, properties that they lose when they become activated; HSCs enclose multiple lipid droplets in the cytoplasm, which contain retinoids, triglycerides, cholesterol and free fatty acids [15]. An important structural feature of HSCs is the presence of bundles of microfilaments, including actin and intermediate filaments such as desmin, vimentin and synemin [16, 17].

PPAR-γ Expression

Quiescent HSCs also express peroxisome proliferator-activated receptor-γ (PPAR-γ), a nuclear receptor transcriptional factor considered to be the fundamental transcriptional regulator for adipogenesis (anabolic pathways), and also believed to have anti-fibrogenic effects [18-20]. It is known that PPAR-γ inhibits type I collagen expression at the level of transcription [18, 21]. In the normal liver, type I collagen is a heterotrimeric protein composed of two α1 (type I) and one α2 (type I) collagen polypeptides encoded by two different genes, and represents about 40-

50% of total collagen proteins although, during cirrhosis, this percentage increases to 60-70% [22].

The PPAR-γ gene, located on chromosome 3 at position 3p25 close to RARβ (retinoic acid receptor) and TRβ (thyroid hormone nuclear receptor) which are at positions 3p24 and 3p21, respectively is expressed in two isoforms, γ1 and γ2, both originating from the same gene [23, 24]; PPAR-γ2 has an additional 30 N-terminal amino acids, relative to PPAR-γ1 [23]; HSCs express predominantly PPAR-γ1 [23]. PPAR-γ forms a heterodimer with another nuclear receptor, (9-*cis* retinoic acid receptor) retinoid X receptor a (RXRa); thereafter, the complex binds to a specific DNA sequence that is located in the promoter region of PPAR-γ target genes and modulates their transcription after binding to specific peroxisome proliferator response elements (PPREs), consisting of a hexameric direct repeat (TGACCT) separated by a single nucleotide [25]. Transcriptional regulation also involves basal transcription factors such as TATA-binding protein, transcription factor for RNA polymerase IIB (TFIIB) and other cofactors known as nuclear transcriptional or histone acetyltransferases, such as CBP/p300 (a cAMP-response-element-binding-protein-binding protein) and NCoA-1/SRC-1 (nuclear receptor coactivator/steroid receptor coactivator) [23] (Fig. **1**).

PPARγ plays an important role in adipogenesis inducing production of several specific adipose markers such as adipocyte fatty acid binding protein (aP2), phosphoenolpyruvate carboxykinase (PEPCK), and LPL [cited in ref 26]. It favours the differentiation and the production of small adipocytes, that are more insulin-sensitive than large adipocytes; PPARγ also induces apoptosis of large adipocytes [26].

Prostacyclin Expression

Prostacyclin, a physiological mediator derived from prostanoids, which are metabolites of arachidonic acid as a result of the action of cyclooxygenase, is produced by HSCs and may exert autocrine effects by means of its nuclear receptor PPAR [10] (Fig. **1**).

Adipogenic Transcription Factor Expression

Quiescent HSCs also express SREBP-1c (sterol-regulatory-element-binding protein 1c), C/EBP-α, β and δ (CCAAT/enhancer binding proteins noticed as the third family of adipocyte-promoting transcription factors that possess both leucine zipper and basic and acidic domains and are active as unliganded homodimers), CREB [cAMP-responsive element (CRE)-binding protein] and liver X receptor-α

(LXRα), which are deemed to be adipogenic transcription factors such as PPAR-γ [26] (Fig. **1**).

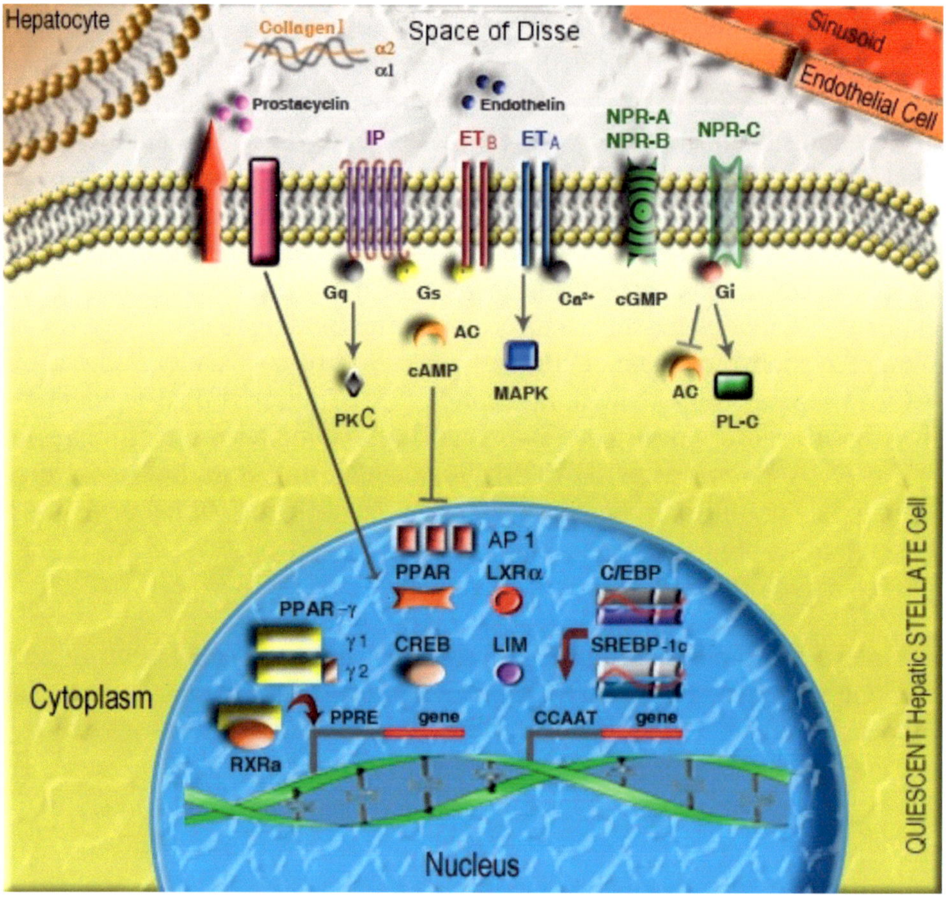

Figure 1: Quiescent HSC. A wide molecular network links HSC to hepatocyte and endothelial cell of the sinusoid, across the space of Disse. Maintenance of the quiescent phenotype of HSC, requires many cellular pathways with anti-fibrogenic effects, such as inhibition of type I collagen expression and prostacyclin and adiponectin secretion. At the nuclear level, quiescent HSC expresses PPAR-γ, CREB, LIM, C/EBP, SREBP-1c and LXRα. The HSC membrane is rich in prostanoid receptor (IP), endothelin receptors A and B (ET_A ET_B) and cardiac natriuretic receptors, A, B and C (NPR-A NPR-B NPR-C). The ↑ symbol means activation. The ┬ symbol means inhibition.

Protein and Cytokine Expression

Under physiological conditions, HSCs express the LIM homeodomain protein Lhx2, which is a transcription factor that is able to conserve their quiescence; in fact, HSCs in the normal liver of Lhx2 knockout mice developed the myofibroblastic phenotype [27].

HSCs also control migratory activity and regulate the expression of hepatocyte growth factor (HGF), TGF-β, insulin like growth factor-I (IGF-I) and other cytokines in an auto- and paracrine manner so as to communicate with all types of adjoining cells [28]. HSCs regulate the expression of IGF-I receptor on their membrane surface, but lack FAS (fatty acid synthase) receptor expression (CD95) [29]. HSCs express various proteins of neural origin (glial fibrillary acidic protein, neuronal growth factor, synaptophysin, RhoN protein, glutamine synthetase, neurotrophin receptors) [1] (Fig. **1**).

Expression of Adiponectin and Receptors

Quiescent HSCs may synthesize the 30-kDa adipocyte complement-related protein adiponectin, a 247 amino-acid protein monomer that forms trimers which further polymerize into larger polymeric complexes that vary in size between 180kDa (hexameres) and 400-600kDa (16-meres) [30, 31]. Furthermore, HSCs express adiponectin receptors such as AdipoR1, AdipoR2 and a short-form leptin receptor (Ob-Ra), whereas the long form-leptin receptor (Ob-Rb) is barely detectable in these cells [30-32] (Fig. **1**).

Other Proteins, Cytokines and Receptors Expressed to Modulate the Liver Microcirculation

HSCs also express glial fibrillary acidic protein, have a low proliferation rate, regulate hepatocyte turnover and development and control the liver microcirculation homeostasis [1, 33] obtained in response to modulation of the liver microcirculation resistance through a mechanism of contraction and relaxation, depending on the type of signal received from vasoconstricting (ET-1) or vasorelaxing (prostacyclin and nitric oxide) mediators [34, 35]. The receptor for these molecules (ET-1 receptor) is formed by a G protein coupled receptor:

- ET_A receptor, which elicits profibrotic and vasoconstricting activity after activation of the mitogen-activated protein kinase (MAPK) and the rapid increase in intracellular Ca^{2+}.

- ET_B receptor, which displays anti-fibrotic and vasorelaxant effects after the rise in intracellular cAMP levels and the reduced activity of extracellular signal-related kinase (Erk) and c-Jun kinase [belonging to the Jun family proto-oncogenes (c-Jun, JunB and JunD), which are critical components of the AP-1 transcription factor] [7, 8, 35].

Furthermore, HSCs express the prostanoid receptor (IP), which is localized on the cell membrane, belongs to the rhodopsin-type G protein-coupled receptor superfamily and contains seven conserved transmembrane domains [36, 37]. The IP receptor is mainly coupled with the Gs protein, but may also couple with Gi and Gq proteins after its initial coupling with Gs and subsequent phosphorylation by cAMP-dependent protein kinase A (PKA) on Ser 357 or protein kinase C (PKC) on Ser 328 [37].

In addition, other vasoconstricting (arginine, vasopressin, angiotensin II) and vasorelaxing (atrial natriuretic peptide, C-type natriuretic peptide) HSC mediators and receptors seem to have interplaying roles in mediating the control of vascular tone and in maintaining a balanced homeostasis [38]. The vasorelaxing cardiac natriuretic hormones, characterized by a peptide ring with a cysteine bridge (considered to be the site of peptidic hormone that binds to its specific receptor), exert a protective effect on endothelial dysfunction by decreasing shear stress, modulating coagulation and fibrinolysis pathways and inhibiting platelet activation [39]. They also share an inhibitory action on some growth factors [39]. HSCs include cardiac natriuretic receptors [40]; in fact, three classes of receptors have been characterized, such as natriuretic peptide receptor-A (NPR-A), which is sensitive to atrial natriuretic peptide [41], natriuretic peptide receptor-B (NPR-B) which is highly specific for C-type natriuretic peptide [42] and natriuretic peptide receptor-C (NPR-C), which binds both natriuretic peptides [41]. NPR-A and NPR-B transduce their biological effects *via* cGMP [41]. In contrast, NPR-C clears natriuretic peptides from the circulation by internalizating them and favouring their lysosomal degradation; furthermore, it also possesses an intracellular G-protein binding domain and can couple to $G_{i/o}$ G-proteins to inhibit cAMP activity and activate phospholipase-Cβ3 (PL-C), thus eliciting vasorelaxant activity [43-46] (Fig. **1**).

LIVER FIBROSIS: MOLECULAR ASPECTS DURING ACTIVATION OF HSCs

Numerous stimuli (viral, ROS, endotoxins, alcool, *etc*) may serve to activate several types of cells (Kupffer cells, macrophages, neutrophils, T cells, natural killer and natural killer T cells) that promote the synthesis of specific cytokines, initially TNF-α and then others (IL-1 α/β, IL-8), all of which are involved in the stepwise initiation and progression of hepatic fibrosis [1, 10], the final step in this cellular and molecular over boost is considered to be caused by HSCs that become activated and then transform themselves into myofibroblasts.

Functional Consequence of HSC Activation

In fact, after activation and transformation, these cells are believed to be the main source of the collagenous and non-collagenous matrix proteins that are deposited in the fibrotic liver. The processes underlying the HSCs transdifferentiation are undoubtedly complex due to the number of cellular, molecular and signaling factors involved. TNF-α promotes the activation and favours the transformation of HSCs into myofibroblasts, thus playing an important function in mediating liver fibrosis [47]; it is also clear that increased production of ECM is a further key point in all forms of hepatic fibrogenesis. When HSCs undergo the so called ''activated'' phenotype, they develop myofibroblast-like cell types, characterized by the loss of lipid droplets, lack of glial fibrillary acidic protein and increased cell proliferation, production of ECM components, increased expression of α-smooth muscle actin fibers and changes in the expression of L-type voltage-operated Ca2+ channels, which mediate Ca2+ influx and regulate cell contraction [48, 49]. This results in excessive ECM deposition, a condition usually observed during fibrosis and cirrhosis. As a consequence, a continuous endothelium and a true basal lamina, finally causing capillarisation of the sinusoids, develops [12]. The current opinion is that the activated phenotype of HSCs is mediated by gene expression, which is in turn regulated by transcription factors (Fig. **2**).

Role of Transcription Factors in Maintaining HSC Activation: FoxO

Quite recently, the Forkhead box gene, group O (FoxO), has been discovered; it is a subfamily of Forkhead transcription factors such as the functionally related proteins FoxO1, FoxO3a and FoxO4, containing an N- and C-terminal protein kinase B (PKB) located in the FoxO domain; they play important roles in metabolism, cell cycle, differentiation, proliferation and protection against oxidative stress, cell death and cell survival [50]. Proteins belonging to the FoxO family contain three highly conserved putative PKB recognition motifs, two located at N- and C-terminal sites, respectively, and one in the forkhead domain [51]. Some kinases (especially phosphoinositide 3-kinase and PKB) that are in turn activated by growth factors (mainly PDGF), regulate the intracellular localization and function of FoxO proteins by phosphorylating them, in a complex manner, within their different intramolecular domains [51-54]. Phosphorylation suppresses transactivation and promotes the translocation of FoxO proteins from the nucleus to the cytosol, reducing the expression of their target genes. In particular, it has been observed that the transcription factor FoxO1 is a crucial player in maintaining the control of transdifferentiation and proliferation of HSCs and liver fibrosis [55]. In fact, its phosphorylation increases liver fibrosis *in vivo* [55] (Fig. **2**).

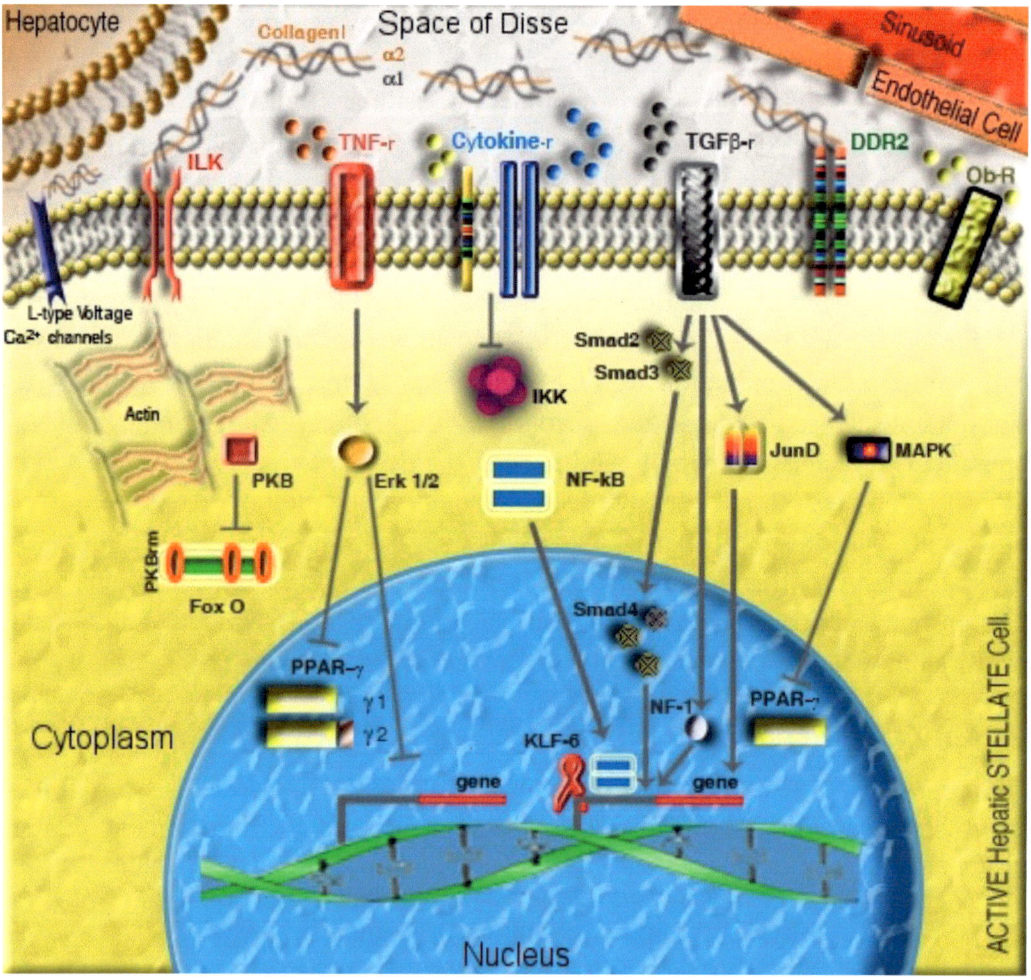

Figure 2: Active HSC. HSC activation leads to upregulation of extracellular matrix (ECM) components, such as type I collagen, creating a fibrotic web in the space of Disse. L-type voltage dependent Ca^{2+} channel (L-type V Ca2+ channel), ILK, which is linked to cytoskeleton and inhibits FoxO and DDR2, which is implied in collagen fibrillogenesis and favors chemotactic migration, are all present in activated HSC. TNFα binds to TNF-r and inhibits the anti-fibrogenetic gene PPAR-γ through Erk ½. Cytokines receptor inhibits the IkB kinase complex (IKK). The TGFβ receptor bond leads to Smad 2 and 3 translocation to the nucleus and formation of a complex with Smad4, JunD and NF-1 which mediate the activation of pro-fibrogenetic gene transcription and PPAR-γ inhibition through p38 MAPK. During activation of HSCs, leptin and leptin receptor (Ob-R) are required. Activated HSCs express NF-kB. The ↑ symbol means activation. The ┬ symbol means inhibition.

ILK

To further complicate the already complex system that regulates the transdifferentiation of HSCs is the significant part attributed to an intracellular serine/threonine protein kinase [integrin-linked kinase (ILK)], the expression of which is upregulated during liver injury [56]. This kinase plays a critical role in HSC activation and fibrogenesis, transducing signals from the extracellular matrix or from growth factors and vasoconstricting peptides (such as TGF-β and ET-1) to the cytoplasm [56]; ILK contains four amino-terminal ankyrin repeats, a carboxyl-terminal kinase domain and a central pleckstrin homology domain which, upon binding to phosphoinositides, is functionally regulated [57]. ILK couples integrins and growth factors to downstream signaling pathways that favour the suppression of apoptosis and promote cell cycle progression. PKB, the major downstream effector of growth factors, cytokines and adhesion receptors in the regulation of cell survival, may be phosphorylated directly by ILK [58, 59]. In this respect, ILK might also play a critical role in the FoxO system, activating the PKB-catalyzed phosphorylation of FoxO1 and increasing liver fibrosis. In addition, ILK has been implicated in integrin-mediated signaling to the actin cytoskeleton, and its overexpression leads to significant changes in HSC adhesion and cell motility through cytoskeletal changes [56] (Fig. **2**).

PPAR-γ

Importantly, it has been elucidated that reduced transcriptional activity of PPAR-γ may result in increased HSC activation and proliferation [38, 60]. Furthermore, there is strong evidence that HSC transformation is accompanied by loss of the adipogenic transcriptional program required for maintenance of HSC quiescence. As demonstrated by an *in vitro* study, TNF-α is able to inhibit PPAR-γ transactivity at a posttranslational level by inhibition of PPAR-γ responsive element binding as well as through phosphorylation of PPAR-γ at the consensus MAPK phosphorylation site (Ser82 of PPAR-γ1), mediated by TNF-α-induced Erk1/2 activation [61] (Fig. **2**).

Profibrotic Genes

Not only do inhibition of genes that maintain HSCs in a condition of homeostasis, but also the overexpression of some genes involved in the synthesis of several molecules, receptors and peptides play a crucial role in favouring liver fibrosis. In fact, HSCs freshly isolated from human cirrhotic livers have been demonstrated to express genes encoding key molecules implicated in liver fibrosis [62]. The activated HSCs produce a number of cytokines and active peptides (TGF-β,

CTGF, ET-1, angiotensin II) that promote their constrictive, proliferative and transformative properties in an autocrine manner, favouring the development of liver fibrosis [1-5, 63, 64] (Fig. **2**).

NF-kB Transcription Factor

One of the most studied transcription factors is surely the nuclear factor-kB (NF-kB); NF-kB, belonging to the Rel family, includes at least five members [p65 (RelA), p50 and p52, RelB and c-Rel] that are able to homo or heterodimerize, translocate to the nucleus and bind to DNA. Its activation is controlled by inhibitors such as IkB proteins which, upon binding to the entire complex in the cytoplasm, obstruct the translocation of NF-kB to the nucleus. Cytokines, through phosphorylation and activation of the IkB kinase complex [(IKK) complex consisting of IKK-a, IKK-h, IKK-g (also called NEMO), IKK-associated protein 1, FIP-3 (type 2 adenovirus E3-14 7- kDa interacting protein), heat shock protein 90 and glutamic acid, leucine, lysine, and serine – abundant protein], lead to the degradation of IkB proteins, favouring NF-kB translocation from the cytoplasm to the nucleus where it may stimulate gene expression [65]. Inhibition of IkB expression may be also mediated by the action of two key molecules [the repressors Cp binding factor 1 (CBF1) and 1methyl-CpG-binding protein-2 (MeCP2)] that seem to favour the methylation of CpG islands in upstream regulatory regions, leading to gene repression [66, 67] (Fig. **2**).

Role of Resistin on NF-kB

Furthermore, resistin (a 114 amino-acid peptide whose inflammatory action is independent of its conformation), which phosphorylates Ser32 and Ser36 residues on the inhibitory protein IkB, may play an important role in allowing active NF-kB to translocate to the nucleus and stimulate gene expression [31, 68], NF-kB was also found to be dependent on the increase in cytosolic calcium concentration induced by resistin, mainly through PL-C activation and release of calcium from intracellular IP-3-sensitive pools, followed by calcium influx from the extracellular environment. This action was found to favour the expression of monocyte chemoattractant protein 1 (MCP-1) and IL-8 [68].

Role of Leptin on NF-kB Transcription Factor and Ob-Rb Receptor

In addition, activated HSCs may express leptin, a 167 amino-acid protein belonging to a cytokine family and encoded by the *ob* gene located within the 7q31.3 locus, which is also able to activate NF-kB and induce chemokine expression [28, 69]. Thus, sustained NF-kB activation endows activated HSCs with a proliferative and antiapoptotic status that may be important for progression of liver fibrogenesis [70]. Evidence of NF-kB action also derives from the fact that NF-kB sites are found in the promoters of genes such as IL-6 and intercellular adhesion molecule-1 (ICAM-1), which are highly induced in activated HSCs during the fibrotic process [71].

Leptin expressed by activated HSCs [69, 72] may bind to the Ob-Rb receptor of Kupffer cells and liver sinusoidal endothelial cells and increase the DNA binding activity of AP-1, thereby enhancing steady- state levels of TGF-beta 1 mRNA in these cells [32]. HSCs also barely express the leptin binding Ob-Rb receptor, which is able to up-regulate MCP-1, vascular endothelial growth factor (VEGF) and angiopoietin-1 expression through increased tyrosine kinase activity, resulting in activation of signal transducer and activator of transcription factor 3 [32, 69]. Leptin can bind to the Ob-Rb receptor in cultured rat HSCs *in vitro* and activate signal transducer and activator of transcription-3 (STAT3) phosphorylation and, thus, increasing total mRNA for the α2 type I collagen gene [73]. Furthermore, leptin-induced phosphorylation of the Ob-Rb receptor *via* Janus kinase-2 (Jak2) activation results in phosphorylation of Erk and Akt, two key signal transduction elements associated with cell growth [74]. In this way, leptin appears to promote HSC survival against apoptosis, and this appears to be mediated by the PI-3 kinase pathway and by inhibition of the expression of the pro-apoptotic HSC mediator, TRAIL/DR5 [74].

KLF-6 Transcription Factor

Another way demonstrating how HSCs may be activated is through the upregulation of Kruppel-like factor-6 (KLF-6), a member of the family of related zinc finger transcription factors that share identical C2H2 C-terminal DNA binding domains [67]; KLF-6 is rapidly induced as an immediate-early gene during HSC activation *in vivo* and in culture [67]. KLF-6 may be alternatively spliced, acting, in the growth-promoting short form, on a number of transcriptional targets such as those of TGF-β1, its receptors, urokinase-type plasminogen activator associated with increased oxidant stress and contractile cytoskeletal proteins [67]; in contrast, the full-length isoform acts as a growth inhibitory and tumor suppressor protein [67] (Fig. **2**).

Profibrotic Mediators: TGF-β ET-1 Other Growth Factors and their Smad Receptors

Among fibrotic mediators, TGF-β is surely one of the most important polypeptides and is constituted by three peptide isoforms (TGF-β1, TGF-β-2, TGF-β-3) [75]. The active forms (dimers of 25kd produced by cleavage of the proform but noncovalently associated to its 75kd propeptide) are composed of two subunits (each one containing two anti-parallel pairs of β-strands and an α-helix stretch) linked by a disulfide-bridge [75, 76]. They may link to the fibronectin in the ECM, thus constituting a reservoir that may be released after ECM activation by proteinases such as plasmin. Inactivation effects are due to two known molecules that function as scavengers, decorin, expressed by HSCs and α2-macroglobulin [77]. In HSCs, TGF-β is involved in the intracellular signaling cascade and transcriptional regulation of certain genes such as Ras, Raf-1, MEK and MAPK [78].

ET-1, a 21 amino acid peptide, also induces HSC activation, invasiveness and fibrogenesis by upregulation of type I collagen gene expression [79]. ET-1 also increases the expression of procollagen α1 during the early, but not the late phase, of HSC activation [80]. Furthermore, it stimulates the expression of MMP-2, favouring the degradation of the normal subendothelial matrix, which is subsequently replaced with a non-functional interstitial extracellular matrix, including type I procollagen. ET-1 also upregulates the expression of TGFβ-1 [80]. The up-regulation of ET-1 synthesis seems to be due to regulation of the enzyme (endothelin-converting enzyme-1) that converts precursor ET-1 to mature ET-1. In this respect, TGF-β seems to play a role, controlling ET-1 production by modulation of the ET-converting enzyme-1 [81]. Activated HSCs then upregulate some receptors, such as those for IL-10 [82], FAS (CD95) [a cellular surface protein with a molecular weight of 42-52 kDa that promotes apoptosis after binding to an inactivated binding (FasL), expressed on the cellular surface of active T cells and NK cells] [29, 83], PDGF, fibroblast growth factor (FGF), VEGF, TGF-β1 [38] and p75 (a member of the TNF receptor superfamily) [84], thus increasing their autocrine effect; on the other hand, they downregulate IGF-I receptor expression [29]. Three TGF-β type receptors and several signaling mediators, commonly named 'Smad proteins', are known [79]. For example, the type 1-TGF-β receptor (a serine/threonine kinases containing a cysteine- rich extracellular domain, a short hydrophobic transmembrane region and a cytoplasmic region harbouring the kinase motif) is an activin receptor-like kinase 5 (ALK5) and increases transcription of genes, such as α2 type I collagen, through the phosphorylation of Smad proteins (Smad2 and Smad3), after forming a hetero-oligomeric complex with Smad4, there is subsequent nuclear translocation of the Smad containing complex [79, 85]. Furthermore, c-Jun N-terminal kinase-dependent Smads2/3 phosphorylation is considered to be the signal through which PDGF and TGF-β1 activate HSCs by *in vivo* and *in vitro* upregulation of MMP activity, increasing cytoskeletal and αSMA organization and favouring their migratory capacity [86-88]. In contrast, inhibitory Smad6 and Smad7 are able to block this cascade to prevent TGF-β-mediated alterations in gene expression [89] (Fig. **2**).

TGF-β and Transcription Factor Stimulation

On the other hand, it is known that TGF-β may also stimulate the transcription of α2 type I collagen by means of transcriptional nuclear factor-I (NF-I) [90]. The TGF-β activating element seems to be located 1.6 kb upstream of the transcription start site; this element contains the 3' portion of the canonical NF-I sequence GCCAAG [91, 92]. However, further studies have demonstrated that transcription factor Sp-1-binding sites of the α2 type I collagen gene, spanning from nucleotides -265 to -241, are necessary for the TGF-β response [93].

In addition, TGF-β is able to induce the expression of TIMP-1 [the prototypic and original ancestral member of the mammalian TIMP family of four proteinase inhibitors (TIMP-1, TIMP-2, TIMP-3 and TIMP-4) that suppress ECM degradation by forming an inhibitory 1:1 complex with the MMPs] [94] through the activator protein-1 transcription factor (AP1) site (5'-TGAGTAA-3') located at the 5' end of the minimal active TIMP-1 promoter (-102 to +60); furthermore, TGF-β is able to repress MMP-1 expression through the AP1 site (5'-TGAGTCA-3') [95]. The AP1 binding site that interacts with the JunD, a member of the Jun family of proto-oncogenes, has an important role in the transcriptional activation of TIMP-1 and IL-6 by means of a TIMP-1 element-1 (UTE-1- regulatory sequence) in association with a 30-kDa nuclear protein [probably the truncated form of full-length RUNX protein (RUNX1A isoform and RUNX2)] [96-98].

The Jun proteins are bZip transcription factors that can form either AP-1 homodimers (Jun/Jun) or AP-1 heterodimers with members of the related bZip protein family of the *fos* proto-oncogene family (c-*fos, fosB, fra1* and *fra2*) or the ATF family (ATF2, ATF3 and ATF4) [99-101].

Overexpression of c-Jun, JunB, Fra2 or FosB in activated HSCs represses TIMP-1 promoter activity; while JunD homodimers are strong stimulators of TIMP-1 gene transcription JunD heterodimers are demonstrated to be less active, In fact, the high level of TIMP-1 expression in activated HSCs seems to be controlled by the balance between JunD homodimers (promoting high rates of TIMP-1 mRNA synthesis) and JunD heterodimers [97]. Furthermore, JunD is implicated in the regulation of HSC activation and liver fibrosis *via* transcriptional regulation of the TIMP-1 gene [102]. On the other hand, activated HSCs produce leptin [69, 72], which increases TIMP-1 transcription *via* phosphorylation of specificity protein 1 (Sp1) and intranuclear pSTAT3 binding to Sp1 [103].

TGF-β, upon binding its receptor, may also stimulate HSC collagen gene expression, inducing activation of the p38 MAPK pathway mediated by the upstream TGF-β activated kinase 1 (TAK1) [78, 104-106]. Remarkably, TGF-β is able to induce phosphorylation of MAPK and subsequent MAPK phosphorylation of PPAR-γ *via* this signaling pathway as a result of decreased PPAR-γ transcriptional activity [107, 108]. In HSC cultures, TGF-β increases the production of ROSs that act as its mediator and, in turn, induce the expression of α1 type I procollagen mRNA [109] (Fig. **2**).

TGF-β and Epithelial to Mesenchymal Transition in Hepatocytes

Recently, TGF-β has also been recognized to be implicated in important pathogenetic pathways, such as those involved in epithelial to mesenchymal transition (EMT) in hepatocytes by inducing snail-1 transcription factor (a zinc-finger transcription factor involved in the degradation of the basement membrane and extracellular matrix), which activates the Mek/Erk and PI3K/Akt survival pathways and tyrosine phosphorylation of Smad2/3 with subsequent recruitment of Smad4 [78, 110, 111]; this results in an increase in the number of fibrogenic cells [112]. In particular, the importance of this discovery lies in the fact that the hepatocytes were able to assume mesenchymal functions. In fact, immunofluorescence showed the colocalization of E-cadherin and vimentin, as well as collagen and albumin, in these cells. Furthermore, type-I collagen was detected intracellularly and extracellularly in albumin-positive hepatocytes, demonstrating that they were the source of synthesized collagen [112]. According to the author of this discovery, the hepatocytes would assume a myofibroblast-like phenotype and function whereas, according to others, they would surely transform themselves into myofibroblasts [113, 114].

In this way, hepatocytes are viewed as perpetuators of hepatic fibrogenesis, directly contributing to liver fibrosis.

On the other hand, mesenchymal cells such as bone marrow-derived circulating cells (also called fibrocytes), portal fibroblasts and fibrogenic cells, respectively, located between the portal tract and the parenchyma or residing around the centrolobular vein, may also assume a myofibroblast-like phenotype and thus favour the development of liver fibrosis [115-117].

Thus, liver fibrosis might be deemed to be a dynamic and plastic process involving several molecular pathways and cellular types with a myofibroblast-like phenotype that cooperates with or is engaged by the principal myofibroblast progenitor (that is again considered to be the transformed HSC) in maintaining and perpetuating the course of fibrosis (Fig. **2**).

Changes of HSC Receptors and Role of MMPs

In HSCs, the amount and relative prevalence of the subtypes of the ET-1 receptor change; in fact, there is an upregulation and sensitization of ET_A receptors, which amplify profibrotic effects [34]. Furthermore, a long-form leptin receptor (Ob-Rb), which contains a longer intracellular domain, is known to activate the JAK-STAT-3 pathway, leading to transcriptional regulation of target genes such as those involved in the production of TGF-β1 [32].

In contrast, HSCs do not change their expression of adiponectin receptors [30]. *In vitro* culture studies demonstrated that HSC activation and proliferation are regulated not only by cytokines and active peptides, but also by ECM components [118]. During activation, HSCs produce MMPs such as MMP-1, MMP-2 (gelatinase A) and MMP-3 (stromelysin), that may degrade the main components of the ECM, such as type IV collagen and laminin. MMPs have a conserved zinc-binding site in their catalytic domain (HEXGHXXGXXH) and the sequence PRCGVPD in the propeptide domain [119]. The cysteine residue serves to anchor the pro-peptide to the catalytic zinc, suppressing enzymatic activity; the absence of this interaction results in proteolytic activity [119]. In particular, HSC upregulation of MMP-2 favours an abnormal degradation of the subendothelial matrix, furthering its replacement by fibrillar collagen [120]. Activated HSCs express the discoidin domain receptor 2 (DDR2), also known as Tyro10 or TKT [121]. DDR-2 is composed of an extracellular discoidin homology (DS) domain consisting of ~150-amino acids followed by a sequence of ~220-amino acids unique to DDRs, a single-span transmembrane domain, a large cytosolic juxtamembrane domain and a C-terminal catalytic tyrosine kinase domain [122-124]. The extracellular domain of DDR2 must be dimerized and or oligomerized to directly bind to the triple-helical structure of fibrillar collagen with high affinity [123, 125]. Thus, DDR2, being able to directly bind to type I collagen and to a lesser extent, types II, III and V collagen, mediates the interaction between fibrillar collagen and HSCs [126, 127]. In this respect, an interesting *in vitro* and *in vivo* study demonstrated that interstitial collagen directly induces HSC activation through specific receptor signaling by DDR2 [128]. DDR-2 was also able to modulate fibrillogenesis of collagen fibrils and fibers and to favour chemotactic migration [121, 125]. Changes of the mechanical properties of the ECM seem to play a fundamental role in the development and maintenance of the myofibroblastic phenotype [129]. In this context, it is known that Smad3-overexpressing HSCs migrate in response to type I collagen [88]; therefore, it seems that both HSCs and fibroblasts may differentiate in myofibroblasts with fully organized stress fibers after the mechanical stiffness-dependent acquisition of α-SMA expression and the TGF-β and Smad3-dependent development of focal adhesions and organization of α-SMA in stress fibers [88] (Fig. **2**).

LIVER FIBROSIS THERAPEUTIC APPROACHES: NEW PERSPECTIVES?

Deeper research in the field of molecular biology have allowed for the possibility to discover and better characterize several elements that are perfectly and dynamically involved in the cause of liver fibrosis: they are the basic cellular actors (principally, activated and transformed HSCs and, second, other non-mesenchymal and mesenchymal cells), ECM proteins, some receptors mediating cell to ECM interactions, molecular factors such as cytokines, integrins, growth factors (principally TGF-β and PDGF) and post-receptor signal regulation (such as some Smads), transcription factors (principally a decreased expression of PPARα, PPARβ/δ in hepatocytes and PPAR-γ in HSCs) and gene response [26, 112-117, 130]. Each one of these elements could be the focus of an anti-fibrogenic therapeutic strategy; however, we'll try to focus our discussion, among the novel therapeutic proposals, on those believed to be of greater scientific interest.

PPARs- Stimulation Strategies

In hepatocytes PPARα regulates the lipid metabolism in the liver controlling the expression of genes involved in the transport, oxidation and export of free fatty acids; it also controls liver inflammation by regulating acute phase protein in hepatocytes and by reducing triglyceride storage and adipose-tissue-derived circulating factors thus increasing insulin sensitivity and ameliorating the level of liver fibrosis [26].

In hepatocytes, PPARβ/δ plays a key role in regulating levels of serum triglycerides in high fat fed mice, by modulating both VLDL production and LPL-mediated catabolism of VLDL-triglycerides; it also suppresses tissue inflammation, increasing oxidative lipid catabolism in the liver, ameliorating serum lipid profiles and global insulin sensitivity, favouring weight loss in obesity [26].

The fact that the expression of PPAR-γ gradually decreases during the passage of quiescent HSCs to activated HSCs strongly indicates its fundamental function in favouring the maintenance of HSCs during a quiescent phase [18].

Several strategies, based on PPARs stimulation and on other compounds (N-acetyl-L-cysteine, PGI2, IL-10, IFN-γ, SB-431542, SB-505124, PGE2, Pirfenidone, PGE1, imatinib, relaxin, IGF-1, adiponectin, antagonist of angiotensin II type 1 receptor) have proven to have potential anti-fibrotic effects.

It is now clear that PPARα agonists (Wy-14,643, Fibrates) upregulate a set of target genes that promote fatty acid breakdown in the liver, thus counteracting hepatic triglyceride and lipid peroxidation accumulation [cited in ref 26]; they also decrease liver fibrosis reducing activation of HSCs and Kupffer cells by means of a diminished release of cytokines by macrophages and other leukocytes. Furthermore, they inhibit the synthesis of apolipoprotein (apo) C-III (that delays the catabolism of triglyceride-rich lipoproteins) and favour the expression of the gene for lipoprotein lipase. Finally, they reduce oxidative stress markers and the level of inflammatory mediator TNF-α [cited in ref 26].

PPARβ/δ agonists (GW501516, L-165041, GW0742, KD3010) ameliorate hepatic insulin sensitivity, serum insulin and leptin levels, glucose and lipid metabolism by means of the increased insulin receptor binding, the increased signaling downstream through PI3K/Akt, which promotes cell survival and the reduced DNA damage and lipid peroxidation [cited in ref 26].

They also have a crucial role in maintaining the glucose-lipid homeostasis by stimulating fatty acid oxidation, inhibiting glucose and VLDL secretion from the liver, increasing circulating HDL levels; furthermore, they regulate the inflammatory activity of macrophages *via* NF-kB suppressing iNOS and COX2 [cited in ref 26].

Finally, they attenuate ethanol-induced hepatic injury and improve liver regeneration with protective liver effects as demonstrated by the important role in maintaining adequate vitamin A storage [cited in ref 26].

PPAR-γ agonists (15-d-PGJ2, GW7845, curcumin, Thiazolidinediones) certainly have an important anti-fibrogenic role [107] that may be deemed as a key therapeutic target to decrease liver fibrosis [18, 19, 21, 61]. As several studies have demonstrated, PPAR-γ expression decreases HSC activation *in vitro* and *in vivo* [18]. PPAR-γ may regulate cell growth and cell differentiation and its mRNA expression is significantly decreased in HSCs isolated from various animal models of liver fibrosis [19]. In fact, it seems that α1 collagen inhibition is mediated by the ability of PPAR-γ to suppress NF-I binding [131]. Furthermore, PPAR-γ suppresses expression of TGF-β, α-SMA and MCP-1 genes as well as chemotaxis and, in animal models of liver fibrosis, prevents the induction of fibrosis and reduces the progression of preexisting fibrosis [18, 19, 21].

In addition, activated HSCs *in vitro* can be phenotypically and functionally reversed to HSCs with a quiescent phenotype by stimulated expression of PPAR-γ [132].

Role of Prostagladin J(2) and GW7845

15-deoxy-delta12,14-prostaglandin J(2) (15-d-PGJ2) **(1)** is a cyclopentenone-type prostaglandin of the J2-series derived from the dehydration of prostaglandin D2 (PGD2), a major cyclooxygenase product in a variety of tissues. Because of the α,β-unsaturated carbonyl moiety present in its cyclopentenone ring structure, 15-d-PGJ2 can act as a Michael reaction acceptor and readily interacts with critical cellular nucleophiles, such as the free sulfhydryls of glutathione and cysteine residues in cellular proteins. Covalent binding of 15-d-PGJ2 to cellular proteins may be one of the mechanisms by which 15d-PGJ2 induces a cellular response. When PPAR-γ is stimulated with the natural ligand 15-deoxy-delta12,14-prostaglandin J(2) (15-d-PGJ2) or the synthetic ligand GW7845 **(2)**, a tyrosine analogue, it is able to inhibit cell growth both through cell cycle arrest and an increase in apoptosis [25]. PGJ2 induces a significant inhibition of phagocytosis, and cytokine expression and an attenuation of liver inflammation and fibrosis.

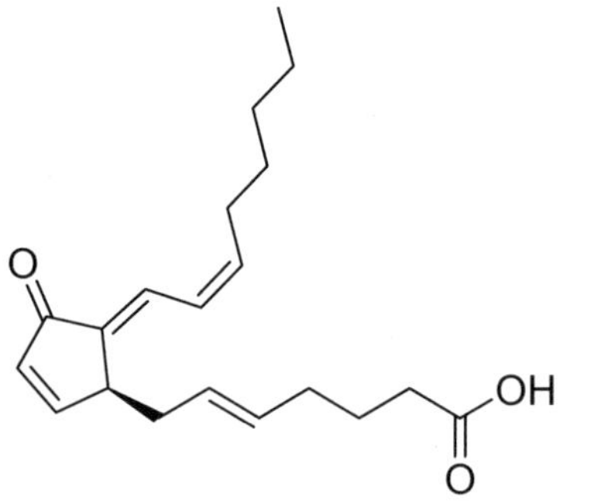

1

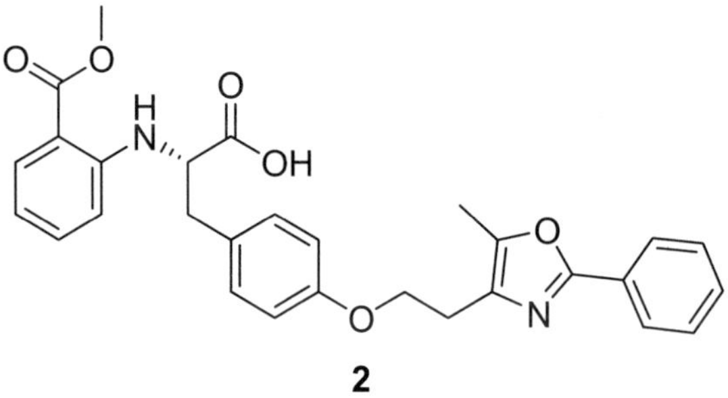

2

Furthermore, both PGJ2 and GW7845 markedly inhibit TGF-β1-induced CTGF expression in HSCs [25]. After 48 h of treatment, G_0/G_1 phase arrest and prevention of cells from entering the S phase was observed. It was also observed that the 15-d-PGJ2 agonist activated other PPAR-γ-independent signaling pathways able to slow down CTGF expression, such as that of the Smads binding element or by the direct inhibition of NF-kB [25]. In fact, PGJ2 is also able to suppress NF-kB activation by directly inhibiting IkB kinase in a PPAR-γ-independent manner. Furthermore, GW7845 inhibition of TGF-β1-dependent induction of ECM genes and activation of Smad3 without induction of toxicity in human HSCs, was found to prevent quiescent HSCs from acquiring a myofibroblastic phenotype during the development of chronic liver injury [133] (Fig. **3**).

Role of Curcumin

According to other authors, induction of the gene expression of endogenous PPAR-γ by means of interruption of PDGF and EGF signaling pathways is also possible using curcumin **(3)**; use of this anti-oxidant polyphenol compound obtained from turmeric, the powdered root of CURCUMA longa (the main yellow pigment of a popular spice, turmeric, widely used as a food colorant) is believed to be another way of counteracting liver fibrosis [134, 135]. In fact, curcumin was demonstrated to suppress the expression of ECM genes, including those encoding α1 type I collagen and fibronectin, probably by interrupting the TGF-β signaling pathway through inhibition of the expression of TGF-β receptors [134]. Other authors demonstrated that curcumin inhibits tyrosine phosphorylation of PDGF-β and EGF receptors by reducing the levels of phosphorylated phosphatidylinositol-3 kinases, Erk and Jun N-terminal kinase [135]. It was also demonstrated that curcumin reduces the amount of cell cycle-stimulating proteins, including cyclin D_1, D_2 and E [136], the expression of which plays a critical role in cell cycle progression, especially at the early G_0/G_1 phase [137]. Moreover, PPAR-γ activation by curcumin was demonstrated to induce apoptosis of activated HSCs *in vitro*, to inhibit transcription factor NF-kB *trans*-activating HSC activity and to reduce α-SMA and fibronectin, as well as mRNA levels of α1 type I collagen [136] (Fig. **3**). Curcumin was not only able to regulate HSCs activation but also the cell cycle progression of damaged hepatocytes inducing them to undergo apoptosis by up-regulating its P53 level expression [cited in ref 26].

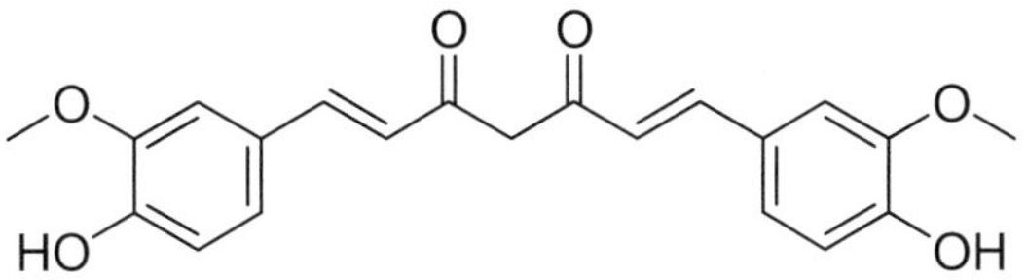

3

In this manner, administration of curcumin is able to attenuate oxidative stress, suppress inflammation and inhibit activation of HSCs.

Role of Thiazolidinediones

Thiazolidinediones (TZDs), a class of antidiabetic drugs able to attenuate insulin resistance and including troglitazone **(4)**, pioglitazone **(5)** and rosiglitazone **(6)**, structurally derive from clofibric acid and selectively bind to two histidine residues, H323 and H449, of the PPAR-γ receptor [138]. TZDs have been demonstrated to reduce TGF-β mRNA expression; consequently, they inhibit TGF-β induced human type I procollagen promoter activity in human HSCs, perhaps by interrupting the inhibition of PPAR-γ transcriptional activity [107, 108] and thus favouring repression of the activity of transcription factors AP-1, STAT and NF-kB [21, 139] (Fig. **3**).

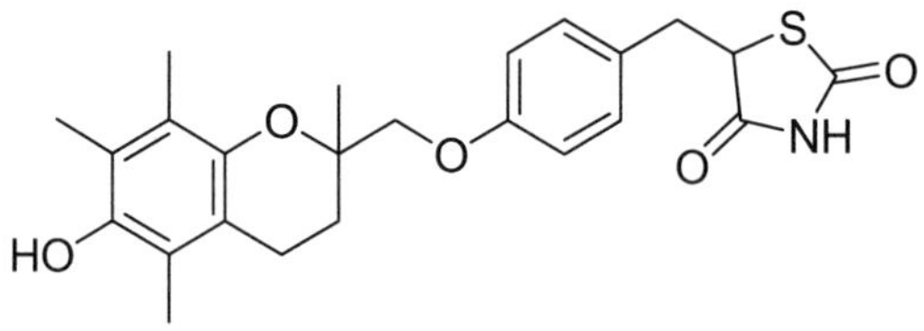

4

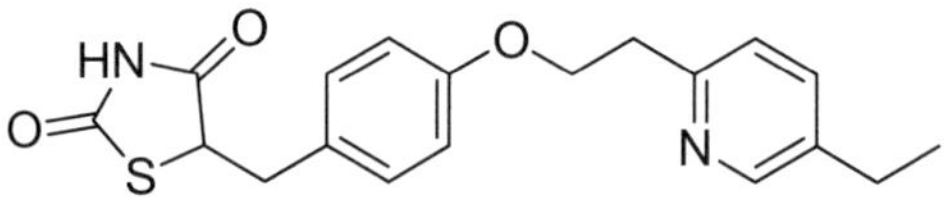

5

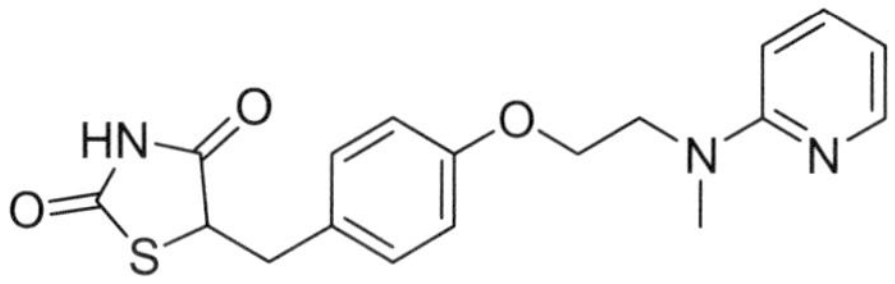

6

TZDs inhibit in dose-dependent mode HSCs proliferation and chemotaxis induced by PDGF and regulate expression of proinflammatory genes; they also suppress upregulation of collagen α1(I) and αSMA [cited in ref 26].

TZDs have demonstrated to have several properties [26].

- Reduction of expression of TNFα and IL-6.
- Decrease of serum ALT level and oxidative stress.
- Down-regulation of the expression of profibrogenic genes of TGF-β, MMP-2 and MMP-9.

- Up-regulation of the expression of adiponectin.
- Protective role against hepatotoxicity through inhibition of HSCs proliferation and inflammation.

It is well known that leptin acts on HSCs by means of its Ob-R receptor, resulting in the phosphorylation of Erk [32, 73, 74]; very recently, TZDs have been also observed to regulate the expression of Ob-R and to reverse Erk activation [140]. This effect is obtained in a PPAR-γ-dependent manner and results in reduction of chemokine expression and HSC proliferation (Fig. **3**).

Role of SC-236

Another compound able to interfere with experimental liver fibrosis is SC-236 **(7)**, a potent, selective, orally active inhibitor of COX-2 with an IC_{50} of 10 nM and approximately 18,000-fold COX-2 selectivity. SC-236, acting through the activation of PPAR-γ, exhibits pro-apoptotic and growth inhibitory properties in HSCs and Kupffer cells [141].

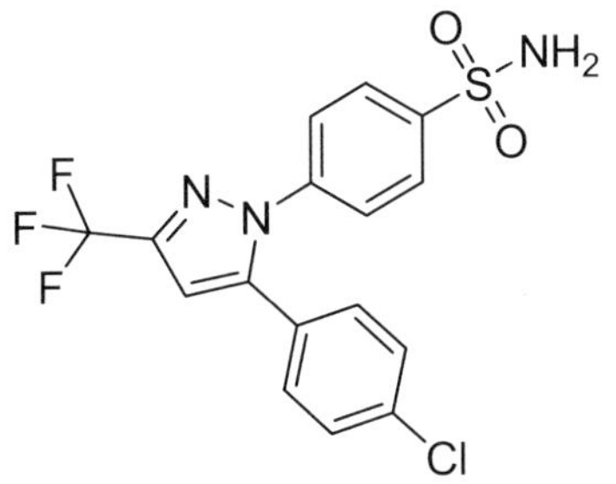

7

Therefore, PPAR-γ stimulation may suppress ECM expression and inhibit cell growth through cell arrest and apoptosis [134-136]. Importantly, the stimulated expression of PPAR-γ in activated HSCs permitted to these cells to restore the unique and specific functions as well as morphologic characteristic of quiescent HSCs. This effect was due to the inhibition of AP-1 binding caused by a PPAR-γ-mediated interference of JunD binding [132] (Fig. **3**).

TGF-β Inhibition Strategies

Pleiotropic fibrogenic effects of TGF-β markedly favour HSC activation and transformation; thus, the possibility of antagonizing or blocking the synthesis of this ligand is surely an exciting therapeutic strategy.

Role of NAC

Recently, N-acetyl-L-cysteine (NAC) **(8)** was recognized as an available drug capable of blocking the TGF-β-dependent Smad pathway signaling in HSCs [109]. NAC is also able to oppose the effect of ROS; in fact, it may directly act on the sulfhydryl group of cellular components and regulate target proteins (such as Raf-1, Mek and Erk) containing reactive cysteine residues that participate in a thiol-disulfide reaction through a redox status [142].

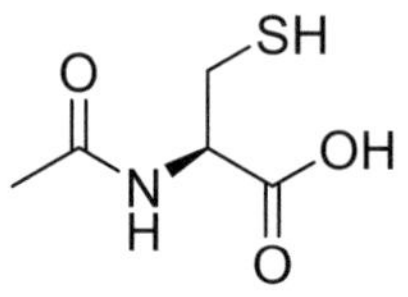

8

In fact, NAC elicits its direct effect reducing activity on HSCs, inducing sustained activation of Erk and leading to phosphorylation of the C terminus of Sp1, which is in turn followed by p21 induction [142]. Furthermore, NAC induces cell cycle arrest, favouring a redox-mediated extracellular proteolysis of PDGF receptor type β [143] (Fig. **3**).

Nac also significantly reduces collagen of the liver by upregulating MMP-13 and MMP-9 and downregulating HSP-60, TIMP-2, TGFβ-1 and collagen-1α [144].

Role of Prostacyclin

Another mediator known to have some effects against TGF-β is prostacyclin **(9)** [145], a physiological mediator derived from prostanoids (metabolites of arachidonic acid through the action of cyclooxygenase) and endowed with anti-platelet and arterial vasodilator abilities [146] (Fig. **4**). It also maintains vascular endothelial homeostasis of the hepatic microcirculation, acting as a local hormone to regulate HSC function by means of specific cellular receptors [(IP, a transmembrane G-protein coupled receptor) and PPARs] [36, 37, 147]. As demonstrated, prostacyclin, acting against TGF-β, might inhibit the activity of CTGF, thus suppressing collagen production [145]. Thanks to its displayed anti-oxidant properties, prostacyclin also suppresses the ROS-mediated expression of α1 type I procollagen mRNA [109]; in fact, it decreases tissue levels of malondialdehyde, increases the hepatic content of glutathione and improves hepatocyte mitochondrial function [148, 149]. Furthermore, it is able to counteract HSC contraction and PDGF-mediated proliferation as well as induce HSC relaxation in opposition to ET-1 by desensitizing the ET_A receptor and upregulating the ET_B receptor through a cAMP/PKA-dependent pathway after binding to the IP receptor [10, 150-153] (Fig. **4**). In addition, activated HSCs both *in vitro* and *in vivo* have Ca^{2+} currents through L-type voltage-operated Ca^{2+} channels, whereas no currents are detected in quiescent HSCs [13, 49]. Moreover, binding studies with (3)H-isradipine, a specific L-type voltage-operated Ca^{2+} channels antagonist, show a large number of binding sites in activated HSCs, while no specific binding is found in quiescent HSCs [49]. An interesting *in vitro* study in activated HSCs, demonstrated that, after administration of prostacyclin in the presence of 3-isobutyl-1-methylxanthine (0 5 mM), there was great elevation in cAMP levels from a basal value of 9.2 +/- 0.8 pmol to 122.2 +/- 12.2 pmol/ 10 min; this induced the disappearance and deaggregation of actin stress fibers in contracted HSCs and promoted their relaxation [151] (Fig. **4**).

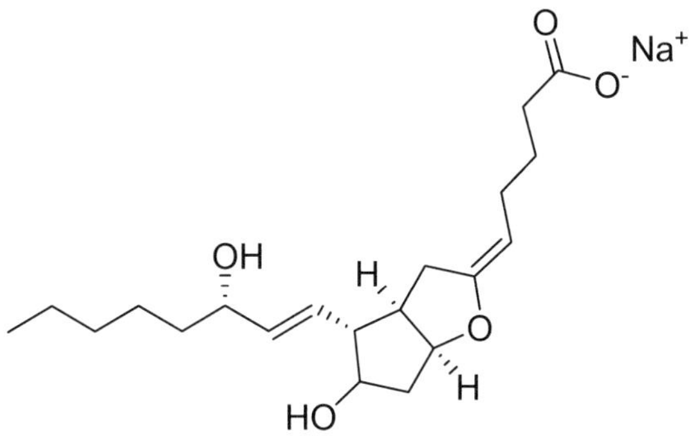

9

It is also known that activation of the IP receptor by prostacyclin stimulates the expression of C/EBP-β and δ by triggering the cAMP/PKA pathway and thus promoting adipogenesis [154]. In particular, in preadipocyte cells, prostacyclin induces the phosphorylation of serine-133 of CREB/ATF1 (members of the leucine zipper class of transcriptional factor cAMP-responsive element binding proteins/activation transcription factor) through the cAMP/PKA pathway, favouring, after the activation of a further Erk pathway, the transactivation of C/EBP-β and δ [154]. Since C/EBP-β and δ are key transcription factors for adipocyte differentiation that are highly expressed in quiescent HSCs and decreased in activated HSCs [26], their stimulation by means of prostacyclin might be helpful in maintaining HSCs in a quiescent phase. In fact, as previously reported, stimulation of the cAMP/PKA/CREB pathway [155] or its parallel pathway Ca^{2+}/ calmodulin kinase-II (CAMK-II) by prostacyclin [156-158] inhibits the proliferation of activated HSCs (Fig. **4**).

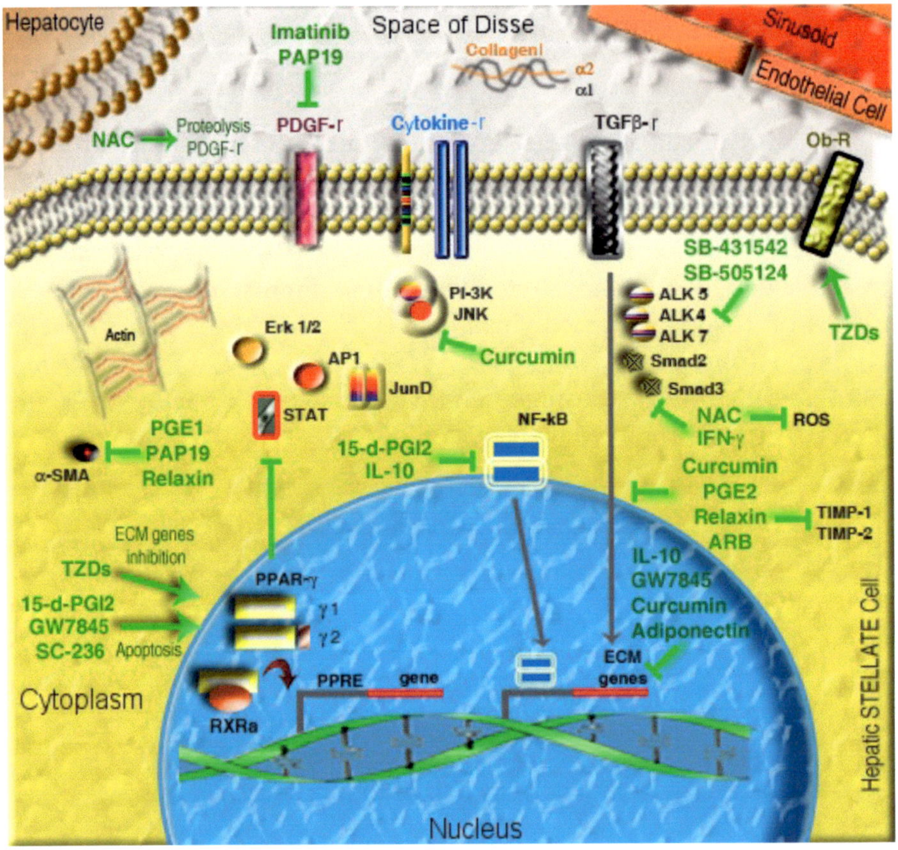

Figure 3: Therapeutic approaches to liver fibrosis. PPAR-γ is stimulated by the natural ligand 15-deoxy-delta12,14-prostaglandin J2 (15-d-PGJ2) or the synthetic ligand GW7845 and by SC-236, a selective COX-2 inhibitor. Thiazolidinediones (TZDs) activate PPAR-γ leading to inhibition of AP-1 STAT and NF-kB, and downregulation of the leptin receptor (Ob-R) and Erk. N-acetyl-L-cystein (NAC) is capable of blocking TGFβ pathways. Interleukin 10 (IL-10) exerts anti-TGFβ action, suppresses NF-kB activation and controls pro-inflammatory cytokines. IFN-γ is able to inhibit TGFβ pathways. SB-431542 and SB-505124, which are potent inhibitors of ALK, interfere with Smad protein transduction activity. PDGF kinase inhibitors exert an anti-fibrotic action. Relaxin has an anti-fibrotic effect and inhibits TGFβ pathways. Sulfasalazine is a potent and selective inhibitor of NF-kB. Adiponectin maintains the quiescent phenotype of HSCs and reverses hepatic fibrosis. Angiotensin II type 1 receptor antagonist suppresses profibrogenetic cytokine synthesis and improves fibrosis. Prostaglandin E2 (PGE2) has anti-TGF action. Prostaglandin E1 (PGE1) suppresses liver fibrogenesis. The ↑ symbol means activation. The ┬ symbol means inhibition.

ONO-1301, a synthetic prostacyclin/IP receptor agonist, increases the expression of hepatic growth factor and decreases the expressions of TGF-β1, CTGF, α-SMA and type-I and type-III collagen, thus suppressing myofibroblast expansion and liver fibrosis [159].

Role of IL-10

Anti-fibrotic cytokine IL-10 is deemed to be a potent anti-TGF-β [75] able to suppress the activation of NF-kB as well as messenger RNA expression of TNF-α and macrophage inflammatory protein-2 (MIP-2) [160]. Therefore, it is capable of controlling proinflammatory cytokines and chemokines that are expressed by NF-kB and of regulating hepatic injury mediated by T cells and macrophages *in vivo* [161]. In accordance with this, an *in vivo* study using a mouse model of liver fibrosis showed that rat IL-10 (rIL-10) had antifibrotic effects; however, the antifibrotic effect of rIL-10 disappeared when mice were irradiated prior to induction of fibrosis, suggesting that the antifibrotic response was lymphocyte-mediated. A better explanation came from the observation that in mice expressing rIL-10, the amount of CD8+ T cells and the production of IL-4 was lower than in mice without rIL-10 expression, suggesting that the decrease in CD8+ T cells in mice might be rIL-10-mediated and that both were fundamental in reducing fibrogenesis [162] (Fig. **3**).

Unfortunetely, in rats with extensive liver fibrosis, exogenous IL-10 rapidly disappears from the circulation due to kidney passage and according to these authors, this is an important limitation for its potential clinical application in liver fibrosis [163].

However, PEGylation of IL-10 to prolong its circulation time and enhance its effectivity might be a good strategy allowing the clinical use of this cytokine [164].

Role of IFN-γ

Another important cytokine with a potent anti-TGF-β effect is IFN-γ [165], which inhibits the TGF-β -induced phosphorylation of Smad3, the association of Smad3 with Smad4, the accumulation of Smad3 in the nucleus and the activation of TGF-β-responsive genes [165]. Furthermore, acting through JAK1 and STAT1, it induces the expression of Smad7 (an effective inhibitor of TGF-β-induced fibrogenesis in experimental models of fibrosis) [166], which antagonizes Smads 2 and 3, preventing the interaction of Smad3 with the TGF-β receptor [89, 165].

In addition, a beneficial effect of a 9-month IFN-γ treatment resulting in decreased fibrosis scores and a reduced number of α-SMA staining HSCs has been observed in a randomized open-labeled multicenter trial with patients affected by chronic HBV infection [167] (Fig. **3**).

A number of 6 targeted sterically stable liposomes (SSLs) modified by a cyclic peptide (pPB) with affinity for the PDGFR-β to deliver IFN-γ to HSCs have been constructed. The study of tissue distribution *via* living-body animal imaging demonstrated that the pPB–SSL–IFN-γ mostly accumulated in the liver until 24 h and caused significant remission of hepatic fibrosis [168].

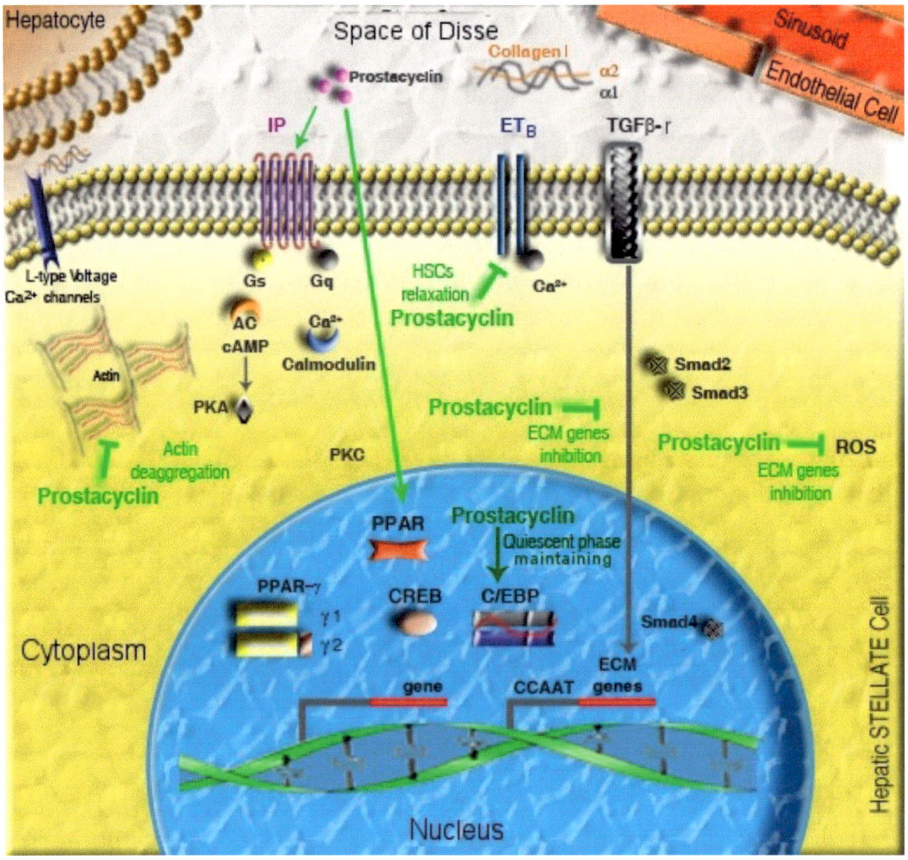

Figure 4: Prostacyclin pathways. Prostacyclin binds to specific cellular receptors such as IP, transmembrane G protein coupled receptors and PPARs. Prostacyclin maintains vascular endothelial homeostasis through activation of endothelin receptor B (ET_B), inhibits TGFβ-dependent collagen synthesis and suppresses ROS-mediated collagen expression. Prostacyclin induces deaggregation of the HSC cytoskeleton. Activation of IP by prostacyclin stimulates the expression of C/EBP and the activation of CREB. The ↑ symbol means activation. The ⊤ symbol means inhibition.

Role of SB-431542 and SB-505124

Further research has resulted in the discovery of a potent inhibitor of ALK5 (SB-431542) **(10)** that acts as a competitive ATP binding site kinase inhibitor [169]; it is able to suppress the activity of ALK5 and potentially ALK4 and ALK7, which are very similar to ALK5 in their kinase domains, resulting in the inhibition of

TGF-β and activin-induced phosphorylation of Smad2 [169]. A similar anti-fibrotic compound was then further developed (SB-505124) **(11)** that, in comparison, was demonstrated to be more potent; SB-505124 retains selective, nontoxic, inhibitor properties of ALK4, ALK5 and perhaps ALK7 [170]. It inhibits activin and TGF-β1-induced phosphorylation of Smad2, TGF-β1 and activin-induced transcriptional activation, TGF-β1-induced activation of JNK/SAP, extracellular signal-regulated kinase 1/2 and p38MAPK and TGF-β1-induced apoptosis [170] (Fig. **3**).

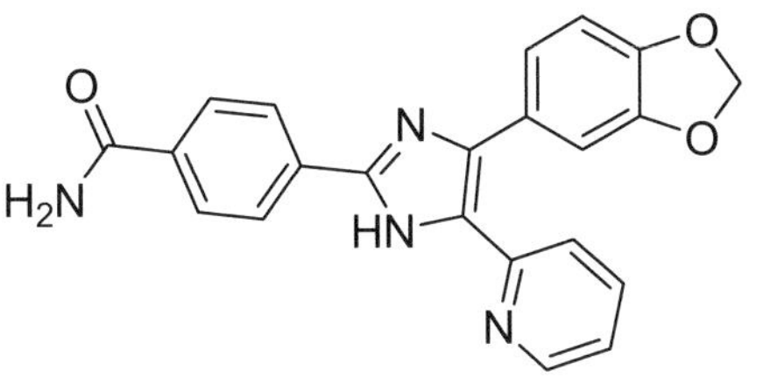

10

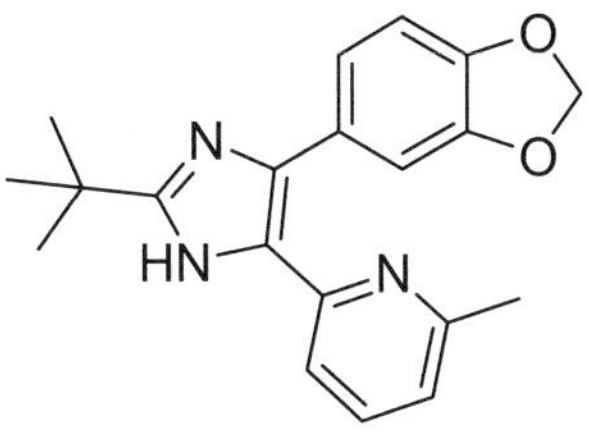

11

Role of PGE2

In vitro experimental administration of prostaglandin E2 (PGE2) **(12)** suppressed TGF-β1-mediated induction of collagen, suggesting that COX-2-derived PGE2 may inhibit both basal and TGF-β1-mediated induction of collagen synthesis by HSCs [171]. However, in an immortalized HSC cell line PGE2 was demonstrated to mediate their proliferative response to PDGF [150] (Fig. **3**).

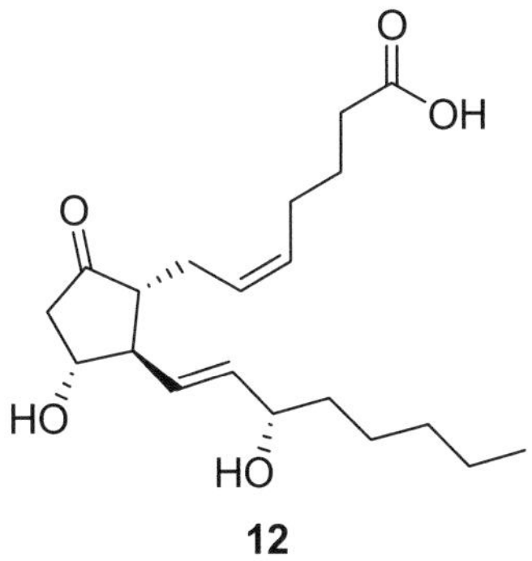

12

Pirfenidone (InterMune) (5 methyl-1-phenyl-2-(1*H*)-pyridone) suppresses TGF-ß1 transcription and subsequent collagen accumulation. In the United States, pirfenidone is in a phase III clinical trial. Recently, the effect of pirfenidone on Th type 1 (Th1) and Th2 responses has been evaluated in experimental liver fibrosis;

Pirfenidone impaired the development of Th2 subpopulation limiting Th2 profibrogenic response [172].

PDG Inhibition Strategies

Inhibiting liver fibrosis *via* a PDGF kinase inhibitor is another attractive strategy.

Role of PGE1

An experiment designed to study the inhibitory effect of exogenous prostaglandin E1 (PGE1) **(13)** on collagen aggregation and activation of HSCs in the liver of rabbits with schistosomiasis, provided convincing evidence of its usefulness against hepatic fibrosis [173]. In the rabbit receiving PGE1, the expression of α-SMA was reduced and limited only to the portal area; a reduction in resistance of the hepatic microcirculation and better blood perfusion in the portal area was also observed [173]. Furthermore, the content of collagen fibrils was clearly decreased, not only due to a downregulation in the expression of the collagen gene, but also an upregulation in the degradation of mature collagen [173]. Importantly, PGE1 prevented the activation of HSCs induced by inflammatory factors, such as PDGF [173] (Fig. **3**).

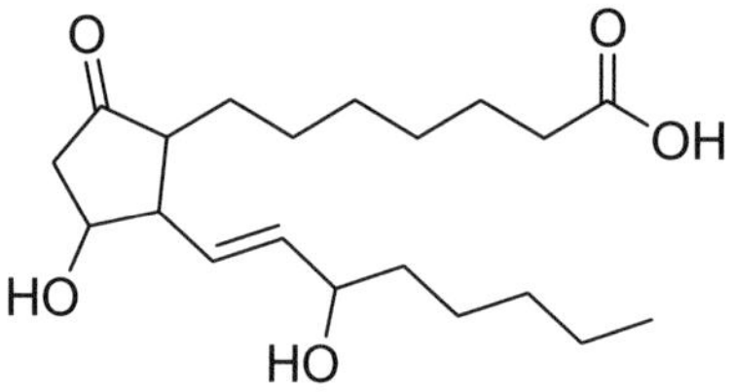

13

Role of Imatinib

Imatinib **(14)**, a drug used in the treatment of chronic myelogenous leukemia and gastrointestinal stromal tumors that demonstrates the ability to inhibit several tyrosine kinases, was shown to also inhibit PDGF-β receptor kinase [174]. This excellent property is potentially useful for attenuating liver fibrosis [175]. Unfortunately, although proven to have antifibrotic effects in cultured HSCs and in animal models of liver fibrosis, this drug has several side effects in other organs (cardiotoxicity) [176] (Fig. **3**). Recently, its ability to inhibit HSCs activation through the enhanced secretion of IL-6, a regenerative cytokine, was demonstrated [177].

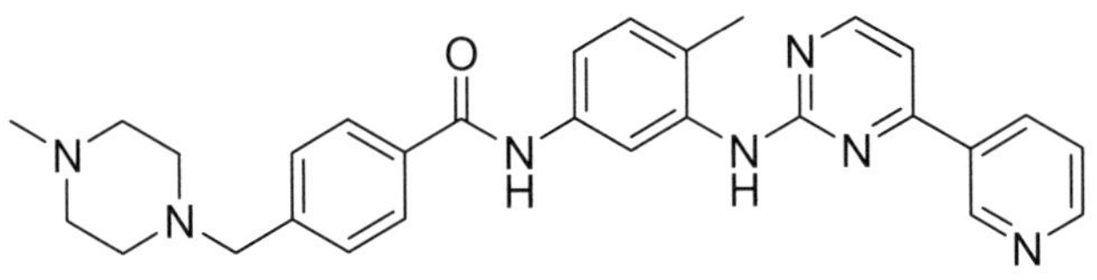

14

Role of M6PHSA

A specific delivery of a PDGF kinase inhibitor to HSCs {4-chloro-*N*-[4-methyl-3-(4-pyridin-3-yl-pyrimidin-2-ylamino)-phenyl]-benzamide (PAP19) **(15)** (an imatinib derivative)} was developed with the goal of inhibiting liver fibrosis without side effects [178]. This compound was obtained by coupling a PDGF kinase inhibitor to a HSCs-directed carrier protein mannose-6-phosphate-human serum albumin (M6PHSA). M6PHSA is a carrier that binds to the M6P/insulin-like growth factor-II receptor on HSCs and rapidly and extensively accumulates in the liver of fibrotic rats. In order to conjugate the PDGF kinase inhibitor to M6PHSA, a novel type of platinum linker chemistry called the universal linkage system (ULS), which allows the stable formation of a platinum-ligand coordination bond, was used [178]. The conjugate compound demonstrated rapid degradation within the lysosomal compartment of HSCs, resulting in quick release of the active drug, which was retained within the fibrotic liver even 2 days after its administration. It was very significant that this therapeutic delivery strategy allowed for lower systemic distribution of the drug, as a much lower dose was administered, which preferentially accumulated within the fibrotic liver and avoided the potential side effects previously observed in nonfibrotic tissues [178]. This compound downregulated the expression of α-SMA and diminished collagen deposition in cultured HSCs whereas, in fibrotic rats, reduced portal-portal bridging in fibrotic areas and α-SMA staining of activated HSCs was observed [178]. Finally, it resulted in a reduction in the amount of activated HSCs [178] (Fig. **3**).

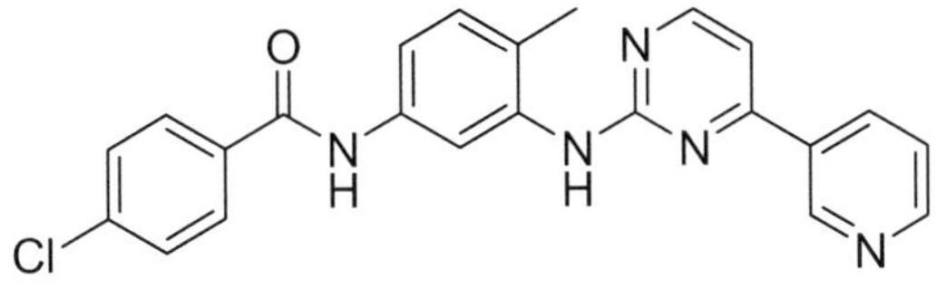

15

Role of Relaxin in Reducing Collagen Deposition

Relaxin. a peptide hormone with a two-chain structure, similar to that of insulin, and encoded by seven genes in humans (RLN1, RLN2 and RLN3 and the insulin-like peptide genes INSL3, INSL4, INSL5 and INSL6) [179] has antifibrotic effects in HSCs [180]. Recombinant human relaxin has been demonstrated to reduce the effective deposition of interstitial collagen both in tissue culture and in an *in vivo* model of hepatic fibrosis [181]. In this study relaxin mediated a reduction in collagen deposition directly, by reducing collagen synthesis, and indirectly, by decreasing TIMP-1 and TIMP-2 expression and thus enhancing matrix degradation [181]. Relaxin was also demonstrated to have anti-TGF-β effects when added to cultured TGF-β1-stimulated rat HSCs by inhibiting collagen overexpression *in vitro* and reducing the level of α-SMA [180, 181]. These effects were mediated through relaxin receptors (LGR7 and LGR8). Both receptors are 7-transmembrane spanning G-protein-coupled receptors that act *via* a pathway involving cAMP. While quiescent HSCs expressed only LGR7, activated HSCs also expressed LGR8 after treatment with relaxin; the fact that this was followed by activation of the cAMP (and this is known to maintain HSCs in a quiescent phase), suggested that relaxin might be useful in the treatment of hepatic fibrosis [180] (Fig. **3**).

Relaxin increases the expression of MMP13 and MMP3, ameliorating the collagen-degrading activity [182].

It also induces a selective and significant decrease of portal pressure in some portal hypertensive models through the increase of intrahepatic NO signaling and a dramatic reduction in contractile filament expression in HSCs [183].

Apoptosis Stimulation Strategies

Apoptosis may be a potential way through which activated HSCs are cleared during resolution of liver injury [184]. In this regard, sulfasalazine a selective and potent inhibitor of NF-kB activation, seems to inhibit the autophosphorylation of IKKα and IKKβ and the subsequent activation of the IKK complex.

Role of Sulfasalazine

Expression of the antiapoptotic factor Gadd45β (able to suppress JNK-induced apoptosis by inhibiting JNK Kinase 2 activation of JNK) decreases after the blockade of IKK and NF-kB by sulfasalazine **(16)** [185]. Thus, sulfasalazine promotes rapid clearance of αSMA positive myofibroblasts, decreases hepatic expression of type I procollagen and TIMP1, increases hepatic MMP activity and accelerates the resolution of liver fibrosis [185] (Fig. **3**).

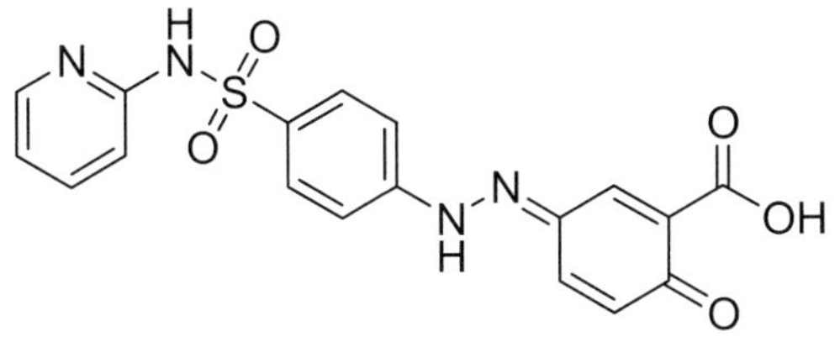

16

Role of IGF-I

Treatment of activated HSCs with IGF-I is also able to induce apoptosis [29] and reduce liver damage and fibrogenesis in cirrhotic rats.

Overexpression of IGF-I by activated HSCs restricts their activation, attenuates fibrogenesis and accelerates liver regeneration [186]. These effects appear to be mediated in part by upregulation of HGF and downregulation of TGFβ1 [186]. These data indicate that IGF-I can modulate the cytokine response to liver injury, facilitating regeneration and reducing fibrosis [29]. In fact, in HSC cultures, IGF-I increased apoptosis in a dose-dependent manner in quiescent HSCs and significantly increased apoptosis in activated HSCs [29]. IGF-I administration induced decreased bcl-2- and bcl-x_L-expression and increased bax gene expression in HSCs [29]. Furthermore, phosphorylation of Erk1/2 was upregulated in quiescent and activated HSCs [29]. By inhibiting Erk1/2 translation, the upregulation of bax and downregulation of bcl-2 was abrogated in quiescent and activated HSCs [29]. According to the authors, since bcl-x_L gene expression was modulated by NF-kB, the proapoptotic effect of IGF-I in HSCs was mediated by ERK1/2 as well as the NF-kB system [29] (Fig. **3**).

Gene transfer of IGF-I (Simian virus 40 vectors encoding IGF-I) to the liver has demonstrated to improve histology and function in animal liver models of cirrhosis; a diminished alpha-SMA expression, up-regulation of MMPs and decreased expression of TIMPS 1 and 2, down-regulation of TGF-β, amphiregulin, PDGF, CTGF and VEGF and induction of HGF were observed [187].

Role of Gliotoxin

Gliotoxin is another compound able to induce HSC apoptosis and attenuate the extent of liver fibrosis through inhibition of NF-kB, as observed in an animal model of liver injury [188]. Furthermore, the addition of gliotoxin to activated rat and human HSCs was also demonstrated to induce apoptosis by means of marked caspase 3 activity, induction of membrane permeability transition and, consequently, release of cytochrome C [189, 190]. However, according to some authors, the clinical use of gliotoxin is limited because due to lack of cell and tissue specificity and a high risk of potentially severe adverse effects [191] (Fig. **3**).

Role of Adiponectin

Adiponectin, an adipokine released by HSCs, upon which it may act in an autocrine fashion through its receptor [30-32], may be critical in maintaining the HSC quiescent phenotype or in reversing hepatic fibrosis by induction of activated HSC apoptosis [30]. In fact, immunoblot studies revealed that adiponectin is only present in quiescent HSCs and is not expressed in activated HSCs; in contrast, its receptors, AdipoR1 and AdipoR2, are present in quiescent and activated HSCs. However, in the latter, RT-qPCR revealed that mRNA expression of AdipoR1 was reduced by nearly 50% compared to quiescent HSCs [30]. Furthermore, immunoblot analysis showed that adiponectin-treated activated HSCs demonstrated increased cleaved caspase 3 (p19), which was not detected in quiescent HSC lysates, demonstrating that, in activated HSCs apoptosis is mediated by caspase 3 [30]. This effect, which was independent of AMP-kinase mediated signalling, was not observed in quiescent HSCs [30]. Moreover, adiponectin knockout leads to severe liver fibrosis [26].

Since adiponectin induces the production of the anti-fibrotic cytokine IL-10 in monocytes and macrophages [31], it has been argued that it may have significant therapeutic implications for the reversal of liver fibrosis (Fig. **3**).

Role of Angiotensin II Type 1 Receptor Antagonist

The identification of the angiotensin II receptor on HSCs [64] and the fact that after its stimulation there is increased synthesis of endothelin, PDGF and chemokines, suggests a possible role for the antagonist of angiotensin II type 1 receptor **(17-19)** in blocking liver fibrosis [192].

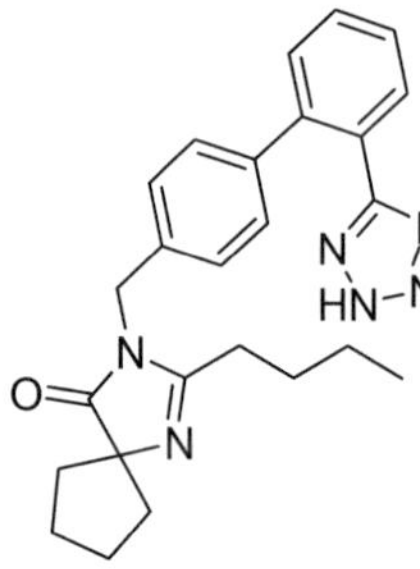

17

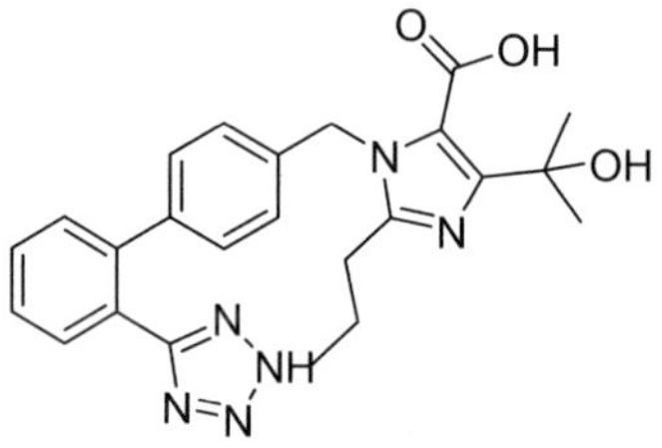

18

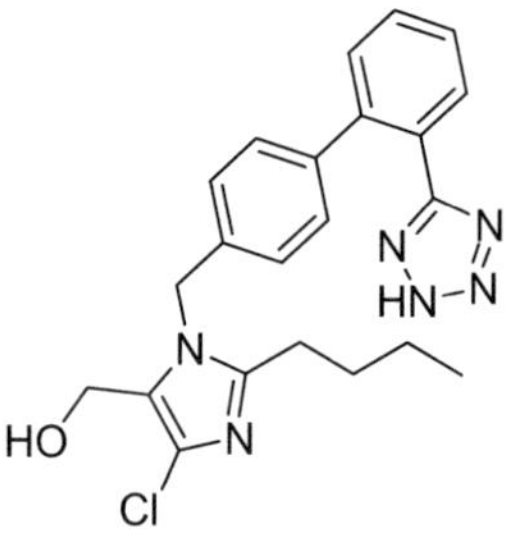

19

In fact, HSC studies *in vitro* and in an animal model of liver fibrosis demonstrated that the angiotensin II type 1 receptor antagonist suppressed fibrogenic responses in activated HSCs including proliferation, collagen synthesis and expression of profibrogenic cytokines (TGF-β1 and CTGF) and improving liver fibrosis in rat models of liver fibrosis [192-194].

In patients with non-alcoholic steatohepatitis, the number of activated HSCs and the ratio of activated/quiescent HSCs were significantly decreased after a 48-week treatment with an angiotensin II type 1 receptor antagonist whereas, quiescent HSCs were slightly increased in parallel with the improvement in liver fibrosis [195]. Two mechanisms of actions were suggested: first, the angiotensin II type 1 receptor antagonist inhibited activated HSCs by blocking angiotensin II type 1 receptors expressed on the surface of HSCs; second, it suppressed the activation of HSCs as a result of the decrease in TGF-β1 [195] (Fig. **3**).

LIVER FIBROSIS: GENE THERAPY

Hepatic fibrosis is clearly a multifactorial event involving several proteins expressed by a large number of genes. HSCs almost certainly play a pivotal role in the unwanted growth of the extracellular matrix. While waiting for a complete understanding of the role of each enzyme and protein involved, many attempts have been made to induce the cessation or the reversal of the process, but all of these showed little activity due to a low uptake at the level of the target cell or major side effects on other cell populations [196]. These attempts have been based on the use of colchicine, interferon [197], biphenyldimethyl-dicarboxylate [198] ,and others. In the face of difficulties with the chemical therapy at the beginning of the new century, gene therapy was assumed as the main subject of this research. Gene transfer by viral vectors relies upon the ultrastructure of the sinusoidal endothelium that, having a non-basal membrane, shows some fenestrae of up to 100 nm, which allow viruses to enter in direct contact with hepatic cells, also known as "Liver Sieve" [199]. Genes are incorporated into vectors that facilitate their entrance and function inside the target cells. Therapeutic genes may be coupled with either viral or non viral vectors, the latter being most frequently used due to their higher transduction efficiency. The specificity, the duration and the inducibility of gene expression relies upon the type of vector [200]. Virtually all liver diseases are amenable to gene therapy. Clinical trials of gene therapy for the treatment of inherited diseases and liver cancer have been initiated, but starting with the early study of Rudolph *et al.* who used the telomerase gene to inhibit induced liver cirrhosis in rats [201], achievement of this goal still seems far away. Genes may be induced in cultured cells and then transplanted into the spleen or the liver [202]. may be physically vehiculated or may be transported by attenuated viruses.

Liver Fibrosis Control *by* Interfering *with the* Immunomodulation

Current strategies to treat hepatic fibrosis are based upon the attempt to stop and possibly reverse the fibrin hyperproduction that occurs as the result of a complex and multi-staged mechanism. This may be done by interfering with the immunomodulation at the site of inflammation.

Role of Transfected Il-18 IL-10 and IFN-γ Genes

Based on the assumed switch of CD4+ T cells by the Th1 to the Th2 condition, Zhang *et al.* induced a reversal of experimental hepatic fibrosis in mice by implanting hepatocytes bearing the IL-18 gene into the spleen. Expression of IL-18 resulted in a Th2 to Th1 switch and, clinically, a significant slowing down of hepatic fibrosis [202]. IL-10 has been investigated by Hung *et al.* Transfection *via* electroporation with a plasmid of the IL-10 gene, in mice with thioacetamide induced hepatic fibrosis, caused a reduction in the disease by acting at several levels: it reduced liver TGF-β1, TNF-α, α1 type I collagen, cell adhesion molecules and TIMPs mRNA [203]. Even transfection with a vector bearing gene coding for IFN-γ proved to be effective in reducing hepatic fibrosis by inhibiting the activation of hepatic stellate cells and decreasing the expression of α-SMA and mRNA of TIMP-1, TGF-β [204].

Since IL-10 has an unfavourable pharmacokinetic profile, which limits its clinical applications, hepatocyte-targeting gene delivery may be an ideal technique for the IL-10 gene therapy of liver fibrosis [205].

Inhibition of HSC Activation by Means of Smad Genetic Block

As evident from the previously reported studies, controlling stellate cell activation is the crucial factor in attempting to inhibit the biochemical cascade leading to liver fibrosis. To this end, Dooley *et al.* proposed the genetic blockade of the TGF-β receptor *via* the Smad7 (an intracellular antagonist of stellate cell TGF-β signaling) [206]. As a consequence, they observed the retention of vitamin A droplets in HSCs and the blockade of the myofibroblast differentiation in the infected cells. They also registered a significant reduction in collagen deposition and a reversal of hepatic fibrosis. Similar results were obtained by Xu *et al.* by transfecting cultured HSCs with a retrovirus carrying the Smad4 nonsense cDNA [207].

Liver Fibrosis Degradation *by* Means *of* Collagen Degrading Enzymes

An additional therapeutic option involves promoting the degradation of liver fibrosis by inducing the activity of collagen degrading enzymes [208]. The means for degradading fibrin may be different; in this regard, the work of Siller-Lopez *et al.* is of particular interest as, in 2 different papers, they caused the reduction of fibrosis by inducing the local synthesis of MMP-8, leading to the reduction of the extracellular matrix and, clinically, a decrease in portal pressure [209].

Similar results were achieved by Roderfeld *et al.* by inducing MMP-9 [210]. The possibility of enhancing the activity of heme-oxigenase 1 (HO-1), an enzyme that in normal liver is expressed by hepatocytes and, to a lesser extent, by a subpopulation of Kupffer cells [211], seems to be promising. During the liver inflammation, HO-1 seems to be a powerful inhibitor of α1 type I pro-collagen transcription and of serum mediated HSCs proliferation [212]. Therefore, enhancing the expression of HO-1 in the fibrotic liver may stop the process as assessed by Tsui *et al.* [213].

In conclusion, in accordance with Prosser *et al.* [214], gene therapy itself is still in its infancy, and ideal gene delivery systems for selective gene transfer with high and prolonged gene expression, as well as decreased cytotoxicity or immunogenicity remain to be developed. Nevertheless, molecular therapeutics has already proven to be a powerful research tool and explorative experimental medicine. They have demonstrated some promise as novel modalities in reducing liver injury, inhibiting HSC activation and promoting the resolution of fibrotic deposition. However, the transition of promising molecular therapeutics from animal models to clinical trials for the validation of their clinical efficacy, adverse effects and target patient population will take a quite long period of time. One may anticipate that, in combination with other more standard therapeutic modalities, these molecular approaches may yield effective treatments for hepatic injury and fibrogenesis in man in the foreseeable future.

CONCLUSION

In vitro studies of HSCs and subsequent studies using animal models of fibrosis have allowed for the evaluation of the antifibrotic effects of some drugs that are likely to block activation and transformation of HSCs and, thus, to improve liver fibrosis.

However, each one of these drugs can have serious drawbacks (short half-life side effects), limiting their administration and fostering further studies to understand how to optimize their beneficial effects [215, 216].

Since animal models of fibrosis are of great importance to understanding the real efficacy of a drug already tested *in vitro* as well as its side effects, it seems of utmost interest to us that light be cast on what model of animal fibrosis is most representative of the human manifestation of this disease. Stated clearly, since human liver fibrosis can be caused by various pathologies (viral, alcoholic, metabolic, *etc.*) and each one of these pathologies might respond better to one drug rather than another, it would be advisable to reproduce animal models of viral, alcoholic and metabolic fibrosis and test various drugs in each one of them.

Another problem is the inability to specifically target some types of cells rather than others and to obtain a suitable drug concentration thus preventing consequent side-effects.

Currently, clinical trials, whose aim is to target fibrosis, use small molecule drugs and monoclonal antibodies; several nanoparticle-based technologies are now in the research phase.

In particular, well-designed nanoparticle-based technologies might offer a safe method to ameliorate delivery, specificity, uptake, stability and release of traditional and novel pharmaceutical agents in the hepatic sites. Nanoparticles seem to be an optimal solution to administer nucleotide-based therapies targeting pro-fibrotic pathways, inflammatory cytokines and growth factors (VEGF CTGF TNF-α and TGF-β1 signaling). They might be very useful to target hepatic fibrosis thanks to natural trafficking of nanoparticles *via* the liver.

We think that the field of genetic engineering is very attracting and may offer many opportunities: however, carefulness is necessary in attempting to modify the expression of certain proteins by manipulating the genetic code. This will require great prudence and deeper studies.

ACKNOWLEDGEMENTS

Declared None.

CONFLICT OF INTEREST

The author confirms that this chapter contents have no conflict of interest.

ABBREVIATIONS

AC	=	Adenylyl cyclase
ALK	=	Activin receptor-like kinase
AP1	=	Activator protein-1 transcription factor
ARB	=	Angiotensin receptor blocker
cAMP	=	Cyclic-adenosine monophosphate
cGMP	=	Cyclic-guanosine monophosphate
C/EBP	=	Adipocyte-promoting transcription factors
cytokine-r	=	Cytokine receptor
CREB	=	[cAMP-responsive element (CRE)-binding protein]
DDR2	=	Discoidin domain receptor
ECM	=	Extracellular matrix
Erk	=	Extracellular signal-related kinase
Gi	=	Protein Gi
Gq	=	Protein Gq
Gs	=	Protein Gs
FoxO	=	Forkhead transcriptional factor
JNK	=	c-Jun N-terminal kinases
HSC	=	Hepatic stellate cell

IFN-γ	=	Interferon-γ
IKK	=	Inhibitor Kinase complex
ILK	=	Integrin linked kinase
IL-10	=	Interleukin-10
IP	=	Prostacyclin receptor
JunD	=	Protein dimers of AP1
KLF-6	=	Kruppel-like factor-6
LIM	=	Transcription factor
LXRα	=	Liver X receptor α
MAPK	=	Mitogen-activated protein kinase
NF-kB	=	Transcriptional nuclear factor kB
NPR-	=	Cardiac natriuretic peptide receptor
PAP19	=	PDGF kinase inhibitor
PDGF	=	Platelet-derived growth factor
PI-3K	=	Phosphatidylinositol-3 kinase
PKA	=	Protein kinase A
PKB	=	Protein kinase B
PKBrm	=	PKB recognition motifs
PKC	=	Protein kinase C
PPAR-	=	Proliferator-activated receptor
PPRE	=	Peroxisome proliferator response elements

RXRa	=	Retinoid X receptor a
SREBP-1c	=	Sterol-regulatory-element-binding protein 1c
STAT	=	Signal transducer and activator of transcription
TGFβ-r	=	Transforming growth factor β receptor
TIMP-	=	Tissue inhibitors of metalloproteinases
TNF-r	=	Tumor necrosis factor receptor

DISCLOSURE

This is an updated report of the earlier review article by Zardi Enrico M "New therapeutic approaches to liver fibrosis: A practicable route?". Curr Med Chem 2008; 15(16): 1628-1644.

REFERENCES

[1] Häussinger D, Kubitz R, Reinehr R, Bode JG, Schliess F. Molecular aspects of medicine: from experimental to clinical hepatology. Mol Aspects Med 2004; 25(3): 221-360.

[2] Castilla A, Prieto J, Fausto N. Transforming growth factors beta 1 and alpha in chronic liver disease. Effects of interferon alfa therapy. N Engl J Med 1991; 324(14): 933-40.

[3] Annoni G, Weiner FR, Zern MA. Increased transforming growth factor-beta1 gene expression in human liver disease. J Hepatol 1992; 14(2-3): 259-64.

[4] Roulot D, Sevcsik AM, Coste T, Strosberg AD, Marullo S. Role of transforming growth factor beta type II receptor in hepatic fibrosis: studies of human chronic hepatitis C and experimental fibrosis in rats. Hepatology 1999; 29(6): 1730-8

[5] Kinnman N, Goria O, Wendum D, *et al.* Hepatic stellate cell proliferation is an early platelet-derived growth factor-mediated cellular event in rat cholestatic liver injury. Lab Invest 2001; 81(12): 1709-16.

[6] Iredale JP. Models of liver fibrosis: exploring the dynamic nature of inflammation and repair in a solid organ. J Clin Invest 2007; 117(3): 539-48.

[7] Sakata K, Eda S, Lee ES, Hara M, Imoto M, Kojima S. Neovessel formation promotes liver fibrosis *via* providing latent transforming growth factor-β. Biochem Biophys Res Commun 2014; 443(3): 950-6.

[8] Rockey DC, Fouassier L, Chung JJ, *et al.* Cellular localization of endothelin-1 and increased production in liver injury in the rat: potential for autocrine and paracrine effects on stellate cells. Hepatology 1998; 27(2): 472-80.

[9] Rockey DC. New therapies in hepatitis C virus and chronic liver disease: antifibrotics. Clin Liver Dis 2006; 10(4): 881-900.

[10] Zardi EM, Dobrina A, Amoroso A, Afeltra A. Prostacyclin in liver disease: a potential therapeutic option. Expert Opinion on Biological Therapy 2007; 7(6): 785-90.

[11] Geerts A, De Bleser P, Hautekeete ML, Niki T, Wisse E. Fat storing (Ito) cell biolog In: Arias IM, Boyer JL, Fausto N (Eds), The liver: Biology and Pathobiology, 3rd ed, Raven Press New York, 1994, pp 819–837.

[12] Guyot C, Lepreux S, Combe C, *et al.* Hepatic fibrosis and cirrhosis: the (myo)fibroblastic cell subpopulations involved. Int J Biochem Cell Biol 2006; 38(2): 135-51.

[13] Kashiwagi S, Suematsu M, Wakabayashi Y, *et al.* Am J Physiol 1997; 272(4 Pt 1): G742-50.

[14] Geerts AJ. On the origin of stellate cells: mesodermal endodermal or neuro-ectodermal? J Hepatol 2004; 40(2): 331-4.

[15] Vogel S, Gamble MV, Blaner WS. Biosynthesis absorption metabolism and transport of retinoids In: Nau H, Blaner WS (Eds.) Retinoids: the biochemical and molecular basis of vitamin A and retinoid action. Springer, Berlin 1999, pp 31–95.

[16] Geerts A, Bouwens L, Wisse E. Ultrastructure and function of hepatic fat-storing and pit cells. Electron Micros Tech 1990; 14(3): 247-56

[17] Uyama N, Zhao L, Van Rossen E, *et al.* Hepatic stellate cells express synemin, a protein bridging intermediate filaments to focal adhesions. Gut 2006; 55(9): 1276-89.

[18] Miyahara T, Schrum L, Rippe R, *et al.* Peroxisome proliferator-activated receptors and hepatic stellate cell activation. J Biol Chem 2000; 275(46): 35715-22.

[19] Marra F, Efsen E, Romanelli R G, *et al.* Ligands of peroxisome proliferator-activated receptor gamma modulate profibrogenic and proinflammatory actions in hepatic stellate cells. Gastroenterology 2000; 119(2): 466-78.

[20] Morrison RF, Farmer SR. J Hormonal signaling and transcriptional control of adipocyte differentiation. J Nutr 2000; 130(12): 3116S-3121S.

[21] Galli A, Crabb DW, Ceni E, *et al.* Antidiabetic thiazolidinediones inhibit collagen synthesis and hepatic stellate cell activation *in vivo* and *in vitro*. Gastroenterology 2002; 122(7): 1924-40

[22] Rojkind M, Giambrone MA, Biempica L. Collagen types in normal and cirrhotic liver. Gastroenterology 1979; 76(4): 710-9.

[23] Mueller E, Drori S, Aiyer A, *et al.* Genetic analysis of adipogenesis through peroxisome proliferator-activated receptor gamma isoforms. J Biol Chem 2002; 277(44): 41925-30.

[24] Greene ME, Blumberg B, McBride OW, *et al.* Isolation of the human peroxisome proliferator activated receptor gamma cDNA: expression in hematopoietic cells and chromosomal mapping. Gene Expr 1995; 4(4-5): 281-99.

[25] Sun K, Wang Q, Huang XH. PPAR gamma inhibits growth of rat hepatic stellate cells and TGF beta-induced connective tissue growth factor expression. Acta Pharmacol Sin 2006; 27(6): 715-23.

[26] Zardi E M Navarini L Sambataro G, *et al.* Hepatic PPARs: their role in liver physiology, fibrosis and treatment. Curr Med Chem 2013; 20(27): 3370-96.

[27] Wandzioch E, Kolterud A, Jacobsson M, *et al.* Lhx2-/- mice develop liver fibrosis. Proc Natl Acad Sci U S A. 2004; 101(47): 16549-54.

[28] Matsumoto K, Nakamura T. Emerging multipotent aspects of hepatocyte growth factor. J Biochem. 1996 Apr; 119(4): 591-600.

[29] Saile B DiRocco P Dudas J, *et al.* IGF-I induces DNA synthesis and apoptosis in rat liver hepatic stellate cells (HSC) but DNA synthesis and proliferation in rat liver myofibroblasts (rMF). Lab Invest 2004; 84(8): 1037-49.

[30] Ding X, Saxena NK, Lin S, *et al.* The roles of leptin and adiponectin: a novel paradigm in adipocytokine regulation of liver fibrosis and stellate cell biology. Am J Pathol 2005; 66(6): 1655-69.

[31] Guzik TJ, Mangalat D, Korbut R. Adipocytokines - novel link between inflammation and vascular function? J Physiol Pharmacol 2006; 57(4): 505-28.

[32] Ikejima K, Lang T, Zhang YJ, *et al.* Expression of leptin receptors in hepatic sinusoidal cells. Comp Hepatol 2004; 3 Suppl 1: S12.

[33] Thimgan MS, Yee HF Jr. Quantitation of rat hepatic stellate cell contraction: stellate cells' contribution to sinusoidal resistance. Am J Physiol 1999; 277(1 Pt 1): G137-43.

[34] Bosch J, Garcia-Pagàn JC. Complications of cirrhosis. I. Portal hypertension. J Hepatol 2000; 32(1 Suppl): 141-56.

[35] Mallat A, Fouassier L, Préaux AM, Mavier P, Lotersztajn S. Antiproliferative effects of ET-1 in human liver Ito cells: an ETB- and a cyclic AMP-mediated pathway. J Cardiovasc Pharmacol. 1995; 26 Suppl 3: S132-4.

[36] Fennekohl A, Schieferdecker HL, Jungermann K, Puschel GP. Differential expression of prostanoid receptors in hepatocytes, Kupffer cells, sinusoidal endothelial cells and stellate cells of rat liver. J Hepatol 1999; 30(1): 38-47.

[37] Ruan KH, Dogne JM. Implications of the molecular basis of prostacyclin biosynthesis and signaling in pharmaceutical designs. Curr Pharm Des 2006; 12(8): 925-41.

[38] Eng F J, Friedman SL. Fibrogenesis I. New insights into hepatic stellate cell activation: the simple becomes complex. Am J Physiol Gastrointest Liver Physiol 2000 Jul; 279(1): G7-G11.

[39] Clerico A, Recchia FA, Passino C, Emdin M. Cardiac endocrine function is an essential component of the homeostatic regulation network: physiological and clinical implications. Am J Physiol Heart Circ Physiol 2006; 290(1): H17-29

[40] Görbig MN, Ginès P, Bataller R, *et al.* Atrial natriuretic peptide antagonizes endothelin-induced calcium increase and cell contraction in cultured human hepatic stellate cells. Hepatology 1999; 30(2): 501-9.

[41] Garbers D L Guanylyl cyclase receptors and their endocrine, paracrine, and autocrine ligands. Cell 1992; Oct 2; 71(1): 1-4.

[42] Koller K J, Lowe D G, Bennett GL, *et al.* Selective activation of the B natriuretic peptide receptor by C-type natriuretic peptide (CNP). Science 1991; 252(5002): 120-3

[43] Anand-Srivastava MB, Sairam MR, Cantin M. Ring-deleted analogs of atrial natriuretic factor inhibit adenylate cyclase/cAMP system. Possible coupling of clearance atrial natriuretic factor receptors to adenylate cyclase/cAMP signal transduction system. J Biol Chem 1990; 265(15): 8566-72.

[44] Maack T, Suzuki M, Almeida FA, *et al.* Physiological role of silent receptors of atrial natriuretic factor. Science 1987; 238(4827): 675-8.

[45] Anand-Srivastava MB, Sehl PD, Lowe DG. Cytoplasmic domain of natriuretic peptide receptor-C inhibits adenylyl cyclase. Involvement of a pertussis toxin-sensitive G protein. J Biol Chem 1996; 271(32): 19324-9.

[46] Villar IC, Panayiotou CM, Sheraz A, *et al.* Definitive role for natriuretic peptide receptor-C in mediating the vasorelaxant activity of C-type natriuretic peptide and endothelium-derived hyperpolarising factor. J Cardiovasc Res 2007; 74(3): 515-25.

[47] Bachem MG, Sell KM, Melchior R, Kropf J, Eller T, Gressner AM. Tumor necrosis factor alpha (TNF alpha) and transforming growth factor beta 1 (TGF beta 1) stimulate fibronectin synthesis and the transdifferentiation of fat-storing cells in the rat liver into myofibroblasts. Virchows Arch B Cell Pathol Incl Mol Pathol. 1993; 63(2): 123-30.

[48] Friedman SL. Seminars in medicine of the Beth Israel Hospital, Boston. The cellular basis of hepatic fibrosis. Mechanisms and treatment strategies. N Engl J Med 1993; 328(25): 1828-35

[49] Bataller R, Gasull X, Ginès P, *et al. In vitro* and *in vivo* activation of rat hepatic stellate cells results in *de novo* expression of L-type voltage-operated calcium channels. Hepatology 2001; 33(4): 956-62.

[50] Accili D, Arden KC. FoxOs at the crossroads of cellular metabolism, differentiation, and transformation. Cell 2004; 117(4): 421-6.

[51] van der Heide LP, Hoekman MFM, Smidt MP. The ins and outs of FoxO shuttling: mechanisms of FoxO translocation and transcriptional regulation. Biochem J 2004; 380(Pt 2): 297-309.

[52] Kaestner KH, Knochel W, Martinez DE. Unified nomenclature for the winged helix/forkhead transcription factors. Genes Dev 2000; 14(2): 142-6.

[53] Brunet A, Bonni A, Zigmond MJ, *et al.* Akt promotes cell survival by phosphorylating and inhibiting a Forkhead transcription factor. Cell 1999; 96(6): 857-68.

[54] Nakae J, Barr V, Accili D. Differential regulation of gene expression by insulin and IGF-1 receptors correlates with phosphorylation of a single amino acid residue in the forkhead transcription factor FKHR. EMBO J 2000; 19(5): 989-96.

[55] Adachi M, Osawa Y, Uchinami H, Kitamura T, Accili D, Brenner DA. The forkhead transcription factor FoxO1 regulates proliferation and transdifferentiation of hepatic stellate cells. Gastroenterology 2007; 132(4): 1434-46.

[56] Shafiei MS, Rockey DC. The role of integrin-linked kinase in liver wound healing. J Biol Chem 2006; 281(34): 24863-72.

[57] Wu C, Dedhar S. Integrin-linked kinase (ILK) and its interactors: a new paradigm for the coupling of extracellular matrix to actin cytoskeleton and signaling complexes. J Cell Biol 2001; 155(4): 505-10.

[58] Downward J. Mechanisms and consequences of activation of protein kinase B/Akt. Curr Opin Cell Biol 1998; 10(2): 262-7.

[59] Persad S, Attwell S, Gray V, *et al.* Regulation of protein kinase B/Akt-serine 473 phosphorylation by integrin-linked kinase: critical roles for kinase activity and amino acids arginine 211 and serine 343. J Biol Chem 2001; 20; 276(29): 27462-9.

[60] Hazra S, Miyahara T, Rippe RA, Tsukamoto H. PPAR Gamma and Hepatic Stellate Cells. Comp Hepatol 2004; 3 Suppl 1: S7.

[61] Sung CK, She H, Xiong S, Tsukamoto H. Tumor necrosis factor-alpha inhibits peroxisome proliferator-activated receptor gamma activity at a posttranslational level in hepatic stellate cells. Am J Physiol Gastrointest Liver Physiol 2004; 286(5): G722-9.
[62] Sancho-Bru P, Bataller R, *et al.* Genomic and functional characterization of stellate cells isolated from human cirrhotic livers. J Hepatol 2005; 43(2): 272-82.
[63] Shao R, Yan W, Rockey DC. Regulation of endothelin-1 synthesis by endothelin-converting enzyme-1 during wound healing. J Biol Chem 1999; 274(5): 3228-34.
[64] Bataller R, Ginès P, Nicolás JM, *et al.* Angiotensin II induces contraction and proliferation of human hepatic stellate cells. Gastroenterology 2000; 118(6): 1149-56.
[65] Ducut Sigala JL, Bottero V, Young DB, Shevchenko A, Mercurio F, Verma IM. Activation of transcription factor NF-kappaB requires ELKS, an IkappaB kinase regulatory subunit. Science 2004; 304(5679): 1963-7.
[66] Mann J, Oakley F, Akiboye F, Elsharkawy A, Thorne AW, Mann DA. Regulation of myofibroblast transdifferentiation by DNA methylation and MeCP2: implications for wound healing and fibrogenesis. Cell Death Differ 2007; 14(2): 275-85
[67] Friedman SL. Transcriptional regulation of stellate cell activation. J Gastroenterol Hepatol 2006; 21 Suppl 3: S79-83.
[68] Bertolani C, Sancho-Bru P, Failli P, *et al.* Resistin as an intrahepatic cytokine: overexpression during chronic injury and induction of proinflammatory actions in hepatic stellate cells. Am J Pathol 2006; 169(6): 2042-53.
[69] Aleffi S, Petrai I, Bertolani C, *et al.* Upregulation of proinflammatory and proangiogenic cytokines by leptin in human hepatic stellate cells. Hepatology 2005; 42(6): 1339-48.
[70] Lang A, Schoonhoven R, Tuvia S, Brenner DA, Rippe RA. Nuclear factor kappaB in proliferation, activation, and apoptosis in rat hepatic stellate cells. J Hepatol 2000; 33(1): 49-58.
[71] Hellerbrand C, Jobin C, Iimuro Y, Licato L, Sartor RB, Brenner DA. Inhibition of NFkappaB in activated rat hepatic stellate cells by proteasome inhibitors and an IkappaB super-repressor. Hepatology 1998; 27(5): 1285-95.
[72] Potter JJ, Womack L, Mezey E, Anania FA. Transdifferentiation of rat hepatic stellate cells results in leptin expression. Biochem Biophys Res Commun 1998; 244(1): 178-82.
[73] Saxena NK, Ikeda K, Rockey DC, Friedman SL, Anania FA. Leptin in hepatic fibrosis: evidence for increased collagen production in stellate cells and lean littermates of ob/ob mice. Hepatology 2002; 35(4): 762-71.
[74] Saxena NK, Titus MA, Ding X, *et al.* Leptin as a novel profibrogenic cytokine in hepatic stellate cells: mitogenesis and inhibition of apoptosis mediated by extracellular regulated kinase (Erk) and Akt phosphorylation. FASEB J 2004; 18(13): 1612-4.
[75] Gressner AM, Weiskirchen R. Modern pathogenetic concepts of liver fibrosis suggest stellate cells and TGF-beta as major players and therapeutic targets. J Cell Mol Med 2006; 10(1): 76-99.
[76] Breitkopf K, Haas S, Wiercinska E, Singer MV, Dooley S. Anti-TGF-beta strategies for the treatment of chronic liver disease. Alcohol Clin Exp Res 2005; 29(11 Suppl): 121S-131S.
[77] Gressner AM, Weiskirchen R, Breitkopf K, Dooley S. Roles of TGF-beta in hepatic fibrosis. Front Biosci 2002; 7: d793-807.
[78] Reimann T, Hempel U, Krautwald S, *et al.* Transforming growth factor-beta1 induces activation of Ras, Raf-1, MEK and MAPK in rat hepatic stellate cells. FEBS Lett 1997; 403(1): 57-60.
[79] Rockey DC, Chung JJ. Endothelin antagonism in experimental hepatic fibrosis. Implications for endothelin in the pathogenesis of wound healing. J Clin Invest 1996; 98(6): 1381-8.
[80] Koda M, Bauer M, Krebs A, Hahn EG, Schuppan D, Murawaki Y. Endothelin-1 enhances fibrogenic gene expression, but does not promote DNA synthesis or apoptosis in hepatic stellate cells. Comp Hepatol 2006; 24; 5: 5.
[81] Shao R, Shi Z, Gotwals PJ, Koteliansky VE, George J, Rockey DC. Cell and molecular regulation of endothelin-1 production during hepatic wound healing. Mol Biol Cell 2003; 14(6): 2327-41.
[82] Mathurin P, Xiong S, Kharbanda KK, *et al.* IL-10 receptor and coreceptor expression in quiescent and activated hepatic stellate cells. Am J Physiol Gastrointest Liver Physiol 2002; 282(6): G981-90.
[83] Taimr P, Higuchi H, Kocova E, Rippe RA, Friedman S, Gores GJ. Activated stellate cells express the TRAIL receptor-2/death receptor-5 and undergo TRAIL-mediated apoptosis. Hepatology 2003; 37(1): 87-95.

[84] Trim N, Morgan S, Evans M, *et al.* Hepatic stellate cells express the low affinity nerve growth factor receptor p75 and undergo apoptosis in response to nerve growth factor stimulation. Am J Pathol 2000; 156(4): 1235-43.
[85] Dooley S, Delvoux B, Streckert M, *et al.* Transforming growth factor beta signal transduction in hepatic stellate cells *via* Smad2/3 phosphorylation, a pathway that is abrogated during *in vitro* progression to myofibroblasts. TGFbeta signal transduction during transdifferentiation of hepatic stellate cells. FEBS Lett 2001; 502(1-2): 4-10.
[86] Yoshida K, Matsuzaki K, Mori S, *et al.* Transforming growth factor-beta and platelet-derived growth factor signal *via* c-Jun N-terminal kinase-dependent Smad2/3 phosphorylation in rat hepatic stellate cells after acute liver injury. Am J Pathol 2005; 166(4): 1029-39.
[87] Yang C, Zeisberg M, Mosterman B, *et al.* Liver fibrosis: insights into migration of hepatic stellate cells in response to extracellular matrix and growth factors. Gastroenterology 2003; 124(1): 147-59.
[88] Uemura M, Swenson ES, Gaça MD, Giordano FJ, Reiss M, Wells RG. Smad2 and Smad3 play different roles in rat hepatic stellate cell function and alpha-smooth muscle actin organization. Mol Biol Cell 2005; 16(9): 4214-24.
[89] Moustakas A, Souchelnytskyi S, Heldin CH. Smad regulation in TGF-beta signal transduction. J Cell Sci 2001; 114(Pt 24): 4359-69.
[90] Schmidt A, Rossi P, de Crombrugghe B. Transcriptional control of the mouse alpha 2(I) collagen gene: functional deletion analysis of the promoter and evidence for cell-specific expression. Mol Cell Biol 1986; 6(2): 347-54.
[91] Jimenez SA, Varga J, Olsen A, *et al.* Functional analysis of human alpha 1(I) procollagen gene promoter. Differential activity in collagen-producing and -nonproducing cells and response to transforming growth factor beta 1. J Biol Chem 1994; 269(17): 12684-91.
[92] Ritzenthaler JD, Goldstein RH, Fine A, Lichtler A, Rowe DW, Smith BD. Transforming-growth-factor-beta activation elements in the distal promoter regions of the rat alpha 1 type I collagen gene. Biochem J 1991; 280 (Pt 1): 157-62.
[93] Chung KY, Agarwal A, Uitto J, Mauviel A. An AP-1 binding sequence is essential for regulation of the human alpha2(I) collagen (COL1A2) promoter activity by transforming growth factor-beta. J Biol Chem 1996; 271(6): 3272-8.
[94] Murphy G, Docherty A. The matrix metalloproteinases and their inhibitors. Am J Respir Cell Mol Biol 1992; 7(2): 120-5.
[95] Hall MC, Young DA, Waters JG, *et al.* The comparative role of activator protein 1 and Smad factors in the regulation of Timp-1 and MMP-1 gene expression by transforming growth factor-beta 1. J Biol Chem 2003; 278(12): 10304-13.
[96] Trim JE, Samra SK, Arthur MJ, *et al.* Upstream tissue inhibitor of metalloproteinases-1 (TIMP-1) element-1, a novel and essential regulatory DNA motif in the human TIMP-1 gene promoter, directly interacts with a 30-kDa nuclear protein. J Biol Chem 2000 Mar 3; 275(9): 6657-63.
[97] Smart DE, Vincent KJ, Arthur MJ, *et al.* JunD regulates transcription of the tissue inhibitor of metalloproteinases-1 and interleukin-6 genes in activated hepatic stellate cells. J Biol Chem 2001; 276(26): 24414-21.
[98] Bertrand-Philippe M, Ruddell RG, Arthur MJ, Thomas J, Mungalsingh N, Mann DA. Regulation of tissue inhibitor of metalloproteinase 1 gene transcription by RUNX1 and RUNX2. J Biol Chem 2004; 279(23): 24530-9.
[99] Karin M, Liu Zg, Zandi E. AP-1 function and regulation. Curr Opin Cell Biol 1997; 9(2): 240-6.
[100] Ziff EB. Transcription factors: a new family gathers at the cAMP response site. Trends Genet 1990; 6(3): 69-72.
[101] Hai T, Curan T. Cross-family dimerization of transcription factors Fos/Jun and ATF/CREB alters DNA binding specificity. Proc Natl Acad Sci USA 1991; 88(9): 3720-4.
[102] Smart DE, Green K, Oakley F, *et al.* JunD is a profibrogenic transcription factor regulated by Jun N-terminal kinase-independent phosphorylation. Hepatology 2006; 44(6): 1432-40.
[103] Lin S, Saxena NK, Ding X, Stein LL, Anania FA. Leptin increases tissue inhibitor of metalloproteinase I (TIMP-1) gene expression by a specificity protein 1/signal transducer and activator of transcription 3 mechanism. Mol Endocrinol 2006; 20(12): 3376-88.
[104] Yamaguchi K, Shirakabe K, Shibuya H, *et al.* Identification of a member of the MAPKKK family as a potential mediator of TGF-beta signal transduction. Science 1995; 270(5244): 2008-11.

[105] Cao Q, Mak KM, Lieber CS. DLPC decreases TGF-beta1-induced collagen mRNA by inhibiting p38 MAPK in hepatic stellate cells. Am J Physiol Gastrointest Liver Physiol 2002; 283(5): G1051-61.
[106] Tsukada S, Westwick JK, Ikejima K, Sato N, Rippe RA. SMAD and p38 MAPK signaling pathways independently regulate alpha1(I) collagen gene expression in unstimulated and transforming growth factor-beta-stimulated hepatic stellate cells. J Biol Chem 2005; 280(11): 10055-64
[107] Galli A, Crabb D, Price D, *et al.* Peroxisome proliferator-activated receptor gamma transcriptional regulation is involved in platelet-derived growth factor-induced proliferation of human hepatic stellate cells. Hepatology 2000; 31(1): 101-8.
[108] Han J, Hajjar DP, Tauras JM, Feng J, Gotto AM Jr, Nicholson AC Transforming growth factor-beta1 (TGF-beta1) and TGF-beta2 decrease expression of CD36, the type B scavenger receptor, through mitogen-activated protein kinase phosphorylation of peroxisome proliferator-activated receptor-gamma. J Biol Chem 2000; 275(2): 1241-6.
[109] Meurer SK, Lahme B, Tihaa L, Weiskirchen R, Gressner AM. N-acetyl-L-cysteine suppresses TGF-beta signaling at distinct molecular steps: the biochemical and biological efficacy of a multifunctional, antifibrotic drug. Biochem Pharmacol. 2005; 70(7): 1026-34.
[110] Vega S, Morales AV, Ocaña OH, Valdés F, Fabregat I, Nieto MA. Snail blocks the cell cycle and confers resistance to cell death. Genes Dev 2004; 18(10): 1131-43.
[111] Nakao A, Imamura T, Souchelnytskyi S, *et al.* TGF-beta receptor-mediated signalling through Smad2, Smad3 and Smad4. Embo J 1997; 16(17): 5353-62.
[112] Zavadil J, Böttinger EP. TGF-beta and epithelial-to-mesenchymal transitions. Oncogene 2005; 24(37): 5764-74.
[113] Kaimori A, Potter J, Kaimori JY, Wang C, Mezey E, Koteish A. Transforming growth factor-beta1 induces an epithelial-to-mesenchymal transition state in mouse hepatocytes *in vitro*. J Biol Chem 2007; 282(30): 22089-101.
[114] Zeisberg M, Yang C, Martino M, *et al.* Fibroblasts derive from hepatocytes in liver fibrosis *via* epithelial to mesenchymal transition. J Biol Chem 2007; 282(32): 23337-47.
[115] Ramadori G, Saile B. Portal tract fibrogenesis in the liver. Lab Invest 2004; 84(2): 153-9.
[116] Cassiman D, Libbrecht L, Desmet V, Denef C, Roskams T. Hepatic stellate cell/myofibroblast subpopulations in fibrotic human and rat livers. J Hepatol 2002; 36(2): 200-9.
[117] Hinz B, Phan SH, Thannickal VJ, Galli A, Bochaton-Piallat ML, Gabbiani G. G The myofibroblast: one function, multiple origins. Am J Pathol 2007; 170(6): 1807-16.
[118] Friedman SL, Roll FJ, Boyles J, Arenson DM, Bissell DM. Maintenance of differentiated phenotype of cultured rat hepatic lipocytes by basement membrane matrix. J Biol Chem 1989; 264(18): 10756-62.
[119] Matrisian LM. Quick guide. Matrix metalloproteinases. Curr Biol 2000; 10(19): R692.
[120] Arthur MJ, Stanley A, Iredale JP, Rafferty JA, Hembry RM, Friedman SL. Secretion of 72 kDa type IV collagenase/gelatinase by cultured human lipocytes. Analysis of gene expression, protein synthesis and proteinase activity. Biochem J 1992; 287 (Pt 3): 701-7.
[121] Karn T, Holtrich U, Bräuninger A, *et al.* Structure, expression and chromosomal mapping of TKT from man and mouse: a new subclass of receptor tyrosine kinases with a factor VIII-like domain. Oncogene 1993; 8(12): 3433-40.
[122] Olaso E, Labrador JP, Wang L, *et al.* Discoidin domain receptor 2 regulates fibroblast proliferation and migration through the extracellular matrix in association with transcriptional activation of matrix metalloproteinase-2. J Biol Chem 2002; 277(5): 3606-13.
[123] Leitinger B. Molecular analysis of collagen binding by the human discoidin domain receptors, DDR1 and DDR2. Identification of collagen binding sites in DDR2. J Biol Chem 2003 May 9; 278(19): 16761-9.
[124] Baumgartner S, Hofmann K, Chiquet-Ehrismann R, Bucher P. The discoidin domain family revisited: new members from prokaryotes and a homology-based fold prediction. Protein Sci 1998; 7(7): 1626-31.
[125] Mihai C, Iscru DF, Druhan LJ, Elton TS, Agarwal G. Discoidin domain receptor 2 inhibits fibrillogenesis of collagen type 1. J Mol Biol 2006; 361(5): 864-76.
[126] Vogel W, Gish GD, Alves F, Pawson T. The discoidin domain receptor tyrosine kinases are activated by collagen. Mol Cell 1997; 1(1): 13-23.
[127] Shrivastava A, Radziejewski C, Campbell E, *et al.* An orphan receptor tyrosine kinase family whose members serve as nonintegrin collagen receptors. Mol Cell 1997; 1(1): 25-34.
[128] Olaso E, Ikeda K, Eng FJ, *et al.* DDR2 receptor promotes MMP-2-mediated proliferation and invasion by hepatic stellate cells. J Clin Invest 2001; 108(9): 1369-78.

[129] Wells RG. The role of matrix stiffness in hepatic stellate cell activation and liver fibrosis. J Clin Gastroenterol 2005; 39(4 Suppl 2): S158-61.

[130] Rockey D C Hepatic fibrosis, stellate cells, and portal hypertension. Clin Liver Dis 2006; 10(3): 459-79, vii-viii.

[131] Hazra S, Xiong S, Wang J, *et al.* Peroxisome proliferator-activated receptor gamma induces a phenotypic switch from activated to quiescent hepatic stellate cells. J Biol Chem 2004; 279(12): 11392-401.

[132] Zhao C, Chen W, Yang L, Chen L, Stimpson SA, Diehl AM. PPARgamma agonists prevent TGFbeta1/Smad3-signaling in human hepatic stellate cells. Biochem Biophys Res Commun 2006; 350(2): 385-91.

[133] Zheng S, Chen A. Activation of PPARgamma is required for curcumin to induce apoptosis and to inhibit the expression of extracellular matrix genes in hepatic stellate cells *in vitro*. Biochem J 2004; 384(Pt 1): 149-57.

[134] Zhou Y, Zheng S, Lin J, Zhang QJ, Chen A. The interruption of the PDGF and EGF signaling pathways by curcumin stimulates gene expression of PPARgamma in rat activated hepatic stellate cell *in vitro*. Lab Invest 2007; 87(5): 488-98.

[135] Nasmyth K. Viewpoint: putting the cell cycle in order. Science 1996; 274(5293): 1643-5.

[136] Xu J, Fu Y, Chen A. Activation of peroxisome proliferator-activated receptor-gamma contributes to the inhibitory effects of curcumin on rat hepatic stellate cell growth. Am J Physiol Gastrointest Liver Physiol 2003; 285(1): G20-30.

[137] Yavrom S, Chen L, Xiong S, Wang J, Rippe RA, Tsukamoto H. Peroxisome proliferator-activated receptor gamma suppresses proximal alpha1(I) collagen promoter *via* inhibition of p300-facilitated NF-I binding to DNA in hepatic stellate cells. J Biol Chem 2005; 280(49): 40650-9.

[138] Nolte RT, Wisely GB, Westin S, *et al.* Ligand binding and co-activator assembly of the peroxisome proliferator-activated receptor-gamma. Nature 1998; 395(6698): 137-43.

[139] Guo YT, Leng XS, Li T, *et al.* Effect of ligand of peroxisome proliferator-activated receptor gamma on the biological characters of hepatic stellate cells. World J Gastroenterol 2005; 11(30): 4735-9.

[140] Lee JI, Paik YH, Lee KS, *et al.* A peroxisome-proliferator activated receptor-gamma ligand could regulate the expression of leptin receptor on human hepatic stellate cells. Histochem Cell Biol 2007; 127(5): 495-502.

[141] Planagumà A, Clària J, Miquel R, *et al.* The selective cyclooxygenase-2 inhibitor SC-236 reduces liver fibrosis by mechanisms involving non-parenchymal cell apoptosis and PPARgamma activation. FASEB J 2005; 19(9): 1120-2.

[142] Kim KY, Rhim T, Choi I, Kim SS. N-acetylcysteine induces cell cycle arrest in hepatic stellate cells through its reducing activity. J Biol Chem 2001; 276(44): 40591-8.

[143] Kawada N, Seki S, Inoue M, Kuroki T. Effect of antioxidants, resveratrol, quercetin, and N-acetylcysteine, on the functions of cultured rat hepatic stellate cells and Kupffer cells. Hepatology 1998; 27(5): 1265-74.

[144] Mazo DF, de Oliveira MG, Pereira IV, *et al.* S-nitroso-N-acetylcysteine attenuates liver fibrosis in experimental nonalcoholic steatohepatitis. Drug Des Devel Ther 2013; 7: 553-63.

[145] Stratton R, Shiwen X, Martini G, *et al.* Iloprost suppresses connective tissue growth factor production in fibroblasts and in the skin of scleroderma patients. J Clin Invest 2001; 108(2): 241-50.

[146] Moncada S, Gryglewski R, Bunting S, Vane JR. An enzyme isolated from arteries transforms prostaglandin endoperoxides to an unstable substance that inhibits platelet aggregation. Nature 1976; 263(5579): 663-5.

[147] Zardi EM, Zardi DM, Cacciapaglia F, *et al.* Endothelial dysfunction and activation as an expression of disease: role of prostacyclin analogs. Int Immunopharmacol 2005; 5(3): 437-59.

[148] Okboy N, Yegen C, Aktan AO, *et al.* The effect of iloprost and NDGA in ischemia reperfusion injury in rat liver. Prostaglandins Leukot Essent Fatty Acids 1992; 47(4): 291-5.

[149] Tanaka J, Malchesky PS, Omokawa S, *et al.* Effects of prostaglandin I2, superoxide dismutase, and catalase on ischemia-reperfusion injury in liver transplantation. ASAIO Trans 1990; 36(3): M600-3.

[150] Hui AY, Cheng AS, Chan HL, *et al.* Effect of prostaglandin E2 and prostaglandin I2 on PDGF-induced proliferation of LI90, a human hepatic stellate cell line. Prostaglandins Leukot Essent Fatty Acids 2004; 71(5): 329-33.

[151] Kawada N, Tran-Thi TA, Klein H, Decker K. The contraction of hepatic stellate (Ito) cells stimulated with vasoactive substances. Possible involvement of endothelin 1 and nitric oxide in the regulation of the sinusoidal tonus. Eur J Biochem 1993; 213(2): 815-23.

[152] Mallat A, Fouassier L, Préaux AM, Mavier P, Lotersztajn S. Antiproliferative effects of ET-1 in human liver Ito cells: an ETB- and a cyclic AMP-mediated pathway. J Cardiovasc Pharmacol 1995; 26 Suppl 3: S132-4.

[153] Mallat A, Préaux AM, Serradeil-Le Gal C, *et al.* Growth inhibitory properties of endothelin-1 in activated human hepatic stellate cells: a cyclic adenosine monophosphate-mediated pathway. Inhibition of both extracellular signal-regulated kinase and c-Jun kinase and upregulation of endothelin B receptors. J Clin Invest 1996; 98(12): 2771-8.

[154] Belmonte N, Phillips BW, Massiera F, *et al.* Activation of extracellular signal-regulated kinases and CREB/ATF-1 mediate the expression of CCAAT/enhancer binding proteins beta and -delta in preadipocytes. Mol Endocrinol 2001; 15(11): 2037-49.

[155] Brindle P, Nakajima T, Montminy M. Multiple protein kinase A-regulated events are required for transcriptional induction by cAMP. Proc Natl Acad Sci USA 1995; 92(23): 10521-5

[156] Sheng M, Thompson MA, Greenberg ME. CREB: a Ca(2+)-regulated transcription factor phosphorylated by calmodulin-dependent kinases. Science. 1991; 252(5011): 1427-30.

[157] Liu F, Thompson MA, Wagner S, Greenberg ME, Green MR. Activating transcription factor-1 can mediate Ca(2+)- and cAMP-inducible transcriptional activation. J Biol Chem 1993; 268(9): 6714-20.

[158] Houglum K, Lee KS, Chojkier M. Proliferation of hepatic stellate cells is inhibited by phosphorylation of CREB on serine 133. J Clin Invest 1997; 99(6): 1322-8.

[159] Xu Q, Sakai K, Suzuki Y, Tambo C, Sakai Y, Matsumoto K. Suppression of fibrogenic gene expression and liver fibrosis using a synthetic prostacyclin agonist. Biomed Res 2013; 34(5): 241-50.

[160] Yoshidome H, Kato A, Edwards MJ, Lentsch AB. Interleukin-10 suppresses hepatic ischemia/reperfusion injury in mice: implications of a central role for nuclear factor kappaB. Hepatology 1999; 30(1): 203-8.

[161] Zhang LJ, Wang XZ. Interleukin-10 and chronic liver disease. World J Gastroenterol 2006; 12(11): 1681-5.

[162] Safadi R, Ohta M, Alvarez CE, *et al.* Immune stimulation of hepatic fibrogenesis by CD8 cells and attenuation by transgenic interleukin-10 from hepatocytes. Gastroenterology 2004; 127(3): 870-82.

[163] Rachmawati H, Beljaars L, Reker-Smit C, *et al.* Pharmacokinetic and biodistribution profile of recombinant human interleukin-10 following intravenous administration in rats with extensive liver fibrosis. Pharm Res 2004; 21(11): 2072-8.

[164] Mattos A, de Jager-Krikken A, de Haan M, Beljaars L, Poelstra K. PEGylation of interleukin-10 improves the pharmacokinetic profile and enhances the antifibrotic effectivity in CCl_4-induced fibrogenesis in mice. J Control Release 2012; 162(1): 84-91

[165] Ulloa L, Doody J, Massagué J. Inhibition of transforming growth factor-beta/SMAD signalling by the interferon-gamma/STAT pathway. Nature 1999; 397(6721): 710-3.

[166] Nakao A, Fujii M, Matsumura R, *et al.* Transient gene transfer and expression of Smad7 prevents bleomycin-induced lung fibrosis in mice. J Clin Invest 1999; 104(1): 5-11.

[167] Weng H, Mertens PR, Gressner AM, Dooley S. IFN-gamma abrogates profibrogenic TGF-beta signaling in liver by targeting expression of inhibitory and receptor Smads. J Hepatol 2007; 46(2): 295-303.

[168] Li Q, Yan Z, Li F, Lu W, Wang J, Guo C. The improving effects on hepatic fibrosis of interferon-γ liposomes targeted to hepatic stellate cells. Nanotechnology 2012; 23(26): 265101.

[169] Inman GJ, Nicolás FJ, Callahan JF, *et al.* SB-431542 is a potent and specific inhibitor of transforming growth factor-beta superfamily type I activin receptor-like kinase (ALK) receptors ALK4, ALK5, and ALK7. Mol Pharmacol 2002; 62(1): 65-74.

[170] DaCosta Byfield S, Major C, Laping NJ, Roberts AB. SB-505124 is a selective inhibitor of transforming growth factor-beta type I receptors ALK4, ALK5, and ALK7. Mol Pharmacol 2004; 65(3): 744-52.

[171] Hui AY, Dannenberg AJ, Sung JJ, *et al.* Prostaglandin E2 inhibits transforming growth factor beta 1-mediated induction of collagen alpha 1(I) in hepatic stellate cells. J Hepatol 2004; 41(2): 251-8.

[172] Navarro-Partida J, Martinez-Rizo AB, Gonzalez-Cuevas J, Arrevillaga-Boni G, Ortiz-Navarrete V, Armendariz-Borunda J. Pirfenidone restricts Th2 differentiation *in vitro* and limits Th2 response in experimental liver fibrosis. Eur J Pharmacol 2012; 678(1-3): 71-7.

[173] Zou WL, Yang Z, Zang YJ, Li DJ, Liang ZP, Shen ZY. Inhibitory effects of prostaglandin E1 on activation of hepatic stellate cells in rabbits with schistosomiasis. Hepatobiliary Pancreat Dis Int 2007; 6(2): 176-81.

[174] von Mehren M. Targeted therapy with imatinib: hits and misses? J Clin Oncol 2005; 23(1): 8-10.
[175] Kinnman N, Goria O, Wendum D, *et al.* Hepatic stellate cell proliferation is an early platelet-derived growth factor-mediated cellular event in rat cholestatic liver injury. Lab Invest 2001; 81(12): 1709-16.
[176] Kerkelä R, Grazette L, Yacobi R, *et al.* Cardiotoxicity of the cancer therapeutic agent imatinib mesylate. Nat Med 2006; 12(8): 908-16.
[177] Kim Y, Fiel MI, Albanis E, *et al.* Anti-fibrotic activity and enhanced interleukin-6 production by hepatic stellate cells in response to imatinib mesylate. Liver Int 2012; 32(6): 1008-17.
[178] Gonzalo T, Beljaars L, van de Bovenkamp M, *et al.* Local inhibition of liver fibrosis by specific delivery of a platelet-derived growth factor kinase inhibitor to hepatic stellate cells. J Pharmacol Exp Ther 2007; 321(3): 856-65.
[179] Samuel CS, Hewitson TD, Unemori EN, Tang ML. Drugs of the future: the hormone relaxin. Cell Mol Life Sci 2007; 64(12): 1539-57.
[180] Bennett RG, Mahan KJ, Gentry-Nielsen MJ, Tuma DJ. Relaxin receptor expression in hepatic stellate cells and in cirrhotic rat liver tissue. Ann N Y Acad Sci 2005; 1041: 185-9.
[181] Williams EJ, Benyon RC, Trim N, *et al.* Relaxin inhibits effective collagen deposition by cultured hepatic stellate cells and decreases rat liver fibrosis *in vivo*. Gut 2001; 49(4): 577-83.
[182] Bennett RG, Heimann DG, Singh S, Simpson RL, Tuma DJ. Relaxin decreases the severity of established hepatic fibrosis in mice. Liver Int. 2013 Jun 16.
[183] Fallowfield JA, Hayden AL, Snowdon VK, *et al.* Relaxin modulates human and rat hepatic myofibroblast function and ameliorates portal hypertension *in vivo*. Hepatology 2014; 59(4): 1492-504.
[184] Iredale JP, Benyon RC, Pickering J, *et al.* Mechanisms of spontaneous resolution of rat liver fibrosis. Hepatic stellate cell apoptosis and reduced hepatic expression of metalloproteinase inhibitors. J Clin Invest 1998; 102(3): 538-49.
[185] Oakley F, Meso M, Iredale JP, *et al.* Inhibition of inhibitor of kappaB kinases stimulates hepatic stellate cell apoptosis and accelerated recovery from rat liver fibrosis. Gastroenterology 2005; 128(1): 108-20.
[186] Sanz S, Pucilowska JB, Liu S, *et al.* Expression of insulin-like growth factor I by activated hepatic stellate cells reduces fibrogenesis and enhances regeneration after liver injury. Gut 2005; 54(1): 134-41.
[187] Sobrevals L, Rodriguez C, Romero-Trevejo JL, *et al.* Insulin-like growth factor I gene transfer to cirrhotic liver induces fibrolysis and reduces fibrogenesis leading to cirrhosis reversion in rats. Hepatology 2010; 51(3): 912-21.
[188] Elsharkawy AM, Wright MC, Hay RT, *et al.* Persistent activation of nuclear factor-kappaB in cultured rat hepatic stellate cells involves the induction of potentially novel Rel-like factors and prolonged changes in the expression of IkappaB family proteins. Hepatology1999; 30(3): 761-9
[189] Wright MC, Issa R, Smart DE, *et al.* Gliotoxin stimulates the apoptosis of human and rat hepatic stellate cells and enhances the resolution of liver fibrosis in rats. Gastroenterology 2001; 121(3): 685-98.
[190] Kweon YO, Paik YH, Schnabl B, Qian T, Lemasters JJ, Brenner DA. Gliotoxin-mediated apoptosis of activated human hepatic stellate cells. J Hepatol 2003; 39(1): 38-46.
[191] Hagens WI, Olinga P, Meijer DK, Groothuis GM, Beljaars L, Poelstra K. Gliotoxin non-selectively induces apoptosis in fibrotic and normal livers. Liver Int 2006; 26(2): 232-9
[192] Friedman SL. The answer: angiotensin II. The question: what do inflammation, oxidant stress and fibrogenesis have in common? J Hepatol 2004; 40(6): 1050-2.
[193] Wei HS, Lu HM, Li DG, *et al.* The regulatory role of AT 1 receptor on activated HSCs in hepatic fibrogenesis: effects of RAS inhibitors on hepatic fibrosis induced by CCl(4). World J Gastroenterol 2000; 6(6): 824-828.
[194] Kurikawa N, Suga M, Kuroda S, Yamada K, Ishikawa H. An angiotensin II type 1 receptor antagonist, olmesartan medoxomil, improves experimental liver fibrosis by suppression of proliferation and collagen synthesis in activated hepatic stellate cells. Br J Pharmacol 2003; 139(6): 1085-94
[195] Yokohama S, Tokusashi Y, Nakamura K, *et al.* Inhibitory effect of angiotensin II receptor antagonist on hepatic stellate cell activation in non-alcoholic steatohepatitis. World J Gastroenterol 2006; 12(2): 322-6.
[196] Beljaars L, Meijer DK, Poelstra K. Targeting hepatic stellate cells for cell-specific treatment of liver fibrosis. Front Biosci 2002; 7: e214-22.
[197] Tasci I, Mas MR, Vural SA, *et al.* Pegylated interferon-alpha plus taurine in treatment of rat liver fibrosis. World J Gastroenterol 2007; 13(23): 3237-44.

[198] Abdel-Salam OM, Sleem AA, Morsy FA. Effects of biphenyldimethyl-dicarboxylate administration alone or combined with silymarin in the CCL4 model of liver fibrosis in rats. ScientificWorldJournal 2007; 7: 1242-55.
[199] Fraser R, Le Couteur DG, Warren A, Cogger VC, Bertolino P, Smith M. The liver sieve and gene therapy. Blood 2003; 101(8): 3338;
[200] Prieto J, Qian C, Hernandez-Alcoceba R, *et al.* Gene therapy of liver diseases. Expert Opin Biol Ther 2004; 4(7): 1073-91.
[201] Rudolph KL, Chang S, Millard M, Schreiber-Agus N, DePinho RA. Inhibition of experimental liver cirrhosis in mice by telomerase gene delivery. Science 2000; 287(5456): 1253-8.
[202] Zhang LH, Pan JP, Yao HP, *et al.* Intrasplenic transplantation of IL-18 gene-modified hepatocytes: an effective approach to reverse hepatic fibrosis in schistosomiasis through induction of dominant Th1 response. Gene Ther 2001; 8(17): 1333-42
[203] Hung KS, Lee TH, Chou WY, *et al.* Interleukin-10 gene therapy reverses thioacetamide-induced liver fibrosis in mice. Biochem Biophys Res Commun 2005; 336(1): 324-31.
[204] Chen M, Wang GJ, Diao Y, *et al.* Adeno-associated virus mediated interferon-gamma inhibits the progression of hepatic fibrosis *in vitro* and *in vivo*. World J Gastroenterol 2005; 11(26): 4045-51.
[205] Chen YX, Huang YH, Zheng WD, Chen ZX, Zhang LJ, Wang XZ. Interleukin-10 gene modification attenuates hepatocyte activation of rat hepatic stellate cells *in vitro*. Mol Med Rep 2013; 7(2): 371-8.
[206] Dooley S, Hamzavi J, Breitkopf K, *et al.* Smad7 prevents activation of hepatic stellate cells and liver fibrosis in rats. Gastroenterology 2003; 125(1): 178-91.
[207] Xu XB, Leng XS, He ZP, *et al.* Inhibitory effect of retroviral vector containing anti-sense Smad4 gene on Ito cell line, LI90. Chin Med J (Engl) 2004; 117(8): 1170-7.
[208] Garcia-Bañuelos J, Siller-Lopez F, Miranda A, Aguilar LK, Aguilar-Cordova E, Armendariz-Borunda J. Cirrhotic rat livers with extensive fibrosis can be safely transduced with clinical-grade adenoviral vectors. Evidence of cirrhosis reversion. Gene Ther 2002; 9(2): 127-34.
[209] Siller-López F, Sandoval A, Salgado S, *et al.* Treatment with human metalloproteinase-8 gene delivery ameliorates experimental rat liver cirrhosis. Gastroenterology 2004; 126(4): 1122-33;
[210] Roderfeld M, Weiskirchen R, Wagner S, *et al.* Inhibition of hepatic fibrogenesis by matrix metalloproteinase-9 mutants in mice. FASEB J 2006; 20(3): 444-54
[211] Bauer I, Wanner GA, Rensing H, *et al.* Expression pattern of heme oxygenase isoenzymes 1 and 2 in normal and stress-exposed rat liver. Hepatology 1998; 27(3): 829-38.
[212] Li L, Grenard P, Nhieu JT, *et al.* Heme oxygenase-1 is an antifibrogenic protein in human hepatic myofibroblasts. Gastroenterology 2003; 125(2): 460-9.
[213] Tsui TY, Lau CK, Ma J, *et al.* Adeno-associated virus-mediated heme oxygenase-1 gene transfer suppresses the progression of micronodular cirrhosis in rats. World J Gastroenterol 2006; 12(13): 2016-23.
[214] Prosser CC, Yen RD, Wu J. Molecular therapy for hepatic injury and fibrosis: where are we? World J Gastroenterol 2006; 12(4): 509-15.
[215] Pinzani M, Rombouts K. Liver fibrosis: from the bench to clinical targets. Dig Liver Dis 2004; 36(4): 231-42.
[216] Bataller R, Brenner DA. Liver fibrosis. J Clin Invest 2005; 115(2): 209-18.

CHAPTER 11

Biologically Active Chitosan Systems for Regenerative Engineering

Tao Jiang[1,2,3], Sangamesh G. Kumbar[2,3,5,6,7], Lakshmi S. Nair[2,3,5,6,7] and Cato T. Laurencin[2,3,4,5,6,7,*]

[1]Department of Medicine, University of Connecticut Health Center, Farmington, CT 06030, USA; [2]Institute for Regenerative Engineering, University of Connecticut Health Center, Farmington, CT 06030, USA; [3]The Raymond and Beverly Sackler Center for Biomedical, Biological, Physical and Engineering Sciences, University of Connecticut Health Center, Farmington, CT 06030, USA; [4]Department of Chemical and Biomolecular Engineering, University of Connecticut, Storrs, CT 06269, USA; [5]Department of Materials Science and Engineering, University of Connecticut, Storrs, CT 06269, USA; [6]Department of Biomedical Engineering, University of Connecticut, Storrs, CT 06269, USA and [7]Department of Orthopaedic Surgery, University of Connecticut Health Center, Farmington, CT 06030, USA

Abstract: Regenerative engineering has recently emerged as the next stage in the evolution of tissue engineering. Advanced materials science is a key component in regenerative engineering. Biologically active materials largely affect cell fate processes and tissue morphogenesis through chemical and biological cues carried by the materials. By physical adsorption of biomolecules on scaffold surface, physical entrapment of biomolecules in polymer microspheres or hydrogels, and chemical immobilization of oligopeptides or proteins on biomaterials, biologically active biomaterials and scaffolds can be derived. These bioactive systems show great potential in regenerative engineering in rendering bioactivity and/or specificity to scaffolds. This review highlights some of the biologically active chitosan systems for regenerative engineering and the associated strategies to develop such bioactive chitosan systems.

Keywords: Advanced materials, bioactive, biocompatibility, biodegradability, biomaterials, biomolecules, chitosan, drug delivery, functionality, growth factor, hydrogel, immobilization, nanoscale, nanotechnology, peptide, polymer, regenerative engineering, scaffolds, surface modification, tissue engineering.

INTRODUCTION

Tissue loss or end-stage organ failure of human body is a severe human health

***Corresponding author Cato T. Laurencin:** Institute for Regenerative Engineering, University of Connecticut Health Center, Farmington, CT 06030, USA; Tel: +1-860-679-6544; Fax: +1-860-679-1533;
E-mail: laurencin@uchc.edu

Atta-ur-Rahman, Allen B. Reitz, M. Iqbal Choudhary and Jizhou Wang (Eds.)

care problem and a major cause of death worldwide. Unlike certain lower species, adult human body lacks the ability to self-repair or regenerate badly damaged or dysfunctional complex body parts due to injury or chronic diseases. Reconstructive surgery and organ transplantation for the permanent replacement of diseased or failed organs, although having saved innumerable lives and improved life quality, suffer from a number of major limitations [1]. In particular in the case of organ transplantation, the shortage of donor tissues or organs is a tremendous obstacle to this therapy and poses an increasing demand with an increase in aging population (Fig. **1**) [2]. Tissue engineering thus was advanced in the late 1980s as an alternative medical therapy aiming to reestablish tissue and organ regeneration *via* a naturally reparative pathway to therapeutically treat diseases. This emerging discipline utilizes techniques from engineering, material science, and stem cell biology with the goal of understanding how cells organize, regenerate and finally form the complex three dimensional structures of tissues and organs [3-6].

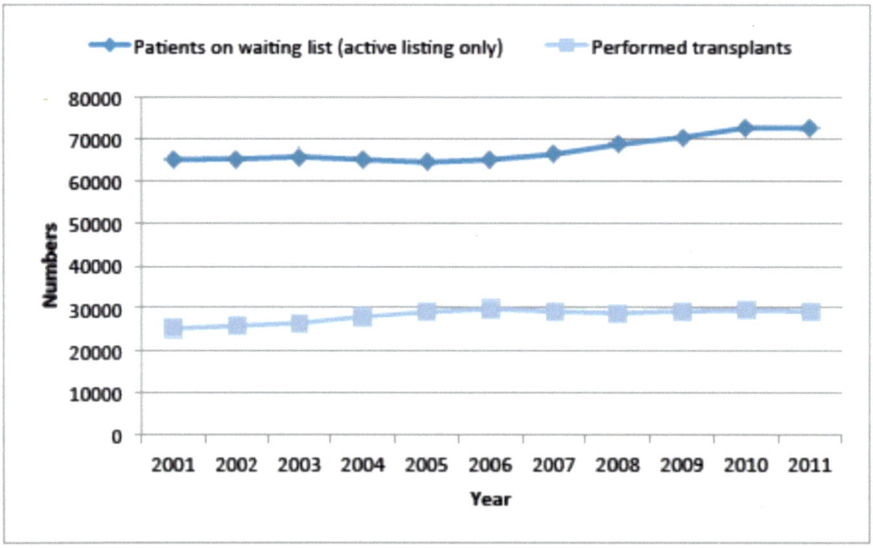

Figure 1: Numbers of the performed organ transplantation and patients actively awaiting an organ. The overwhelming number of patients necessitates alternative therapies such as regenerative engineering to the current organ transplantation. Graph is plotted based on the data presented in Ref. 2.

Tissue engineering, as defined by Laurencin *et al.* applies biological, chemical, and engineering principles towards the repair, restoration or regeneration of living tissues using biomaterials, cells, and factors alone or in combination [7]. In this therapeutic approach, biomaterials, usually in the form of three dimensional porous scaffolds, play multiple significant roles to provide structural maintenance of the shape of a defect, and void volume for vascularization, and also serve as an initial temporary extracellular matrix (ECM) for cell adhesion, proliferation, differentiation and maturation. In addition, scaffolds can act as a delivery vehicle for loaded bioactive molecules, growth factors, and cells to the defect site for tissue morphogenesis and defect healing [8-10]. Ideally, tissue engineering scaffolds degrade in a controllable manner so as to match the natural ECM formation, gradually transfer the structural and functional roles to the newly formed tissue, and finally get resorbed and metabolized. Cells, often tissue-specific cell types, progenitors or stem cells that are capable of differentiating into multiple phenotypes, are critical for a tissue-engineered system to perform and succeed in many cases to compensate for a deficiency of normally functional cell population [4, 9, 11]. Growth factors, typically acting as signaling molecules *via* binding to specific transmembrane receptors on the target cells, regulate a number of cell behaviors such as cell adhesion, proliferation, migration and differentiation [12, 13]. They play an essential role in tissue morphogenesis and neo-tissue formation in tissue engineering. Thus growth factors are important components in tissue engineering systems that can stimulate cells to alter cellular behaviors. In addition to the types of growth factors, the spatial distribution, concentration gradient of the growth factors on scaffolds, as well as the release kinetics are crucial for them to exert their biological activity on the target cells [14, 15]. Although it is not indispensable that all three components are present in any tissue engineering approach, it is undoubted that a combinational approach may augment the therapeutic effect.

In recent years, the importance of convergence research integrating tissue engineering, advanced materials, and stem cells with seemingly disparate disciplines of physical sciences and developmental biology has been recognized, and a new research direction has been proposed and termed as "regenerative engineering" [16]. While tissue engineering focuses on the ability to repair a specific tissue, regenerative engineering aims to regenerate or reconstruct complex tissues and biological systems, such as the whole human limb. Advanced materials science, nanotechnology, stem cell science, and developmental biology are important fields within this new multidisciplinary paradigm to engineer complex tissues [16, 17].

Advanced materials science is an essential component in regenerative engineering. Traditionally, biomaterials were designed to be biocompatible so that they do not elicit severe immune response and biodegradable so that the macromolecular chains can be broken down and the small degradation products are non-toxic and can be absorbed by a biological system [18]. More recently, with the advances in the biomaterials field, the development of localized/targeted drug delivery, and especially the understanding and investigation of cell-biomaterial interactions, much focus has been put on to develop advanced bioactive biomaterial systems *via* two approaches. One is the development of biomaterials as a delivery vehicle for growth factors, genes and cells. This approach has incorporated drug delivery concept into tissue engineering in order to achieve accelerated tissue regeneration. For example, single or multiple growth factors were loaded onto various tissue-engineered constructs for the repair of skin, blood vessel, nerve, cartilage and bone [19-21]. The other is to develop chemically modified bioactive biomaterials for use as extracellular matrices for tissue regeneration [22, 23]. Biomaterial surface modifications by immobilization of a variety of bioactive agents such as oligopeptides, oligosaccharides, or growth factors provide biomaterials required bioactivity. These biomaterial surfaces that exhibit recognition and specificity may trigger specific biological reactions and thus precisely control cell behavior [24, 25].

Another key concept in regenerative engineering is the nanotechnology. The application of nanoscale materials is crucial in various approaches to regenerating human tissues for several reasons. First, many natural human tissues themselves are composed of finely organized ECM in the nanoscale. For example, natural bone itself is a nanocomposite material composed mainly of hydroxyapatite nanocrystallites in an organic collagen-rich matrix [26]. The individual, helical collagen chains are 10nm in length and self-assemble into organized collagen fibers, measuring approximately 500nm in length. Hydroxyapatite, the mineralized inorganic component of bone, exists as plate-like nanocrystals, measuring 20-80nm in length. The triple helical structure of collagen gives bone a structural framework with high tensile strength and flexibility, while the nanocrystalline hydroxyapatite provides rigidity and high compressive strength [27]. Second, nanotopographical features, when interacting with cells, have been proven to control and regulate cellular processes as important signaling modalities [28, 29]. Nanoscale materials can be fabricated into different forms such as nanoparticles, nanofibers, nanospheres, nanotubes, nanogels, nanocapsules, and surfaces with nanotopography. These nanostructures can be prepared using a variety of biofabrication methods such as electrospinning, spray drying, phase separation, molecule self-assembly, chemical vapor deposition, and nano-imprinting.

Many biomaterials have been investigated for tissue engineering applications. These include a wide range of materials from metals to ceramics and polymers, from bio-inert polymers to biodegradable ones, from synthetic biomaterials to natural ones [30]. Among them chitosan is a naturally occurring unique biopolymer that received much research interest recently [31, 32]. Chitosan, the N-deacetylated derivative of chitin which forms the exoskeleton of crustacean shells, has structural similarity to glycosaminoglycan. Chitosan shows excellent biocompatibility [33], controllable degradation rate dependent on its degree of deacetylation (DDA) [34], antimicrobial and antifungal activities [35], making it highly capable of being used as a tissue engineering substrate material. In addition, chitosan macromolecular chains carry numerous reactive sites in the form of hydroxyl and amino groups, which can further be modified with biological factors to present functional domains that selectively regulate cell growth in a controlled manner.

This review focuses on the use of advanced biologically active chitosan systems for regenerative engineering. Various methods and efforts to introduce bioactive functionalities to attain cell specificity in tissue engineering and regenerative medicines will be discussed. These include the development of chitosan scaffolds simultaneously serving as drug delivery vehicles for tissue engineering, and immobilization of biomolecules on chitosan attained either by physical adsorption or chemical immobilization *via* its reactive sites. Furthermore, the application of finely designed micro- and nanoscale chitosan-based structures for regenerative engineering is also discussed.

STRUCTURE AND CHEMISTRY OF CHITOSAN

Chitosan occurs as a copolymer of N-acetyl-glucosamine and N-glucosamine units randomly or block distributed throughout the biopolymer chain (Fig. **2a**). When the number of N-glucosamine units (DDA) is higher than 50%, the biopolymer is termed chitosan [36]. Chitosan is soluble in dilute acidic solutions such as acetic acid or hydrochloric acid, in which chitosan molecular chains are protonated through the amino groups (Fig. **2b**). Chitosan aqueous solution can readily be fabricated into different forms such as films, tubes, sponges and other porous structures for various tissue engineering and drug delivery applications [37, 38]. Moreover, the cationic nature of protonated chitosan was found to be of significant biological importance in a number of situations. It has been demonstrated that the cationic polymer chains can complex with certain growth factors and negatively charged DNA and thereby protect them from heat inactivation, proteolysis, and nuclease degradation [39, 40].

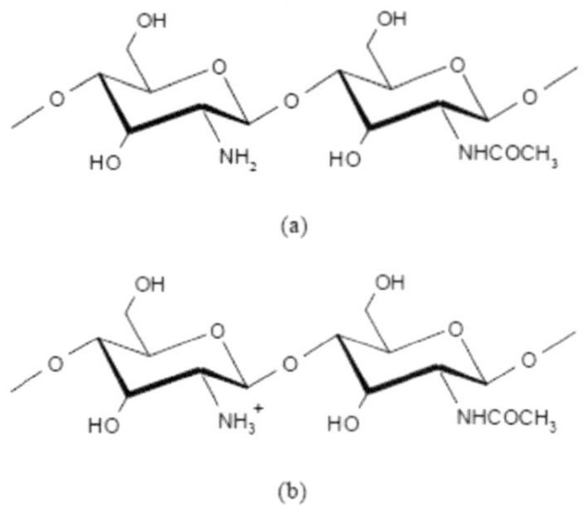

Figure 2: Molecular structures of (a) chitosan free base and (b) protonated chitosan.

Another interesting feature of chitosan is the presence of abundant reactive hydroxyl and amino functionalities that provides opportunities for multiple physical and chemical modifications. Thus desired biological activities can be introduced to chitosan to perform specific functions. Three different approaches have been investigated to introduce beneficial bioactivity to chitosan by using the reactive hydroxyl and amine groups of chitosan. The first approach is based on ionic complexation between positively charged chitosan and negatively charged bioactive agent of interest. For example, negatively charged heparin molecule can be introduced to chitosan systems by ionically interacting with positively charged chitosan molecules [41]. The second approach is based on physical adsorption or entrapment of biomolecules on chemically modified chitosan surface or within chitosan substrates. This strategy sometimes involves chemical modification of chitosan to alter its physicochemical properties through introduction of certain structural moieties to chitosan polymer chains. Chitosan can be converted into water-soluble form by introducing acidic functionalities, for instance modifying with lactobionic acid. The modified chitosan in combination with heparin hydrogels were used to immobilize fibroblast growth factor-2 (FGF-2) to attain neovascularization and fibrous tissue formation in a mouse model [42]. Chitosan was further modified with multiple functionalities such as lactose and azide *via* sequential condensation reaction at the chitosan amine sites. This resulting polymer can be photocrosslinked and avoids the use of other toxic crosslinking agents during scaffold fabrication and biomolecules incorporation. These chitosan derivatives showed great potential in incorporating growth factors under mild conditions for a variety of biomedical applications such as for wound healing and controlled angiogenesis [43]. Chemical modification of featured biomolecules on chitosan renders more diversity to this biomaterial. Beneficial bioactivity can be introduced through the incorporation of covalently bound oligopeptides to chitosan. Oligopeptide sequences of interest can be directly bound to chitosan through the reaction between carboxyl groups on the peptide and amine groups on chitosan to form imide bonding [44]. This reaction often involves the use of carbodiimide such as 1-ethyl-3-(3-dimethylaminopropyl) carbodiimide (EDC) and N-hydroxysuccinimide (NHS) to activate the carboxyl groups on peptides and decrease side reactions. Scaffold architecture, surface topography, biochemical and mechanical cues have been shown to significantly improve cellular events and *in vivo* tissue regeneration. Specifically, electrospun nanofiber matrices have gained tremendous interests due to their intrinsic structural resemblance to native tissue ECM. Electrospun nanofiber matrices of poly(ε-caprolactone)/chitosan (PCL/chitosan) blends were immobilized with type I collagen *via* EDC coupling to regulate the stem cell differentiation into osteogenic lineage [45]. The amine functionalities of chitosan were utilized to facilitate this reaction. Many synthetic polymers without proper functionalities for such modifications can benefit by this approach. For instance, EDC collagen coupling resulted in 0.120±0.016 micro g of collagen immobilization onto a 1 cm^2 area of the PCL/chitosan nanofibers, which was 2.6-folds higher than the amount of collagen that can be retained by physical adsorption. These collagen-functionalized matrices significantly improved adhesion, spreading, proliferation and osteogenic differentiation of bone marrow derived stromal cells as compared to PCL nanofiber matrices. Such an approach can extremely benefit synthetic polymers based nanostructures to improve their bioactivity for regenerative engineering [45]. Another approach is to prepare a reactive chitosan intermediate, for example thiolated chitosan, carboxylated chitosan, or maleimide activated chitosan through the amine groups [46-50]. These active chitosan intermediates can further react with sulphydryl or amino groups of some peptide sequences or whole protein molecules to form covalent immobilization. Other chemical modification methods of chitosan through its hydroxyl groups for biological purposes have also been reported. For example, carboxylation can also be achieved through the hydroxyl groups on chitosan followed by imide bonding formation with peptides [51]. In addition, photochemical reactions can occur between the hydroxyl groups on chitosan and azido derivative of peptides to form covalent bonding [52, 53]. All the aforementioned modification methods take advantage of the amine and hydroxyl functionalities of chitosan, to develop biologically active chitosans with cell recognition and specificities for various tissue engineering and regenerative medicine applications. Table **1** summarizes some major chemical modifications reported in the literature to immobilize various biomolecules on chitosan in an effort to develop biologically active chitosan derivatives.

Table 1: Selected chemistry towards the biomolecule immobilization on chitosan backbone

Substrate function *	Coupling agent	Active intermediate	Major reacting groups on peptides or proteins
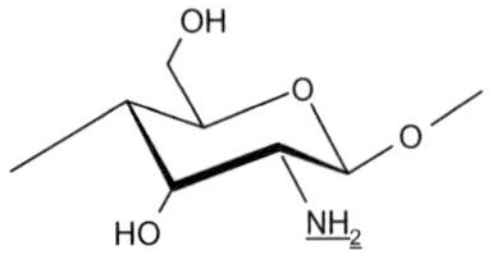	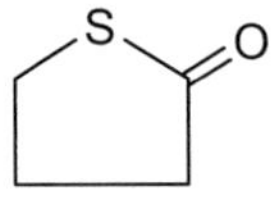	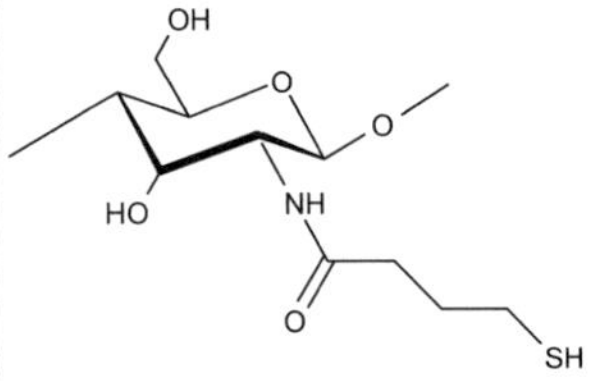	– SH
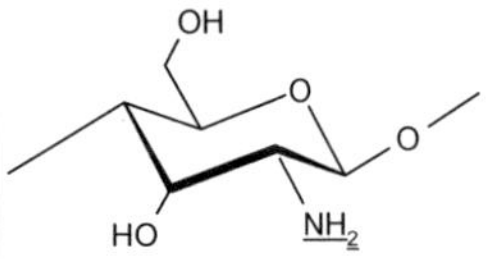	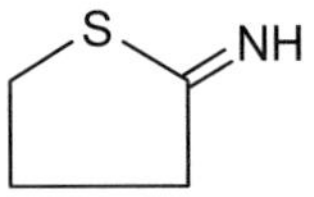	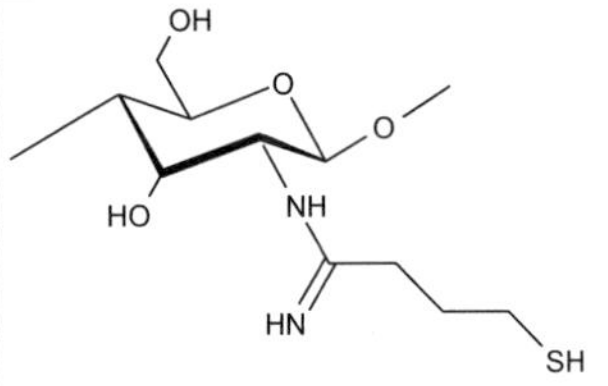	– SH
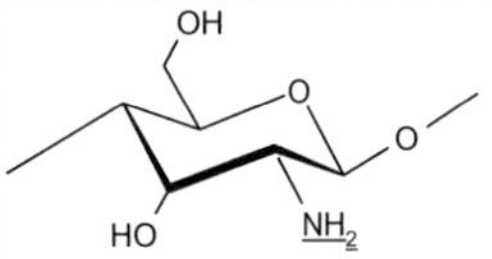	HS-CH_2COOH	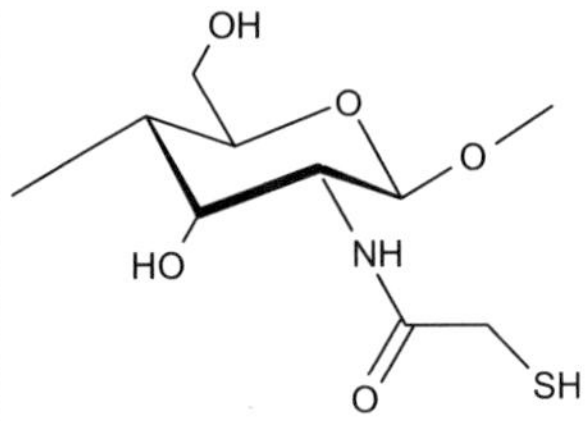	– SH
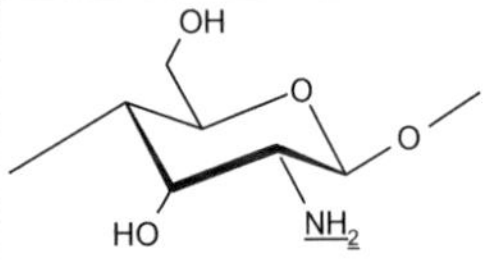	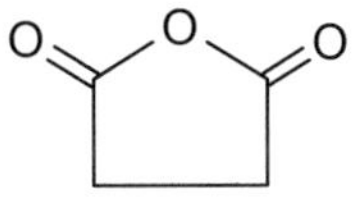	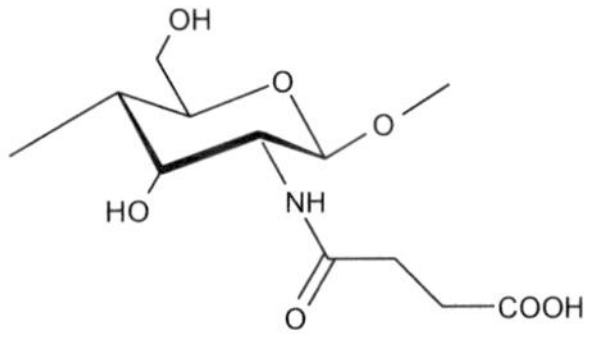	– NH_2
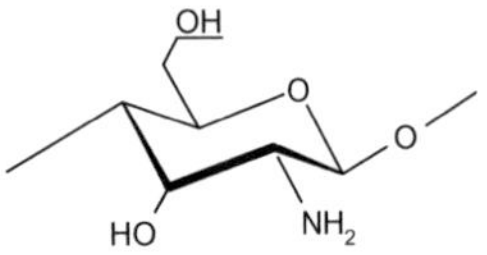	Cl-CH_2COOH	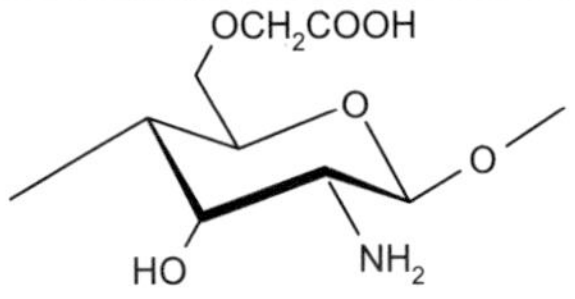	– NH_2
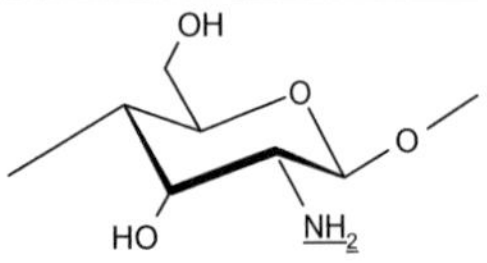	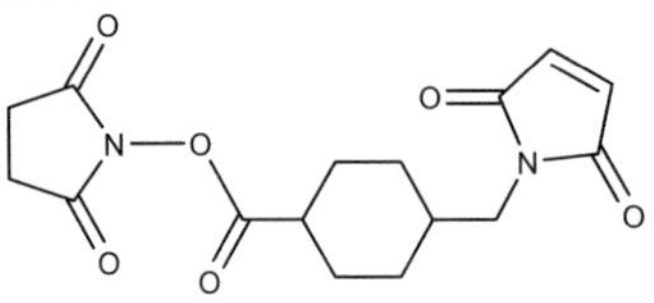	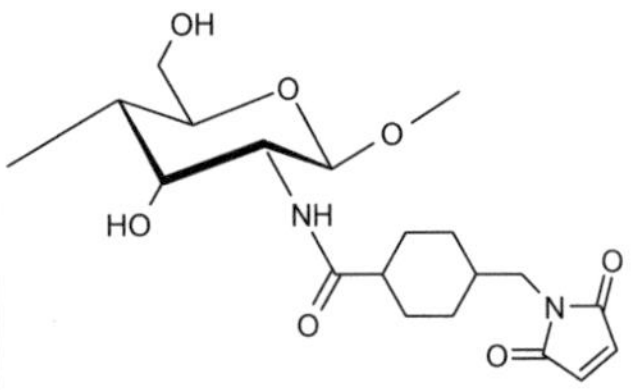	– SH

*: Underlined group represents the reacting group.

CHITOSAN DELIVERY SYSTEMS

Biomaterials, cells and growth factors are three key components in tissue engineering. To achieve rapid tissue regeneration, biomaterials, in the form of tissue specific scaffolds should be utilized in combination with cells and/or growth factors to augment the regenerative potential of a tissue-engineered system. In such systems, the loaded tissue-specific cells with a well-defined three-dimensional distribution on scaffolds, proliferate, differentiate, and function to induce tissue regeneration therein. Growth factors play another crucial role in tissue morphogenesis and regeneration. Direct injection of growth factors for tissue regeneration, although is straightforward and may succeed to some extent, has significant drawbacks. They may quickly diffuse away from the injection site, and the short half lives of growth factors requires repeated and large amount injections to achieve a therapeutic effect [54]. In addition, the un-localized delivery and distribution of these factors may lead to undesirable systemic effect and toxicity [14]. Therefore, controlled growth factor delivery from tissue engineering scaffolds emerges as a more meaningful approach in tissue engineering and regenerative medicine, which aims to mimic the orchestrated endogenous growth factor production during natural tissue morphogenesis [15].

The natural polymer chitosan possesses the intrinsic advantages of being used as a delivery vehicle. Chitosan and its derivatives require only mild processing conditions, thus can avoid growth factor inactivation under harsh processing conditions. Chitosan can be fabricated into different forms and shapes to incorporate growth factors and other biomolecules for various *in vivo* applications. In addition, chitosan is biocompatible and non-toxic to cells, which thus can be used as a suitable cell delivery vehicle for *in vivo* tissue regeneration therapies. For instance, injectable hydrogels have attracted a great deal of attention as cell carriers and bioactive agents in regenerative engineering due to their ability to fill complex three-dimensional (3D) tissue gaps and relative ease of *in vivo* administration. A variety of chitosan-based injectable polysaccharide hydrogels have been developed for cell and protein delivery. Among the various approaches chitosan crosslinking involving amine functionalities has been popular to cause gelation using various crosslinking agents. For instance, cross-linked hydrogels were produced by the reaction between the amino functionality of chitosan and the aldehyde of dextran aldehyde resulting in Schiff's base formation. This approach eliminated the use of additional crosslinking agents, which may pose undesired side effects regarding cytotoxicity and biocompatibility. This approach will also allow tuning their mechanical properties by altering the crosslinking density with using higher aldehyde content. Bovine serum albumin release followed a steady release pattern from this gel and the release was dependent on the extent of hydrogel crosslinking. These gels encapsulated with human fetal osteoblasts supported their survival and continued to proliferate under *in vitro* culture [55]. Table **2** summarizes chitosan and chitosan-based bioactive delivery systems for tissue regeneration.

Table 2: Chitosan based biological factor or cell delivery systems for tissue regeneration

Incorporated Substance	Actions	Substrate Material	Application	References
Bone morphogenetic protein - 2 (BMP-2)	Bone morphogenesis	Chitosan Chitosan - poly(glutamic acid	Bone regeneration Osseointegration	[56, 141]
Platelet-derived growth factor - BB (PDGF-BB)	Mitogenic Bone formation and remodeling	Chitosan Chitosan - chondroitin sulfate Chitosan - poly(L-lactide)	Bone regeneration Periodontics Orthopaedics Plastic surgery	[57, 59, 142]
Transforming growth factor - β (TGF-β)	Multifunctional	Chitosan Chitosan - collagen Chitosan - tricalcium phosphate Chitosan - collagen - chondroitin sulfate	Bone and cartilage regeneration	[58, 61, 143, 144]
Fibroblast growth factor - 2 (FGF-2)	Mitogenic Angiogenic Regulates cell differentiation	Chitosan - lactose hydrogel Photocrosslinkable azide-chitosan-lactose hydrogel	Angiogenesis Wound healing	[42, 43, 65-68]
Epidermal growth factor (EGF)	Cell proliferation ECM synthesis	Chitosan	Wound healing	[145]
Vascular endothelial growth factor (VEGF)	Angiogenic Endothelial cell proliferation	Chitosan - dextran sulfate nanoparticles	Angiogenesis	[62]
RGD - containing protein	Cell adhesion and migration	Chitosan - alginate - hyaluronate	Cartilage regeneration	[146]
Growth hormone (GH)	Cell proliferation and differentiation	Chitosan	Bone regeneration	[60]
Plasmid DNA encoding PDGF	Expression of PDGF	Chitosan - coral	Periodontal tissue regeneration	[147]
Plasmid DNA encoding TGF-β	Expression of TGF-β	Chitosan - collagen Chitosan - gelatin	Periodontal tissue regeneration Cartilage regeneration	[148, 149]

Table 2: contd...

Silver sulfadiazine	Antibiotic	Chitosan	Wound healing	[150]
Chondrocytes	Initial cell source	Chitosan - β- glycerol phosphate -glycosamine	Cartilage regeneration	[151]
Chondrocytes and meniscus cells	Initial cell source	Chiotsan-g-poly(N-isopropylacrylamide)	Cartilage and meniscus regeneration	[73]

As a straightforward approach, chitosan films, porous sponges, or microgranules have been fabricated using a variety of techniques such as solvent evaporation, freeze drying with ionic or chemical crosslinking, and solution coagulation [56-58]. Growth factors were then loaded onto the scaffolds by soaking them in growth factor solutions. The factor-loaded bioactive scaffolds promoted cell growth and functions thus stimulated tissue regeneration as compared to chitosan scaffolds alone. Nevertheless, this type of factor loading generally resulted in an initial burst release followed by a short period of slow release for a few days, which is undesirable in many cases that requires a long time sustained factor release. Efforts were made to improve growth factor release profile by adding negatively charged component into chitosan system. Such an addition can potentially interact with the positively charged residues of growth factors and provide stability for the system and retard growth factor delivery [59]. Chitosan micro- or nanospheres were also developed as an alternative carrier system for growth factor delivery [60-64]. The growth factors were more uniformly distributed within these particles, and the release was primarily controlled by diffusion and more sustainable. These growth factor encapsulated micro- and nanosphere systems can be used directly for bone and cartilage tissue engineering, angiogenesis, or they can be loaded to other forms of porous scaffolds for tissue regeneration.

Chitosan-based hydrogel carriers with specific growth factors show great potential in skin wound healing, angiogenesis and vascularization. A number of chemically modified forms of chitosan were used to deliver bioactive agents to attain desired effects. Chitosan macromolecules with lactose moiety attached to its backbone can be prepared through the condensation reaction between carboxyl groups of lactose and amine groups of chitosan [42, 65]. This lactose-introduced chitosan hydrogel in the form of injectable provides a non-invasive means of delivering growth factors. Several studies showed the feasibility of introducing photoreactive azide functionality to chitosan along with lactose moiety [43, 66-68]. Thus azide-chitosan-lactose systems can be photocrosslinked under a brief UV irradiation to produce hydrogels. Such systems are promising since growth factors can be dispersed into aqueous azide-chitosan-lactose solution, applied to an *in vivo* site, and readily crosslinked to gel and perform the functions. Additionally, chitosan-based hydrogels stabilize growth factors and preserve their biological activity. Growth factors then are released as the hydrogels degrade *in vivo*.

Physically entrapped growth factor delivery from chitosan carriers, although achieved successes to certain extent, has several associated limitations. This is because sensitive growth factors are prone to undergo denaturation during the loading process, and their typically short half lives in an *in vivo* milieu make them less therapeutically effective. Thus, gene delivery has emerged as a promising technique to achieve localized, sustained expression of growth factors. Since plasmid DNA is more stable than recombinant growth factors, it seems to be a more favorable candidate to obtain sustained delivery systems. Chitosan has the intrinsic advantage of forming polyelectrolyte complex with negatively charged species that can be potentially used to deliver genes of interest. Chitosan in solution has protonated structure and exhibits excellent affinity towards negatively charged plasmid DNA thus favoring complex formation. These polyelectrolyte complexes formed by the condensation of DNA with chitosan are capable of delivering viable DNA to cells, which thereafter express specific growth factors at the local site. Research has been performed to fabricate chitosan gene delivery matrices for a variety of tissue engineering applications and a detailed account of such systems were reviewed elsewhere [69].

Cell-based therapy is particularly important to regenerate damaged tissues that have little capacity of spontaneous healing such as cartilage, ligament and intervertebral disc [70, 71]. Therefore, a biodegradable polymeric carrier with autologous cells represents an important approach to regenerate these tissues. Chitosan has shown great promise as a cell delivery carrier due to its proven biocompatibility, biodegradability and non-toxicity. A smart thermo-responsive chitosan-poly(N-isopropylacrylamide) (CH-PNIPAM) has been developed and applied as a carrier for chondrocytes and meniscus cells [72, 73]. This comb-like CH-PNIPAM system showed a sol-gel transition at a critical temperature below 37 °C. Cell suspension thus can be mixed with CH-PNIPAM solution at room temperature. When such a cell-containing CH-PNIPAM solution is transferred to an *in vivo* environment, a sol-gel transition occurs and results in a three dimensional porous gel within which cells can adhere and proliferate.

BIOACTIVE CHITOSANS *VIA* SURFACE IMMOBILIZATION

Biological recognition has been considered crucial to accomplish appropriate biological performances of biomaterials through pre-designed cell/tissue biomaterial interactions [22]. Surface or bulk modification of biomaterials is an important tool to design biomaterials that exhibit specificity to trigger defined and desired biological reactions. Physical immobilization such as adsorption and ionic interaction, and chemical immobilization such as covalent coupling are two useful means of introducing bioactivity to biomaterials. Table **3** summarizes research performed on chitosan surface modifications to develop functional chitosan-based substrates.

Table 3: Functional chitosan-based biomaterials obtained by surface modification

Biomaterial	Immobilized Molecule *	Immobilization method	Functions	References
HAp-coated chitosan	Laminin-1 CDPG**YIGSR** CSRARKQAAS**IKVAV**SAD	Physical adsorption	Cell adhesion and migration Neurite outgrowth	[77]
Chitosan	BMP-2	Physical adsorption	Osseointegration	[56]
Chitosan	Heparin	Ionic interaction	Protection of locally produced growth factors	[74, 75]
Chitosan	G**RGD** **RGD**S G**RGD**S **RGD**SGGC	Covalent coupling	Cell adhesion	[44, 46, 49, 52, 53]
Chitosan	**YIGSR** **IKVAV** CDPG**YIGSR** CSRARKQAAS**IKVAV**SAD	Covalent coupling	Cell adhesion	[47, 48, 51]
Chitosan	**BMP-2**	Covalent coupling	Cell proliferation and differentiation	[50]

*:Letters in bold represent the identified bioactive peptide sequences.

Biomolecule immobilization on chitosan through adsorption or ionic interaction is a simple and effective technique to achieve favorable properties. Biomolecule adsorbed chitosan surfaces have been shown to be capable of promoting biomaterial surface functionality and enhancing cell/tissue growth [48, 67]. In addition, protonated form of chitosan selectively binds to negatively charged biomolecules and performs their functions. Thus produced chitosan-heparin complex has been shown to stimulate human skin wound healing [74, 75].

Covalent immobilization of bioactive molecules is a popular method of chitosan surface modifications, which essentially makes use of the functional groups of chitosan. Growth factors such as BMP-2 were chemically immobilized on a nanofibrous chitosan substrate as a guided bone regenerative membrane [50] in contrast to the physical adsorption method illustrated previously [56]. This method involved using a bi-functional reagent, which can react with chitosan on one end and BMP-2 on the other. The chemically bound BMP-2 remained bioactive and was more effective in stimulating osteoblastic cell proliferation and differentiation as compared to the BMP-2 adsorbed chitosan membrane. While the complete protein can be used, it seems to be more advantageous to work with short peptide sequences derived from their parental proteins. The reasons reside on the facts that short peptide sequences preserving the bioactivity of their deriving protein are easier to build up, that they are chemically well-defined and can be obtained in highly purified form and relatively large quantity [76], that they show fewer antigenic reactions *in vivo* [51], and that they are more efficient since they can make all of them available and active for binding to cell-surface receptors by adopting appropriate arrangement [22]. In this regard, oligopeptides containing RGD (Arg-Gly-Asp) sequences from fibronectin have been covalently coupled on chitosan substrates [44, 46, 49, 52, 53]. Other adhesion factors such as oligopeptides containing YIGSR (Tyr-Ile-Gly-Ser-Arg) and IKVAV (Ile-Lys-Val-Ala-Val) sequences from laminin have been covalently immobilized onto chitosan using various chemical methods [47, 48, 51]. These peptide sequences are the recognition sites for many integrins in their adhesion protein ligands. These immobilized peptides thus function to promote cell attachment *via* their binding to specific integrins. It has been shown that the modified chitosan can exhibit desired bioactivity as indicated by the enhance cell adhesion, proliferation, and differentiation.

It should be noted that although physical and chemical immobilization are different options to achieve functionality, different methods can be used for the same biomolecule to be immobilized on a polymeric substrate. For instance, the aforementioned YIGSR and IKVAV containing peptides were both physically adsorbed and covalently coupled on chitosan for nerve regeneration [48, 77]. It is interesting to note that in some cases the physically adsorbed peptides or proteins performed better [47], but in other cases the covalently coupled ones functioned more efficiently [50]. In general, immobilization through physical adsorption or ionic interaction is simpler with lower cost as compared to chemical modification which is often a more involved process. However, immobilization through covalent binding provides a means to achieve high loading efficiency, low loss of biomolecules, and more controllable and sustainable release of bioactive biomolecules within a relatively long time frame [78]. The selection of immobilization methods should depend on specific purposes.

BIOACTIVE SINTERED MICROSPHERE SCAFFOLDS FOR BONE TISSUE ENGINEERING

Bone reconstruction and regeneration has become a worldwide problem due to trauma-related bone fractures, osteoporotic fractures and joint diseases among the aging population. Bone regeneration, in many situations, requires scaffolds that not only can provide a temporary space for cell and tissue ingrowth, but need to serve as a mechanically competent matrix to bear load initially. Thus, a functional scaffold which can elicit appropriate biological responses and is biomechanically compatible with its surrounding milieu is highly promising in bone tissue engineering.

Human trabecular bone has a compressive modulus in the range of 10 – 2000 MPa [79] and a compressive strength in the range of 2 – 180 MPa [80] dependent on the anatomic locations in the body. In addition to the 3-D porous architecture of scaffolds, it is important to match the mechanical properties of the scaffolds to their *in vivo* microenvironment. Evidence suggests that implants are more biocompatible when the material properties approximate those of the local tissue [81]. Our laboratory has pioneered sintered microsphere based three dimensional porous scaffolds for bone tissue engineering [82, 83]. In such scaffolds the regular packing of poly(lactic acid-glycolic acid) (PLAGA) microspheres gives interconnected porous structure with sufficient porosity for cell and tissue ingrowth. The sintering between adjacent polymeric microspheres provides excellent mechanical strength to the scaffolds. It has been shown that the mechanical properties of the sintered PLAGA microsphere scaffolds can be manipulated by varying the sintering temperature or sintering time [82]. Previous studies have demonstrated great potential of using this scaffold for load-bearing bone regeneration [84]. We further demonstrated that a second polymer component chitosan can be introduced to the PLAGA microsphere system [85]. The newly developed chitosan/PLAGA sintered microsphere scaffold, while remains to be biocompatible, biodegradable and mechanically robust (compressive modulus 200 – 450 MPa and compressive strength 3.5 – 16 MPa, both in the range of trabecular bone), possesses functional groups attribute to chitosan [85]. These functional groups as illustrated in the former sections are tremendously beneficial for physical binding or chemical coupling with biomolecules.

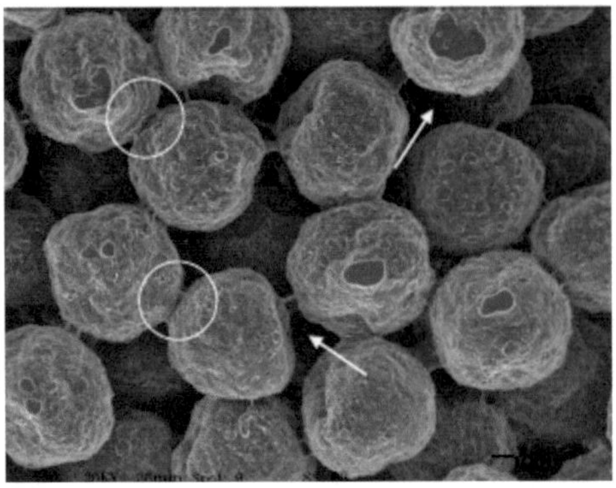

Figure 3: Scanning electron micrograph of heparin immobilized chitosan/PLAGA sintered microspere scaffold. Circles indicate fusion between adjacent microspheres; arrows indicate void spaces among microspheres.

These scaffolds were modified by heparin allowing negatively charged heparin to bind to protonated chitosan/PLAGA scaffolds *via* ionic interaction. Fig. (**3**) shows a typical scanning electron micrograph of heparin immobilized chitosan/PLAGA scaffold. The fusion between adjacent microspheres which contributes to the mechanical strength of the scaffold, and the void spaces among microspheres caused by microsphere packing which give rise to the total porosity are demonstrated in the figure. The possible underlying complexing mechanism between chitosan and heparin is presented in Fig. (**4**) [73]. Heparin modification was confirmed by immobilization of fluorescence conjugated heparin and visualized under a confocal microscope. Fig. (**5**) shows the confocal micrographs of heparin immobilized chitosan/PLAGA sintered microsphere scaffolds. The green fluorescence indicated the uniform distribution of heparin molecules on the scaffolds. This heparinized scaffold, while preserved the mechanical properties of the original chitosan/PLAGA scaffold, supported the proliferation and migration of MC3T3-E1 osteoblast-like cells as shown in Fig. (**6**). In addition, heparin immobilized chitosan/PLAGA scaffolds stimulated MC3T3-E1 cell differentiation as evidence by the enhanced alkaline phosphatase activity and osteocalcin expression of cells on the heparinized scaffold [86]. Such heparin modification is advantageous as they can regulate cell growth and function possibly through its interactions with various growth factors [87, 88]. Also such heparin modified scaffolds can potentially bind with a number of growth factors with heparin-bind domains such as BMP-2, FGFs, and VEGF [89]. These growth factors are actively involved in the regulation of cell functions of osteoblastic lineage and vascularization, which are critical to new bone formation. Our ongoing research is focusing on evaluating the ability of the heparinized scaffold to bind and release BMP-2 and its *in vivo* bone regeneration capacity in a rabbit ulna critical-sized defect model.

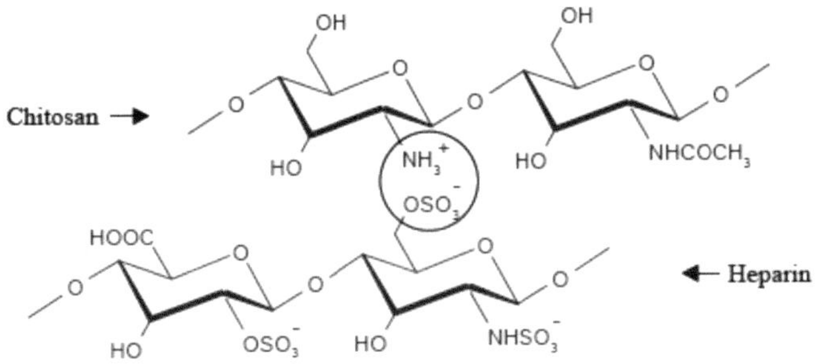

Figure 4: Possible ionic complexing mechanism between chitosan and heparin.

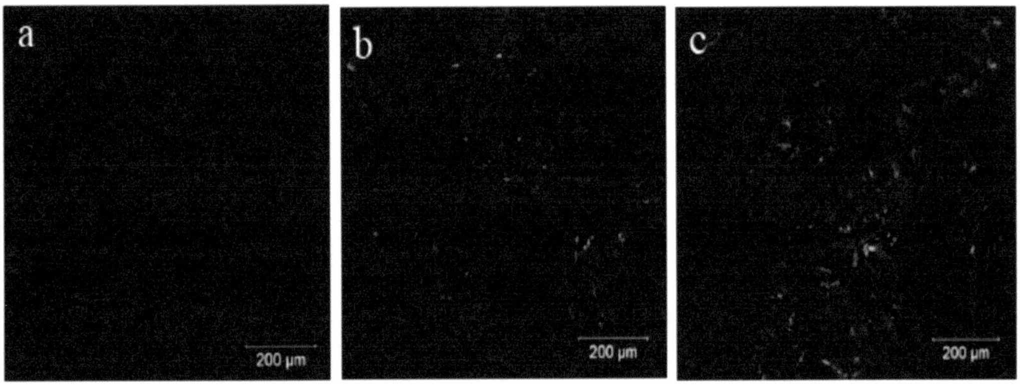

Figure 5: Confocal fluorescent micrographs of fluorescein heparin immobilized chitosan/PLAGA scaffolds after treated with heparin solutions at the concentration of **(a)** 0 μg/ml (negative control); **(b)** 40μg/ml; and **(c)** 200μg/ml. Heparin distribution was uniform throughout the scaffolds. Magnification: 10×.

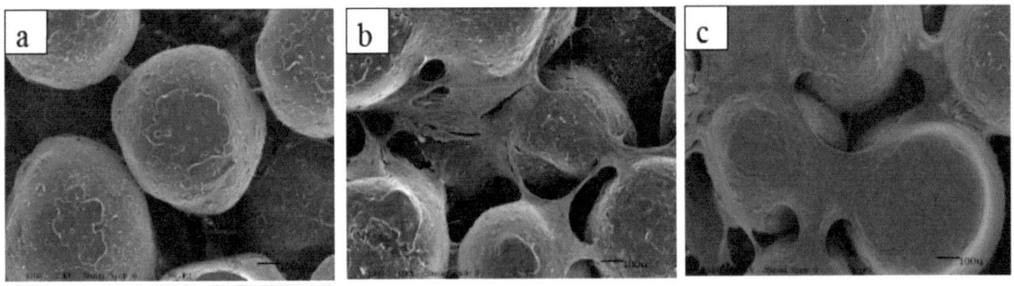

Figure 6: Scanning electron micrographs showing MC3T3-E1 cell proliferation on heparin immobilized chitosan/PLAGA scaffolds at **(a)** 7 days; **(b)** 14 days; and **(c)** 21 days after cell seeding. Cells were found to attach and proliferate on the scaffold primarily around the microsphere adjoining areas after 7 days. Cell proliferation propagated to microsphere surfaces and cells migrated from one microsphere to another or to the porous structure after 14 days of culture. Magnification: 100×.

NANOSCALE CHITOSAN STRUCTURES FOR REGENERATIVE ENGINEERING

Nanoscale surface topographical features have shown to play a significant role in regulating cell functions by providing cells with mechanotransductive cues. For example, surface nanostructures of square array of nanoscale pits were able to retain mesenchymal stem cell phenotype and multipotency [90]; while a nanografting surface induced mesenchymal stem cell differentiation [91]. In the natural *in vivo* environment, tissue specific nanotopographical features and nanostructures are ubiquitous. These nanostructures such as the nanotopographies of endothelial basement membranes [92] and the aligned collagen nanofibers in ligament and tendon tissues [93] play critical biological and mechanical roles. The fabrication of chitosan nanostructures incorporating bioactive agents for regenerative engineering is a relatively new area but has attracted great interests over the past decade.

Layer-by-Layer Self-Assembly

The layer-by-layer (LbL) self-assembly technique involves the preparation of multilayer films that are composed of polyelectrolyte molecules [94]. The process is based on sequential LbL adsorption of thin polyion films from solutions based on the attractive electrostatic force between a charged surface and an oppositely charged polyelectrolyte [95]. Polyelectrolytes are macromolecules that have a large number of charged or chargeable groups when dissolved in polar solvents, primarily water [96, 97]. When dissolved in water, the polyelectrolytes will dissociate into a macromolecular ion and counter ions and the charge of each is controlled by the pH of the solution. These electrostatic interactions are the driving force leading to their deposition at interfaces. The charge of the polyelectrolyte must alternate from layer to layer, and due to the electrostatic attraction results in the build-up of defined multilayer macromolecule films on the surface. The charged macromolecule must have a critical number of binding sites to be adsorbed onto the previously adsorbed layer, and also retain an adequate number of bonds for the adsorption of the subsequent layer. In addition to electrostatic forces, Coulombic interactions, van der Waals and entropic effects are also important for the stability of the system.

In acidic pH, the amine groups of chitosan are positively charged, turning chitosan into a water-soluble cationic polyelectrolyte. The pK_a of chitosan is near neutrality, and the soluble–insoluble transition occurs at a pH between 6.46 and 7.32, a suitable range for biological applications [98]. Based on their novel properties, LbL films have been used for applications such as implantable device coatings, drug delivery systems, gene therapy, and 3D biomimetic cell/biomaterial constructs [99, 100]. LbL assembly has been particularly useful in modifying biomaterial surfaces with great control over surface composition and coating thickness. In one study, chitosan was combined with poly(γ-glutamic acid) (γ-PGA), a microbial produced polyanion having properties of low immunogenic reaction and low cytotoxicity [101]. Multilayered ultrathin films were assembled by LbL, with a maximum of six layers. Chitosan interacts electrostatically with γ-PGA, forming polyelectrolyte multilayered films that function as model substrates to study cell–biomaterial interactions in tissue regeneration. A number of other biomacromolecules carrying negative charges have also been investigated in forming polyelectrolyte multilayers (PEMs) with chitosan. Kipper *et al.* deposited PEMs of heparin (polyanions) and chitosan (polycations) onto glass substrates [102]. A biomolecule with the highest negative charge density among any known biological macromolecules, heparin is known to be able to bind with a number of heparin-binding growth factors that are involved in the regulation of many cell fate processes. Fibroblast growth factor - 2 (FGF-2) was adsorbed on the PEMs coated glass substrate, which significantly increased mesenchymal stem cell proliferation than when FGF-2 was delivered in solution, possibly because the layer of heparin can stabilize and potentiate the activity of FGF-2 [103]. The same group also developed a procedure to modify surfaces with PEMs with either negatively or positively charged polyelectrolyte complex nanoparticles (PCNs) adsorbed to or embedded in the top layer. The PCNs thus could be used as a means to introduce nanoscale topography and varying surface chemistry into the surface coatings with PEMs [104, 105]. In an effort to integrate nanostructured coating with bioactive small molecules in the PEM approach, PEMs of a recombinant elastin-like recombinamer (ELR) containing peptide sequence arginine-glycine-aspartic acid (RGD) and chitosan were deposited to the surface of gold-coated crystals [106]. Enhanced osteoblastic cell adhesion and differentiation were observed on the chitosan/ELR-RGD coated surface.

In addition to coating 2D substrates with PEMs, multilayered coatings with polyelectrolytes could also be applied toward 3D structures to create functional scaffolds for regenerative engineering. In one study, PEMs of chitosan and heparin were coated on chitosan nanofibrous scaffolds [107]. The presence of heparin might serve as a reservoir to present active heparin-binding growth factors in an effective manner to cells and tissues *in vivo* for successful morphogenesis. PEMs of chitosan and heparin were also deposited onto sheep cortical bone, mimicking the periosteum by presenting osteoinductive factors, osteogenic cells, and antimicrobial factors [108]. It was suggested that PEMs of chitosan and heparin might improve the performance of large segmental bone allografts. Furthermore, stand-alone 3D PEM scaffolds have also been fabricated. Silva *et al.* deposited multiple layers of chitosan and chondroitin sulfate onto 3D packet of paraffin spheres. The PEM scaffolds consisting of chitosan and chondroitin sulfate were obtained after depositing 10 double layers and dissolving the paraffin spheres [109]. The porous PEM scaffolds were able to support the growth and chondrogenic differentiation of human mesenchymal stem cells.

Electrospinning

Polymeric nanofibers due to their similarity to natural ECM are of great interest for biomedical applications [110]. They are characterized by ultra-thin continuous fibers, high surface-to-volume ratio and porosity, and variable pore size distribution. Electrospinning is an attractive technique to fabricate nanofibers from various biodegradable polymers due to the ease of fabrication, efficient control over the process, and easy scale-up [111]. Zeleny first studied the electrostatic spinning in 1914 [112], but it was not until 1934 when Formhals *et al.* [113] reported the first patent describing the process of developing polymeric nanofibers *via* electrospinning. In electrospinning, polymeric nanofibers result from a jet of polymer solution under the influence of applied electrical field between an ejecting needle and a collector [110]. A suitable polymer solution viscosity is critical to fabricate nanofibers without any beads or beads-on-a-string appearance. For example, increase in viscosity or polymer concentration results in the increase of fiber diameter. By changing polymer concentration alone, it is possible to fabricate nanofibers with diameters in the range of a few nanometers to several micrometers while keeping other electrospinning parameters constant. However, if the viscosity of polymer solutions becomes too low, the ejected jet may break down into droplets and result in electrospraying [114]. The applied electrical potential is another important factor affecting the fiber diameter. Other electrospinning parameters include polymer solution flow rate, distance between spinneret and collector, motion of the grounded target, and ambient conditions. A semi-quantitative method, known as response surface methodology, was used to evaluate the possible quantitative relationship between electrospinning parameters and average fiber diameter [115]. A study by Agarwal *et al.* [115] suggested that response surface methodology, coupled with a second-order model and non-linear regression analysis, could be a useful tool to predict chitosan-polylactide electrospun nanofiber diameter prepared under various parameters. Laurencin and colleagues were the first to demonstrate that electrospun nanofibrous structure could serve as a novel scaffold to engineer tissues [116]. This pioneering work has been recently featured as one of the top 25 papers published by the Journal of Biomedical Materials Research over the past 50 years. Laurencin *et al.* have then investigated a number of electrospun biodegradable nanofiberous scaffolds using polyphosphazenes, PLAGA, and poly (epsilon-caprolactone) for wound healing [117], drug delivery [118], and regenerating soft [119-125] and hard tissues [126]. Furthermore, polyphosphazene-nanohydroxyapatite composite [127] and polyphosphazene-PLAGA blend [128] nanofiber scaffolds as well as composite scaffolds comprising PLLA nanofibers and sintered microspheres [129] have been investigated for bone tissue engineering applications.

The electrospinning of chitosan remains challenging due to its rigid crystalline structure, insufficient viscosity, and limited solubility in common organic solvents. Hence chitosan is often electrospun as a polymeric blend with several synthetic components. Recently we have reported a smart chitosan modification methodology that allows electrospinning chitosan irrespective of molecular weight or deacetylation value and without blending with synthetic polymers [130]. For instance, 2-nitrobenzyl-chitosan, a novel chitosan derivative at various molar compositions of chitosan:2-nitrobenzaldehyde, was synthesized by the reaction between amino groups of chitosan and aldehyde groups of 2-nitrobenzaldehyde. In this Schiff base, 2-nitrobenzaldehyde protects the amine functionalities of chitosan and improves its solubility in trifluoroacetic acid. 2-nitrobenzaldehyde is a photoactivatable-caged compound that cleaves off from iminochitosan on ultraviolet exposure yielding neat chitosan. Derivatives showed improved solubility in trifluoroacetic acid and dynamic viscosities in the range of 1.34 ± 0.7 to 12 ± 0.5 Pa·s based on the degree of substitution and concentration. Bead-free nanofibers were produced from high viscous solutions with diameters range of 100–600 nm and photolysis did not alter fiber morphology. Newly synthesized iminochitosan derivatives were very effective against microorganisms including bacteria (Gram-positive and Gram-negative), fungi, and yeast. These nanofiber matrices may serve as scaffolds for a variety of tissue healing and factor delivery applications [130]. Electrospinning allows for the generation of different chitosan structures such as random nanofibers, aligned nanofibers, polymer/ceramic nanocomposites, micro/nanofiber composites, core-shell nanofibers, and 3D structures [131-135]. Bioactive chitosan nanofibrous structures can be fabricated through different approaches. Physical adsorption of biological factors onto chitosan nanofibers is the most straightforward methods to introduce bioactivity. In one study, fibronectin was adsorbed onto the surface of chitosan nanofiber mat for cardiac tissue regeneration. Such surface modification with fibronectin was shown to enhance cellular adhesion to the nanofibers and cell migration into the interfibrous milieu [136]. Heparin, due to its high negative charge, anticoagulant properties, and ability to bind with many growth factors with heparin-binding domains, has been frequently used to modify chitosan nanofibrous scaffolds for vascular regenerative engineering [107, 137, 138]. Du *et al.* fabricated graded chitosan/ PCL blend nanofibrous scaffolds exhibiting chitosan-rich and PCL-rich portions by sequentially co-electrospinning chitosan and PCL solutions. Since chitosan is positively charged while PCL is neutral, negatively charged heparin and heparin-binding vascular endothelial growth factor (VEGF) can be immobilized onto the electrospun scaffolds with a vertical gradient as well [138]. Comparing to the uniform scaffold, the VEGF-immobilized gradient chitosan/PCL scaffold showed more stable and sustained VEGF release as well as enhanced adhesion and proliferation of human umbilical vein endothelial cells. Rather than surface immobilization through heparin binding, growth factors can also be incorporated into the bulk of the chitosan nanofibers. Xie *et al.* developed a nanoparticle-containing chitosan-poly(ethylene-oxide) (PEO) nanofibrous system with dual growth factor delivery capability. Vascular endothelial growth factor (VEGF) was loaded within the chitosan-PEO nanofibers to promote angiogenesis in the short term; while platelet-derived growth factor-BB (PDGF-BB) encapsulated PLAGA nanoparticles were embedded inside nanofibers for a sustained release of PDGF-BB for accelerated tissue regeneration and remodeling [139].

CONCLUSION AND FUTURE DIRECTIONS

Chitosan is an attractive candidate biomaterial that shows great potential in the area of tissue engineering and regenerative medicine due to its biocompatibility, controlled biodegradability and functionality. With the advances in the areas of controlled delivery and bioactive biomaterials, biologically active chitosan systems have been increasingly developed and investigated. These systems, either by incorporating various growth factors or by physically or chemically attaching biomolecules or peptide sequences, exhibit specific bioactivity attribute to the loaded or attached factors. Although much progress has been achieved to apply bioactive chitosan systems in regenerative engineering, a few limitations of the current systems need to be overcome in order to develop more clinically meaningful chitosan substrates for various tissue regeneration. Most of the current chitosan-based growth factor delivery systems use either spongy scaffolds with adsorbed growth factors, or individual growth factor encapsulated micro- or nanospheres lacking overall structural integrity. The former systems show poorly control over the growth factor release and the mechanical properties are generally low. The latter systems, although have a better spatial distribution of growth factors and better release kinetics, can only be used as a filling material without providing any structural support and function. Therefore, it is promising to develop bioactive chitosan systems with controlled delivery capacity that also shows tissue compatible mechanical properties. On the other hand, modification of chitosan by introducing various recognition molecules or domains represents an intriguing research direction. With the understanding of the immobilization processes and cellular responses to the modified two dimensional films, it is essential to develop three dimensional structures exhibiting recognition and specificity for tissue regeneration. In addition, considering that chitosan is biodegradable *in vivo* and the biomolecule modified surface will lose the bioactivity after being degraded, chitosan bulk modification with particular biomolecules will be of interest and importance. Furthermore, the use of biologically active small molecules is emerging in the area of regenerative engineering because small molecules do not possess many of the downsides associated with protein-based growth factors [140]. A convergence of research incorporating advanced chitosan structures, small molecule drugs, and stem cells becomes an important future direction.

ACKNOWLEDGEMENTS

Authors gratefully acknowledge funding from the Raymond and Beverly Sackler Center for Biomedical, Biological, Physical and Engineering Sciences. Authors also acknowledge the funding from National Science Foundation Award Number IIP-1311907, INT0115595 and EFRI-1332329.

CONFLICT OF INTEREST

The authors confirm that this chapter contents have no conflict of interest.

DISCLOSURE

This is an updated report of the earlier review article by T. Jiang, S.G. Kumbar, L.S. Nair, and C.T. Laurencin "Biologically active chitosan systems for tissue engineering and regenerative medicine (review)" Curr Top Med Chem 2008; 8(4): 354-364.

ABBREVIATIONS

BMP = Bone morphogenetic protein

DDA	=	Degree of deacetylation
ECM	=	Extracellular matrix
EGF	=	Epidermal growth factor
FGF	=	Fibroblast growth factor
GH	=	Growth hormone
Hap	=	Hydroxyapatite
PCL	=	Poly(ε-caprolactone)
PDGF	=	Platelet-derived growth factor
PLAGA	=	Poly(lactic-co-glycolic acid)
PLLA	=	Poly(l-lactic acid)
TGF-β	=	Transforming growth factor – β
VEGF	=	Vascular endothelial growth factor

REFERENCES

[1] Langer R, Vacanti JP. Tissue Engineering. Science 1993; 260: 920-6.
[2] 2011 Annual Report of the U.S. Organ Procurement and Transplantation Network and the Scientific Registry of Transplant Recipients. US Department of Health and Human Services, Health Resources and Services Administration, Washington, DC. 2012.
[3] Gurtner GC, Callaghan MJ, Longaker MT. Progress and potential for regenerative medicine. Ann Rev Med 2007; 58: 299-312.
[4] Atala A. Tissue engineering and regenerative medicine: Concepts for clinical application. Rejuvenation Res 2004; 7: 15-31.
[5] Sipe JD. Tissue engineering and reparative medicine. Ann NY Acad Sci 2002; 961: 1-9.
[6] Griffith LG, Naughton G. Tissue engineering - Current challenges and expanding opportunities. Science 2002; 295: 1009-1014.
[7] Laurencin CT, Ambrosio AMA, Borden MD, Cooper JA. Tissue engineering: Orthopedic applications. Ann Rev Biomed Eng 1999; 1: 19-46.
[8] Liu XH, Ma PX. Polymeric scaffolds for bone tissue engineering. Ann Biomed Eng 2004; 32: 477-86.
[9] Muschler GE, Nakamoto C, Griffith LG. Engineering principles of clinical cell-based tissue engineering. J Bone Joint Surg Am 2004; 86A: 1541-58.
[10] Spector M. Biomaterials-based tissue engineering and regenerative medicine solutions to musculoskeletal problems. Swiss Med Wkly 2006; 136: 293-301.
[11] Lavik E, Langer R. Tissue engineering: current state and perspectives. Appl Microbiol Biotechnol 2004; 65: 1-8.

[12] Le Roith D, Blakesley VA. Biology of growth factors. In: Canalis E, Eds. Skeletal growth factors. Philadelphia: Lippincott Williams & Wilkins 2000; pp. 31-50.
[13] Hubbell JA. Matrix-bound growth factors in tissue repair. Swiss Med Wkly 2006; 136: 387-91.
[14] Chen RR, Mooney DJ. Polymeric growth factor delivery strategies for tissue engineering. Pharm Res 2003; 20: 1103-12.
[15] Whitaker MJ, Quirk RA, Howdle SM, Shakesheff KM. Growth factor release from tissue engineering scaffolds. J Pharm Pharmacol 2001; 53: 1427-37.
[16] Laurencin CT, Khan Y. Regenerative engineering. Sci Transl Med 2012; 4: 160ed9/1-3.
[17] Deng M, James R, Laurencin CT, Kumbar SG. Nanostructured polymeric scaffolds for orthopaedic regenerative engineering. IEEE Trans NanoBioscience 2012; 11: 3-14.
[18] Black J. Biological performance of materials. New York: Marcel Dekker 1999; pp. 3-12.
[19] Tabata Y. Tissue regeneration based on growth factor release. Tissue Eng 2003; 9: S5-S15.
[20] Simmons CA, Alsberg E, Hsiong S, Kim WJ, Mooney DJ. Dual growth factor delivery and controlled scaffold degradation enhance *in vivo* bone formation by transplanted bone marrow stromal cells. Bone 2004; 35: 562-9.
[21] Richardson TP, Peters MC, Ennett AB, Mooney DJ. Polymeric system for dual growth factor delivery. Nat Biotechnol 2001; 19: 1029-34.
[22] Hubbell JA. Bioactive biomaterials. Curr Opin Biotechnol 1999; 10: 123-9.
[23] Ratner BD. The engineering of biomaterials exhibiting recognition and specificity. J Mol Recogn 1996; 9: 617-25.
[24] Lutolf MP, Hubbell JA. Synthetic biomaterials as instructive extracellular microenvironments for morphogenesis in tissue engineering. Nat Biotechnol 2005; 23: 47-55.
[25] Hubbell JA. Materials as morphogenetic guides in tissue engineering. Curr Opin Biotechnol 2003; 14: 551-8.
[26] Rho J-Y, Kuhn-Spearing L, Zioupos P. Mechanical properties and the hierarchical structure of bone. Med Eng Phys 1998; 20: 92-102.
[27] Laurencin CT, Kumbar SG, Nukavarapu SP. Nanotechnology and orthopedics: a personal perspective. Wiley Interdiscip Rev Nanomed Nanobiotechnol 2009; 1: 6-10.
[28] Bettinger CJ, Langer R, Borenstein JT. Engineering substrate topography at the micro - and nanoscale to control cell function. Angew Chem Inter Ed 2009; 48: 5406-15.
[29] Li J-R, Shi L, Deng Z, Lo SH, Liu G-Y. Nanostructures of designed geometry and functionality enable regulation of cellular signaling processes. Biochemistry 2012; 51: 5876-93.
[30] Ratner BD, Bryant SJ. Biomaterials: Where we have been and where we are going. Ann Rev Biomed Eng 2004; 6: 41-75.
[31] Khor E, Lim LY. Implantable applications of chitin and chitosan. Biomaterials 2003; 24: 2339-49.
[32] Di Martino A, Sittinger M, Risbud MV. Chitosan: A versatile biopolymer for orthopaedic tissue-engineering. Biomaterials 2005; 26: 5983-90.
[33] VandeVord PJ, Matthew HWT, DeSilva SP, Mayton L, Wu B, Wooley PH. Evaluation of the biocompatibility of a chitosan scaffold in mice. J Biomed Mater Res 2002; 59: 585-90.
[34] Freier T, Koh HS, Kazazian K, Shoichet MS. Controlling cell adhesion and degradation of chitosan films by N-acetylation. Biomaterials 2005; 26: 5872-8.
[35] Choi BK, Kim KY, Yoo YJ, Oh SJ, Choi JH, Kim CY. *In vitro* antimicrobial activity of a chitooligosaccharide mixture against Actinobacillus actinomycetemcomitans and Streptococcus mutans. Inter J Antimicrob Agents 2001; 18: 553-7.
[36] Khor E. Chitin: fulfilling a biomaterials promise. Oxford: Elsevier 2001; pp. 1-8.
[37] Madihally SV, Matthew HWT. Porous chitosan scaffolds for tissue engineering. Biomaterials 1999; 20: 1133-42.
[38] Jiang T, Pilane CM, Laurencin CT. In Nanoscale materials science in biology and medicine; Laurencin CT, Botchwey EA, Eds. Warrendale, Pennsylvania: Materials Research Society 2004; Vol.845, pp. 187-92.
[39] Masuoka K, Ishihara M, Asazuma T, *et al.* The interaction of chitosan with fibroblast growth factor-2 and its protection from inactivation. Biomaterials 2005; 26: 3277-84.
[40] Richardson SCW, Kolbe HJV, Duncan R. Potential of low molecular mass chitosan as a DNA delivery system: biocompatibility, body distribution and ability to complex and protect DNA. Int J Pharm 1999; 178: 231-43.

[41] Kweon DK, Song SB, Park YY. Preparation of water-soluble chitosan/heparin complex and its application as wound healing accelerator. Biomaterials 2003; 24: 1595-601.
[42] Fujita M, Ishihara M, Simizu M, *et al.* Vascularization *in vivo* caused by the controlled release of fibroblast growth factor-2 from an injectable chitosan/non-anticoagulant heparin hydrogel. Biomaterials 2004; 25: 699-706.
[43] Ishihara M, Obara K, Ishizuka T, *et al.* Controlled release of fibroblast growth factors and heparin from photocrosslinked chitosan hydrogels and subsequent effect on *in vivo* vascularization. J Biomed Mater Res A 2003; 64A: 551-9.
[44] Ho MH, Wang DM, Hsieh HJ, *et al.* Preparation and characterization of RGD-immobilized chitosan scaffolds. Biomaterials 2005; 26: 3197-206.
[45] Cheng Y, Ramos D, Lee P, Liang D, Yu X, Kumbar SG. Collagen Functionalized Bioactive Nanofiber Matrices for Osteogenic Differentiation of Mesenchymal Stem Cells: Bone Tissue Engineering. J Biomed Nanotechnol 2014; 10: 287-98.
[46] Masuko T, Iwasaki N, Yamane S, *et al.* Chitosan-RGDSGGC conjugate as a scaffold material for musculoskeletal tissue engineering. Biomaterials 2005; 26: 5339-47.
[47] Itoh S, Matsuda A, Kobayashi H, Ichinose S, Shinomiya K, Tanaka J. Effects of a laminin peptide (YIGSR) immobilized on crab-tendon chitosan tubes on nerve regeneration. J Biomed Mater Res B Appl Biomater 2005; 73B: 375-82.
[48] Matsuda A, Kobayashi H, Itoh S, Kataoka K, Tanaka J. Immobilization of laminin peptide in molecularly aligned chitosan by covalent bonding. Biomaterials 2005; 26: 2273-9.
[49] Li J, Yun H, Gong Y, Zhao N, zhang X. Investigation of MC3T3-E1 cell behavior on surface of GRGDS-coupled chitosan. Biomacromolecules 2006; 7: 1112-23.
[50] Park YJ, Kim KH, Lee JY, *et al.* Immobilization of bone morphogenetic protein-2 on a nanofibrous chitosan membrane for enhanced guided bone regeneration. Biotechnol Appl Biochem 2006; 43: 17-24.
[51] Suzuki M, Itoh S, Yamaguchi I, *et al.* Tendon chitosan tubes covalently coupled with synthesized laminin peptides facilitate nerve regeneration *in vivo*. J Neurosci Res 2003; 72: 646-59.
[52] Chung TW, Lu YF, Wang HY, *et al.* Growth of human endothelial cells on different concentrations of Gly-Arg-Gly-Asp grafted chitosan surface. Artif Organs 2003; 27: 155-61.
[53] Chung TW, Lu YF, Wang SS, Lin YS, Chu SH. Growth of human endothelial cells on photochemically grafted Gly-Arg-Gly-Asp (GRGD) chitosans. Biomaterials 2002; 23: 4803-9.
[54] Tabata Y. Significance of release technology in tissue engineering. Drug Discov Today 2005; 10: 1639-46.
[55] Cheng Y, Nada AA, Valmikinathan CM, *et al.* In situ gelling polysaccharide - based hydrogel for cell and drug delivery in tissue engineering. J Appl Polym Sci 2014; 131: 39934/1-11.
[56] Lopez-Lacomba JL, Garcia-Cantalejo JM, Casado JVS, Abarrategi A, Magana VC, Ramos V. Use of rhBMP-2 activated chitosan films to improve osseointegration. Biomacromolecules 2006; 7: 792-8.
[57] Park YJ, Lee YM, Park SN, Sheen SY, Chung CP, Lee SJ. Platelet derived growth factor releasing chitosan sponge for periodontal bone regeneration. Biomaterials 2000; 21: 153-9.
[58] Lee JY, Kim KH, Shin SY, *et al.* Enhanced bone formation by transforming growth factor-beta 1-releasing collagen/chitosan microgranules. J Biomed Mater Res A 2006; 76A: 530-9.
[59] Park YJ, Lee YM, Lee JY, Seol YJ, Chung CP, Lee SJ. Controlled release of platelet-derived growth factor-BB from chondroitin sulfate-chitosan sponge for guided bone regeneration. J Control Release 2000; 67: 385-94.
[60] Cho BC, Kim JY, Lee JH, *et al.* The bone regenerative effect of chitosan microsphere-encapsulated growth hormone on bony consolidation in mandibular distraction osteogenesis in a dog model. J Craniofac Surg 2004; 15: 299-311.
[61] Lee JE, Kim SE, Kwon IC, *et al.* Effects of a chitosan scaffold containing TGF-beta 1 encapsulated chitosan microspheres on *in vitro* chondrocyte culture. Artif Organs 2004; 28: 829-39.
[62] Huang M, Vitharana SN, Peek LJ, Coop T, Berkland C. Polyelectrolyte complexes stabilize and controllably release vascular endothelial growth factor. Biomacromolecules 2007; 8: 1606-14.
[63] Kim J-H, Kim Y-S, Park K, *et al.* Self-assembled glycol chitosan nanoparticles for the sustained and prolonged delivery of antiangiogenic small peptide drugs in cancer therapy. Biomaterials 2008; 29: 1920-30.

[64] Niu X, Feng Q, Wang M, Guo X, Zheng Q. Porous nano-HA/collagen/PLLA scaffold containing chitosan microspheres for controlled delivery of synthetic peptide derived from BMP-2. J Control Release 2009; 134: 111-7.
[65] Fujita M, Ishihara M, Shimizu M, *et al.* Therapeutic angiogenesis induced by controlled release of fibroblast growth factor-2 from injectable chitosan/non-anticoagulant heparin hydrogel in a rat hindlimb ischemia model. Wound Repair Regen 2007; 15: 58-65.
[66] Ono K, Saito Y, Yura H, *et al.* Photocrosslinkable chitosan as a biological adhesive. J Biomed Mater Res 2000; 49: 289-95.
[67] Obara K, Ishihara M, Ishizuka T, *et al.* Photocrosslinkable chitosan hydrogel containing fibroblast growth factor-2 stimulates wound healing in healing-impaired db/db mice. Biomaterials 2003; 24: 3437-44.
[68] Obara K, Ishihara M, Fujita M, *et al.* Acceleration of wound healing in healing-impaired db/db mice with a photocrosslinkable chitosan hydrogel containing fibroblast growth factor-2. Wound Repair Regen 2005; 13: 390-7.
[69] Jiang T, Nair LS, Laurencin CT. Chitosan composites for tissue engineering: bone tissue engineering scaffolds. Asian Chitin J 2006; 2: 1-10.
[70] Sittinger M, Hutmacher DW, Risbud MV. Current strategies for cell delivery in cartilage and bone regeneration. Curr Opin Biotechnol 2004; 15: 411-8.
[71] Roughley P, Hoemann C, DesRosiers E, Mwale F, Antoniou J, Alini M. The potential of chitosan-based gels containing intervertebral disc cells for nucleus pulposus supplementation. Biomaterials 2006; 27: 388-96.
[72] Lee SB, Ha DI, Cho SK, Kim SJ, Lee YM. Temperature/pH-sensitive comb-type graft hydrogels composed of chitosan and poly(N-isopropylacrylamide). J Appl Polym Sci 2004; 92: 2612-20.
[73] Chen JP, Cheng TH. Thermo-responsive chitosan-graft-poly(N-isopropylacrylamide) injectable hydrogel for cultivation of chondrocytes and meniscus cells. Macromol Biosci 2006; 6: 1026-39.
[74] Kratz G, Arnander C, Swedenborg J, *et al.* Heparin-chitosan complexes stimulate wound healing in human skin. Scand J Plast Reconstr Surg Hand Surg 1997; 31: 119-23.
[75] Kratz G, Back M, Arnander C, Larm O. Immobilised heparin accelerates the healing of human wounds *in vivo*. Scand J Plast Reconstr Surg Hand Surg 1998; 32: 381-5.
[76] Blumenthal DK. Use of synthetic peptides to study substrate specificity of protein phosphatases. In: Kemp BE, Eds. Peptides and protein phosphorylation. Boca Raton, Florida: CRC press 1990; pp. 135-44.
[77] Itoh S, Yamaguchi I, Suzuki M, *et al.* Hydroxyapatite-coated tendon chitosan tubes with adsorbed laminin peptides facilitate nerve regeneration *in vivo*. Brain Res 2003; 993: 111-23.
[78] Hoffman AS, Hubbell JA. Surface-immobilized biomolecules. In: Ratner BD, Hoffman AS, Schoen FJ, Lemons JE, Eds. Biomaterials science: an introduction to materials in medicine. San Diego, California: Elsevier Academic Press 2004; pp. 225-33.
[79] Keaveny TM, Hayes WC. Mechanical properties of cortical and trabecular bone. In: Hall BK, Eds. Bone A treatise, volume 7: bone growth-B. Boca Raton, Florida: CRC press 1993; pp. 285-344.
[80] Goldstein SA. The Mechanical-Properties of Trabecular Bone - Dependence on Anatomic Location and Function. J Biomech 1987; 20: 1055-61.
[81] Palsson BO, Bhatia SN. Tissue engineering. Upper Saddle River, New Jersey: Pearson Prentice Hall 2004.
[82] Borden M, Attawia M, Khan Y, Laurencin CT. Tissue engineered microsphere-based matrices for bone repair: design and evaluation. Biomaterials 2002; 23: 551-9.
[83] Borden M, El-Amin SF, Attawia M, Laurencin CT. Structural and human cellular assessment of a novel microsphere-based tissue engineered scaffold for bone repair. Biomaterials 2003; 24: 597-609.
[84] Borden M, Attawia M, Khan Y, El-Amin SE, Laurencin CT. Tissue-engineered bone formation *in vivo* using a novel sintered polymeric microsphere matrix. J Bone Joint Surg Br 2004; 86B: 1200-8.
[85] Jiang T, Abdel-Fattah WI, Laurencin CT. *In vitro* evaluation of chitosan/poly(lactic acid-glycolic acid) sintered microsphere scaffolds for bone tissue engineering. Biomaterials 2006; 27: 4894-903.
[86] Jiang T, Khan Y, Nair LS, Abdel-Fattach WI, Laurencin CT. Functionalization of chitosan/poly(lactic acid-glycolic acid) sintered microsphere scaffolds *via* surface heparinization for bone tissue engineering. J Biomed Mater Res A 2010; 93A: 1193-208.

[87] Hausser HJ, Brenner RE. Low doses and high doses of heparin have different effects on osteoblast-like Saos-2 cells *in vitro*. J Cell Biochem 2004; 91: 1062-73.
[88] Blanquaert F, Barritault D, Caruelle JP. Effects of heparan-like polymers associated with growth factors on osteoblast proliferation and phenotype expression. J Biomed Mater Res 1999; 44: 63-72.
[89] Capila I, Linhardt RJ. Heparin - protein interactions. Angew Chem Int Ed 2002; 41: 391-412.
[90] McMurray RJ, Gadegaard N, Tsimbouri PM, *et al.* Nanoscale surfaces for the long-term maintenance of mesenchymal stem cell phenotype and multipotency. Nat Mater 2011; 10: 637-44.
[91] Yim EK, Pang SW, Leong KW. Synthetic nanostructures inducing differentiation of human mesenchymal stem cells into neuronal lineage. Exp Cell Res 2007; 313: 1820-9.
[92] Liliensiek SJ, Nealey P, Murphy CJ. Characterization of endothelial basement membrane nanotopography in rhesus macaque as a guide for vessel tissue engineering. Tissue Eng Part A 2009; 15: 2643-51.
[93] Goh JC-H, Ouyang H-W, Teoh S-H, Chan CK, Lee E-H. Tissue-engineering approach to the repair and regeneration of tendons and ligaments. Tissue Eng 2003; 9: 31-44.
[94] Decher G. Polyelectrolyte multilayers. In: Schlenoff JB, Decher G, Schlenoff JB, Eds. Multilayer thin films: sequential assembly of nanocomposite materials, Weinheim: Wiley-VCH 2003; pp. 1-46.
[95] Groth T, Lendlein A. Layer-by-layer deposition of polyelectrolytes-a versatile tool for the *in vivo* repair of blood vessels. Angew Chem Int Ed 2004; 43: 926-8.
[96] Dautzenberg H, Jaeger W, Kötz J, Philipp B, Seidel C, Stscherbina D. Polyelectrolytes: formation, characterization and application, Munich: Carl Hanser Verlag 1994.
[97] Netz RR, Andelman D. Neutral and charged polymers at interfaces. Phys Rep 2003; 380: 1-95.
[98] Sogias IA, Khutoryanskiy VV, Williams AC. Exploring the factors affecting the solubility of chitosan in water. Macromol Chem Phys 2010; 211: 426-33.
[99] Zankovych S, Diefenbeck M, Bossert J, *et al.* The effect of polyelectrolyte multilayer coated titanium alloy surfaces on implant anchorage in rats. Acta Biomater 2013; 9: 4926-34.
[100] Yuxi L, Jing Y, Ziqi Z, Junjie L, Rui Z, Fanglian Y. Formation and characterization of natural polysaccharide hollow nanocapsules *via* template layer-by-layer self-assembly. J Colloid Interface Sci 2012; 379: 130-40.
[101] Antunes JC, Pereira CL, Molinos M, *et al.* Layer-by-layer self-assembly of Chitosan and Poly (γ-glutamic acid) into polyelectrolyte complexes. Biomacromolecules 2011; 12: 4183-95.
[102] Boddohi S, Killingsworth CE, Kipper MJ. Polyelectrolyte multilayer assembly as a function of pH and ionic strength using the polysaccharides chitosan and heparin. Biomacromolecules 2008; 9: 2021-8.
[103] Almodóvar J, Bacon S, Gogolski J, Kisiday JD, Kipper MJ. Polysaccharide-based polyelectrolyte multilayer surface coatings can enhance mesenchymal stem cell response to adsorbed growth factors. Biomacromolecules 2010; 11: 2629-39.
[104] Boddohi S, Almodóvar J, Zhang H, Johnson PA, Kipper MJ. Layer-by-layer assembly of polysaccharide-based nanostructured surfaces containing polyelectrolyte complex nanoparticles. Colloids Surf B Biointerfaces 2010; 77: 60-8.
[105] Boddohi S, Moore N, Johnson PA, Kipper MJ. Polysaccharide-based polyelectrolyte complex nanoparticles from chitosan, heparin, and hyaluronan. Biomacromolecules 2009; 10: 1402-9.
[106] Costa RR, Custodio CA, Arias FJ, Rodriguez-Cabello JC, Mano JF. Layer-by-layer assembly of chitosan and recombinant biopolymers into biomimetic coatings with multiple stimuli-responsive properties. Small 2011; 7: 2640-9.
[107] Almodóvar J, Kipper MJ. Coating electrospun chitosan nanofibers with polyelectrolyte multilayers using the polysaccharides heparin and N, N, N-trimethyl chitosan. Macromol Biosci 2011; 11: 72-6.
[108] Almodóvar J, Mower J, Banerjee A, Sarkar AK, Ehrhart NP, Kipper MJ. Chitosan-heparin polyelectrolyte multilayers on cortical bone: periosteum-mimetic, cytophilic, antibacterial coatings. Biotechnol Bioeng 2013; 110: 609-18.
[109] Silva JM GN, Costa R, Sher P, Reis RL, Van Blitterswijk CA, Karperien M, Mano JF. Nanostructured 3D constructs based on chitosan and chondroitin sulphate multilayers for cartilage tissue engineering. PLOS One 2013; 8: e55451/1-11.
[110] Deng M, James R, Laurencin CT, Kumbar SG. Nanostructured polymeric scaffolds for orthopaedic regenerative engineering. IEEE Trans Nanobioscience 2012; 11: 3-14.

[111] Huang ZM, Zhang YZ, Kotaki M, Ramakrishna S. A review on polymer nanofibers by electrospinning and their applications in nanocomposites. Composites Sci Technol 2003; 63: 2223-53.
[112] Zeleny J. The electrical discharge from liquid points, and a hydrostatic method of measuring the electric intensity at their surfaces. Phys Rev 1914; 3: 69-91.
[113] Formhals A. Process and apparatus fob pbepabing. US Patent 1975504, 1934.
[114] Kumbar SG, Bhattacharyya S, Sethuraman S, Laurencin CT. A preliminary report on a novel electrospray technique for nanoparticle based biomedical implants coating: precision electrospraying. J Biomed Mater Res B Appl Biomater 2007; 81: 91-103.
[115] Agarwal P, Mishra P, Srivastava P. Statistical optimization of the electrospinning process for chitosan/polylactide nanofabrication using response surface methodology. J Mater Sci 2012; 47: 4262-9.
[116] Li WJ, Laurencin CT, Caterson EJ, Tuan RS, Ko FK. Electrospun nanofibrous structure: a novel scaffold for tissue engineering. J Biomed Mater Res 2002; 60: 613-21.
[117] Merrell JG, McLaughlin SW, Tie L, Laurencin CT, Chen AF, Nair LS. Curcumin-loaded poly(epsilon-caprolactone) nanofibres: diabetic wound dressing with anti-oxidant and anti-inflammatory properties. Clin Exp Pharmacol Physiol 2009; 36: 1149-56.
[118] Katti DS, Robinson KW, Ko FK, Laurencin CT. Bioresorbable nanofiber-based systems for wound healing and drug delivery: optimization of fabrication parameters. J Biomed Mater Res B Appl Biomater 2004; 70: 286-96.
[119] Kumbar SG, James R, Nukavarapu SP, Laurencin CT. Electrospun nanofiber scaffolds: engineering soft tissues. Biomed Mater 2008; 3: 034002/1-15.
[120] Kumbar SG, Nukavarapu SP, James R, Nair LS, Laurencin CT. Electrospun poly(lactic acid-co-glycolic acid) scaffolds for skin tissue engineering. Biomaterials 2008; 29: 4100-7.
[121] Taylor ED, Nair LS, Nukavarapu SP, McLaughlin S, Laurencin CT. Novel nanostructured scaffolds as therapeutic replacement options for rotator cuff disease. J Bone Joint Surg Am 2010; 92: 170-179.
[122] James R, Kumbar S, Laurencin C, Balian G, Chhabra A. Tendon tissue engineering: adipose-derived stem cell and GDF-5 mediated regeneration using electrospun matrix systems. Biomed Mater 2011; 6: 025011/1-13.
[123] James R, Toti US, Laurencin CT, Kumbar SG. Electrospun nanofibrous scaffolds for engineering soft connective tissues. Methods Mol Biol 2011; 726: 243-58.
[124] Peach MS, James R, Toti US, *et al.* Polyphosphazene functionalized polyester fiber matrices for tendon tissue engineering: *in vitro* evaluation with human mesenchymal stem cells. Biomed Mater 2012; 7: 045016/1-13.
[125] Peach MS, Kumbar SG, James R, *et al.* Design and optimization of polyphosphazene functionalized fiber matrices for soft tissue regeneration. J Biomed Nanotechnol 2012; 8: 107-24.
[126] Wang J, Valmikinathan CM, Liu W, Laurencin CT, Yu X. Spiral-structured, nanofibrous, 3D scaffolds for bone tissue engineering. J Biomed Mater Res A 2010; 93: 753-62.
[127] Bhattacharyya S, Kumbar SG, Khan YM, *et al.* Biodegradable polyphosphazene-nanohydroxyapatite composite nanofibers: scaffolds for bone tissue engineering. J Biomed Nanotechnol 2009; 5: 69-75.
[128] Deng M, Kumbar SG, Nair LS, Weikel AL, Allcock HR, Laurencin CT. Biomimetic structures: biological implications of dipeptide-substituted polyphosphazene-polyester blend nanofiber matrices for load-bearing bone regeneration. Adv Funct Mater 2011; 21: 2641-51.
[129] Brown JL, Peach MS, Nair LS, Kumbar SG, Laurencin CT. Composite scaffolds: bridging nanofiber and microsphere architectures to improve bioactivity of mechanically competent constructs. J Biomed Mater Res A 2010; 95: 1150-8.
[130] Nada AA, James R, Shelke NB, *et al.* A smart methodology to fabricate electrospun chitosan nanofiber matrices for regenerative engineering applications. Polym Adv Technol 2014; 25: 507-15.
[131] Homayoni H, Ravandi SAH, Valizadeh M. Electrospinning of chitosan nanofibers: Processing optimization. Carbohydr Polym 2009; 77: 656-61.
[132] Wang W, Itoh S, Konno K, *et al.* Effects of Schwann cell alignment along the oriented electrospun chitosan nanofibers on nerve regeneration. J Biomed Mater Res A 2009; 91: 994-1005.
[133] Zhang Y, Venugopal JR, El-Turki A, Ramakrishna S, Su B, Lim CT. Electrospun biomimetic nanocomposite nanofibers of hydroxyapatite/chitosan for bone tissue engineering. Biomaterials 2008; 29: 4314-22.

[134] Shim IK, Suh WH, Lee SY, *et al.* Chitosan nano-/microfibrous double-layered membrane with rolled-up three-dimensional structures for chondrocyte cultivation. J Biomed Mater Res A 2009; 90: 595-602.

[135] Zhang J-F, Yang D-Z, Xu F, Zhang Z-P, Yin R-X, Nie J. Electrospun core– shell structure nanofibers from homogeneous solution of poly (ethylene oxide)/chitosan. Macromolecules 2009; 42: 5278-84.

[136] Hussain A, Collins G, Yip D, Cho CH. Functional 3-D cardiac co-culture model using bioactive chitosan nanofiber scaffolds. Biotechnol Bioeng 2013; 110: 637-47.

[137] Wang T, Ji X, Jin L, *et al.* Fabrication and characterization of heparin-grafted poly-l-lactic acid–chitosan core–shell nanofibers scaffold for vascular gasket. ACS Appl Mater Interfaces 2013; 5: 3757-63.

[138] Du F, Wang H, Zhao W, *et al.* Gradient nanofibrous chitosan/poly ε-caprolactone scaffolds as extracellular microenvironments for vascular tissue engineering. Biomaterials 2012; 33: 762-70.

[139] Xie Z, Paras CB, Weng H, *et al.* Dual growth factor releasing multi-functional nanofibers for wound healing. Acta Biomater 2013; 9: 9351-9.

[140] Lo KW-H, Jiang T, Gagnon KA, Nelson C, Laurencin CT. Small-molecule based musculoskeletal regenerative engineering. Trends Biotechnol 2014; 32: 74-81.

[141] Hsieh CY, Hsieh HJ, Liu HC, Wang DM, Hou LT. Fabrication and release behavior of a novel freeze-gelled chitosan/gamma-PGA scaffold as a carrier for rhBMP-2. Dent Mater 2006; 22: 622-9.

[142] Lee JY, Nam SH, Im SY, *et al.* Enhanced bone formation by controlled growth factor delivery from chitosan-based biomaterials. J Control Release 2002; 78: 187-97.

[143] Lee JY, Seol YJ, Kim KH, *et al.* Transforming growth factor (TGF)-beta 1 releasing tricalcium phosphate/chitosan microgranules as bone substitutes. Pharm Res 2004; 21: 1790-6.

[144] Lee JE, Kim KE, Kwon IC, *et al.* Effects of the controlled-released TGF-beta 1 from chitosan microspheres on chondrocytes cultured in a collagen/chitosan/glycosaminoglycan scaffold. Biomaterials 2004; 25: 4163-73.

[145] Alemdaroglu C, Degim Z, Celebi N, Zor F, Ozturk S, Erdogan D. An investigation on burn wound healing in rats with chitosan gel formulation containing epidermal growth factor. Burns 2006; 32: 319-27.

[146] Hsu SH, Whu SW, Hsieh SC, Tsai CL, Chen DC, Tan TS. Evaluation of chitosan-alginate-hyaluronate complexes modified by an RGD-containing protein as tissue-engineering scaffolds for cartilage regeneration. Artif Organs 2004; 28: 693-703.

[147] Zhang YF, Wang YN, Shi B, Cheng XR. A platelet-derived growth factor releasing chitosan/coral composite scaffold for periodontal tissue engineering. Biomaterials 2007; 28: 1515-22.

[148] Zhang YF, Cheng XR, Wang JW, *et al.* Novel chitosan/collagen scaffold containing transforming growth factor-beta 1 DNA for periodontal tissue engineering. Biochem Biophys Res Commun 2006; 344: 362-9.

[149] Guo T, Zhao JN, Chang JB, *et al.* Porous chitosan-gelatin scaffold containing plasmid DNA encoding transforming growth factor-beta 1 for chondrocytes proliferation. Biomaterials 2006; 27: 1095-103.

[150] Mi FL, Wu YB, Shyu SS, *et al.* Control of wound infections using a bilayer chitosan wound dressing with sustainable antibiotic delivery. J Biomed Mater Res 2002; 59: 438-49.

[151] Hoemann CD, Sun J, Legare A, McKee MD, Buschmann MD. Tissue engineering of cartilage using an injectable and adhesive chitosan-based cell-delivery vehicle. Osteoarthritis Cartilage 2005; 13: 318-29.

Subject Index

A

Atta-ur-Rahman, Allen B. Reitz, M. Iqbal Choudhary and Jizhou Wang (Eds.)

B

C

D

E

F

G

V

X

Z

Made in the USA
Las Vegas, NV
03 November 2026

57424493R00257